Avian Incubation

Oxford Ornithology Series
Edited by C. M. Perrins

1. *Bird Population Studies: Relevance to Conservation and Management* (1991)
   Edited by C. M. Perrins, J.-D. Lebreton, and G. J. M. Hirons
2. *Bird-Parasite Interactions: Ecology, Evolution, and Behaviour* (1991)
   Edited by J. E. Loye and M. Zuk
3. *Bird Migration: A General Survey* (1993)
   Peter Berthold
4. *The Snow Geese of La Pérouse Bay: Natural Selection in the Wild* (1995)
   Fred Cooke, Robert F. Rockwell, and David B. Lank
5. *The Zebra Finch: A Synthesis of Field and Laboratory Studies* (1996)
   Richard A. Zann
6. *Partnerships in Birds: The Study of Monogomy* (1996)
   Edited by Jeffery M. Black
7. *The Oystercatcher: From Individuals to Populations* (1996)
   Edited by John D. Goss-Custard
8. *Avian Growth and Development: Evolution within the Altricial-Precocial Spectrum* (1997)
   Edited by J. M. Starck and R. E. Ricklefs
9. *Parasitic Birds and Their Hosts* (1998)
   Edited by Stephen I. Rothstein and Scott K. Robinson
10. *The evolution of Avian Breeding Systems* (1999)
    J. David Ligon
11. *Harriers of the world: Their Behaviour and Ecology* (2000)
    Robert E. Simmons
12. *Bird Migration* 2e (2001)
    Peter Berthold
13. *Avian Incubation: Behaviour, Environment, and Evolution* (2002)
    Edited by D. C. Deeming

Avian Incubation

BEHAVIOUR, ENVIRONMENT, AND EVOLUTION

Edited by
D.C. DEEMING
Hatchery Consulting and Research,
9 Eagle Drive,
Welton,
Lincoln,
UK

UNIVERSITY PRESS

OXFORD
UNIVERSITY PRESS

Great Clarendon Street, Oxford OX2 6DP

Oxford University Press is a department of the University of Oxford.
It furthers the University's objective of excellence in research, scholarship,
and education by publishing worldwide in

Oxford New York

Athens Auckland Bangkok Bogotá Buenos Aires Cape Town
Chennai Dar es Salaam Delhi Florence Hong Kong Istanbul Karachi
Kolkata Kuala Lumpur Madrid Melbourne Mexico City Mumbai Nairobi
Paris São Paulo Shanghai Singapore Taipei Tokyo Toronto Warsaw

with associated companies in Berlin Ibadan

Oxford is a registered trade mark of Oxford University Press
in the UK and in certain other countries

Published in the United States
by Oxford University Press Inc., New York

First published 2002

British Library Cataloguing in Publication Data
Data available

Library of Congress Cataloging in Publication Data
Data available

ISBN 0 19 850810 7 (Hbk)

1 3 5 7 9 10 8 6 4 2

Typeset by
Newgen Imaging Systems (P) Ltd., Chennai, India
Printed in Great Britain
on acid-free paper by
Biddles Ltd, Guildford & King's Lynn

Preface

My interest in avian incubation has extended over 20 years and has been largely based on laboratory and commercial studies of the effects of incubation conditions on development of poultry and ostriches. Back in the late 1980s my research on the physiological basis of egg turning generated an idea that rates of egg turning would correlate with the relative albumen content of the egg and so altricial species (with a higher albumen content) would require more turning than precocial species. This idea remained untested primarily because existing reviews of incubation (Freeman and Vince 1974; Drent 1975; Skutch 1976) failed to describe any reports of the rate of turning during natural incubation. However, during 1999 I happened upon a couple of reports where rates of egg turning in nests were reported and so I set about a literature search in order to see if these were isolated instances. The 82 data points from 61 species I identified allowed me to test my hypothesis (see Chapter 11).

During the course of this search I encountered more and more papers on incubation in wild birds, which over the years I had not really had need to consult. It seemed to me a shame that this information had not been brought together in a book solely concerned with incubation in nests. I set about developing the idea of such a book and contacting colleagues, old and new, in order to find authors for those chapters I did not feel fully qualified to write. *Avian Incubation* is the end result of these endeavours and I believe that our efforts have produced an invaluable reference text for years to come.

This book is split into sections that reflect the diverse biology of natural incubation. My introductory chapter aims to describe what little we know about the evolution of avian reproduction. The second section describes, albeit briefly, fundamental aspects of avian reproduction: nests, eggs and embryonic development. Factors affecting the behaviour of incubating birds and their embryos during incubation are considered in the third section. Chapters in the next section describe how the incubation environment is maintained and as well as considering temperature, gas exchange and turning, there is also a review of the microbiology of nests. The diversity of nesting environments in birds is celebrated in the next chapter with chapters on underground nesting in megapodes, the incubation environment of hummingbirds, a model of intermittent incubation of passerines, incubation in species nesting in extreme climates, and strategies of brood parasites. The penultimate section considers ecological and evolutionary aspects of incubation, including the energetic and fitness costs of incubation. My final chapter aims to bring together the concepts described in the book highlighting areas of interest where incubation scientists could perhaps focus their research efforts in the future.

This book has brought together a wide range of experts in their diverse fields from the United Kingdom, United States of America, Canada, Australia, The Netherlands and Israel. Editing a book can be a difficult task when there is such a variety of authors but on the whole my task was interesting and stress-free. The large number of contributors meant that I decided to adopt the third person throughout the book and some styles of writing may have been sacrificed to ensure that the book is clear and accessible to the widest possible readership. Indeed, it is my sincere hope that *Avian Incubation* will be of interest not only to those people interested in avian reproduction but also to those interested in artificial incubation of bird eggs, and to ornithologists, professional and amateur alike.

My sincere thanks go out to all of the contributors for all their efforts in making this book the valuable resource that it is. Many thanks to Richard Tibbitts and Simone End of *AntBits Illustration* for providing excellent illustrations for my own chapters and the front cover. I would like to thank Dr Linda Birch of the *Alexander Library*, at the *Edward Grey Institute of Field Ornithology*, University of Oxford, for her invaluable assistance in researching my own chapters. I thank Cathy Kennedy of *OUP* for sticking with the idea; I hope the end result is worthy. Finally, many thanks to my wife Roslyn, and daughters Katherine and Emily (born in January 2001), for their patience and love during the course of this project.

Finally, I would like to dedicate this book to my good friend Amos Ar for his outstanding contribution to improving our understanding of avian incubation in nests and incubators alike, which, as Chapter 10 shows, continues to this day.

Lincoln D. C. D.
August 2001

Contents

Contributors xii

1 **Importance and evolution of incubation in avian reproduction** 1
D. C. Deeming

The bird–nest incubation unit 1
Evolution of avian incubation 2
Evolutionary development of avian incubation 6

2 **Location, structure and function of incubation sites** 8
M. H. Hansell and D. C. Deeming

Incubation sites 9
Nest shapes 13
Nest construction 15
The evolution of nest building 24
Function of nests 25

3 **Functional characteristics of eggs** 28
D. C. Deeming

Egg morphology and composition 28
Egg dimensions 33
The shell 35
The avian egg: a remarkable evolutionary design 41

4 **Embryonic development and utilisation of egg components** 43
D. C. Deeming

Basic patterns of embryonic development 43
Changes to egg components 50
Embryonic development in non-commercial species 53

5 **Hormonal control of incubation behaviour** 54
C. M. Vleck

Species and gender differences in prolactin levels and parental behaviour 55
Role of steroid hormones and prolactin in initiating incubation 57
Behavioural effects on prolactin secretion 59
Prolactin and incubation behaviour 62

6 Behaviour patterns during incubation 63
D. C. Deeming

Patterns of behaviour 63
Activity of birds during incubation 67
Constancy of incubation 73
Evolution of incubation behaviour 78
%Attentiveness and eggmass 85

7 Parent–embryo interactions 88
R. B. Brua

Embryo–embryo interactions 88
Parent–embryo interactions 91
Importance of parent–embryo interactions 99

8 The brood patch 100
R. W. Lea and H. Klandorf

Morphology 101
Occurrence of brood patches 105
The endocrine basis for brood patch development 108
Mechanisms of action of the brood patch: evidence of importance of tactile sensitivity 112
Indirect action of the brood patch 113
Functional activity of the brood patch 115
Further research into brood patch form and function 118

9 Maintenance of egg temperature 119
J. S. Turner

Maintenance of egg temperature: steady state 120
Maintenance of egg temperature: unsteady state 129
Limitations of the biophysical approach to maintenance of egg temperature 138
What is 'egg temperature'? 141

10 Nest microclimate during incubation 143
A. Ar and Y. Sidis

Definition of nest microclimate 145
Temperature regulation of the eggs 145
Thermal considerations of the incubating adult 152
Nest humidity 154
Nest gas composition 156
Egg turning 159
The nest microclimate: further research 160

11 **Patterns and significance of egg turning** 161
D. C. Deeming

The process of egg turning 161
Reasons for egg turning during incubation 171
Evolutionary aspects of egg turning 175
Areas for further research into turning during incubation 177

12 **Microbiology of natural incubation** 179
G. K. Baggott and K. Graeme-Cook

Fungal associations of bird nests 180
The nest bacterial community 183
Behavioural control of potential pathogens 184
Microbes on the egg surface 186
Improving our understanding of the microbiology of natural incubation 189

13 **Underground nesting in the megapodes** 192
D. T. Booth and D. N. Jones

Sources of incubation heat 192
Incubation technique 194
Nest incubation conditions 197
Physiology of embryonic development 200
Hatching and escaping the nest 202
Areas for future research in megapode incubation 205

14 **Characteristics and constraints of incubation in hummingbirds** 207
W. A. Calder III

Basic breeding biology 209
Hummingbirds and incubation: a scaling perspective 210
Nest site selection 213
Incubation behaviour 215
Temperature regulation of females and eggs during incubation 216
Incubation to fitness: hummingbird persistence was unlikely! 220
Future research 221

15 **Intermittent incubation: predictions and tests for time and heat allocations** 223
F. R. Hainsworth and M. A. Voss

Modelling of time and heat allocations during intermittent incubation 224
Percent time allocated to foraging during an incubation cycle 228
Predictions from the model 230
Tests of the model predictions for maximal % time foraging for juncos 233
Significance of the test 235

16 **Incubation in extreme environments** 238
C. Carey

Cold environments 239
High altitude environments 242
Desert environments 247
Wet-nesters 251
Prospects for birds breeding in extreme environments 252

17 **Tactics of obligate brood parasites to secure suitable incubators** 254
S. G. Sealy, D. G. McMaster and B. D. Peer

Suitable hosts 255
Tactics to ensure hosts accept the parasite's egg 256
Tactics to lay in nests before incubation has advanced too far 257
Tactics to ensure adequate incubation of the parasite's egg 261
Tactics to ensure the parasite's egghatches first 265
Priorities for future research 268

18 **Ecological factors affecting initiation of incubation behaviour** 270
P. N. Hébert

Physio-ethology of the onset of incubation 270
Patterns of onset of incubation 271
Ecological factors affecting the onset of incubation 274
Direction of future studies 279

19 **Adaptive significance of egg coloration** 280
T. J. Underwood and S. G. Sealy

Crypsis 280
Egg/nest recognition 284
Filtering solar radiation 292
Eggshell strength 293
Aposematism 293
Intraclutch variability 294
White eggs 296
Why so blue? 296
Priorities for future research 297

20 **Energetics of incubation** 299
J. M. Tinbergen and J. B. Williams

Terminology 300
Energy constraints during incubation and clutch size 303
Energetic costs of incubation in terrestrial birds 304
Energetic cost of incubation in seabirds 311
Future perspectives for research into energetics of incubation 313

21 Incubation and the costs of reproduction 314
J. M. Reid, P. Monaghan and R. G. Nager

Why might incubation be costly? 315
Occurrence and distribution of incubation costs 321

22 Perspectives in avian incubation 326
D. C. Deeming

Areas for further research into incubation 326
The future for incubation research? 328

References 330

Index 417

Contributors

AMOS AR,
Department of Zoology, Faculty of Life Sciences,
Tel Aviv University,
Ramat Aviv 69978,
Israel.

GLENN K. BAGGOTT,
School of Biological and Chemical Sciences,
Birkbeck College,
Malet St,
London WC1E 7HX,
UK.

DAVID T. BOOTH,
Department of Zoology and Entomology,
The University of Queensland,
Brisbane,
Queensland 4072,
Australia.

ROBERT W. BRUA,
Canadian Wildlife Service,
Prairie and Northern Wildlife Research Centre,
115 Perimeter Road,
Saskatoon,
Saskatchewan S7N 0X4,
Canada.

BILL A. CALDER, III,
Deparment of Ecology & Evolutionary Biology,
Box 210088,
University of Arizona,
Tucson,
Arizona 85721,
USA.

CYNTHIA CAREY,
Department of EPO Biology,
Campus Box 334,
University of Colorado,
Boulder,
Colorado 80309-0334,
USA.

D. CHARLES DEEMING,
Hatchery Consulting & Research,
9 Eagle Drive,
Welton,
Lincoln,
UK.

KATE GRAEME-COOK,
School of Biological and Chemical Sciences,
Birkbeck College,
Malet St,
London WC1E 7HX,
UK.

F. REED HAINSWORTH,
Department of Biology,
Syracuse University,
Syracuse,
New York 13244-1270,
USA.

MIKE H. HANSELL,
Division of Environmental & Evolutionary Biology,
Institute of Biomedical and Life Sciences,
Graham Kerr Building,
University of Glasgow,
Glasgow G12 8QQ,
UK.

PERCY N. HEBÉRT,
Department of Wildlife,
Humboldt State University,
Arcata,
California 95521,
USA.

DARRYL N. JONES,
Australian School of Environmental Studies,
Griffith University,
Nathan,
Queensland 4111,
Australia.

HILLAR KLANDORF,
Division of Animal and Veterinary Sciences,
College of Agriculture and Forestry,
West Virginia University,
PO BOX 6108,
Morgantown,
West Virginia 26506-6108,
USA.

BOB W. LEA,
Department of Biological Sciences,
University of Central Lancashire,
Preston PR1 2HE,
UK.

D. GLEN McMASTER,
Saskatchewan Wetland Conservation Corporation,
101-2022 Cornwall Street,
Regina,
Saskatchewan S4P 2K5,
Canada.

PAT MONAGHAN,
Division of Environmental & Evolutionary Biology,
Graham Kerr Building,
University of Glasgow,
Glasgow G12 8QQ,
UK.

RUDI G. NAGER,
Division of Environmental & Evolutionary Biology,
Graham Kerr Building,
University of Glasgow,
Glasgow G12 8QQ,
UK.

BRIAN D. PEER,
Department of Biology,
Lawrence University,
Appleton,
Wisconsin 54912,
USA.

JANE M. REID,
Division of Environmental & Evolutionary Biology,
Graham Kerr Building,
University of Glasgow,
Glasgow G12 8QQ,
UK.

SPENCER G. SEALY,
Department of Zoology,
University of Manitoba,
Winnipeg,
Manitoba R3T 2N2,
Canada.

YISRAEL SIDIS,
Department of Zoology,
Faculty of Life Sciences,
Tel Aviv University,
Ramat Aviv 69978,
Israel.

JOOST M. TINBERGEN,
Zoological Laboratory,
Kerklaan 30,
9750 AA Haren,
The Netherlands.

J. SCOTT TURNER,
Department of Environmental & Forest Biology,
State University of New York College of Environmental Science & Forestry,
Syracuse,
New York 13210,
USA.

TODD J. UNDERWOOD,
Department of Zoology,
University of Manitoba,
Winnipeg,
Manitoba R3T 2N2,
Canada.

CAROL M. VLECK,
Department of Zoology and Genetics,
Iowa State University,
Ames,
Iowa 50011,
USA.

MARGARET A. VOSS,
Department of Biology,
State University of New York College at Potsdam,
44 Pierrepont Avenue,
Potsdam,
New York 13676-2294,
USA.

JOE B. WILLIAMS,
Department of Evolution, Ecology, and Organismal Biology,
Ohio State University,
1735 Neil Ave,
Columbus,
Ohio 43210,
USA.

1 *Importance and evolution of incubation in avian reproduction*

D. C. Deeming

Reproduction in birds is extremely conservative with the vast majority of the 8,900 species adopting bird–egg contact incubation to maintain an appropriate environment for embryonic development. The exact conditions adopted by different species vary considerably, but factors such as initial egg mass, incubation period, eggshell conductance to gases and weight loss during incubation are all interrelated in a wide range of avian species (Rahn and Paganelli 1990; Chapter 3). Man has exploited this uniformity in avian incubation to use birds from one species, usually the domestic fowl (*Gallus gallus*) to incubate the eggs of a second species (e.g. raptor eggs), and to build artificial incubators that mimic the incubation environment for up to 120 000 fowl eggs at a time.

Previous reviews of incubation in birds have concentrated on specific aspects of avian reproduction including the nest (Hansell 2000), attentiveness at the nest (Kendeigh 1952), the egg (Romanoff ands Romanoff 1949), embryonic development (Romanoff 1960), embryonic physiology (Romanoff 1967; Freeman and Vince 1974; Seymour 1984a; Metcalfe *et al.* 1987; Deeming and Ferguson 1991a) or commercial aspects of incubation (Tullett 1991). A lot of incubation research has focused on the development of embryos in incubators or on the processes of artificial incubation. The aim of this book is to bring together the wealth of information about incubation by birds in their natural environment and so provide a comprehensive review of current understanding of the natural incubation environment of as many birds as is possible. This first chapter defines incubation and discusses some of the evolutionary history of avian reproduction. Thereafter the book is split into five sections that deal with fundamental aspects of avian reproduction, behaviour of adults and embryos, the incubation environment, unusual aspects of incubation in specific groups of birds, and evolutionary aspects of incubation in birds. In each chapter it is hoped that areas for further research will be suggested and so the concluding chapter highlights areas where research on incubation may proceed.

The bird–nest incubation unit

Incubation is a relatively simple process, but plays a crucial role in the reproduction of birds. A useful working definition of avian incubation is the process by which eggs are kept at temperatures suitable for embryonic development, in a humid environment that is regularly changed to allow for exchange of respiratory gases, and during which the eggs are turned regularly. Further refinement of this definition, to better distinguish it from reptilian incubation, is that embryo temperature is generally maintained within 37–38°C with most of the heat energy being supplied by the incubating bird and not coming from the environment. The temperature of the embryo and egg recorded in nests (Rahn 1991) are not usually well matched because contact incubation can cause temperature gradients within the egg (Turner 1991; Chapter 9). Nest humidity is maintained above that of the adjacent air, but rarely close to saturation, conditions that result in 10–20% weight loss from the egg during incubation (Rahn and Paganelli 1990). With few exceptions (e.g. megapodes; Chapter 13), birds incubate their eggs in an open nesting environment free from packing by a substrate, unlike the nests of most reptiles. Of course, reptiles do not turn their eggs (Deeming 1991).

How a bird achieves the environmental consequences of incubation (i.e. high temperature, a specific gaseous environment and egg movement) is the focus of this book. Most birds create an incubation environment by laying eggs in a nest and sitting above and in contact with them. Thus, the 'incubator' is not the bird alone but a combination of the nest and the bird (see also Chapter 10); it is the bird-nest unit that allows successful incubation. The location of the incubation site, the construction of the nest and the persistency of incubation behaviour all contribute to an ideal incubation environment in individual species. How these conditions vary are explored throughout this book.

Evolution of avian incubation

How could the bird-nest incubation unit have come about? Despite our relatively good knowledge of the evolution of different types of birds (Chatterjee 1997; Dingus and Rowe 1998; Feduccia 1999), there is almost no evidence of their reproductive behaviour. Although many fossil eggshells have been attributed to avian species (Mikhailov *et al.* 1994; Chatterjee 1997) only a few eggs have been found containing the remains of avian embryos. These include the elantiornithine *Gobipteryx minuta* and the neornithine *Gobipipus reshetovi*, late Cretaceous birds from the Gobi Desert, Mongolia (Elzanowski 1981; Chatterjee 1997). Apparently, eggs of *G. reshetovi* have even been found in 'clutches' sitting vertically or obliquely with their narrower ends down in the sediment (Chatterjee 1997). Despite the report of a few adult hind limb bone fragments associated with *G. minuta* eggs (Sabbath 1991), there is no clear evidence of the incubation behaviour of extinct birds.

By contrast, discoveries of embryos over the past 20 years have meant that many claims have been made about the reproductive habits of dinosaurs (Carpenter *et al.* 1994; Carpenter 1999). To date, the best example of potential nesting behaviour in a dinosaur has been in the theropod *Oviraptor philoceratops* whose fossils appear to indicate that this species practised contact incubation (Norell *et al.* 1995; Clark *et al.* 1999). This fossil evidence is used to strengthen the evolutionary link between theropods and birds (Clark *et al.* 1999). However, how good is the evidence that theropods practised contact incubation?

Did *Oviraptor* incubate its eggs?

The original description of *Oviraptor* was based of its discovery in Mongolia lying on top of a nest containing eggs assumed to be laid by *Protoceratops*. It was suggested that, at its demise, the animal had been raiding the nest and eating the eggs (Osborn 1924), but more recent evidence identifies the eggs as those of *Oviraptor* (Norell *et al.* 1994). Moreover, recent discoveries in Mongolia and China have found other adult *Oviraptor*s lying on top of nests of oviraptoroid eggs (Norell *et al.* 1995; Dong and Currie 1996; Clark *et al.* 1999). It is now generally concluded that *Oviraptor* was brooding the eggs in a similar way to that seen in extant birds (Norell *et al.* 1995; Dong and Currie 1996; Clark *et al.* 1999). Norell *et al.* (1995) described 'brooding' as simply 'the behaviour of sitting on nests', despite it being more properly associated with the provision of heat by an adult to eggs or young animals to maintain a high temperature for development of embryos or thermoregulation of neonates. Norell *et al.* (1995) used examples of brooding from extant birds and python snakes, which implicitly reinforced the idea that this *Oviraptor* was providing the eggs with heat. However, does all of the evidence about *Oviraptor* fossils support this idea?

Oviraptor eggs measure 18 × 6–7.5 cm (Clark *et al.* 1995) and 15 × 5.5 cm (Dong and Currie 1996) and they strongly resemble those of the Elongatoolithidae described by Mikhailov *et al.* (1994). Ultrastructure of the shells of *Oviraptor* eggs (Dong and Currie 1996) is also similar to that described for elongatoolithid eggs (Mikhailov *et al.* 1994). This similarity suggests that *Oviraptor* eggs will have shell porosity characteristics similar to that of elongatoolithid eggs. Hence, the water vapour conductance (Ar *et al.* 1974) values of elongatoolithid shells (Sabbath 1991; Mikhailov *et al.* 1994) are 2–5 times greater than that of bird eggs of equivalent weights (Figure 1.1). These relative values are similar to those found for modern crocodilian eggs (Deeming and Thompson 1991) and suggest that the eggs were incubated in a high humidity environment. Therefore, the eggs were not only buried in but were also covered by a substrate.

The *Oviraptor* described by Norell *et al.* (1995) and Clark *et al.* (1999) is lying across a nest but the eggs adopt a toroidal shape lying to the sides of the body remains and in particular, lateral to the pes (Figure 1.2). Of the presumed clutch of 30 eggs almost half would lie uncovered by the body. Dong and Currie (1996) also suggest that their animal was lying with its feet within the circle of eggs. Furthermore, the eggs in one nest lie inclined at low angles (13–16°) to the ground sloping away from the centre of the nest (Dong and Currie 1996) and in the other nest two layers of eggs are described (Norell *et al.* 1995; Clark *et al.* 1999).

This orientation of the eggs in the nest and in relation to the animal undermines the idea that *Oviraptor* was 'brooding' the eggs. Avian incubation relies upon close physical contact between the skin and the egg surface and so bird eggs form only one layer within the nest. Eggs in the lower layer would not contact the brood patch and may not have received sufficient heat to allow normal embryonic development. Similarly, in one nest at least the *Oviraptor* eggs are lying end-on in a toroidal pattern meaning that animal–egg contact would be

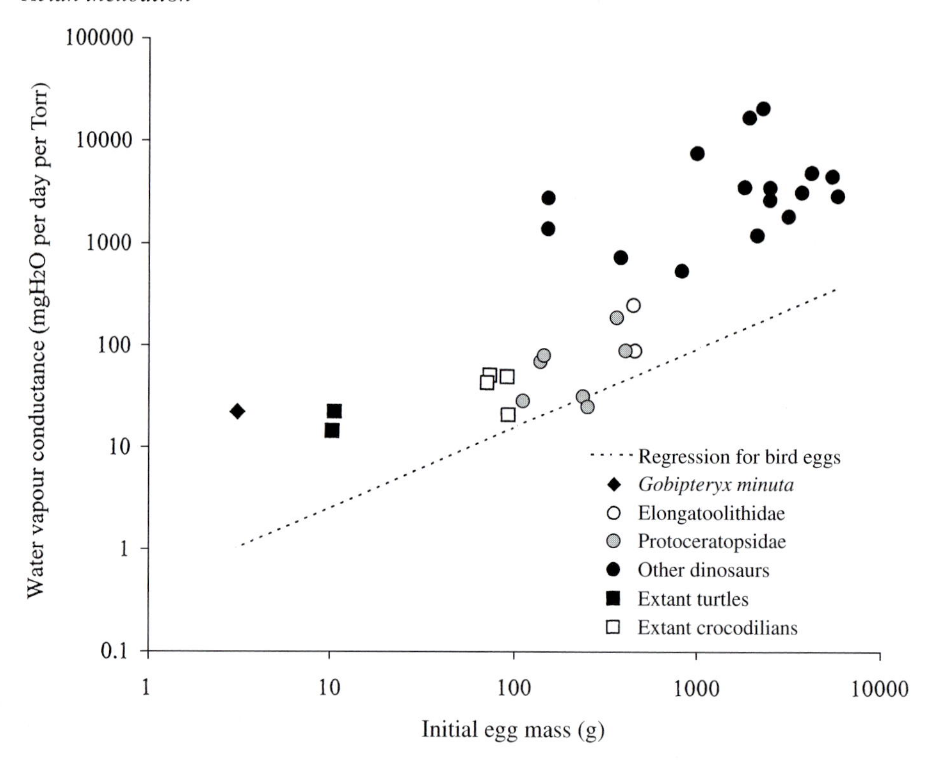

Fig. 1.1 The relationship between initial egg mass and water vapour conductance for a species of an elantiornithine bird (*Gobipteryx minuta*), eggs attributed to a variety of dinosaurs and rigid-shelled eggs of extant crocodilians and turtles. Regression line is for the relationship between water vapour conductance and initial egg mass in extant birds: $G_{H_2O} = 0.432\,M^{0.78}$ (Ar *et al.* 1974). Data from Seymour (1979), Williams *et al.* (1984), Sabbath (1991), Deeming and Thompson (1991), Grigorescu *et al.* (1994), Mikhailov *et al.* (1994), Sahni *et al.* (1994) and Kern and Ferguson (1997).

minimal because the eggs present only a small surface area to the animal's skin (Figure 1.2). Furthermore, the eggs have their large ends in the centre and slope away from the centre of the nest forming a mound rather than a cup (Figure 1.2). This is difficult to sit upon without disturbing the arrangement. Unless *Oviraptor* possessed some form of insulation layer which extended well beyond the skin most of the eggs would have been exposed to the elements. Unless they were buried in a substrate with their high shell porosities the eggs would have quickly dried out.

Unlike Carpenter (1999), I suggest that eggshell characteristics strongly reflect the incubation environment. In *Oviraptor* the eggs were almost certainly completely covered by a substrate, most likely sand, and the animal was lying on top of the substrate forming the nest mound rather than sitting directly on top of the eggs (Figure 1.2). Brooding in terms of avian contact incubation was not possible and this nesting pattern is more typically of extant reptiles than birds. To support

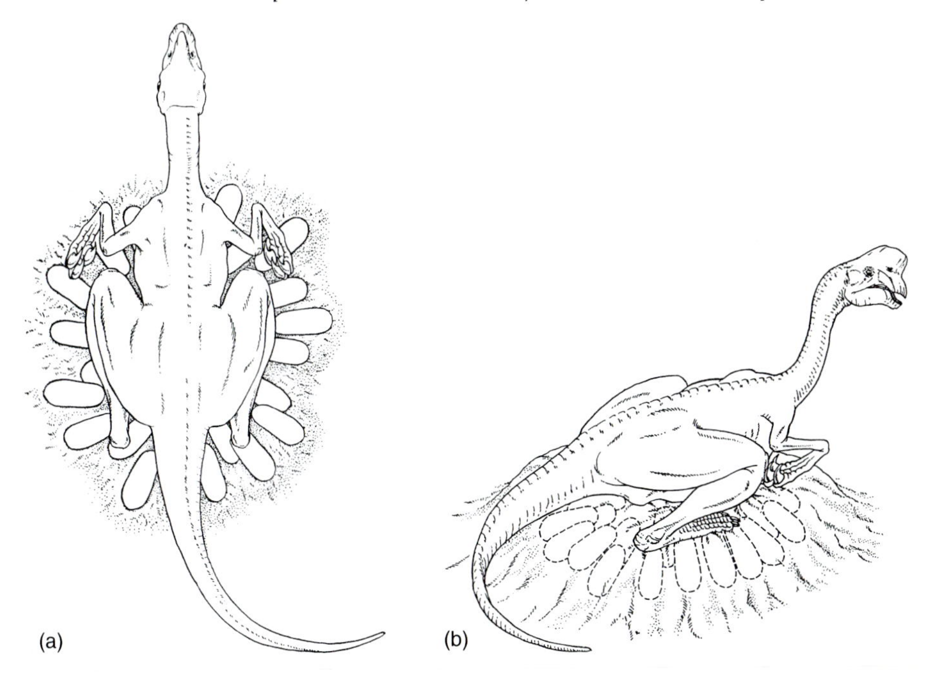

Fig. 1.2 (a) A dorsal view of an *Oviraptor philaceratops* on top of its nest. Note the circle of eggs which lie end on to, but not covered by, the body of the animal. (b) A side view of an *Oviraptor* sitting upon a nest mound. Note that the eggs lie buried in a toroidal ring with their long axis pointing upwards towards the centre.

this contention I predict that when the water vapour conductance of *Oviraptor* eggshells is determined it will be found that it is high compared with avian eggs of an equivalent mass.

This attachment of *Oviraptors* to their covered nests was so strong that it remained unbroken by the events that killed them. Protective guarding of nests from predation is known in modern crocodilians, which lie on top of nests and do not feed for several months (Ferguson 1985; Webb and Cooper-Preston 1989). By contrast, birds will often simply abandon a nest if threatened by predation (Chapter 6). As is the case in crocodilians, perhaps the adult *Oviraptor* was also waiting to assist with hatching and escape from the nest mound and to offer protection of the hatchlings during their first few days of life.

Most of the available fossil evidence suggests that all dinosaur eggs described to date were buried (Seymour 1979; Williams *et al.* 1984; Carpenter 1999) and so incubation conditions were more similar to modern reptiles than to birds. Simply using the physical association of the *Oviraptor* with the eggs to suggest a reproductive strategy, and not examining all of the available evidence, could lead to the wrong conclusions being drawn about the evolution of reproductive biology of this species and other dinosaurs and birds alike.

Evolutionary development of avian incubation

Presumably contact incubation in birds evolved from a habit of burying and guarding eggs and so it remains possible that the intimate relationship between adult *Oviraptors* and their nests may have been a key step in the evolution of avian incubation. Any departure from the behaviour of abandoning eggs to their fate probably had some selective advantage particularly if reproductive effort was high (e.g. a large number of energy rich eggs). Parental care of eggs has evolved several times in extant reptiles with close contact between adults and eggs during incubation although in almost all species there is no exchange of heat between the adults and the eggs (Shine 1988). In early birds, development of a higher body temperature may have combined with egg–adult contact during incubation to elevate egg temperature and to increase the rate of embryonic development. This reduced the period of incubation and hence the risk of predation of the adult and the eggs. Contact incubation probably provided greater control of the environment for embryonic development and selection pressure acted to reinforce and further refine this behaviour such that it became a peculiarly conservative avian behaviour.

In modern birds, only in Megapodes is there any divergence from the basic pattern of contact incubation. These birds may bury their eggs like reptiles but they exhibit considerable care of the nest mounds in order to maintain the correct incubation environment (Chapter 13). This habit of burying eggs is not a primitive trait and it evolved from avian contact incubation perhaps from burrow nesting behaviour (Jones *et al.* 1995) or perhaps the egg covering behaviour described by Maclean (1974).

The evolutionary history of birds during the Mesozoic remains sketchy but it is clear that only modern bird types (e.g. Gaviiformes and Charadriiformes) crossed the Cretaceous–Tertiary (K–T) boundary whereas a wide range of other birds (the Enantiornithes, Hespornithiformes, and Ichthyornithiformes) failed to do so (Chatterjee 1997; Dingus and Rowe 1998). Despite considerable debate about why only neornithine birds survived, few people have suggested a mechanism which could allow neornithines to survive the K–T boundary. I would like to suggest that the evolution of functional contact incubation only occurred in neornithine birds and not in other Cretaceous birds, which relied on the more reptilian-like incubation environment. This idea is supported by water vapour conductance values for the eggshells of the elantiornithine *G. minuta* which have a water vapour conductance of 22.4 $mgH_2O\,d^{-1}\,Torr^{-1}$ compared with a predicted value of 2.7 $mgH_2O\,d^{-1}\,Torr^{-1}$ based on egg mass derived from relationships in modern birds (Figure 1.1; Sabbath 1991). Therefore, elantiornithoid (and other non-neornithine) birds may not have had eggshells typical of modern birds and some of the fossil eggs currently attributed to dinosaurs may have been avian. It is curious to note that fossil eggs of dinosaurs are very common, but those of birds are rare both before and after the K–T Boundary. Perhaps this reflects the habit of incubating eggs in an open nest above the ground, which makes them less likely to fossilise. Whatever the cause of the mass extinction at the K–T boundary the evolution of true contact incubation may have been a critical factor in allowing neornithine to survive beyond the Mesozoic (Deeming and Ferguson 1989).

Acknowledgements

Many thanks go to Mike Thompson for constructive comments on an earlier draft of this chapter and to Richard Tibbitts for the superb illustrations of *Oviraptor*.

2 *Location, structure and function of incubation sites*

M. H. Hansell and D. C. Deeming

It is only in a minority of species of birds that the role of caring for eggs is the sole responsibility of the parents, instead most birds build some kind of nest to which they devolve some of the role of parental care. Indeed the interaction between the bird-nest unit is central to successful incubation (Chapter 1). This requires an investment of time and energy into the construction of a nest, with the benefit to parents of removing later costs, giving them greater flexibility in allocating time and energy to other activities.

Animals build constructions in order to gain a measure of control over the environment (Hansell 2000). Only a few bird species build nests especially for adult use (Skutch 1961) and so birds build nests to provide a more controlled environment for the eggs. Nests of many species also provide protection for chicks, although those of families that have precocial chicks only have to care for the eggs since chicks leave the nest shortly after hatching. For those species which have altricial young, e.g. all of more than 5,700 species of passerines (Sibley and Monroe 1990), nests also exhibit adaptive features concerned with the protection of chicks as well as the eggs. As regulation of the temperature and humidity of the nest environment will be less important for chicks than for eggs, the design of the nest may not exhibit features that provide optimal conditions for incubation (Table 2.1).

Nests protect broods in a variety of ways and so the design of the nest may be constrained by trade-offs between conflicting functional roles of which the control of the physical environment of the incubating eggs is just one. Priorities in the rival functions of the nest, e.g. between regulation of egg temperature, protection of the chicks from disease, or the nest from discovery by a predator, may in some species result in a nest that apparently has little role in regulating the incubation environment. Nests of pigeons and doves (Columbidae) which are generally minimal twig platforms located in trees, rock ledges or cavities, provide examples of this (Goodwin 1983).

Descriptions of nests and nest building behaviour can be found in various books (Skutch 1976; Collias and Collias 1984; Hansell 2000). This chapter is concerned with not only the structure and function of incubation sites but also it will emphasise the role of nests in controlling the incubation environment, exploring how this may be compromised by rival functions. Hence, incubation sites are considered first followed by morphology and construction of nests. Following sections investigate patterns of evolutionary change in nest design, and conclude with

Table 2.1 Multiple possible roles of the nest in control of the environment of eggs and chicks. (+ symbol indicates a possible role for that feature of the nest design in the control of the factor indicated.)

Factor		Nest characteristic	Means of control	Egg phase	Chick
Physical	Temperature	Site	Exposure to sun and wind	+	+
		Shape	Roof/no roof	+	+
		Materials	Amount and type	+	+
	Humidity	Site	Exposure to sun and wind	+	+
		Shape	Roof/no roof	+	+
		Materials	Water absorption, penetration and evaporation	+	+
Biological	Predation	Site	Accessibility of nest	+	+
		Shape	Restriction of access to nest cavity	+	+
		Materials	Concealment of nest or giving it misleading identity	+	+
	Parasites	Site	Wasps and spiders may deter arthropod parasitoids	–	+
		Shape	Better ventilation may reduce nest parasites	–	+
		Materials	Open nest fabric may be better ventilated and reduce parasite hiding places. Materials with anti microbial and insecticidal properties	–	+
	Parental access to chicks	Shape	Roofed nest may limit control by parent to feed chick of choice	–	+

a consideration of nest design in relation to its varied functions, with an emphasis on how this may affect the nest cavity environment during the period of egg incubation.

Incubation sites

Site categories

Incubation sites in birds are generally fixed locations within the environment where the eggs remain whether the incubating adult is absent or present. They are usually associated with some kind of nest structure, the form of which can vary considerably. The only exceptions are the emperor and king penguins (*Aptenodytes forsteri* and *Aptenodytes patagonica* respectively) where incubation takes place on the feet of the adult bird (Handrich 1989a; Williams 1995). After laying, the single egg is carefully manoeuvred on to the feet of the bird and a brood pouch is formed by a flap of bare skin, the upper part of the feet and a feathered flap of skin of the abdomen. With the egg tucked away in this pouch the adult (the male in the emperor and both genders in the king penguin) has to remain standing during incubation but it is able to move around whilst keeping the egg off the ground or ice.

All other species of birds locate their nests in one of a wide variety of sites. Hansell (2000) recognised eight broad categories of nest location:

(1) *tree/bush* – nests located in branches at any height above the ground;
(2) *grass/reeds* – nests located above the ground, secured to this vegetation;
(3) *on the ground* – nests located on the ground or upon small piles of vegetation on the ground;
(4) *tree hole/cavity* – nests in cavities in trees either excavated or adopted by the adults (giant cacti can be classed as trees in this instance);
(5) *ground hole/cavity* – nests found in natural or excavated cavities on the ground;
(6) *on wall* – nests attached to cliff faces or the walls of buildings;
(7) *on ledge* – nests on walls or cliff faces but supported from below; and
(8) *on water* – construction of nests allows them to float on water.

Within these broad categories it is possible to recognise more detailed sub-categories. For example considering the tree/bush category alone, nests may vary greatly in their height above the ground, or in their position relative to the centre or periphery of the tree. Koepke (1972), in a study of nest sites used by birds in Peruvian rainforest, describes the nests of the lined forest falcon (*Micrastur gilvicollis*) as in the upper branches of canopy trees, while those of the hoazin (*Opisthocomus hoazin*) are in bushes, within 2–3 m of the ground. The nest of the varied sitella (*Daphoenositta chrysoptera*) is situated in the angle of substantial tree limbs (Campbell 1901), that of the red-capped manakin (*Pipra mentalis*; Tyrannidae) slung between a fine horizontal forked twig towards the end of a branch (Skutch 1969), while the multi-chambered nest of the rufous-fronted thornbird (*Phacellodomus rufifrons*; Furnariidae) and the yellow-olive flycatcher (*Tolmoyias sulphurescens*; Tyrannidae) hang down from the ends of slender branches (Skutch 1969).

The sites chosen by birds to incubate their eggs are very varied and this will have an impact on the incubation environment of the eggs. At the gross level there are differences in climate reflecting differences in season and latitude, this is then modulated in ways that reflect the choice of nest site and at the microclimate level by the structure of the nest itself.

Factors affecting incubation site location

Factors which affect the location of a nest have only been described for a small number of species but must include: availability of nest material, provision of optimal environmental conditions, avoidance of predation, and proximity to food resources. These factors apply to adults, and/or eggs, and/or hatched chicks.

Selection of nest material can often be species specific (Hansell 2000). Tomialojc (1992) suggested that this has determined the breeding range of the song thrush (*Turdus philomelos*) in Europe. Tomialojc (1992) compared the nest linings over the different parts of the song thrush's range and suggested that, despite the presence of suitable prey items, dry areas and urban sites lack the wet conditions and rotten wood, as well as suitable replacement materials such as mud or dung, which are required for the typical nest lining. Availability of nest material did not seem to impose any restrictions on the range of the closely related blackbird (*Turdus merula*).

Exposure of the nest and its occupants to fluctuations in the physical environment will be different in each of the categories of nest site. Nests located in the more exposed sites will experience greater exposure to fluctuations of climate (temperature, rainfall, wind) than those located within foliage. Nests of the orange-tufted sunbird (*Nectarinia osea*) in urban sites in Israel were commonly suspended within 2 m of a sheltering structure (often a building), more often located on its eastern aspect (Sidis *et al.* 1994). Around 80% of the entrance holes were directed away from the prevailing westerly wind direction. Experimental analysis showed that wind increased nest wall heat conductance by 20% and cooling of unattended eggs by 10%. Furthermore, almost 70% of nests were located in fully-shaded positions with almost all other nests having only up to 2 hours exposure to full sun which served to prevent overheating of the nest environment (Sidis *et al.* 1994).

Many other species exhibit specificity in the choice of nest location apparently based on environmental constraints. Sage sparrows (*Amphispiza belli*) in Idaho place their nests 34 cm above the ground some 21 cm from the perimeter of the bush and located so as to avoid the south-west sector. This part of the bush is exposed to the afternoon sun and to the periodic strong winds (Petersen and Best 1985a). The pinon jay (*Gymnorhinus cyanocephalus*), which nests in the canopy of pine trees, selects the southern sector of the tree in 83% of observed cases, apparently to allow additional sunlight to reach the nest and so save incubation costs in this early nesting species. The superb fairy wren (*Malurus cyareus*) chooses the northern sector of a bush and the white-winged fairy wren (*Malurus leucopterus*) the north-east although it is unclear why these positions are chosen (Tideman and Marples 1988).

Avoidance of low night-time temperatures seems to be an objective of the location of broad-tailed hummingbirds (*Selasphorus platycephalus*) nesting at about 3,000 m in the Rocky Mountains. Nests are located near to the canopy but close under an overhead branch (Figure 14.1). Nests in these preferred sites were found to loose less heat through convection, conduction, and radiation than those at sites without overhead protection (Calder 1973b). For the rufous hummingbird (*Selasphorus rufus*) in British Colombia, during the spring, low nest sites, are more protected from environmental variations than higher sites, but later in the season nest sites are higher and sheltered so as to match the environmental conditions experienced in the spring (Horváth 1964).

Changing seasonal environmental conditions can apparently alter the choice of nest sites in the yellowhammer (*Emberiza citrinella*) and Brewer's sparrows (*Spiszella breweri*); both show an increase in the height above ground that the nest is placed as the breeding season progresses (Peakall 1960; Peterson and Best 1985b). The change in nest height over the season may not be an adaptation to changing temperature because changes in the biological environment could, for example, alter the nature of predatory risks. Further research is needed to clarify this point.

Equally in the absence of experimental evidence, nest sites apparently chosen for their biological features may also be affected by other factors. Many species of birds from at least eight families, mainly kingfishers and parrots, regularly locate their nests in the terrestrial mounds or arboreal nests of termites (Hansell 2000). This may provide the bird clutches with some protection against predation but equally the moderated climate of the interior of the termite nest might favourably control the environment of the nest cavity.

Skutch (1976) considered that nest location in relation to predation risk was based on four principles: invisibility, inaccessibility, impregnability, and invincibility. Invisibility is usually attained by placing the nest in dense cover or a cavity although some ground nesting species rely on cryptic coloration to conceal the adult birds during incubation or the exposed eggs during the absence of the parent. Inaccessibility is achieved by location of the nest in a site free from terrestrial predators, e.g. seabirds nesting on islands, or difficult to reach by predators, the floating nests of grebes (Podicipediformes) or horned coot (*Fulica cornuta*). Siting a nest on or in a cliff wall has much the same effect. Building nests which hang far below branches can also serve to deter predation. The red-headed weaver (*Anaplectes rubriceps*) goes further and removes the foliage from the nest anchor, making access to the nest by snakes more difficult. Location of a nest in thorn trees can also be an effective deterrent. Impregnability involves construction of nest structures designed to resist attack. In particular, hornbills (Bucerotidae) almost completely block the entrance hole to their nest cavity with the adult female imprisoned within and reliant on the male bird to provision her and the hatched birds (Kemp 1995). Invincibility seems to be a minor and relatively unsuccessful strategy, reliant as it is on direct defence of the nest by the adults (Skutch 1976).

Although not all agree (e.g. Andrén 1991) many recent studies demonstrate that predation is a critical factor in determining nest site location (reviewed by Hansell 2000). Factors such as habitat type, level of vegetation cover, density of nests and variety of species in the habitat are all important in determining nest predation. Colonial nesting can have benefits as it can provide effective nest defence by sheer weight of adult numbers. Associations between two bird species, usually a predatory type and another perhaps prey species, has been shown to have advantages to the secondary species through improved defence of the nesting areas from potential predators.

There are also many instances of birds nesting in association with arthropods ranging from caterpillars and spiders through to various social insects including termites, ants bees and wasps (reviewed by Hansell 2000). It has not always been demonstrated that these associations are necessarily beneficial to the bird. However, nesting attempts by rufous-naped wrens (*Campylorhynchos rufinucha*) were more successful, in terms of fledging young, when a wasp (*Polybia rejecta*) nest was present (Joyce 1993). The vigorous defence of the wasps for their own nest is seen as being adopted by the birds. A potential disadvantage of this association is that the birds are also vulnerable to attack from the insects themselves (Hansell 2000).

In general, nest sites tend to located in close proximity to sources of food and water for the incubating adult. Certainly for those species demonstrating low %attentiveness (Chapter 6), the need to be close to a food source is critical. In some species however, particularly seabirds, the scarcity of suitable nest sites means that the adult has to travel a great distance from the nest location to a feeding site. Extreme examples of this are the emperor penguin and the snow petrel (*Pagodroma nivea*) which nest far from the sea in Antarctica (Maher 1962; Williams 1995).

The remaining influences on the climate enveloping the eggs are the nest itself and the nature of the source of heat. The nature of heat transfer from parent to egg is not the concern of this chapter (see Chapter 9), however, a mean core egg temperature of 35.7°C during incubation (Rahn 1991) means that egg temperatures are usually above ambient (although this is not always so; Chapter 16). A major role of the nest in incubation is to conserve heat although if ambient temperature is higher than the egg temperature then the nest also acts to help insulate the eggs from excessive heat (Rahn 1991). Similarly the nest can assist in maintaining a suitable humidity during incubation (Rahn and Paganelli 1990). These functions are achieved largely through nest shape and composition. However, in the following sections it can be seen, not only that varied nest architecture and composition is influenced by considerations of egg incubation but also by other, sometimes conflicting, requirements for the successful fledging of the brood.

Nest shapes

Shape categories

Hansell (2000) recognised eight nest shapes based on the examination of nests of more than 500 species in museum collections (Figure 2.1):

(1) *cup* – a nest with a distinct concavity to hold eggs, e.g. the song thrush;

(2) *dome* – a roofed nest with a side entrance such as that of the long-tailed tit (*Aegithalos caudatus*);

(3) *dome and tube* – a roofed nest with an additional antechamber or entrance tube, seen in an extreme form in the nest of the weaver (*Malimbus rubriceps*);

(4) *plate* – an above-ground platform with a shallow, indistinct concavity for eggs, e.g. the woodpigeon (*Columba palumbus*);

(5) *bed* – a ground nest with a shallow, indistinct concavity for eggs, e.g. the lapwing (*Vanellus vanellus*);

(6) *scrape* – a shallow depression in the ground with little or no gathered material, as made by a number of species nesting on coastal sand such as the Arctic tern (*Sterna paradisea*);

(7) *mound* – a heap of material typical of megapodes (Chapter 13);

(8) *burrow* – a cavity excavated in the ground, e.g. the bee-eater (*Merops apiaster*); or in a tree, e.g. green woodpecker (*Picus viridis*). Where cavity nesters employ

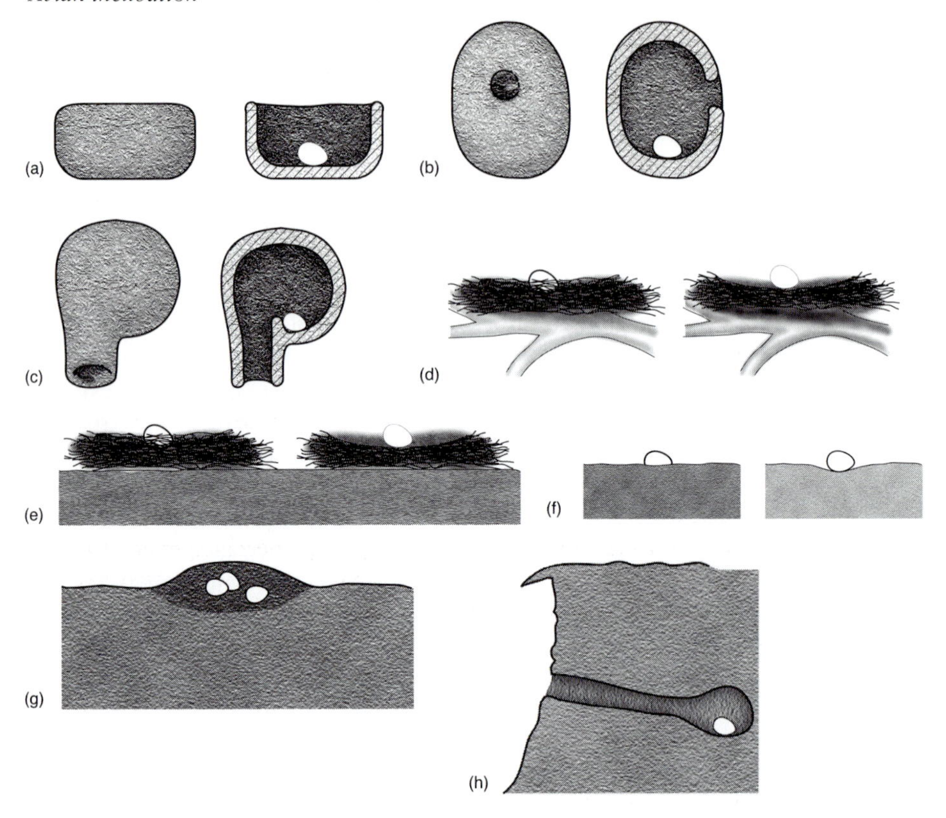

Fig. 2.1 Nest shapes can be assigned to one of eight possible categories: (a) cup; (b) dome; (c) dome and tube; (d) plate; (e) bed; (f) scrape; (g) mound; (h) burrow. Re-drawn after Hansell (2000).

ready-made sites, Hansell (2000) described nest shape on the basis of the structure assembled within, i.e. *cup, bed, etc.*

Not all nest shapes are possible in each type of nest location. The first four categories above are usually located above the ground whereas the remainder are located on, or in (i.e. in a burrow) the ground. For arboreal nests a main distinction is usually between enclosed or roofed nests (with or without a tube) and open nests. The distinction between cup and plate is that between a nest which is deep and has steep sides and the near flat platform. Therefore, plate, cup, and dome nests could be seen as representing sites which allow for different degrees of exposure of the eggs to ambient conditions, and as such could be seen as representing a continuum in the degree of responsibility of the parents for maintenance of egg temperature.

Open nests of cup or similar design are found in 63% of families of arboreal nesting passerines (Collias 1997), yet their design and composition vary greatly between species showing that, even considering a single nest shape, species will differ in the degree of responsibility of the nest for the maintenance of egg temperature and humidity. The cup nest of the Florida jay (*Aphelocoma coerulescens*) for example consists of a structural layer of sticks lined with rootlets, whereas that of the blue vanga (*Cyanolanius madagascarinus*) has two categories of plant material and two types of silk in the structural layer, decorated on the outside with lichen and lined with hair fungi.

Another aspect of variation even within the cup nest design is that of the weight of the nest in relation to the species building it. Unsurprisingly, there is positive correlation between bird weight and nest weight, which can for example be seen in the extended family Corvidae as recognised by Sibley and Monroe (1990). Species in this family overwhelmingly build nests of cup shape, although embracing species both at the lower and higher range of body masses found in the passerines. However, the smallest species tend to build nests that are smaller in proportion to their weight than the largest species such as crows *(Corvus* spp.). The nest of the brown-throated wattle-eye (*Platysteira cyanea*), at a mere 2.5 g, is considerably lighter than the 13.4 g bird, whereas that of Stella's jay (*Cyanocitta stelleri*), at 272 g, is more than twice the weight of the 128 g bird, (nest weight data Hansell unpublished data; bird weights: Dunning 1993). A possible explanation for this relationship between bird weight and nest weight is predation pressure. This may select for more concealed nests in smaller species, giving small nest size an advantage. Nests of large species, on the other hand, are hard to conceal and are selected for large size perhaps to provide better protection against predation or to be better insulated.

Bed and scrape shapes of nests, respectively a simple platform on the ground and a shallow depression on the ground, also represent nest structures with a limited role in maintaining egg temperature. Burrows or adapted cavities in the ground, or in trees, provide opportunities for regulation of the incubation environment similar to that of domed constructed nests. In the very long (1.8 m) burrow nests of the bee-eater respiration by the incubating birds, eggs, and nestlings affects the gas composition of the nest chamber (Ar and Piontkewitz 1992). In order to prevent a deleterious decrease in nest partial pressure for oxygen the maximal nest ventilation has to be 0.8 L min^{-1} which is easily achieved by the feeding visits of the birds (which convect 1.2 L per visit). The location of the tree hole nests of some species on particular sides of the trunk shows that nest site choice can probably be used to influence nest climate.

The varied shapes of nests again illustrate the multiplicity of selection factors operating on nest design. Choice of nest site dictates aspects of nest design. Arboreal nests need structural integrity, and may need specialised attachment. Birds of different sizes use different principles to protect their nests from predation which affect nest design, and cavity nesters devolve some of the nest functions including insulation, to the wall of the cavity itself.

Nest construction

Materials

After the choice of nest site and the species-typical nest shape, the selection of materials with which to build a nest is the third level at which parent birds can construct an incubation environment. Although there are an infinite number of kinds of materials that could be recognised, Hansell (2000) considered that at the most basic level there are only three: *animal*, '*vegetation*', and *mineral*. There were other materials that were uncommonly employed in nest construction including cow dung and synthetic material, such as paper.

Animal material includes salivary mucus used by swifts (Apodidae). More common is feathers, which can be derived from the parent, the female eider (*Somateria mollissima*) pulls breast feathers out to line her nest, or derived from other species and collected from the general environment. Similarly, fur and hair of mammals is collected in the same manner. Cast snakeskin is employed on the outer layer of some nests or within the nest cavity of others. Arthropod silk is also a very important material of animal origin particularly for arboreal-nesting species building nests of less than 30 g. Microscopic examination allowed Hansell (2000) to recognise three types of arthropod silk in nests: lepidopteran, spider cocoon, and spider web/retreat.

Hansell (2000) recognised 24 types of nest material derived from plants, fungi and lichens based on taxonomic criteria and the part of the plant involved. Broadly speaking these include various types of leaves, including pine needles. Broad leaves can be employed intact or as skeleton remains retrieved from leaf litter. Woody material was classified into fine woody stems less than 2 mm in thickness compared with sticks which were thicker and stiffer. Vine tendrils were characterised by a corkscrew form, and rootlets were distinguished from woody stems by their lack or rigidity and fine side branches. Other materials included bark and plant fibres. Grass material was split into narrow and broad leaves, stems and heads. Palm fronds were distinguished from palm fibres. Most of these materials are collected as dead plant material, however, green plant material is used in some nests (see also Chapter 12) as are flower heads. Other categories of 'vegetation' material included: horsehair fungus, fern stem scales, moss, lichen, rushes and seaweed.

Of mineral materials there are essentially two, stones and mud. Stones are relatively unimportant in nest structures although gentoo penguins (*Pygoscelis papua*) use substantial numbers of stones to raise their eggs above flat ground (Williams 1995). By contrast, mud is a vital component of many nests. Rowley (1970) found that 5% of bird species employ mud in a variety of ways and forms.

Hansell (2000) chose these various categories of nest material after the inspection of nests of several hundred bird species to reflect what birds are apparently trying to achieve in nest construction: simplicity in both construction behaviour and in the completed structure. Through choice of a small number of specific nest materials, construction behaviour can be relatively simple, allowing repetition of the same building movements for each piece added. This also has the effect of creating a nest of rather regular structure which is consequently less likely to fall apart.

Recognition of the differences in behaviour required for handling each type of material helps in the recognition of useful categories of building material. This is illustrated by grass, a plant Hansell (2000) differentiated into leaf, stem, and inflorescence. All these components are found in birds nests, and one may be included in the nest of many birds largely to the exclusion of the other two. This is because each has distinctive properties and requires a different building technique. Grass leaves for example are ribbon-like and flexible, grass stems are more rigid, hollow beams. The former can be bent around a bird to make a nest cup, the latter must be buckled at intervals to make polygons built up round the bird to make a cup.

Nest differentiation

Nests have multiple functions and certain materials are more suited to fulfilling one function rather than another, consequently nests are differentiated structures. Hansell (2000) described the composition of a nest in terms of four functionally distinct zones: *attachment* (any materials employed to secure the nest in place), *outer* (decorative) layer (material placed on the outside of the structure altering its appearance without affecting strength or attachment), *structural layer* (provides integrity to nest shape preventing distortion or it falling apart). The fourth nest zone, the *lining*, is a layer having no obvious structural role which lies inside the structural layer making contact with the eggs (Figure 2.2).

Not all four nest zones are necessarily present in all nests. For example, the simplest form of ground nest is a scrape, which has no materials in any nest zone. The nest of the ostrich (*Struthio camelus*) is no more than a bowl-like depression in the ground (Deeming and Ar 1999) and in many Charadriiformes nests are simply depressions on pebble beaches. Arboreal nests must possess the mechanical integrity not to fall apart. They must therefore have a structural layer of materials that creates a nest wall of whatever shape (dome, cup or plate) that will keep the eggs and parent secure in the nest site. One or more of the other three nest zones may, however, be absent. Only nests not fully supported from below will require attachment materials to secure the nest. The outer decorative layer is typical only of the nests of small birds, and generally absent from those of large ones (Hansell 2000). The nest lining layer is also absent from the nests of many species, leaving the materials of the structural layer in contact with the eggs.

Wherever a nest is placed it needs to remain secure for the duration of its use. Nests placed on the ground need no special design features to secure them in position, but many, although not all nests raised above the ground do. Materials of the attachment zone are those that secure the nest to the nest site. The position of the attachment falls into a number of fairly distinct categories (Figure 2.3). *Top* attachment is usually associated with the nest being suspended from its top to the nest site directly above it. *Top-lip* attachment typically describes the attachment of an open nest slung like a hammock between horizontal, supporting branches; typically, as in the nests of vireos (Vireonidae) and white-eyes (Zosteropidae) the nest is sited in the angle of a forked twig with the attachment materials looped over the branches lying either side of it. The nests of hummingbirds such as the long-tailed hermit (*Phaethornis superciliosus*) have a *top-side* attachment to the underside of a leaf where the attachment materials are wrapped around the nest and leaf.

Many nest types of bottom-supported nests are entirely supported from below by a branch or branches, others, like the nest of the Costa's hummingbird (*Calypte costae*), are balanced on a sturdy horizontal branch to which it is attached by silken 'guys' to prevent it from toppling off. Where nests are located in reeds or tall grasses, attachment materials need to be employed to prevent the nest from slipping down the narrow vertical plant stems, an attachment type described as *bottom multiple (vertical)* attachment and seen for example in the attachment of the nest of the reed warbler (*Acrocephalus scirpaceus*). The nests of many

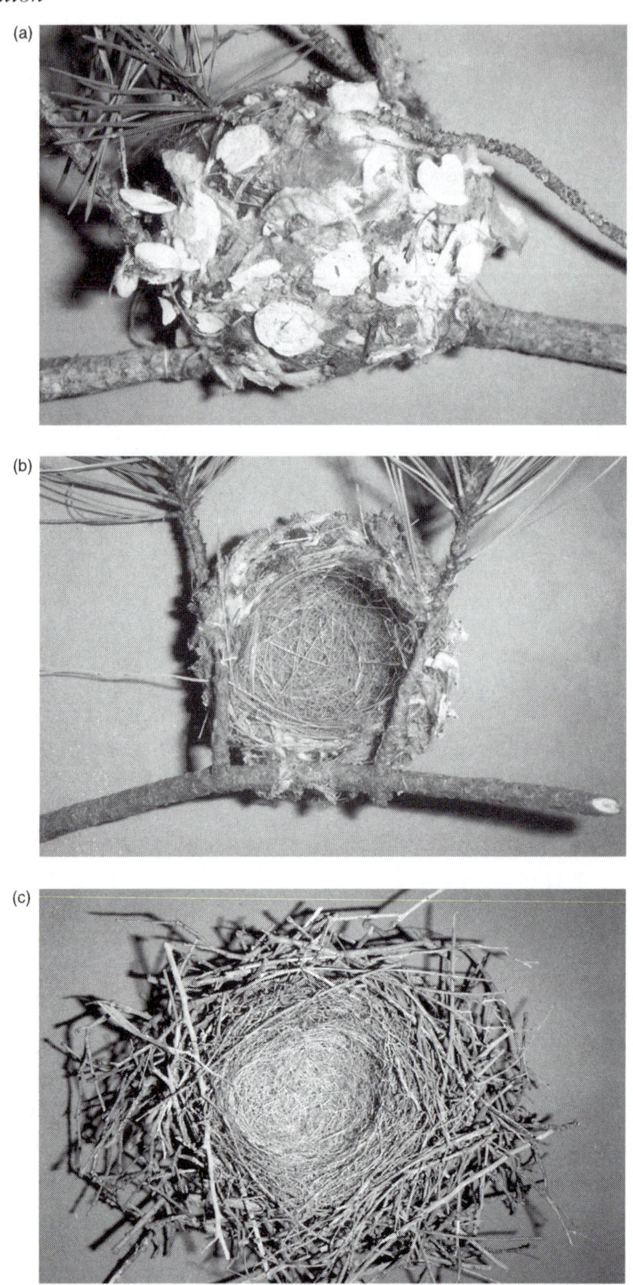

Fig. 2.2 The four nest zones: (a) Outer layer. The underside of the cup nest of the western warbling vireo (*Vireo [gilvus] swainsonii*) covered in white, papery spider cocoons. (b) Attachment. Above view of the nest of the plumbeus vireo (*Vireo solitarius plumbeus*) showing top lip attachment to twigs by means of arthropod silk. (c) Structural and lining layers shown in above view of the cup nest of the white-throated magpie-jay (*Calocitta formosa*). Photos by M. H. Hansell.

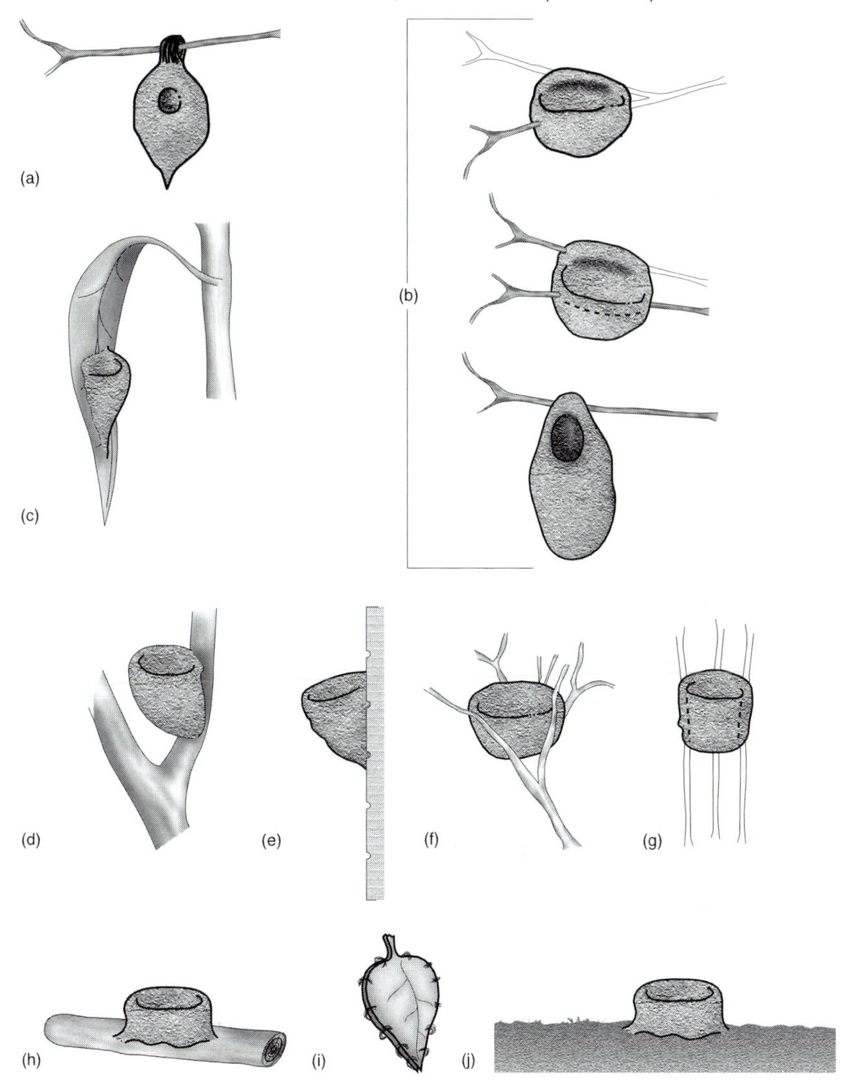

Fig. 2.3 Nest attachment can be assigned to one of ten possible categories: (a) top; (b) top lip; (c) top side; (d) bottom side; (e) wall; (f) bottom multiple (branched); (g) bottom multiple (vertical); (h) bottom; (i) leaf purse; (j) on ground. Re-drawn after Hansell (2000).

arboreal species are placed in the angle of upwardly directed forking branches. Their attachment is described as *bottom multiple (branched)*. Nests in this position are frequently held very securely in position by their own weight and so require little in the way of specialised attachment although the supporting branches may to a degree be embedded in the body of the nest. Attachment to the vertical walls of cliffs or buildings requires an adhesive material such as salivary mucus in the case of the edible-nest swiftlet (*Collocalia fuciphaga*) or, in the case of the cliff swallow (*Hirundo pyrrhonota*), mud. Attachment zone materials are varied however, and

Hansell (2000) found them in virtually all nests where they were present, to be a sub-set of materials used in the structural layer; consequently nest attachments can generally be regarded as specialised extensions of the structural wall of the nest.

Materials of the outer nest zone are attached to the outside of the structural wall of the nest and appear to serve to alter the nest's appearance. For that reason they may be called 'decorative' materials, although it is possible that they could serve additional or alternative functions. In general, Oniki (1985) showed that pale nest materials were associated with open sites and dark materials with shaded sites. The outer layer is particularly a feature of nests of smaller arboreal species and the expectation that an outer nest layer is more characteristic of small rather than large nests was tested for three passerine families (Tyrannidae, Corvidae and Fringillidae). There was a significant negative effect of nest cup diameter on the presence of outer layer materials (Hansell 2000). The most commonly used material in outer layers is lichen flakes (41% of sample), followed by moss (30%), and pieces of white membranous silk or white spider cocoons (27%).

Hansell (2000) investigated the various ideas that have been invoked to explain the presence of lichen as an outer layer. These include: mimicry of the general nest environment (mimicry), resemblance of a specific part of the nest environment, e.g. a knot of a branch (masquerade), and an attempt to blend into the back ground (crypsis). There was little association with the amount of lichen on the nest and the adjacent nest environment suggesting that a mimicry function was not particularly important.

The diameter of the nests was much greater than the diameter of the branches supporting them which suggested that outer layers do not provide a significant masquerade role. Most of the evidence suggested that application of lichen and other white material to the external surface of the nest led to disruption of the nest outline and hence crypsis is the likely to be the most significant function. A further possible additional function of lichen on the outer surface may be as a form of water-proofing. The nest of the long-tailed tit has over 2,000 flakes of lichen interspersed with 150 small pieces of membranous spider cocoon (Hansell 2000). This may serve to provide an initial water-proofing layer and it would be interesting to investigate the microclimate within this nest type under a variety of environmental conditions.

Nests must retain their structural integrity during their role as a site for holding eggs and chicks. All nests, other than ground and cavity nesters, have a structural layer which can often form most of its bulk. For arboreal nests, the structural materials are dominant in terms of nest weight, and may well be the only zone represented in the nest design. Nests supported by the bottom multiple method do not need specialised attachment materials, larger nests generally lack an outer (decorative) layer and many nests lack specialised lining materials. In many arboreal nest designs therefore, the main or even only insulative layer between the nest chamber and external conditions is the structural wall of the nest. This must therefore have multiple roles. There are the structural roles of supporting the eggs and, since the chicks of most arboreal nesters are not precocial, of providing a secure platform for unfledged young. Also, where the structural layer provides all or most of the material between the clutch and the outside world, there is the role of assisting nest chamber insulation.

To serve the mechanical role, different materials are appropriate for different sizes of nest. This can be seen in the broad relationship between nest size and the use of three major groups of structural materials, arthropod silk, grass and sticks. The use of arthropod silk as a structural material, usually mixed with vegetation such as moss to form a 'Velcro' fabric, is widespread among small birds of at least 25 of the 45 passerine families identified by Sibley & Monroe (1990), but these birds are overwhelmingly small species (Hansell 2000). Nests using grass either in the form of grass stems or grass leaves as the dominant material in the structural zone are generally built by somewhat larger birds, while stick nests are typical of larger arboreal passerines and large non-passerine species such as ibises (Threskiornithidae) and storks (Ciconiidae).

The fourth nest zone is the lining layer. When present, this lies inside the structural layer in contact with the eggs. Not all nests have a lining layer. For example, in a survey of the nests of the Tyrannidae, Corvidae, and Fringillidae, respectively 35, 38 and 33% of nests were found to be without lining materials (Hansell 2000). A wide variety of materials are employed in the nest lining layer and Hansell (2000) showed that the commonest were: grass stems, feathers, rootlets, plant down, fur/hair, and palm fibres.

With the exception of feather and fur/hair, each of these are commonly used in other layers of the nest construction. The primary role of lining material is seen as insulation for the incubating bird and the eggs. The effectiveness at trapping air of the different materials does affect the insulative properties of a nest. Feathers are more likely to used to line nests of early breeding species than late breeders (Møller 1984). Fur provided good insulation in nests of the eastern white-crowned sparrow (*Zonotrichia leucophrys oriantha*) but the rootlet lining of the rose-breasted grosbeak (*Pheucticus ludovicianus*) was much less effective (Skowron and Kern 1980). Møller (1987b) investigated the possibility that the use of feathers in the nests of the barn swallow (*Hirundo rustica*) might have a role in the control of haematophagous nest parasites but failed to find a relationship between parasite load in the nest and the use of feathers. There is, however, both correlational and experimental evidence that green plant materials rich in secondary compounds in the nest lining of common starlings (*Sturnus vulgaris*) have the function of controlling nest ectoparasites (Clark & Mason 1985, 1988). The serious problems raised by disease and parasitism may account for the very open designs of the nests of some species. The open twig platforms of species of doves may provide little insulation but also little refuge for mites and fleas. The balance of this trade-off between well insulated or poorly insulated nests could account for the open texture of the fabric of the nests in some species and the absence of a specialised lining in others.

Relationships between nest composition and construction

Materials not only give characteristic properties to the structures they are built out of, they also to a large extent dictate the method of construction. Birds do not depend upon specialised building apparatus for nest building. This can be recognised from the fact that, whereas the diet of a bird may be quite evident from its visible anatomy, the type of nest it builds cannot be predicted in this way. Birds depend upon specialised behaviour not specialised anatomy to build nests, and the behaviour must be appropriate to the materials being used. Hansell (2000) identified a variety of construction techniques each of which relates to nest construction using certain kinds of material. They are: (a) *Sculpting*; (b) *Moulding*; (c) *Sticking together*; (d) *Piling up*; (e) *Interlocking*, and (f) *Weaving*. These identify the methods by which the structural wall of the nest is created, after which non-structural layers may be added to either the outer or inner surfaces.

Sculpting is distinct from other nest construction techniques as it does not involve any form of assembly in the process of nest-building. The results of the analysis by Hansell (2000) are briefly described here. Sculpting involves the beak being used to excavate a cavity in a tree or the ground, which is large enough to house the incubating bird and its offspring. Woodpeckers (Picidae) are the ultimate sculptors of cavities in trees and large cacti. Holding on to the trunk with strong claws the bird uses its beak to chisel away at the wood. This technique can be seen as an extension of the feeding behaviour in this group. Excavation of a cavity in the ground, or arboreal termite nest, by kingfishers (Alcedinidae), however, requires the beak to be used in a way that is quite different from the feeding pattern in these species. Cavities can also be gouged out by the beak and in the parrots (Psittacidae) the powerful beak is used to bite away at the substrate.

The moulded mud nests of swallows and martins (Hirundinidae) and saliva nests of swifts (Apodidae) use the principle that the building material can be worked into a shape when initially malleable yet soon harden to produce a hard, rigid and relatively impervious wall. Moulding behaviour for nest-building could be described as nest shaping but this ignores the complex construction behaviours involved in making a nest. Nest construction in swallows (*Hirundo* spp.) involves collection of mud in and on top of the beak which, upon application, appears to help avoid air cavities developing in the nest wall, so preventing cracks or points of weakness. This problem is also minimised by a vibrating movements of the builder's head which appears to partially liquefy mud by the phenomenon of thixotropy so fusing the surfaces of separate beak loads together as the nest wall is built up. Inclusion of grasses or hair onto the mud wall of the nest as it grows helps to add tensile strength to a nest which lacks any support from below. The use of mud in vegetation-based nests is usually to line the cup and a smooth finish is achieved by shaping movements of the breast, a technique seen in birds employing other construction materials.

Some species employ a technique of using mud or mucus to stick together other nest materials (Hansell 2000). Here the amount of vegetation is the significant part of the nest and is only held together by the adhesive material. The white-winged chough (*Corcorax melanorhamphos*) produces a bottom supported mud cup with vertical walls which can support vertical forces, and grass and strips of bark embedded in the cup can take radial strains. Both the white-throated swift (*Aeronautes saxatalis*) and the cliff flycatcher (*Hirundinea ferruginea*) use salivary mucus to stick together vegetation. The nest of the chimney swift (*Chaetura pelagica*) is a bracket of straight sticks held together and attached to a vertical surface by a salivary mucus (Pearson and Burroughs 1936).

The piling up technique is a simple process by which a platform is built up at the nest site. Typically, piling up involves the birds placing sticks on top of one another to produce a nest platform in a tree. This technique is employed in many species of storks (Cicionidae), pelicans (Pelicanidae), raptors (Accipitidae), and pigeons (Columbidae). The piling up technique can be modified by the birds, commonly types of crow (Corvidae), working sticks more precisely into contact with one another in order strengthen the structure. The magpie goose (*Anseranus semipalmata*) creates a piled up nest structure in marshy ground simply by bending down rooted rush stems toward the central nest cup thereby raising the nest site above the water (Davies 1962).

Interlocking is the major category covering nest construction and could be considered as an extension of piling-up behaviour. Hansell (2000) recognised three key means by which birds interlock their nest materials: (a) *entangle*; (b) *stitches* and *pop rivets*; and (c) *Velcro*. Entangle is a construction type involving a wide variety of techniques that broadly achieve a method for ensuring that the nest materials stay together. These can be simply pulling and pushing twigs together as in the case of the castlebuilders (*Synallaxis* spp.). When grass and other flexible materials are used then there is usually some element of intertwining of the materials, however, grassy materials may also be 'felted' together (Skutch 1960) by additional strands of grass being inserted into the existing nest wall; both these techniques are apparently employed in the construction of the grass and rootlet nest of the white-winged snowfinch (*Montifringilla nivalis*; Aichorn 1989). Interlocking vine tendrils producing a very open nest fabric, seen in the insubstantial yet sturdy platform constructed by the rufous piha (*Lipaugus unirufus*).

Some species are able to stitch together their nest materials to form a nest structure (Hansell 2000). The common tailorbird (*Orthotomus sutorius*) forms a purse by stitching together the margins of green leaves. The thread employed includes lepidopteran and spider silk and plant down and stitches are made by driving the thread through leaf, grasping it on the other side and then driving through the leaf again. Coarseness of the thread and the elasticity of the living leaves prevents the structure from coming apart. Spiderhunters (*Arachnothera* spp.) suspend a nest chamber below a leaf using fine threads of silk (Madge 1970). These threads are driven through the leaf and the thickened end is trapped by the leaf.

Hansell (2000) considered 'Velcro' to be the most effective technique for interlocking nest materials. The Velcro process is based on the hook-and-eye principle and in nest building involves entanglement of vegetation or lichen in threads of silk. This form of construction is relatively widespread in birds as it is a simple behaviour that allows quite complex structures to be produced. An example of this is the wall of the nest of the long-tailed tit which uses predominantly fluffy spider egg cocoon silk entangled with small-leaved mosses.

Weaving is a construction method using long strips of vegetation and is typified by the generally suspended roofed nests of the Old World ploceine weavers such as the red-billed weaver (*Quelea quelea*) and the elongated hanging bags of the much larger New World orioles and oropendolas like the crested oropendola (*Icterus nigrogularis*). Both these groups of birds use combinations of loops, hitches, loop tucks, spiral coils, and other devices to weave together the long flexible vegetation strands to create the nest wall (Collias & Collias 1964; Heath & Hansell 2001).

The evolution of nest building

The steps that lead to the change of nests through evolution can broadly be divided into two types, innovations of design and of technology. In the former, the same materials are used to produce a nest with new architectural features; in the latter the key biological change is the selection of a new material. The latter is likely to have the more profound effects, as the new material may open up a whole range of new design options leading to a radiation in architecture and possibly facilitating speciation (Hansell 2000). An example of such a technological innovation could have been the inclusion of arthropod silk as a structural material in the nest, since it provides a very simple mechanism for holding vegetation nest materials together and is strong enough to support the whole weight of nest, clutch, and incubating parent. This is dramatically shown by the nest of the rock warbler (*Origma solitaria*), which is attached by silk to a cave roof. The use of arthropod silk has arisen several families of passerines independently, but we know too little of their detailed phylogeny to understand its consequences.

Winkler and Sheldon (1993), having determined the phylogeny of the swallows and martins (Hirundinidae) using DNA hybridisation, were able to show that one technological change, from excavating burrows or cavities to the construction of nests with collected mud, led to a diversification of nest materials, nest design, and nest site. These wall-attached mud nests range from the simple cup of the barn swallow to the enclosed chamber with entrance tube built by the cliff swallow. The most obvious consequence of this innovation has been the ability to colonise new habitats and attach nests in less accessible sites on the walls of cliffs or buildings. Such changes in the nest design may also have affected the incubation environment, although this does not appear to have been the primary selection pressure.

Eberhard (1998) in a study of the cavity nesting African parrot genus *Agapornis*, showed that from species that built a simple cup nest in a tree cavity there evolved species that elaborated the cup into a dome that fully lined the cavity, more obviously a change in design than technology in this case. In the Furnariidae, cavity nesting seems to have been the ancestral condition (Zyskowski & Prum 1999), yet the family exhibits one of the most diverse range of nest designs of any among the passerines. It includes the domed mud nest of *Furnarius* spp. and enclosed stick nests of thornbirds (*Phacellodomus* spp.). In both families the new nest designs have allowed opportunities for better security from predation or foraging opportunities. These designs may also reflect the evolution of new methods of controlling nest cavity environment but the evidence for this is currently lacking.

Collias (1997) on the basis of classifying nests into just three basic types (*hole, open,* and *domed*) pointed out that all are present in the few families that are considered to be the most primitive of the passerines i.e. Eurylaimidae (broadbills), Rhinocryptidae (tapaculos) and Menuridae (Australian lyrebirds). He concluded that diversity of nest design in the passerines evolved early. In some families nest design is apparently conservative, for instance in the Columbidae (pigeons and doves) where a simple twig platform is almost universal regardless of nest site (Goodwin 1983). Other families, notably the Furnariidae but also the Tyrannidae, show great nest variability. Hansell (2000) showed that in the Tyrannidae changes in nest design through evolution may have been frequent. Attaching nest designs to a phylogeny of the Phylloscartes group, produced by Lanyon (1988a) from skeletal evidence, showed that within the nine genera there were species building domed nests (both bottom supported and suspended) as well as open nests (both bottom supported or top lip attached). In the 10 genera of the Flatbill and Tody-tyrant assemblage (Lanyon 1988b), there is a species building a bottom multiple supported cup nest among species with suspended nests, some globular, some greatly elongated and some with a downwardly-directed entrance tube.

This evident ability at least in some families to respond readily to selection pressures does not however tell us what the selection pressures are and therefore what the nests are designed for. Crook (1960, 1963) in pioneering comparative studies on the weaver birds (Ploceinae), tried to understand the functional design of nests by relating them to habitat type even though the sequence of evolutionary events giving rise to them was poorly known in order. On the basis of this he concluded, amongst other things, that the innovation of a dense 'middle roof layer' seen in the nest of the brown-throated weaver (*Ploceus castanops*) evolved to adapt this species to areas of higher rainfall. The unique building technique of the red-headed weaver (*Anaplectes rubriceps*), in which green sticks are bound together with the tails of bark projecting from their ends he interprets as having evolved from an ancestor finding a replacement for strips of green grass as it moved from a wetter to a dryer breeding habitat.

Molecular genetic techniques will give us more and more detailed phylogenies. From this the pattern of evolutionary changes in nest design and composition will be seen more clearly. However, this will need to be accompanied by field and laboratory experiments to understand more fully what functions such nest designs serve.

Function of nests

The nest is a critical part of the bird-nest incubation unit (Chapter 1) and it can have an important role in maintaining the incubation environment. For instance, the temperature and humidity of the nest air is usually higher than the ambient environment (Rahn 1991). However, nests have a range of other functions including restraining the eggs in an inaccessible nest site and helping in their concealment prior to incubation. In many instances nests also act to contain the hatched brood for a variable time leading up to fledging, maintaining them as far as is possible in a healthy, parasite-free environment. Nests, therefore, protect the eggs and brood from the physical and biological hazards that surround them (Table 2.1). This function is assisted by the attentions of a parent (Chapter 6). Certain nest materials and construction techniques may be ideal for reducing these risks but the nest may be designed to provide protection from more than one of them.

Differentiation of the nest into different zones of distinctive materials can therefore be understood as the distribution of different nest functions to separate zones. However, since the nest may be expressed as a collection of zones, materials in one zone could compromise the effectiveness of materials in another. It is logical to expect trade-offs between the use of materials concerned with different functions. Consequently in trying to understand the selection of materials to control the temperature and humidity of eggs during the incubation period it is important to

Table 2.2 Possible trade-offs in the functional design of nests differentiated into two or more distinct layers. (Note: Most of these influences are speculative due to the absence of experimental evidence.) For each of the three nest zones (1. Outer; 2. Structural; 3. Lining) the primary adaptive role is indicated, then possible additional effects of controlling or exacerbating biological and physical hazards to the brood are listed, in particular influences on the incubation environment of the eggs.

1. OUTER NEST LAYER	
Primary role (Biological)	Conceals nest from predators (disruptive camouflage/crypsis)
Other possible effects:	
Biological	None.
Incubation control	Reflects light, diminishing nest heating by sunlight.
	Reduces water penetration, diminishing conductive heat loss.
	Reduces water evaporation, assisting/reducing conductive heat loss.
2. STRUCTURAL NEST LAYER	
Primary role (Mechanical)	Insures integrity of nest structure.
Other possible effects:	
Nest roof	
Biological	Keeps predators out.
	Conceals brood from predators.
	Restricts adult escape from predators.
	Reduces adult control of chick feeding.
Incubation control	Enhances insulation.
Amount of nest material	
Biological	Nest more detectable by predators.
Mechanical	Nest stronger.
Incubation control	Insulation improved.
Density of fabric	
Biological	Increases nest ectoparasite population.
Incubation control	Reduces rainwater penetration.
	Reduces nest-drying rate.
	Increases/reduces nest cavity humidity.
	Reduces/increases heat loss.
3. NEST LINING LAYER	
Primary role (Depends upon type of material):	
Feathers and similar materials:	
Biological	Enhances/reduces nest ectoparasites.
Incubation control	Increases insulation.
	Reduces nest-drying rate.
	Increases/reduces nest cavity humidity.
Green Plant materials	
Biological	Controls parasitism and disease of brood.
Incubation control	Increases nest cavity humidity.
Fine grasses and similar materials	
Biological	Reduces damage or abrasion of eggs and chicks.
Incubation control	Increases/reduces insulation.
	Increases/reduces nest cavity humidity.

consider how this might be affected by selection for optimising other nest functions (Table 2.2). Indeed some materials may be good for controlling the incubation environment in certain circumstances but poor in others. Currently little is known about this but some observations suggest that more detailed studies on functional design would repay investigation.

An illustration of this concerns the function of the lining layer. It has already been indicated that this layer is absent in many nest designs. However, the question that should be asked is not why this occurs, but why it is there at all. It could be argued that the nests of many species could be made to provide adequate insulation if they were of the same material only more of it. However, the inclusion of a lining layer of feathers for example might produce the same effect for less material. Møller (1984) has shown that the use of feather linings is related to cold climate, while Hilton *et al.* (unpublished data) have confirmed the greater insulative power of feathers compared with grass and moss. This could provide us with an estimate of just how big a nest built entirely of moss would have to be to provide the insulative power of moss walled nest lined with feathers. What would be the disadvantages of a large all-moss nest? Two possible candidates can be suggested: cost and predation.

So little is currently known about the costs of nest building to judge whether such a cost difference would be biologically significant. However, Møller (1990) was able to show that the level of predation on unoccupied blackbird nests baited with eggs was positively correlated with nest size. With a selection pressure such as this, the inclusion of a specialised insulative nest lining could be considered an anti-predator adaptation. Slagsvold (1982a, 1982b) showed that fieldfares (*Turdus pilaris*) provided with larger nests had higher nesting success and identified three possible reasons to explain their relatively small size: cost, predation, and the rate of nest drying. Slagsvold (1989a, 1989b) was able to show that smaller nests did indeed dry out faster and Hilton *et al.* (unpublished data) showed that not only does dampness dramatically reduce the insulative properties of a range on nest materials, it reduces the effectiveness of good insulative materials like feathers to the level of the poorer ones. This leads to the counter-intuitive prediction that in a wet environment, a better ventilated nest might provide overall better insulation than a less ventilated one. Slagsvold (1989b) suggested that in the case of the fieldfare this might also affect the choice of nest materials, with a more absorbent material being more suitable than grass in a sheltered nest site but less suitable in a more exposed site where a grass nest would not become so saturated in rain and would dry out faster.

Bird nests are generally highly characteristic within species but vary greatly between species. This chapter has shown that nests can perform multiple functional roles one of which may be to regulate the physical environment surrounding the eggs. Much of the between-species variability could be explained by the relative importance of these roles, however, factors such as the availability of nest materials and the size of the bird may give rise to different solutions to the same problem between species. Experimental work on nest function now needs to be pursued alongside careful descriptions of the anatomy and composition of the nests themselves.

3 *Functional characteristics of eggs*

D. C. Deeming

The egg is a key feature of avian reproductive biology and is the *raison d'etre* for the bird-nest incubation unit. As a consequence there has been a lot of interest in oology and egg composition and morphology have been extensively studied in a variety of species although, because it's commercial significance, there has been a lot of emphasis on the eggs of the domestic fowl (*Gallus gallus*). This chapter only briefly reviews the morphology and composition of eggs from, where possible, non-commercial species; more extensive reviews of avian oology are provided by Romanoff and Romanoff (1949), Romanoff (1967) and Burley and Vadhera (1989). This chapter also describes the relationships between egg characteristics, embryonic development and various aspects of the incubation environment.

Egg morphology and composition

Egg formation

A detailed description of the process of egg formation in the domestic fowl can be obtained from Gilbert (1971). Briefly, a female bird hatches with all of the ova (egg cells) she is capable of laying in her lifetime already within her single ovary found on the left side adjacent to the kidney. The right ovary has no function and should the left ovary be lost, it will develop into a testis-like structure without any reproductive function. When the hen becomes sexually mature hormonal changes cause some of the tiny ova to begin to develop in sequence. Proteins and lipids are produced in the liver and transported as vitellogellin and lipoprotein via the blood to be used by ovarian tissue to increase the diameter of each ovum and as it enlarges it becomes recognisable as the yolk (White 1991). In the ovary of a sexually mature female several ova (up to 15) of varying sizes and stages of development are present.

A mature ovum is released from the tissue in the ovary that surrounds it and moves into the mouth of the oviduct (infundibulum) located close by in the abdominal cavity. It is here that fertilisation takes place if sperm are present. Irrespective of whether it has been fertilised the yolk moves into and down the oviduct in preparation of the formation of albumen and the shell. In the magnum dehydrated albumen (egg white) proteins are laid down as a capsule around the yolk. The two shell membranes are deposited around this albumen layer whilst the egg is in the isthmus. In the shell gland there is secretion of fluid which hydrates the albumen proteins making them swell (often called 'plumping') and stretching the shell membranes allowing the egg to assume its normal shape. The hard shell then forms on top of the fibres of the outer shell membrane by crystallisation of calcite (a type of calcium carbonate, $CaCO_3$) into cones and then blocks that abut each other forming a solid layer (Bradfield 1951; Board and Sparks 1991). Once the shell is completed shell accessory material may be deposited on the shell surface and the egg is voided through the cloaca.

Descriptions of egg formation in species other than the domestic fowl are uncommon. Patterns of yolk deposition are described for various seabirds (Grau 1984; Astheimer *et al.* 1985; Astheimer and Grau 1990; Warham 1990). In general, rate of yolk deposition is correlated with initial egg mass, 7–8 days for the 7–8 g yolk of the fairy tern (*Gygis alba*) through to 25 days for the 90 g yolk of albatrosses (*Diomedea* spp.), although the rate of deposition varies between Procellariformes, other seabirds and non-seabirds (Astheimer and Grau 1990). In the common guillemot (*Uria aalge*) yolk deposition takes place over 11 days for first eggs but even though yolk deposition was faster in replacement eggs eventual yolk mass was significantly smaller (Birkhead and Del Nevo 1987).

Egg morphology

Egg morphology is largely described based on that of the domestic fowl (Romanoff and Romanoff 1949; Burley and Vadhera 1989) and it is assumed that this is applicable to all species of bird. Given differences in the relative composition of eggs from birds exhibiting different developmental modes (see below), it would be interesting to investigate whether this assumption is true. Avian eggs consist of three main components: shell, albumen, and yolk.

The shell is an inflexible layer of inorganic material, mainly calcite crystals deposited on the outer surface of the fibrous outer shell membrane which in turn is deposited on the inner shell membrane (Tullett 1987; Board and Sparks 1991; Solomon *et al.* 1994). Each membrane consists of layers of fibres running parallel to the shell surface. The fibres in the outer membrane are shorter and have a wider diameter (1.3–3 μm), forming a more open mesh, than the longer but narrower fibres (0.9 μm) of the inner shell membrane. A limiting membrane (2.7 μm thick) that is in contact with, and covers the fibres of the inner shell membrane seems to serve to separate albumen from the membranes. The external surface of the shell may, or may not, be covered by a layer of accessory material (organic cuticle or inorganic cover).

A fibrous vitelline membrane surrounds the yolk. This membrane is less robust than the shell membranes and is deposited as an inner peri-vitelline layer before ovulation and an outer peri-vitelline layer initiated some 15 minutes post ovulation as the ovum enters the magnum (Wishart and Staines 1999). Yolk forms a spherical body comprised of spheres and granules. Most of the yolk consists of pigmented yellow yolk although there is a colourless component to the yolk. This white yolk is found in the centre of the sphere, the latebra and as a tube (the neck of the latebra) extending to the nucleus of Pander immediately below the blastoderm. White yolk is less dense than yellow yolk and so the upper hemisphere of the yolk has a slightly lower density than the lower hemisphere (Rahn 1991). Use of fat-soluble dyes has shown that yolk deposition exhibits a diurnal pattern and so the yolk is multi-layered (Grau 1976).

Albumen forms a multi-layered structure with distinct zones. The outer thin albumen is a narrow fluid layer adjacent to the limiting membrane. The outer thick albumen is a gel that forms the centre of the albumen and is attached to the limiting membrane at the poles of the egg. The inner thin albumen is another fluid layer surrounding the yolk whereas the inner thick albumen is a thin viscous layer immediately adjacent to the vitelline membrane around the yolk. This inner thick albumen is continuous with the chalazae, twisted proteinaceous fibres that appear to suspend the yolk within the albumen layers.

Egg composition

Whole yolk (yellow and white combined) of fowl eggs consists of 47.5% water, 33.0% lipid, and 17.4% protein (Burley and Vadhera 1989). Free carbohydrate is only 0.2% of the yolk whereas inorganic compounds form 1.11% and other compounds form 0.8% of the yolk (Burley and Vadhera 1989). The lipid fraction is very high in avian eggs (60% of dry yolk; Burley and Vadhera 1989) but free lipid is absent in fresh yolk and lipids are found combined with proteins as lipoproteins. The predominant lipids fractions (Speake *et al.* 1998) are triacylglycerols (66.1% of total lipids) followed by phospholipids, especially lecithins (24.7%) and cholesterol (1.3%). Within each class of lipid the fatty acid residues have saturated and unsaturated hydrocarbon chain lengths of between 14–22 carbon atoms although C16 and C18 chain lengths predominate (Speake *et al.* 1998).

The fatty acid profile of yolk lipid is a function of both physiology and nutrition of the female. The major saturated and monounsaturated fatty acids can be synthesised in the maternal liver as well as obtained from the diet but the 18-carbon polyunsaturated fatty acids (linoleic and a-linolenic) can only be derived from the diet (Speake and Thompson 1999). The composition of dietary lipid has quite important effects on the fatty acid profiles of yolk depending on the species. Grains and seeds are rich in linoleic acids, vegetation is rich in a-linolenic and fish is rich in long-chain polyunsaturated fatty acids. The predominant component of the diet influences the fatty acid profile of the egg yolk. Comparison of yolks from wild-derived and captive-derived eggs do reveal significant differences in fatty acid profiles reflecting the differing diets (Speake *et al.* 1998) but the significance of these differences in yolk composition on embryonic development and hatchability have yet to be fully determined.

Whole albumen in the fowl egg consists of 88.5% water and 11.5% solids (Burley and Vadhera 1989). Proteins constitute 10.5% of whole albumen compared with 0.5% carbohydrates (free glucose), 0.5% inorganic ions and 0.02% lipids. There are at least 13 individual proteins within the albumen of fowl eggs with ovalbumin forming 54% of the total (Burley and Vadhera 1989). Ovotransferrin and ovomucoid form the next biggest fractions (12 and 11% respectively) with ovoglobulins G2 and G3 forming 4% each. Lysozyme forms 3.5% of the total protein with ovomucin and ovoinhibitor both forming 1.5% of the total. The other proteins form less than 1% of the total. The composition of the different layers of albumen reflect in part their different chemical composition. In particular, the outer thick layer of albumen is five times richer in ovomucin than the layers of thin albumen and the chalazae are particularly rich in ovomucin (Burley and Vadhera 1989).

Investigation of the albumen of eggs from other species has revealed that whilst the major protein fractions are usually present in all species their relative proportions vary considerably and can be species dependent (Osuga and Feeney 1968; Sibley and Ahlquist 1972). For instance, compared with the fowl egg, ovomucoid forms a higher proportion of emu (*Dromaius novaehollandiae*) eggs.

Albumen has many roles in avian development. It is the major source of water and mineral ions within the egg required as the embryo develops (Chapter 4). Albumen proteins are a key component of the anti-microbial defence of the yolk prior to, and during early incubation. The various proteins are enzyme inhibitors, can iron, and vitamins, and include lysozyme, a potent bactericide (Tranter and Board 1982; Burley and Vadhera 1989). Removal of albumen from the egg can reduce embryonic growth (Chapter 4) indicating that it is a significant source of nutrition for the embryo.

Altricial–precocial relationships between components of egg contents

Avian hatchlings display variation in apparent maturity from blind, helpless altricial types through to fully active, independent precocial types (Nice 1962; Starck and Ricklefs 1998). It has long been recognised that the initial composition of eggs reflects the degree of hatchling maturity (Romanoff and Romanoff 1949; Ar and Yom-Tov 1978; Carey *et al.* 1980; Ar *et al.* 1987; Vleck and Vleck 1987; Sotherland and Rahn 1987). Eggs from species producing altricial young have a higher albumen content than eggs from those species producing precocial species with semi-altricial and semi-precocial species having intermediate values (Figure 3.1). These differences in composition are indicative of the different water and energy contents in these groups of eggs (Figure 3.2). Altricial species produce water-rich but energy-poor eggs compared with the water-poor, energy-rich eggs of precocial species (Vleck and Vleck 1987). This in turn correlates with the incubation periods required by these different types of hatchlings; for a given mass altricial species have eggs with lower energy content and shorter incubation periods than other groups (Vleck and Vleck 1987). There is also a close correlation between the water content of the egg and of the hatchling (Table 3.1).

These differences in egg composition are also reflected in the composition of the lipid fraction of the yolk (Speake and Thompson 1999). Eggs from precocial species contain yolks high in energy-rich tri-acylglycerides (TAG) but in a semi-precocial lesser black-backed gull (*Larus fuscus*) there is a decrease in the amount of TAG and an increase in lower energy phospholipids (PL). Eggs of the pigeon (*Columba livia*) have even lower levels of TAG and higher levels of PL and cholesterol esters. Yolks of eggs of the semi-altricial emperor penguin (*Aptenodytes forsteri*) have a lipid composition comparable to that of the precocial fowl but Speake and Thompson (1999) consider this as an adaptation to the pattern of embryonic development in this species. These data suggest that the composition

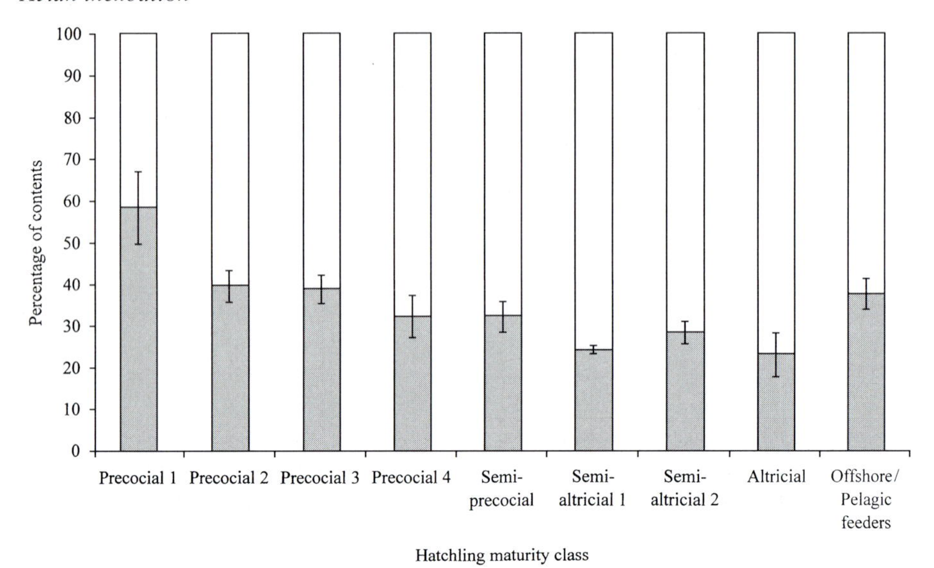

Fig. 3.1 Proportions of albumen (white) and yolk (grey) in the contents of avian eggs from species exhibiting different hatchling maturity. Values are mean percentages (±SD) derived from Sotherland and Rahn (1987).

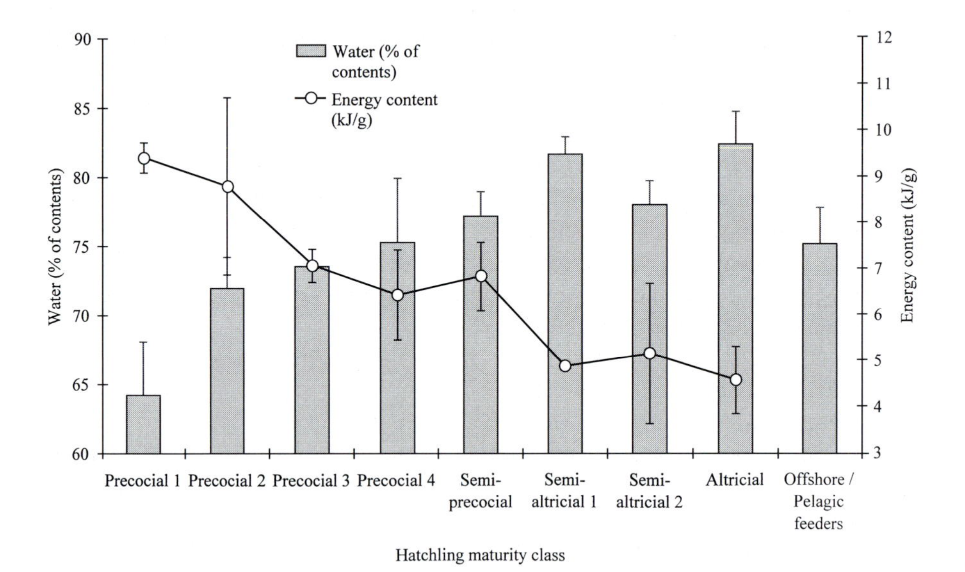

Fig. 3.2 Percentage water content of avian eggs (mean + SD) and energy content (mean ± SD; kJ g^{-1}) of avian eggs from species exhibiting different hatchling maturity. Data from Sotherland and Rahn (1987) and Vleck and Vleck (1987) respectively.

Table 3.1 The relationship between mean values for percentage water content of eggs at oviposition and of hatchlings for different groups of hatchling type. Data from Sotherland and Rahn (1987).

% Water in hatchling type	Egg	Hatchling
Megapodes	66.0	67.5
Precocial	73.4	72.4
Semi-precocial	76.9	77.9
Semi-altricial	81.7	83.6
Altricial	83.6	83.9

of eggs is determined by the energy demands of the embryo, although it is not known how this is achieved, and further work on a wider range of species is needed to ensure that this pattern holds true for most bird species.

Egg dimensions

Egg size

The size of eggs has long interested researchers (Heinroth 1922) and in particular Schönwetter (1960–1985) compiled tables of the mass, linear dimensions and shell thickness of bird eggs from over 8,000 species. This superb resource has been utilised to examine Class-wide relationships for egg dimensions. Rahn and Paganelli (1988a) summarised the data in Schönwetter's tables on the basis of orders of birds (Figure 3.3) and showed that egg mass in passerines and non-passerines both showed lognormal distributions. For passerines mean egg mass derived from the lognormal distribution was 2.85 g (SD ± 1.99) compared with a normal mean value of 3.72 g (SD ± 3.45). For non-passerines the mean derived from the lognormal distribution was 19.3 (SD ± 3.56) which was much lower than the mean derived from untransformed data (42.8 g, SD ± 88.7). The range of egg mass passerines was highly restricted compared with non-passerine species (Figure 3.3; Rahn and Paganelli 1988a).

Egg mass is a function of adult body mass (Rahn *et al.* 1975). The slope (i.e. power function b) for the relationship between body mass and egg mass is 0.67 and common to all orders of birds but the value of the proportionality constant (a) varies and is characteristic of each group. Therefore, for any given body mass, initial egg mass varies between birds of different Orders. For example, in Procellariformes a 100 g bird will produce a 21 g egg compared with a 4.5 g egg for parasitic Cuculiformes. A similar effect is seen in Passeriformes where relative egg mass can vary between different families.

Egg shape

Avian eggs adopt a variety of shapes usually described in terms of elongation (i.e. e = length/breadth) although more mathematical descriptions of shape

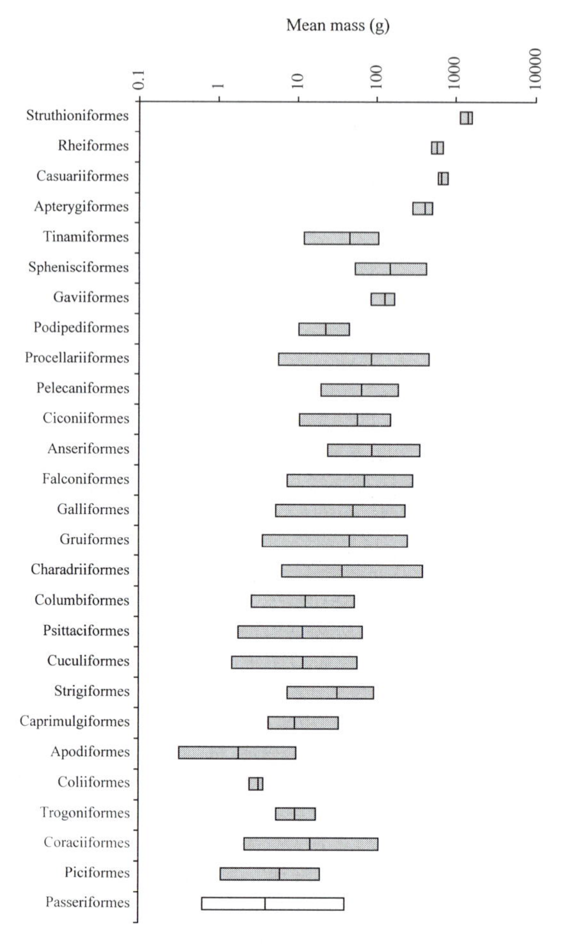

Fig. 3.3 The mean (centre line) and range of initial egg mass for the different orders of birds (data from Rahn and Paganelli 1988a). Note the log scale on the y-axis.

are available (Smart 1991). The range of shapes is from the elongated eggs of Apterygiformes and Gaviiformes ($e = 1.61$) through the more rounded eggs of the ostrich (*Struthio camelus*) and the Strigiformes with an elongation of 1.21 (Rahn and Paganelli 1988b). Elongation has a significant regression with egg mass but r^2 is negligible and Class-wide analysis masks relationships between elongation and mass within different families: although elongation often increases with egg mass in some families it decreases (Rahn and Paganelli 1988b).

Smart (1991) reviewed the factors affecting egg shape. The broad end of the egg may be formed in the zone of relaxation behind the leading edge of a peristaltic wave that moves the egg down the oviduct (Smart 1991). Warham (1990) reviewed the relationship between egg shape and pelvis shape for petrels (Procellariformes). It was found that there were significant relationships between egg shape index (breadth/length) and pelvis width and depth. The species laying the most spherical eggs had a deeper pelvis which were more spread caudally and those species laying elongated eggs had a shallow pelvis that narrowed caudally.

The role of egg shape may be related to increasing egg (and hatchling) mass for a given diameter of the oviduct, increasing shell conductance in the broad end of the egg or matching the egg to a particular nest environment (Smart 1991). For example, Tschanz *et al.* (1969) showed that the highly pointed eggs (e = 1.63) of the guillemot (*Uria aalge*) tend to roll in a tight circle thereby minimising the risk of the egg falling from the nesting ledge. By contrast, analysis by Barta and Székeley (1997) suggested that the shape of eggs was related to clutch size although their assumption of a circular brood patch was not realistic.

Relationships between egg composition, mass and shape

Mass of an egg at oviposition is a function of the combined mass of the yolk, albumen, and shell and not only do these vary between species but these components also contribute to the degree of variability in egg mass within a species. Relationships between initial egg mass and the mass of the egg components have been investigated in only a few species of different developmental modes (Williams 1994). In some species, albumen mass increases faster than egg mass whereas other species it is yolk mass which increases faster and in further species there are no differences between the components. There are no general patterns regarding the relative contributions of the yolk and albumen to egg mass for any developmental mode and there is little evidence to suggest that larger eggs are associated with higher protein contents. Shell mass increases faster than egg mass in passerines and non-passerines alike (Table 3.2).

Egg mass is correlated with the linear dimensions in non-passerines and passerines alike (Rahn and Paganelli 1988b). In both groups as egg mass increases there is a higher rate of increase in maximum length (L) than for maximum breadth (B; Table 3.2) and the degree of variation in length is twice that for breadth. Hoyt (1979) showed that initial egg mass could be predicted from linear dimensions (Table 3.2). The mean value of the constant, k, varies for non-passerines and passerines (0.542 and 0.522 respectively) and the range of values within a group is large and this variation means that it usually better to calculate k for individual species (Rahn and Paganelli 1988b).

Relationships between egg composition and linear dimensions have not been fully explored. Anderson and Deeming (2001) showed that in the houbara (*Chlamydotis undulata*) wet albumen mass was significantly correlated with both L and B but wet yolk mass was only significantly correlated with B. This, together with increased variation for L, probably reflects the constraints imposed on B by the diameter of the oviduct. Such relationships are important because the contents of the egg determine the physical dimensions of the shell, which then constrains the volume available for embryonic development. The volume of albumen may well prove to be more important than yolk in determining egg size and rates of embryonic growth (Chapter 4).

The shell

The eggshell is a remarkable structure performing a variety of functions during the lifetime of an egg. The shell is the medium through which heat is conducted

Table 3.2 Summary of the relationships between various egg parameters, initial egg mass and incubation period.

Parameters	Relationship	Reference
Non-passeriformes: Length : egg mass	$L = 14.7\ W^{0.341}$	1
Non-passeriformes: Breadth : egg mass	$B = 11.3\ W^{0.327}$	1
Passeriformes: Length : egg mass	$L = 15.1\ W^{0.345}$	1
Passeriformes: Breadth : egg mass	$B = 11.3\ W^{0.325}$	1
Egg mass : linear dimensions	$W = 5.48 \times 10^4\ LB^2$	2
Volume: linear dimensions	$V = 5.07 \times 10^4\ LB^2$	2
Shell area : egg mass	$A = 4.835\ W^{0.662}$	3
Volume : egg mass	$V = 4.951\ W^{0.666}$	3
Density : egg mass	$p = 1.038\ W^{0.006}$	3
Non-passeriformes: Shell mass : egg mass	$SM = 0.0546\ W^{1.113}$	4
Passeriformes: Shell mass : egg mass	$SM = 0.0553\ W^{1.024}$	4
Water vapour conductance : egg mass (n = 29)	$G_{H_2O} = 0.432\ W^{0.780}$	5
Water vapour conductance : egg mass (n = 90)	$G_{H_2O} = 0.384\ W^{0.814}$	6
Total functional pore area : egg mass	$Ap = 9.2 \times 10^{-5}\ W^{1.236}$	5
Shell thickness : egg mass (n = 367)	$L = 5.126 \times 10^{-3}\ W^{0.456}$	5
Shell thickness : egg mass (n = 3217)	$L = 0.0546\ W^{0.441}$	4
Pore numbers : egg mass	$N = 304\ W^{0.767}$	7
Incubation period : egg mass	$I_p = 12.03\ W^{0.217}$	8
Incubation period : egg mass	$I_p = 11.64\ W^{0.221}$	6
Rate of water loss : egg mass	$M_{H_2O} = 13.243\ W^{0.754}$	9
Pre-internal pipping oxygen consumption : egg mass	$M_{O_2} = 23.5\ W^{0.734}$	10

Where: A = surface area of shell (cm^2); Ap = Functional pore area (cm^2); B = maximum breadth (mm); G_{H_2O} = water vapour conductance (mg H_2O $Torr^{-1}$ d^{-1}); I_p = incubation period (d); L = maximum length (mm); M_{H_2O} = Rate of water loss (g d^{-1}); M_{O_2} = pre-internal pipping oxygen consumption (ml d^{-1}); n = samples size; N = pore number; p = egg density (g cm^{-3}); SM = shell mass (g); V = egg volume (cm^3); W = initial egg mass (g). References: 1 – Rahn and Paganelli (1988b); 2 – Hoyt (1979); 3 – Paganelli *et al.* (1974); 4 – Rahn and Paganelli (1989); 5 – Ar *et al.* (1974); 6 – Ar and Rahn (1978); 7 – Ar and Rahn (1985); 8 – Rahn and Ar (1974); 9 – Ar and Rahn (1980); 10 – Rahn and Paganelli (1990).

during contact incubation and it provides physical protection to the egg contents but its coloration can also act to help camouflage the egg within the nest (Chapter 20). It also acts as a reservoir of calcium and magnesium ions for the developing embryo (Chapter 4). Permeated by microscopic pores the shell acts as an effective mediating barrier for gaseous exchange between the egg contents and the ambient environment (Paganelli 1991). Extensive research has revealed the inter-relationships between pore number, shell thickness, functional gas conductance and incubation parameters.

Pore morphology

Although total pore number in an eggshell scales with egg mass (Table 3.2) the distribution of pore across the eggshell is not uniform. The shell over the air space has a higher pore density than the rest of the eggshell (Paganelli 1991). Moreover, pore morphology varies between different species. Board *et al.* (1977) and Board

Table 3.3 Classification of pore systems in avian eggshells (modified from Board *et al.* 1977 and Board and Scott 1980).

Pores	Pore canal	Shell accessory material (SAM) type	Example
1. Outer orifice open *(No shell accessory material [SAM])*	a. Unbranched		Wood pigeon (*Columba palumbus*)
	b. Unbranched and branched		Ostrich (*Struthio camelus*)
2. Outer orifice occluded *(Amorphous SAM found as skin on shell surface with cracks over the outer pore opening)*	a. Unbranched		Herring gull (*Larus argentatus*)
	b. Unbranched and branched		
3. Outer orifice plugged *(SAM forms a plug within the outer opening of the pore)*	a. Unbranched		Lily-trotter (*Micropara capensis*)
	b. Unbranched and branched		Rhea (*Rhea americana*)
4. Outer orifice capped *(SAM covers the entire shell surface, capping and sometimes blocking the pore opening)*	a. Unbranched	i. Organic	Black swan (*Cygnus atratus*)
		ii. Inorganic	Cormorant (*Phalacrocorax carbo*)
	b. Unbranched and branched	i. Organic	Emperor penguin (*Aptenodytes forsteri*)
		ii. Inorganic	None
5. Outer orifice reticulate *(No SAM, pores open into a calcareous stratum containing a plexus of tubules)*	a. Unbranched		White stork (*Ciconia alba*)
	b. Unbranched and branched		Emu (*Dromaius novaehollandiae*)

and Scott (1980) developed an arbitrary classification of pore systems in eggshells largely based around the presence of external covering material and the shape of the pores (Table 3.3). In general, pore morphology consists of unbranched canals extending from the upper surface of the mammilary cone layer through the palisade layer opening into a broad funnel shape at the outer surface, thereby resembling an elongated post horn (Figure 3.4). Only in thick shells of large eggs of swans (*Cygnus* spp.) and the ostrich (Figure 3.4), does the pore canal branch within the palisade layer. The chemical type and morphology of the shell accessory material (SAM) was a further criterion for pore classification.

Functional aspects of shell conductance to gases

Movement of gases, typically water vapour, oxygen and carbon dioxide, across the shell is by diffusion through the pore canals (Rahn *et al.* 1979; Tullett 1984) and the rate of diffusion is related to the functional porosity of the shell and the partial pressure gradient across the eggshell. Therefore, the rate of loss of

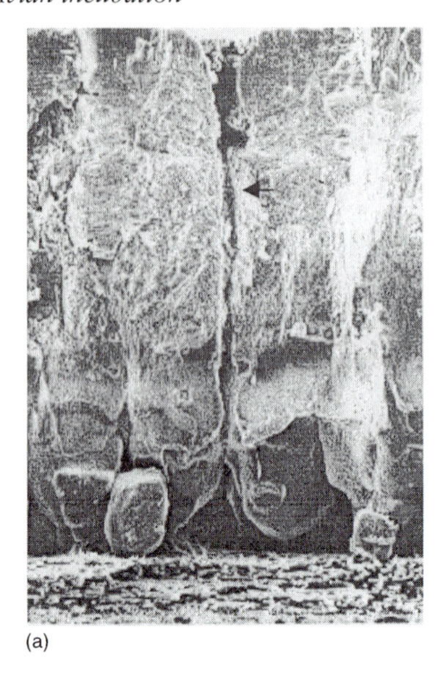

(a)

(b)

Fig. 3.4 Scanning electron micrographs of radial fracture sections of eggshell from (a) domestic fowl (*Gallus gallus*) and (b) ostrich (*Struthio camelus*). Arrows indicate the single pore canal in (a) and the multiple pore canals in (b). Photographs courtesy of Glenn Baggott and Nick Sparks respectively.

water vapour across the eggshell can be determined using a modification of Fick's law:

$$M_{H_2O} = G_{H_2O}\,(P_{iH_2O} - P_{eH_2O}), \tag{3.1}$$

where M_{H_2O} = daily rate of water loss (mg d^{-1}), G_{H_2O} = water vapour conductance (mg H_2O Torr^{-1} d^{-1}), P_{iH_2O} = water vapour pressure of water vapour inside the shell (Torr), and P_{eH_2O} = water vapour pressure of water vapour outside the egg (Torr) (Ar *et al.* 1974). Under normal circumstances (but see below) G_{H_2O} is fixed at oviposition and P_{iH_2O} is always at 100% saturation (Ar *et al.* 1974) so that variation in M_{H_2O} is a function of P_{eH_2O}, i.e. humidity in the nest air. With a respiratory quotient of 0.73, the mass gain of oxygen equals the mass loss of carbon dioxide and so loss of weight during incubation can assumed to be due to water vapour alone (Rahn and Paganelli 1990). G_{H_2O} is a numerical measure of shell porosity and is a function of total functional pore area (Ap, cm^2) and the length of the pores, effectively shell thickness (L, cm):

$$G_{H_2O} = 23.42\ \mathrm{Ap}\ L^{-1}. \tag{3.2}$$

These equations apply equally to oxygen and carbon dioxide and values for G_{O_2} and G_{CO_2} can be calculated from G_{H_2O} (Tullett 1984).

In bird eggs G_{H_2O}, Ap, L, and pore numbers all scale to initial egg mass (Table 3.2). Moreover, incubation period (Ip) also scales to egg mass (Table 3.2) and there is an inter-relationship between water vapour conductance, egg mass, and incubation period (Ar and Rahn 1978) where:

$$(G_{H_2O}\ I_p)W^{-1} = 5.13\,\text{mg}(\text{g Torr})^{-1}. \quad (3.3)$$

The rate of water loss from eggs during incubation also scales to egg mass (Table 3.2) although there is also a good relationship if Ip is also considered (Ar and Rahn 1980):

$$M_{H_2O}\ I_p = 151.9\ W^{0.992}. \quad (3.4)$$

On average fractional water loss during incubation is a constant, i.e. 0.15 ± 0.025, irrespective of egg mass (Ar and Rahn 1980). This means that in all bird species water vapour conductance of the eggshell and the humidity are linked with incubation period in order to ensure that there is a standard rate of water loss from the egg (Ar and Rahn 1980). This is required to maintain an appropriate level of tissue hydration in the hatchling and to ensure a sufficiently large air space in the egg at internal pipping which allows normal expansion of the lungs and air sacs (Ar 1991a). Hence, the wedge-tailed shearwater (*Puffinus pacificus chlororhynchus*) has an incubation period twice that expected from egg mass and this is compensated for by a water vapour conductance some 50% of that predicted from egg mass (Ackerman *et al.* 1980).

Embryonic metabolism is also affected by egg mass and shell characteristics with pre-internal pipping oxygen consumption scaling with egg mass (Table 3.2). Rahn and Paganelli (1990) concluded that the average daily flux for each pore was independent of egg mass and had values of 69, 50 and 49 ml d^{-1} for O_2, CO_2 and H_2O respectively. These values predicted estimates for the gas tensions within the air cell immediately before internal pipping which were close to the actual values observed (Rahn and Paganelli 1990).

Low shell conductance has been shown to affect oxygen consumption in the fowl and other domestic species (Tullett and Deeming, 1982; Burton and Tullett 1983). Below average values for conductance restrict oxygen consumption by the embryo which has to conform to the amount of oxygen which the shell will allow across. By contrast, above average conductance allows the embryo to regulate its oxygen consumption. The metabolic restriction imposed by low shell conductance has adverse effects on rates of embryonic growth and embryonic mortality is high in low conductance eggs because of hypoxia.

Changes in water vapour conductance during incubation

Eggshell conductance is usually considered to be fixed at oviposition, which means that gas exchange across the shell is dependent solely on the partial pressure of gases on either side of the pore (Rahn *et al.* 1974). However, there are several reports where rates of weight loss increase significantly with increasing length of incubation (Carey 1979; Hanka *et al.* 1979; Grant *et al.* 1982; Rahn *et al.* 1983). In some of these and other instances water vapour conductance has been shown to increase during incubation (Table 3.4). In some species these changes can be quite considerable, i.e. of the order of 40–100% of the initial conductance

Table 3.4 Mean values (± SE) for water vapour conductance (mg H_2O $Torr^{-1}$ d^{-1}) for various species of birds showing changes observed during incubation.

Species	Before[#]	After[#]	% Change	Reference
Hirundo rustica[A]	0.223 ± 0.015	0.458 ± 0.016	105.4	1
Riparia riparia[A]	0.317 ± 0.032	0.509 ± 0.052	60.6	1
Turdus migratorius[B]	0.98 ± 0.10	1.45 ± 0.08	48.0	2
Agelaius phoeniceus	0.97 ± 0.04	1.42 ± 0.11	146.4	2
Petrochelidon pyrrhonota[A, C]	–	–	41	3
Serinus canarius (infertile)[D]	0.440 ± 0.010	0.656 ± 0.065	49.1	4
Serinus canarius (fertile)[D]	0.492 ± 0.045	0.971 ± 0.172	97.4	4
Ficedula hypoleuca	0.377 ± 0.067	0.688 ± 0.155	182.5	5
Aix galericulata[D]	3.35 ± 0.21	6.13 ± 0.44	83.0	6
Gallus gallus[E]	15.1	16.1	6.6	7
Aptenodytes patagonica	28.5 ± 1.4	33.1 ± 1.5	16.1	8
Cygnus olor[E]	95.0[F]	105.8[F]	11.4	9
Leipoa ocellata[E]	20.2 ± 3.8	64.2 ± 8.6	217.8	10

[#]Treatment: [A]Before and after development of the chorio-allantoic membrane; [B]Fresh eggs and eggs containing embryos of 0.25 g or more; [C]Actual values for conductance not reported; [D]Fresh eggs and during incubation in nests; [E]Fresh and hatched eggs; [F]Values for whole egg surface derived from data presented so no SE are reported. [E]No SE values reported. References: 1 – Birchard and Kilgore (1980); 2 – Carey (1979); 3 – Sotherland *et al.* (1980); 4 – Kern (1986); 5 – Kern *et al.* (1992); 6 – Baggott and Graeme-Cook (personal communication); 7 – Booth and Rahn (1990); 8 – Handrich (1989); 9 – Booth (1989); 10 – Booth and Seymour (1987).

(Table 3.4) and occur during the early part of incubation, e.g. the red-winged blackbird (*Agelaius phoeniceus*; Carey 1979) although no explanation for the mechanism of this change has been offered. Other increases in conductance, are associated with erosion of the mammilary cones of the inner surface of the eggshell and this seems to be more significant in megapodes. Booth and Seymour (1987) reported 3-fold increases in G_{H_2O} during development (Table 3.4) in eggs of the malleefowl (*Leipoa ocellata*). This was due to the erosion of the shell leading to shortening of the pore (by 12–21%) and removal of the narrowest part of the pore next to the outer shell membrane (Booth and Seymour 1987; Booth and Thompson 1991).

Of the ten reports of increases in water vapour conductance shown in Table 3.4, only three (Carey 1979; Kern 1986; Baggott and Graeme-Cook 1997) use sequential sampling of G_{H_2O} in individual eggs at various stages during development to demonstrate true changes in conductance early in incubation. Booth and Rahn (1990) also repeatedly weighed eggs of domestic species but were only able to demonstrate an increase in G_{H_2O} by the end of incubation in fertile but not infertile fowl eggs. No effect was observed in any eggs of the turkey (*Meleagris gallopavo*) or the quail (*Coturnix coturnix*). The other reports in Table 3.4 usually describe studies where different samples of eggs were studied at various stages of incubation. There is a strong possibility, therefore, that sampling error affected the results because of the natural variation for G_{H_2O} within a population of eggs. Handrich (1989a) reported that G_{H_2O} of king penguin (*Aptenodytes patagonica*) eggs averaged at 28.1 mg H_2O $Torr^{-1}$ d^{-1} but the range for individual eggs was 10.3–37.8 mg H_2O $Torr^{-1}$ d^{-1}. Carey (1986) reported ranges for water vapour conductance for red-winged blackbird eggs where the highest values were 5–11 times higher than the lowest values. In the ostrich, five fold increases are also seen for the range for mass specific water vapour conductance (Deeming and Ar 1999). Hence, the possibility exists that the apparent increases in G_{H_2O} in a variety of species may have been the product of limiting sampling within the population of eggs. Substantially more data of the appropriate kind are required on a lot more species before the effects of incubation on water vapour conductance can be understood.

Experimental removal of the cuticle can also increase the water vapour conductance of some species of domestic species but not others (Deeming 1987). Changes in conductance of pied flycatcher (*Ficedula hypoleuca*) eggs were not due to loss of cuticle (Kern *et al.* 1992) but in the mandarin duck, microbial degradation of the cuticle may be responsible for increases in shell conductance (Chapter 12). Handrich (1989a) estimated that during incubation around 50% of the cuticle was rubbed from the shells of king penguin eggs showing a 16% rise in G_{H_2O} during incubation (Table 3.4). Artificial removal of the whole cuticle increased shell conductance by 18% (Handrich 1989a). Thompson and Goldie (1990) also showed that removal of the cuticle from eggs of the adelie penguin (*Pygoscelis adelie*) increased G_{H_2O} by over 20% raising the possibility of this occurring naturally. In some species at least the cuticle does provide a significant barrier to loss of water vapour by contributing to the overall conductance of the shell.

The avian egg: a remarkable evolutionary design

The egg was a key evolutionary step in avian reproduction and has proved to be a remarkable structure with very elegant adaptations to the environment in which they exist. There are few structures as self-contained as the egg that exchanges only heat, oxygen, carbon dioxide, and water vapour with its environment. Over the past thirty years the study of oology has been greatly enhanced by the collected work of Hermann Rahn, Amos Ar, and Charles Paganelli, and their colleagues around the world, which has elucidated the inter-relationships between egg mass, shell conductance, rates of water loss, and incubation period. The egg has proved to be a relatively conservative structure whose morphology and function has become inextricably linked to the bird-nest incubation unit and this is largely true for the majority of bird species. There are differences between some birds, e.g. in gross composition of the egg contents between species of different hatchling maturity, and yet these factors are also linked with the incubation period experienced by the egg.

Future research should aim to investigate potential species differences in egg composition and how these affect egg size and embryonic development through to hatching. There is also a need to determine why water vapour conductance should change during incubation and whether with this widespread phenomenon. Moreover, as we become more aware of the bird-nest incubation unit (Chapter 1), and the symbiosis between the adult and the embryo (Chapter 9), there may be a need to re-examine the incubation environment in terms of size and other characteristics of the clutch as a whole rather than for individual eggs.

Acknowledgements

Many thanks to Glenn Baggott for his constructive comments on a previous draft of this chapter and to Glenn and Nick Sparks for allowing me to use micrographs of shell structure.

4 *Embryonic development and utilisation of egg components*

D. C. Deeming

Development of bird embryos of domestic species is well reported but given the diversity of bird species there are relatively few descriptive reports of development in non-commercial species (Ricklefs and Starck 1998). Patterns of embryonic development in different species have been found to be largely similar and the domestic fowl (*Gallus gallus*) embryo has proved to be a useful general model for avian development. In this chapter space prevents an in-depth description of the pattern of embryonic development in any particular species and this would be largely redundant given previous publications (Romanoff 1960; Freeman and Vince 1974; Bellairs and Osmond 1998; Ricklefs and Starck 1998). Rather, an attempt is made to provide a basic description of embryonic development through to hatching, with emphasis on the changes in extra-embryonic environment within the shell. Freeman and Vince (1974) describe behavioural aspects of development and interactions between the developing embryo and the incubating adult as described in Chapter 7.

Basic patterns of embryonic development

The process of converting yolk and albumen into an embryo can be split into two parts: differentiation and growth. Deeming and Ferguson (1991b) argued that these two phases correspond to the embryonic and foetal stages of development observed in mammals. During the first phase, the bulk of the embryonic tissues are formed such that 90% of the bodily organs are present at day 12 of incubation in the fowl. There is also development of extensive extra-embryonic tissues that are essential for normal development. Some but not all of these tissues are incorporated into the bird at hatching. The second phase involves the increase in size of the embryo and the maturation of tissues up to a point that it is ready to hatch.

Differentiation

Despite the superficial differences in maturity between hatchlings of different developmental modes (i.e. altricial, semi-altricial, semi-precocial, and precocial), it was concluded by Ricklefs and Starck (1998) that embryonic development in birds could be described by the same series of stages proposed by Hamburger and Hamilton (1951) for the domestic fowl. The pattern during the phase of differentiation of embryonic development is listed very briefly here using the domestic fowl as a model and readers are directed to other texts (Hamburger and Hamilton 1951; Romanoff 1960; Freeman and Vince 1974; Ricklefs and Starck 1998) for more detail.

Avian embryonic development is characterised by development of a three dimensional structure from a plate of cells on top of a large yolk. In amphibians and mammals, the embryo develops from a round ball of cells which include a very small amount to yolk. After formation of the zygote, early embryonic development occurs in the oviduct during the process of egg formation so that at oviposition the blastoderm is at the gastrula stage of development consisting of ~60 000 cells. Further development is suspended until after the onset of continuous incubation.

The avian embryo initially develops from a flat blastoderm comprised of a dorsal layer of the epiblast (the future ectoderm), and a ventral hypoblast (the future endoderm). During gastrulation mesoblast cells invade between these layers and as the embryo begins to develop a recognisable form this layer splits into a dorsal somatic mesoderm associated with ectoderm and a ventral splanchnic mesoderm associated with endoderm.

Embryonic development in vertebrates generally proceeds from the anterior end with development of the head happening before that of the tail. By 24 hours of incubation in the fowl (stage 6) the head fold begins to develop closely followed by somite formation which progresses down the body only reaching the tip of tail by $3\frac{1}{2}$ days (stage 22). Therefore, the brain and heart develop early in development (from ~30 hours stage 8/9) with recognisable eyes at 48 hours (stage 12). By stage 19 (72 hours) the embryo is lying on its left side with a well developed heart, brain, and eyes. Development of the alimentary tract and respiratory system is initiated on day 3 of incubation and morphogenesis continues during the first half of incubation so that a fully formed, albeit small, bird is recognisable at day 12 of incubation.

Growth

Embryonic growth in birds has been the subject of considerable research with the commonest method to describe its pattern being the parabolic growth model (Ricklefs and Starck 1998). After analysis on growth rates of 35 species of bird Ricklefs and Starck (1998) concluded that all bird species exhibit similar rates of growth regardless of hatchling maturity and variations are due to differences in the initial egg mass and the length of incubation. Indeed apparent differences in growth rates of altricial and precocial species are related to differences in relative egg mass rather than developmental mode (Ricklefs and Starck 1998).

If it is the case that egg mass is a controlling factor in avian embryonic growth then two important questions arise: (1) how the developing embryo is able to assess how large its egg is? and (2) how is the embryo to know long its incubation period will be? Deeming and Ar (1999) highlighted this problem using the example of the ostrich (*Struthio camelus*) embryo. Variation in egg mass is considerable in this species (range: 1–2 kg) yet incubation periods of the eggs at the extreme of this range only differ by only 2–3 days. Furthermore, it is accepted that hatchling mass is a function of egg mass (Tullett and Burton 1982). Therefore, to produce hatchlings of the same relative size (% of initial egg mass) within roughly the same time period, an ostrich embryo in a 1800 g egg will have to grow almost 1.5 times faster than an embryo in a 1200 g egg. As Deeming and Ar (1999) pointed out: 'how is the rate of embryonic growth controlled?' and 'how does an embryo "know" how large its egg is (and hence its final body mass)?' In the domestic fowl, egg mass has no effect on embryo mass at 12 days of incubation yet at 15 days embryos in larger eggs are significantly larger in smaller embryos (Burton and Tullett 1985) although space in the shell is not limited at this time.

Some studies perhaps give some hint of the factors affecting embryo mass. Deeming (1989) showed in the fowl that removal of 5 g of albumen at 3 days of incubation significantly reduced embryo mass at 14 days of incubation. Furthermore, Finkler *et al.* (1998) have shown that wet embryo mass was significantly reduced by removal of albumen but unaffected by removal of yolk. There is certainly scope for more research into this interesting area of embryonic development.

Descriptions of embryonic growth other than mass have been provided for some seabirds (Haycock and Threlfall 1975; Maunder and Threlfall 1972; Mahoney and Threlfall 1981). Morphometric measurements of around 10 anatomical features have been made for embryos of the herring gull (*Larus argentatus*), black-legged kittiwake (*Rissa tridactyla*), and the common guillemot (*Uria aalge*). The patterns of growth through time were similar in all three species with the culmen and most limb and torso measurements exhibiting linear growth whereas growth of the head and orbit was curvilinear. Relationships between growth of body parts and embryonic stage have yet to be fully investigated in birds.

Development of extra-embryonic membranes

Embryogenesis does not occur in isolation within the egg and a number of membranes and fluid compartments allow the embryo to survive within the confines of an egg. A brief outline of the pattern of development and role of the membranes is provided here. Romanoff (1960) provides further details.

The amnion is formed from a layer of ectoderm and somatic mesoderm that are continuous with the presumptive body wall (and neural tube). By 30 hours of incubation in the fowl embryo (stage 9) there is the beginning of the head-fold of the amnion which gradually expands and by 48 hours has started to grow over the embryo (stage 12). A tail fold arises at this time and grows over the tail region of the embryo. The two folds meet at a point (the sero-amniotic connection) nearer the tail-end around 72 hours of incubation (stage 20). The fusion of the amniotic folds produces a fluid filled sac lined with ectoderm. Amniotic mesoderm lines the extra-embryonic coelom (EEC). The role of the amnion is primarily protective forming a fluid-filled environment in which the embryo can develop and so with embryonic growth the amniotic sac increases in size reaching a maximum volume at day 15 (Figure 4.1).

The amnion is avascular but is highly muscular exhibiting waves of muscular contractions from early in incubation. These become important in the later stages

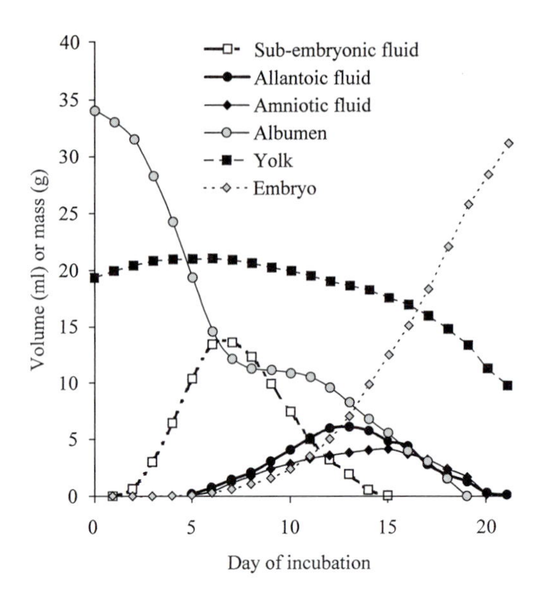

Fig. 4.1 The pattern of changes in the various fluid compartments within a domestic fowl (*Gallus gallus*) egg during the incubation period. Data from Romanoff (1967).

of incubation during the muscular retraction of the yolk sac into the body cavity. The amnion immediately next to the body wall forms the navel ring observed in hatchlings. The amnion lying over the body of the embryo degenerates a few hours prior to internal pipping into the air space.

The chorion is continuous with the amnion but becomes distinguished from it as the latter folds over the embryo. It is external to the amnion with its ectoderm external to the somatic mesoderm that lines a lot of the EEC. As development proceeds the chorion continues to grow and in combination with the outer allantoic membrane it forms the chorio-allantoic membrane (CAM) and lines the inner surface of the inner shell membrane. At its furthest point from the embryo the chorionic somatic mesoderm remains attached to the splanchnic mesoderm of the yolk sac membrane and forms the limit of the EEC. Chorionic ectoderm represents the outer limit of all living embryonic tissue and it forms the walls of the albumen sac.

The yolk sac membrane (YSM) consists of a layer of endoderm adjacent to the sub-embryonic fluid and yolk and a layer of splanchnic mesoderm that lines the EEC. It arises as an extension of the alimentary tract and later in development, it is attached to the small intestine at the junction of the jejunum and the ileum via a narrow tube, the yolk stalk. During the first half of development in the chick the YSM rapidly grows over the surface of the yolk by moving along the inner surface of the vitelline membrane (VM). After the fourth day of incubation, when the VM is dissolved away by the embryo, it is the attachment of the YSM that maintains the integrity of the sub-embryonic fluid/yolk compartment (bounded by the YSM above and the VM below). The YSM eventually surrounds the yolk replacing the VM except for a small area at the former vegetal pole of the yolk adjacent to the albumen sac. The YSM is a highly vascular membrane with an extensive network of blood vessels. As development proceeds the YSM epithelium becomes more complex in morphology and extensive folding of the membrane causes it to protrude into the yolk. During the first third of incubation the YSM also acts as the major respiratory membrane until the CAM develops. The YSM is also the key site for ion transport between the albumen and the yolk and is critical in formation of sub-embryonic fluid. The major role of the YSM is the transfer of nutrients in the yolk to the blood stream and it has a key role in lipid metabolism (Speake *et al.* 1998).

The allantois arises at around 65–69 hours of incubation in the fowl (stage 18) as a diverticulum of the hindgut to form a sac lined with endoderm with splanchnic mesoderm lining the EEC. As the allantois expands it comes to fill the EEC lining the mesodermal surfaces of the chorion, amnion and YSM and surrounding the amniotic sac, the yolk, and the albumen. The allantoic sac is fluid-filled acting as a repository for the waste products from the kidneys. It reaches its maximum volume around day 12 and declines during the growth phase of development (Figure 4.1). The outer aspect of the allantois comes to lie next to the chorion forming the highly vascular CAM, which can be seen lining the inner surface of the inner shell membrane beyond hatching. The CAM is the primary respiratory membrane for the embryo during its growth phase and it has an important role in absorption of calcium ions from the eggshell. The inner allantoic membrane covers much of the amnion and the YSM. It has a role in yolk sac retraction and in part is drawn into the body cavity with the residual yolk.

Patterns of embryonic development in the altricial–precocial spectrum

Despite apparent external differences between precocial and altricial species (Portman 1959), Ricklefs and Starck (1998) concluded that there were few morphological differences between embryos of different developmental modes. Therefore, a single description of the morphological changes during development will fit the 30 species (penguins to passerines) for which developmental charts are published (see list in Ricklefs and Starck 1998). However, as Ricklefs and Starck (1998) point out there are few reports that give the time periods for the different developmental stages in different species. Although information is limited it seems that the duration of early embryonic stages (differentiation phase) is relatively uniform in different species but there is greater variation in the length of later stages (growth phase). It is these that are then correlated with the length of incubation. Hence, Hamburger and Hamilton stage 39 lasts 48 hours in the fowl, incubation period of 21 days, but only 18–24 hours in the budgerigar (*Melopsittacus undulatus*) and rice finch (*Lonchura oryzivora*), with incubation periods of 18 days (Ricklefs and Starck 1998). However, this relationship is not simple because the quail (*Coturnix coturnix*) also has an incubation period of 18 days but stage 39 lasts 24–36 hours (Ricklefs and Starck 1998).

This situation is highlighted by a comparison of the embryonic development of the domestic fowl and the red-winged blackbird (*Agelaius phoeniceus*). These two species (with precocial and altricial hatchlings respectively) differ in the length of incubation (21 and 12 days respectively) but Daniel (1957) showed that the pattern of development in real time was very similar. Indeed at 12 days the two species showed similar patterns of development and Daniel (1957) suggested that the fowl embryo was in a state which could allow it to hatch. However, ossification of the scapula and coracoid starts at day 12 (57% of incubation period) in the fowl but at day 9 in the red-winged blackbird (75% of the incubation period). Compared with the fowl embryo, many stages in embryonic development in the ostrich occur at the same relative period of the full incubation period (Ar and Gefen 1998). Other developmental features (e.g. appearance of scales on the leg and growth of the eyelid, stage 38) develop significantly early in the later stages of development in the ostrich embryo. Relative embryonic growth is relatively slow for the first two-thirds of incubation but very much faster during the last third of incubation. This may be related to the short incubation period of the ostrich relative to egg mass (Deeming and Ar 1999).

Although Ricklefs and Starck (1998) suggest that stage length, particularly during the later stages, may determine developmental differences in altricial and precocial species, there has been little research into the comparative development of birds. As Richardson *et al.* (1997) point out, birds are considered to show the least embryonic variation of vertebrates but there needs to be more study in a greater range of birds to ensure that there is a general rule for all species.

Hatching

Hatching is a critical phase of embryonic development because it leads to the emergence of the hatchling from the egg. The process starts with the foetal chick manoeuvring its body to lie with its long axis parallel to the long axis of the egg and with its neck bent so that its head lies on its breast. The beak is moved under the right wing so that its tip comes to lie close to the CAM adjacent to the inner shell membrane delimiting the air space. During this process the residual yolk is being pulled into the abdominal cavity. Bond *et al.* (1988b) suggested that all birds, except for the megapodes, adopted this pattern although Deeming (1995) has shown that the ostrich is an exception. The next phase in hatching is the partial draining of the blood vessels in the CAM so as to allow the egg tooth on the upper mandible to pierce the inner shell membrane thereby allowing the bird access to air. This act of internal pipping allows filling of the lungs and air sacs (Duncker 1978) prior to the first external pip hole in the eggshell.

The time interval between internal and external pipping is variable and is a function of the partial pressure of carbon dioxide and oxygen (p_{O_2} and p_{CO_2} respectively) in the air of the air space. In the domestic fowl, p_{CO_2} in the air space is an important stimulus for external pipping (Visschedijk 1968a). Decreasing p_{O_2} in the air space accelerated the time of pipping but increasing p_{CO_2} was twice as effective. Applying paraffin to the shell covering the air space, lowered the p_{O_2} of the air space, increased p_{CO_2} content and brought forward the time of pipping (Visschedijk 1968b). By contrast, drilling a hole into the shell over the air space retarded external pipping; neither of these treatments, however, affected hatch time (Visschedijk 1968b). Burton *et al.* (1989) showed that air space p_{CO_2}

increased markedly after internal pipping into the air space reaching values of 75 and 60 mmHg in low and high porosity eggshells respectively. Furthermore, internal and external pipping occurred at progressively later times with increasing shell porosity but hatching time was unaffected by shell porosity (Burton *et al.* 1989). After internal and external pipping, pulmonary respiration allows a decrease in p_{CO_2} of the embryonic blood that continues in the first few days of post-hatching life (Tazawa *et al.* 1983).

Several authors have described climax of hatching in attempts to define differences between different species of bird. Oppenheim (1972) suggested that, with the exception of the megapodes, there was only one method for hatching which produced a symmetrical eggshell after hatching climax. This conclusion was challenged by Bond *et al.* (1988b) who examined hatching in around 150 species of bird and found that whilst many species did adopt the symmetrical method of hatching there were a significant proportion which produced an asymmetrical eggshell.

In the symmetrical pattern of hatching climax the chick makes many co-ordinated movements which have three main components (Bond *et al.* 1988b): (1) vigorous oblique back thrusts of the beak against the eggshell; (2) simultaneous pressure of the upper spine and tarsal joints against the shell; and (3) a subsequent leftward rotation of the body within the shell. The beak thrusts effectively produce a series of pip holes in the shell, which trace the circumference of the egg and produce a cap which can be pushed off the egg by thrusting movements by the chick. Having effectively loosened the end of the egg the chick is able to push its way out of the shell. Differences in this process between species lie in the extent to which there is rotation within the eggshell before the shell cap can be pushed off. The rhea (*Rhea americana*) shows rotation angles of less than 90° compared with 240–270° in the domestic fowl and ring-necked pheasant (*Phasianus colchinus*) and 300–360° in some pigeons (Columbiformes) and owls (Strigiformes). It may not be significant but semi-altricial and altricial species did seem to require a greater degree of rotation before hatching.

Internal pipping in the ostrich is achieved by the chick rubbing its beak against the shell membranes at a position located away from the air space (Deeming 1995). No egg tooth is present and the beak is covered by an amorphous layer which is eroded by the rubbing action (Richardson *et al.* 1998). Internal pipping is achieved by the bird pushing against the inner shell membrane, which distorts the air space and pulls the membrane over the beak, which is held against the shell. External pipping creates a large pip hole which is enlarged by the beak and right foot during a rotation of only 90°.

The asymmetrical method of hatching involves producing a pip hole lower down the eggs and then an oblique movement of the beak upwards towards the blunt pole of the egg, which produces a line of pip holes up to 90° of the circumference of the shell (Bond *et al.* 1988b). This type of hatching is largely adopted by those species with long beaks such as the avocet (*Recurvirostra avosetta*), curlew (*Numenius arquata*) and godwits (*Limosa* spp.) as well as puffins (*Fratercula* spp.) but it is not clear whether other long-billed groups (e.g. Pelecaniformes, Gruiformes or Ciconiformes) employ the same techniques.

Although attempts have been made to correlate the different hatching strategies with shell strength, thickness or hardness (see Bond *et al.* 1988b), Bond *et al.* (1986) showed that there is a good correlation with the toughness of the shell and shell membranes combined and the degree of rotation during hatching climax. Hence repeated chipping is required in those shells where cracks do not extend a long way (e.g. quail) but in the hard, brittle shell of the ostrich cracks extend a long way and relatively few breaks in the shell are needed before hatching is possible (Deeming 1995).

Timing of hatching is also subject to variation (Chapter 18). In some species incubation appears to be initiated with the first egg in the clutch so that subsequent eggs are always a day or two behind (Drent 1975). Hatching is asynchronous because the resulting hatchlings are different ages. The other extreme is the hatching synchrony, observed in many precocial species, where all of the birds hatch more-or-less together. This synchrony is often the result of considerable vocal communication between individual birds (Chapter 7). It is often assumed that with synchronous hatching incubation starts once the clutch is complete but in some species age of embryonic development can vary by 1–2 days (Afton 1979). In the rhea, embryonic development can be accelerated by up to 7 days by the presence of hatching chicks (Bruning 1973). The question arises of, if the start of incubation is the same for all of the eggs, why is there a need for vocal communication between hatching birds. It is always assumed that the rate of embryonic development is the same in synchronous species but this has yet to be confirmed in wild nests.

Changes to egg components

Whilst the composition of bird eggs is relatively well described (Chapter 3) descriptions of the utilisation of egg components are largely confined to that of the commercially important (precocial) species (Romanoff 1967). Through necessity, the following descriptions are mainly based on the domestic fowl although other species will be discussed where appropriate. There is considerable scope for further investigation of the patterns of utilisation of the egg components in different birds of the various developmental modes.

Albumen

Albumen has numerous roles in supporting embryogenesis, e.g. protection from microbial infection and, through its mass, aiding to maintain thermal inertia of the egg contents, but its main role during incubation is to provide water, mineral ions, and proteins for the developing embryo. During the course of pre-incubation and incubation it undergoes considerable changes in composition and by the time of hatching it is absent from the egg.

Prior to incubation many bird eggs undergo a period of storage and in the domestic fowl there are changes in the composition of the albumen during this time. There is a loss of dissolved carbon dioxide across the eggshell leading to an increase in pH from around 7.3–7.5 up to 9.3–9.5 in eggs exposed to prolonged storage (Meuller 1958). As storage proceeds there is a general thinning of the thick albumen capsule around the yolk (Oosterwoud 1987). In the fowl these changes can affect embryonic viability once incubation is started but no data are available for wild species. This would certainly be of an area for further research given the periods which some eggs have to endure before incubation is started. For example, the clutch in the ostrich is around 8–14 eggs which take up to 28 days to lay during which time they lie in an open nest exposed to the sun (Bertram 1992).

Once incubation is initiated then the developing embryo changes the composition of the albumen. From day 2 in the fowl, there is a measurable amount of sub-embryonic fluid (SEF) a watery solution found in the cavity below the embryo and on top of, but not physically separated from, the yolk (Romanoff 1967). Formation of SEF involves the transport of sodium ions across the embryonic tissue from the albumen into the yolk. The electro-osmotic gradient draws chloride ions and water across the membrane and a salt-solution forms under the embryonic tissue (Howard 1957; Elias 1964; Deeming *et al.* 1987; Deeming 1989; Babiker and Baggott 1992, 1995). SEF also dissolves yolk material giving it the look of skimmed milk. As SEF is formed then there is a rapid drop in the mass of the albumen which slows as the amount of SEF being formed reduces (Figure 4.1). Fluid transport is an active process requiring energy and poisoning the membrane prevents it (Babiker and Baggott 1992; Latter and Baggott 2000a). The rate of this process can be diffusion limited (Deeming *et al.* 1987; Babiker and Baggott 1991; Latter and Baggott 1996) and can be affected by incubation temperature and rate of egg turning (Deeming 1989a, 1989b). In effect, early in development the embryo moves its water reservoir from the albumen to the yolk where it is easily accessible for the later formation of allantoic and amniotic fluids, and hydration of its own tissues. Decline in SEF volume reflects the movement of water and mineral ions to the embryo and other extra-embryonic fluid compartments. In precocial bird species the volume of SEF is a reflection of egg mass (Romanoff 1967) but to date SEF volume in species from other developmental modes has not been reported. Presumably the higher albumen content of altricial eggs (Chapter 3) would allow a higher rate of SEF formation but this needs further investigation.

Around day 12 in the fowl there is a second phase of albumen utilisation (Figure 4.1) as there is a transfer of albumen proteins into the amniotic fluid via the sero-amniotic connection (SAC). At 12 days of incubation the ectodermal cell layer of the SAC perforates to form a crescent of oval holes (Hirota 1894; Romanoff 1960; Deeming 1991). Over the next 3–4 days albumen proteins move through the holes into the amniotic fluid causing a rise in the protein level of the fluid (Deeming 1991). The proteins are taken into the alimentary tract by the embryo swallowing the amniotic fluid and eventually albumen proteins are found in the yolk sac. As a result protein concentration of the yolk is actually higher at the end of development than at the start (Romanoff 1967). The albumen sac has normally disappeared by day 16 but this process can be slowed or even prevented by failure to turn eggs during incubation (Deeming 1991). Witschi (1949) showed that movement of the proteins occurs in the sparrow (*Passer* sp.) from day 9 through to day 11 of the 13-day incubation period. Compared to the fowl, this is earlier in time but relatively later in the incubation period. It is unclear whether this broadly reflects the development in altricial eggs. Given the variation in the composition of albumen between different developmental modes (Figure 3.1), there is considerable scope for further investigation of the process of albumen utilisation in other birds. Moreover, despite Witschi's (1949) claim that there were muscular contractions of the allantoic membrane, the process by which the albumen proteins move through into the amniotic fluid has yet to be fully investigated.

Yolk

Yolk lipids form the primary energy source and yolk is an important source of proteins for the embryo. Absorption of yolk material is achieved by the highly vascular YSM which acts in a similar way to the embryonic gut by transferring the yolk components to the blood stream (Freeman and Vince 1974). Proteins are digested by proteolytic enzymes secreted by the YSM with the products being moved across the membrane by an active amino acid transfer system or by phagocytosis. Lipid utilisation by the embryo involves three phases mediated by the YSM (Speake *et al.* 1998): (1) uptake of lipids by phagocytosis; (2) re-modelling of lipid structures; and (3) assembly of modified lipids into lipoproteins for secretion into the circulation. Mass of yolk does not drastically change during embryonic development (Figure 4.1) and a considerable amount of yolk is left at hatching. This results from dynamic changes in the components of the yolk with the protein concentration increasing towards the end of development, whereas the lipid content is significantly depleted at hatching (Romanoff 1967). Patterns of yolk utilisation are largely based on the domestic fowl but there has been a recent move to investigate yolk composition (Chapter 3) and metabolism in other species, including penguins (Speake *et al.* 1999a), gulls (Royle *et al.* 1999), waterfowl and pheasants (Speake *et al.* 1999b).

The eggshell

The eggshell plays many roles in avian incubation, not least as physical protection for the contents against microbial contamination and a mediating barrier for gaseous diffusion (Board 1980). Shell is also a major source of calcium and magnesium for the developing embryo (Bond *et al.* 1988a). The embryo utilises these ions by secretion of a carbonic acid from the external surface of the CAM which then diffuses to the surface of the mammillary cones of the shell and erodes them away. The released calcium ions then diffuse back to the CAM and are absorbed and transported away by the blood. Although most of the calcium ions are used in the skeleton of the embryo some are stored in the yolk (Johnston and Comar 1955). By contrast, in the embryonic yellow-headed blackbird (*Xanthocephalus xanthocephalus*) most of the calcium ions removed from the eggshell are stored in the yolk rather than in the embryo suggesting that in this altricial species at least, ossification of the skeleton occurs after hatching (Packard 1994).

Bond *et al.* (1988a) found in hatched eggshells from altricial species that there was less change in the morphology of the mammillary cones, associated with erosion of the calcite crystals observed in shells from precocial species. These findings conflict with the work of Starck (1996) who found few differences between altricial and precocial species in the sequence and extent of ossification of skeletal cartilage. Unfortunately, there was no overlap in the species being studied in these different reports and further research is required to correlate calcium removal from the eggshell and ossification in hatchlings of different levels of maturity.

Embryonic development in non-commercial species

Our understanding of development of the fowl embryo is well advanced both in terms of embryogenesis and in the utilisation of the egg components. Whilst the fowl has proved to be a useful model for all birds, it is clear that diversity in developmental modes and egg composition may be reflected in differences in the pattern of development. Further research could examine how embryo growth is achieved and how final hatchling mass is determined within eggs of different mass. A better understanding of the degree of variation in rates of development in wild nests would perhaps help clarify whether hatching synchrony is established at the start or the end of incubation. Also a study of the water and electrolyte balance of altricial and semi-altricial embryos could give some insight into the role of the large amounts of albumen in these eggs. It is hoped that future studies of embryonic development and physiology will begin to better reflect the diversity of developmental patterns in birds and hence improve our understanding of the reproductive biology of this group.

Acknowledgement

Many thanks to Nick French for his constructive comments on a previous draft of this chapter.

5 *Hormonal control of incubation behaviour*

C. M. Vleck

Contact incubation is displayed by nearly all birds to maintain egg temperature within acceptable limits for embryonic development. The behaviours involved have been studied in exquisite detail in a few species such as herring gulls (*Larus argentatus*; Baerends and Drent 1970). This work analysed nest building activity, incubation, retrieval of eggs, and re-settling behaviour of adult gulls, and how the tendency to sit quietly on the eggs varied over both the reproductive season and over a single bout of incubation. Changes in behaviour were attributed to changes in tendency to incubate, but without a clear understanding of the underlying factors influencing behaviour.

In the ensuing 30 years much has been learnt about the proximate control of incubation behaviour. It is now known that it is influenced by a combination of hormonal and other physiological signals, environmental inputs from the eggs or nest (Hall 1987; Evans 1990d), and social cues such as presence or absence of a mate (Silver and Gibson 1984; Vowles and Lea 1986; Bédécarrats *et al.* 1997). Previous breeding experience seems to increase responsiveness of an individual to hormonal and environmental signals for incubation (Cloues *et al.* 1990; Wang and Buntin 1999). The pattern of incubation behaviour can be affected by ambient temperatures (Conway and Martin 2000a), and incubation is terminated under adverse environmental conditions or when the parent's energy reserves become too low (Ainley *et al.* 1983; Davis and McCaffrey 1986; Astheimer *et al.* 1995; Vleck and Vleck 2001). Incubating birds may also neglect viable eggs for self-preservation, which may not kill the embryo, but will prolong incubation. This is particularly true of seabirds that forage offshore such as Procellariiformes (Boersma *et al.* 1980) and alcids (Sealy 1984; Astheimer 1991).

In this chapter the focus is on the hormonal factors that initiate incubation and maintain birds in a 'broody' state. Broodiness in the poultry literature is defined as that condition following a period of egg laying characterised by ovarian regression, incubation of eggs (if present), attentiveness to a nest and care of young (Burke and Dennison 1980). As such it defines a series of parental behaviours that are in contrast to the sexual and reproductive behaviours (courtship, aggression, mate defence, etc.) that usually proceed it and are directed toward a mate rather than toward offspring.

Species and gender differences in prolactin levels and parental behaviour

The neural and hormonal basis of incubation behaviour has been studied extensively in a few domesticated species (reviewed in Buntin 1996), including the domestic fowl (*Gallus gallus*), turkeys (*Meleagris gallopavo*) and domesticated ring doves (*Streptopelia risoria*). Much of this work has focused on the hypothalamus–pituitary axis and the secretion of the hormone prolactin. In galliforms, the general goal has been to understand how broodiness can be inhibited in order to prevent the egg production losses that accompany broodiness (Sharp 1997). Parental behaviour of galliform birds is not typical of most birds, however, because males provide little or no parental care (Chapter 6). In addition many of these poultry species have been artificially selected for precocial maturation, high egg production, reduced seasonality, and reduced broodiness. Thus, control of parental behaviour and prolactin secretion in these domesticated species may not be fully representative of the majority of other bird species. The other well-studied group, the Columbiformes, provide bi-parental care, but doves and pigeons secrete prolactin to stimulate maturation and secretions of the highly specialised crop gland, an adaptation for providing nutrition to the hatchlings that is not found in any other order of birds. Much of the work on incubation in free-living birds has focused on the correlations between plasma levels of prolactin and parental behaviour, and how these vary with social system, ecology and life-history traits (Dawson and Goldsmith 1983; Oring *et al.* 1986b; Hiatt *et al.* 1987; Wingfield and Goldsmith 1990; Silverin 1991; Vleck *et al.* 1991; Brown and Vleck 1998). As yet, however, there is no clear understanding of how the diversity in incubation patterns (Chapter 6) arises from the underlying physiological mechanisms.

A rise in prolactin is associated with the onset and maintenance of egg incubation and care of young in all species that have been examined (Goldsmith 1991; Buntin 1996). The rise may occur abruptly at the time of egg laying and incubation, or prolactin may rise more slowly and peak during mid-incubation. These differences may result from selection for different reproductive strategies including differences in clutch size, laying interval and at what point within the laying sequence incubation normally begins (see Chapter 18).

In species with precocial young, prolactin levels usually decline rapidly after the chicks hatch, and the presence of the chicks can modify this rate of decline (Dittami 1981; Silverin and Goldsmith 1983; Opel and Proudman 1989). Prolactin levels generally decline more slowly in species with altricial young, presumably because parental brooding of non-thermoregulating chicks is required for their survival, just as it is for embryo survival. Prolactin levels often begin to decline only after the chicks achieve thermal independence and no longer require constant brooding (Goldsmith 1991). The waning of importance of prolactin after hatching and the transition to other factors that control parental behaviour, including sensory input from the chicks themselves, varies between species, suggesting patterns have been subject to selection. In off-shore feeding species such as the king

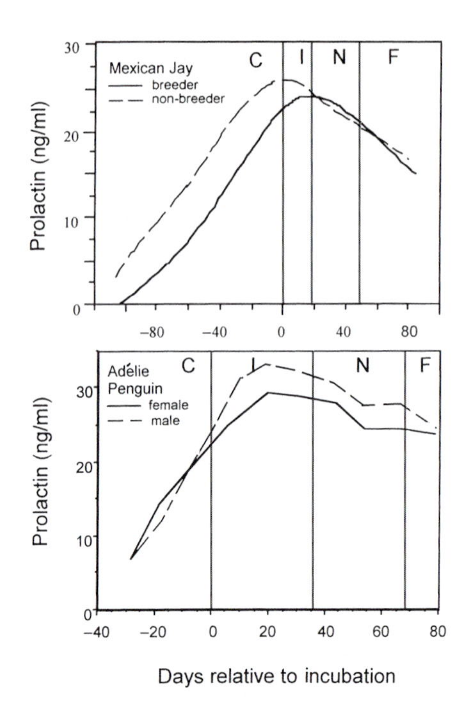

Fig. 5.1 Plasma prolactin levels across the breeding season in two species in which prolactin is elevated well past the incubation stage. Data are plotted relative to the day incubation begins. C = courtship stage, I = incubation stage, N = nestling stage, and F = fledgling stage. Mexican jays (*Aphelocoma ultramarina*) live in large social groups with multiple asynchronous nests; both breeders and non-breeders feed incubating females and chicks. Adélie penguins (*Pygoscelis adeliae*) have an extended period of incubation and chick care during which they must make lengthy trips between the breeding and foraging grounds. Figures based on data in Brown and Vleck (1998) and Vleck *et al.* (2000).

penguin (*Aptenodytes patagonicus*) and Adélie penguin (*Pygoscelis adeliae*), in which chicks spend long intervals dependent on parental feeding, prolactin levels in the parents remain elevated for weeks to months after hatching (Cherel *et al.* 1994; Vleck *et al.* 1999; Figure 5.1). A lengthy interval of elevated plasma prolactin after eggs hatch may be necessary to stimulate continued return of the parents to the breeding colonies from distant, off-shore feeding sites. Likewise, in co-operatively breeding Mexican jays (*Aphelocoma ultramarina*), prolactin levels in all group members remain elevated for many weeks after the start of incubation rather than declining to baseline within a few days of hatching (Figure 5.1; Brown and Vleck 1998). In this species there may be up to five asynchronous nests over a three month breeding season, and essentially all members of the group feed incubating females and care for the young.

There are also differences in prolactin profiles between males and females, even when both members of the pair incubate. Levels of prolactin are usually higher in females than in males, whether or not the male participates in incubation

(Goldsmith and Williams 1980; Dawson and Goldsmith 1982; Hiatt *et al.* 1987; Lormée *et al.* 2000; but see Gratto-Trevor 1990). For example, prolactin in incubating female Harris' hawks (*Parabuteo unicinctus*) is more than double the level found in the males, even though the latter do participate in incubation (Vleck *et al.* 1991). In Mexican jays, peak prolactin levels are only about 20% higher in females than in males (Brown and Vleck 1998), although the males do not incubate, but rather feed the incubating females who seldom leaves the nest. Species in which the male incubates exclusively or considerably more than the female, however, have higher prolactin levels in males than in females (Oring *et al.* 1986a, 1988; Gratto-Trevor *et al.* 1990; Buntin *et al.* 1998). It would be interesting to know whether there is a correlation between elevated prolactin levels in males and the extent to which males of that species feed the incubating female on the nest. Given the correlation between chick feeding behaviour and prolactin in many species, such a relationship would not be unexpected.

Although elevated prolactin titre is highly correlated with incubation behaviour, it may not always be necessary for the display of parental behaviour. In domesticated species, broody females deprived of their nest show a decline in prolactin levels, but will begin incubating as soon as they are returned to their nests. This can occur even though prolactin levels do not rise immediately (El Halawani *et al.* 1980; Zadworny *et al.* 1985; Lea and Sharp 1989). Such behavioural effects, however, could be attributed to a 'carry-over' effect of the previous prolactin elevation that may have primed the behavioural display (Buntin 1996). In poultry species, forced confinement of non-breeding adults with chicks can induce parental behaviour, even in the absence of any change in prolactin (Richard-Yris *et al.* 1995). In addition, parental behaviour does not seem to be fine-tuned by prolactin. Variation in the intensity of parental care is usually not tightly correlated with variation in the level of prolactin (Silverin and Goldsmith 1984; Janik and Buntin 1985; Ketterson *et al.* 1990; Vleck *et al.* 1991; Buntin *et al.* 1996; Schoech *et al.* 1996). This suggests that the level of parental effort is not modulated in any precise way by the level of prolactin secretion (or *vice-versa*).

Role of steroid hormones and prolactin in initiating incubation

In most species full-blown incubation behaviour does not begin with the first laid egg. In domestic ducks (*Anas platyrhynchos*) females spend increasing numbers of short bouts in a nest box over the course of laying 10 eggs, and at the end of clutch formation there is sudden increase in nest occupation (Hall 1987). Likewise in canaries (*Serinus canaria*), the percent of time spent in the nest rises with days after onset of laying. Initiation of incubation (defined as 50% of time spent in the nest) can vary from 0 to 4 days after the initiation of laying (Sockman and Schwabl 1999).

Several studies in domesticated species have indicated that exogenous prolactin administration induces sitting behaviour (e.g. Youngren *et al.* 1991), but may require interaction with ovarian steroids as well (reviewed in Buntin 1996). Turkeys pre-treated with estradiol benzoate and then given exogenous progesterone express interest in nest boxes, which can then be maintained by prolactin injections alone. Once incubation is fully established, endogenous prolactin levels are elevated and no exogenous hormone treatment is needed for sustained incubation (El Halawani *et al.* 1986). Induction of brood patch formation and broody behaviour in many species can be achieved only if prolactin treatments are accompanied by steroids (Jones 1971; Hutchison 1975; El Halawani *et al.* 1986; Chapter 8). The sharp drop in gonadal steroids after egg production, however, suggests sex steroids are not involved in incubation maintenance. Indeed, in domesticated species, gonad removal or steroid inactivation does not disrupt incubation that has already been established (Zadworny and Etches 1987; Ramesh *et al.* 1995; Lea *et al.* 1996). Such experimentation would obviously be difficult in wild birds.

Columbiformes differ from other species in that prolactin levels are not typically elevated until incubation is well established. In these species synergistic action of estradiol and progesterone is important to initiate incubation (Cheng and Silver 1975), although prolactin may be important in maintaining incubation behaviour and parental behaviour. Ring doves, separated from their nests but given exogenous prolactin, maintain nest attachment for up to 10 days and resume incubation given the opportunity, whereas controls do not (Lehrman 1964; Janik and Buntin 1985).

The initiation of incubation affects hatching synchrony (see Chapter 18), but may also limit clutch size in some indeterminate or semi-determinate laying species (reviewed in Haywood 1993a). For example, in the zebra finch (*Taeniopygia guttata*), contact with eggs in the nest on the second or third day of the egg-laying period appears to disrupt ovarian follicular growth and thus limit clutch size (Haywood 1993b). These types of observations led to the attractive hypothesis that in females increasing sensitivity to tactile contact with eggs through the laying cycle could elevate prolactin (see below) which could in turn have anti-gonadotropic effects that limit clutch size (Mead and Morton 1985). Individual variation in prolactin level is negatively correlated with clutch size in American kestrels (*Falco sparverius*; Sockman *et al.* 2000). This could be explained if those females that are most sensitive to the stimuli from eggs in the nest respond with the highest rate prolactin increase, which accelerates the onset of incubation behaviour and cessation of laying. This hypothesis further predicts that rising prolactin should be accompanied by decreasing sex steroids through the lay cycle, and in fact change in the ratio of these hormones could be what triggers the initiation of full incubation behaviour. Recent work in canaries supports this idea (Sockman and Schwabl 1999). Estradiol decreases by about three-fold on the day that incubation begins and progesterone decreases to a lesser extent. These decreases occur with incubation initiation irrespective of clutch size or when in the laying sequence incubation begins (Sockman and Schwabl 1999). Unfortunately, prolactin levels were not reported in this study.

In males similar relationships between androgens and prolactin may be important to initiate incubation. In the spotted sandpiper (*Actitis macularis*), a polyandrous, sex-role reversed species, testosterone decreases with the onset of incubation at the same time that prolactin increases (Fivizzani and Oring 1986;

Oring *et al.* 1986b). Furthermore, anti-androgens can bring about a premature onset of incubation (Oring and Fivizzani 1991). Interactions between hormone levels and their effects on incubation behaviour may vary with whether or not species are determinate or indeterminate layer, and when within the lay sequence incubation normally begins. Clearly both correlative and manipulative experimentation will be needed to identify patterns, which should aid in addressing the evolutionary significance of these patterns.

In males that develop brood patches it may be interaction between testosterone and prolactin that initiates brood patch formation and incubation. In general, however, high levels of testosterone are thought to be incompatible with the display of male parental behaviour. Male birds that engage in parental behaviour may show elevated levels of testosterone during courtship and mating stages, but these levels generally plummet when the birds enter the parental phase of reproduction (Wingfield *et al.* 1990). Androgen supplementation is known to disrupt or decrease parental behaviour in several free-living species (Silverin 1980; Hegner and Wingfield 1987; Oring *et al.* 1989; Ketterson *et al.* 1992; Chandler *et al.* 1997). Testosterone in these species does not necessarily block the display of parental behaviour, but rather seems to increase the likelihood that the birds will engage in other behaviours (e.g. courtship or territorial defence) at the expense of parental behaviour.

Behavioural effects on prolactin secretion

Hormone levels promote incubation behaviour, but the act of incubation also affects hormone titres. Cause and effect relationships between incubation behaviour and prolactin can be studied in species in which essentially continuous tactile input from the eggs is required to maintain elevated prolactin and broodiness (El Halawani *et al.* 1980; Hall *et al.* 1986; Hall 1987; Sharp *et al.* 1988). In galliforms and anseriforms, the effect of incubation on prolactin is known to be mediated by contact between the eggs or chicks and the brood patch (Opel and Proudman 1988; Richard-Yris *et al.* 1998b; Chapter 8). When the brood patch is anaesthetised or denervated, prolactin levels fall and incubation behaviour decreases, even when the eggs are still present in the nest (Hall 1987; Book *et al.* 1991). Brood patch denervation in female turkeys decreases prolactin levels compared to surgical controls and decreases the time they spend in nest boxes (Figure 5.2). These birds also do not become broody and have higher egg production rates than controls (Book *et al.* 1991).

Not all species, however, display a tight relationship between continuous brood patch stimulation, elevated prolactin secretion and incubation behaviour. In many species in which both members of the pair incubate, plasma prolactin remains high even when birds are away from their nest for hours to weeks. In these species the birds return to the nest to incubate even if they have not had recent tactile input from the eggs (Hector and Goldsmith 1985; Hall 1986; Lea *et al.* 1986; Garcia *et al.* 1996; Jouventin and Mauget 1996; Lormée *et al.* 2000; Vleck *et al.* 2000). In the emperor penguin (*Aptenodytes forsteri*), the female leaves the colony after

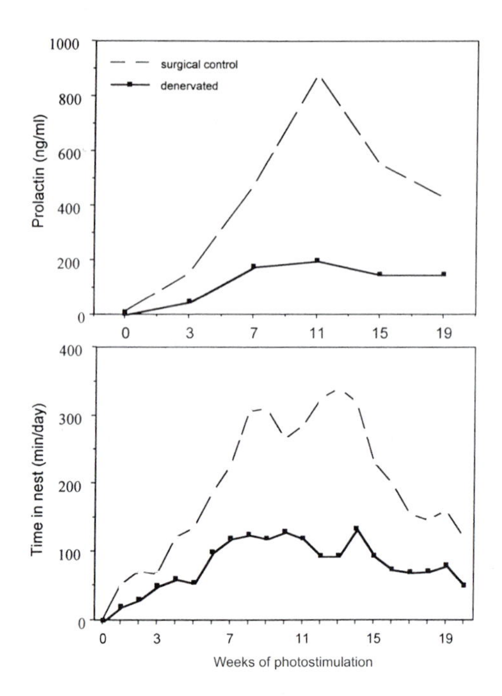

Fig. 5.2 The effect of surgical denervation of the brood patch in female turkeys (*Meleagris gallopavo*) on photoinduced prolactin secretion and time spent in a nest box. The dashed line indicates data from surgical controls and the solid line comes from brood patch denervated birds. Modified from Book *et al.* (1991), with permission from Physiology and Behaviour.

egg laying and the male incubates while the females forages. Yet female prolactin levels are higher when they finally return to feed and brood the newly hatched chicks than when they left approximately 2 months earlier (Lormée *et al.* 1999).

Manipulation of incubation period

One way to study the cause and effect relationship between prolactin, incubation behaviour and sensory input is through manipulative experimentation. In domesticated species there are multiple studies that indicate tactile and vocal input from precocial chicks to incubating birds decreases prolactin secretion and disrupts incubation behaviour (Richard-Yris *et al.* 1987; Opel and Proudman 1988; Leboucher *et al.* 1990; Lea *et al.* 1996; Richard-Yris *et al.* 1998b). A similar 'experiment' occurs in nature when eggs of brood parasites hatch before the natural eggs of the parasitised incubator. In yellow warblers (*Dendroica petechia*), however, the presence of a brown-headed cowbird chick (*Molothrus ater*) does not alter female nest attentiveness patterns (McMaster and Sealy 1999).

In free-living species, eggs and chicks can be switched between nests to artificially extend or terminate the incubation period. If such manipulation affects prolactin levels it suggests sensory input is important in controlling changes in prolactin secretion. In *Diomedia* albatrosses (Hector and Goldsmith 1985), the pied flycatcher (*Ficedula hypoleuca*; Silverin and Goldsmith 1984) and Adélie penguins (Vleck *et al.* 2000), lengthening the incubation period does not alter the timing of the post-hatching decline in prolactin levels. In pied flycatchers, however, individuals whose incubation period has been shortened do show an early decline in prolactin (Silverin and Goldsmith 1984). In Wilson's phalaropes (*Phalaropus tricolor*) extending the incubation period lengthens the period of prolactin elevation and shortening it has the opposite effect (Oring *et al.* 1988). Oring *et al.* (1988) suggested that in species, such the Wilson's phalarope, in which the natural incubation period is variable, the drop in prolactin secretion that normally occurs at hatching should be cued by the hatching event. Conversely, endogenous control of prolactin secretion, independent of hatching events, should be adaptive in species with little variation in the incubation period, such as the pied flycatcher, because this would decrease the time a bird spends incubating non-viable eggs. In species in which prolactin levels and parental behaviour is relatively independent of exposure of the bird to nest and eggs, prolactin levels may be determined by an endogenous schedule or by other external cues such as day length (Burke and Dennison 1980; Ebling *et al.* 1982; Silverin and Goldsmith 1997). In these species, pairs that breed late in the season may abandon their chicks at a relatively young age because parental prolactin levels are falling. Such a pattern may accommodate a late-season trade-off between parental self-maintenance and further chick investment (Moreno *et al.* 1997; Vleck *et al.* 2000).

In many species prolactin levels decrease dramatically within 24 to 48 hours of nest loss or egg removal (Etches *et al.* 1979; El Halawani *et al.* 1980; Goldsmith *et al.* 1984; Ramsey *et al.* 1985; Hall 1986, 1987; Sharp *et al.* 1988; Lea and Sharp 1989; Lea *et al.* 1996; Leboucher *et al.* 1996; Richard-Yris *et al.* 1998a). A tight feedback loop between tactile input from eggs or chicks and continued prolactin secretion should be adaptive in species that can re-nest after premature loss of eggs or chicks. Otherwise, protracted prolactin secretion after nest loss might hinder the rise in gonadotropins and sex steroids necessary for re-initiation of another nest cycle (reviewed in Sharp *et al.* 1998).

In many high latitude seabirds, however, plasma prolactin does not change rapidly after nest loss (Garcia *et al.* 1996; Jouventin and Mauget 1996; Vleck *et al.* 2000). In emperor penguins, prolactin levels are elevated for up to 3 months after failure (Lormée *et al.* 1999). In species, which cannot re-nest within a season, continued prolactin secretion after nest loss may contribute to their propensity to remain in the nesting colony and to attempt to steal and incubate the eggs of other birds (Vleck personal observation; Fowler *et al.* 1994). Adélie penguins will incubate infertile eggs for nearly four weeks past the normal hatch date (Vleck *et al.* 2000). Such persistence in incubation probably reflects the prolonged elevation in prolactin after egg hatching (Figure 5.2). This may in turn, be the result of selection to maintain a hormone milieu conducive to promoting return of ocean-foraging birds to the colony to care for eggs and chicks during the extended nesting season.

Manipulation of prolactin levels

Direct evidence for prolactin involvement in incubation behaviour comes from experiments that alter prolactin levels. Exogenous prolactin has been shown to affect incubation behaviour in a number of species (reviewed in Buntin 1996). For example, ovine prolactin from implanted osmotic pumps increases incubation behaviour about 2.5 fold relative to controls in captive, laying American kestrels (Sockman *et al.* 2000). Free-living female willow ptarmigan (*Lagopus lagopus*) receiving supplemental prolactin display more intense defence of chicks compared to controls, and chick survival is also increased (Pedersen 1989). Additional experimentation with wild birds awaits less invasive ways to increase prolactin levels. Pharmacological and immunological treatments that block prolactin secretion will limit broody behaviour in domestic species (reviewed in Sharp 1997), but have not yet been shown to be effective in any wild birds. Reduction of prolactin activity via inhibition of the prolactin releasing hormone, vasoactive intestinal peptide, reduces incubation behaviour in domestic fowl, and the effect can be prevented by simultaneous injection of prolactin (Sharp *et al.* 1989.) In ring doves, however, this treatment reduces prolactin levels, but not sitting behaviour (Lea *et al.* 1991). Immunisation against prolactin decreases its activity and reduces incubation behaviour in female turkeys (Crisóstomo *et al.* 1997) and in bantam fowl (March *et al.* 1994). The development of ways to block prolactin activity in free-living birds would be useful to establish the importance of prolactin in maintaining incubation behaviour in other birds.

Prolactin and incubation behaviour

A variety of social, environmental, physiological and experiential factors may influence incubation behaviour, but the most important proximate factor that maintains incubation is elevated prolactin. It may be that multiple, redundant mechanisms have evolved to reinforce parental behaviour because of its direct tie to fitness. In the process of domestication, breeders have had limited success in blocking incubation behaviour, probably because of its importance to fitness and the multiple stimuli that can affect it. On the other hand, brood parasites lay eggs and exhibit no parental behaviour past egg laying. Despite this fact, one obligate brood parasite, the brown-headed cowbird displays a seasonal rise in plasma prolactin (Dufty *et al.* 1987). This suggests that sensitivity to prolactin can be decreased and the behaviour eliminated, possibly by decreasing prolactin binding activity within the brain (Ball *et al.* 1988). Prolactin secretion and parental behaviour appear to be mutually reinforcing such that is it difficult to tease apart cause and effect, although the distinction for wild birds may have little biological importance. Despite general similarities in the control of incubation behaviour, there are obvious species differences associated with different life history strategies. Additional studies of free-living birds under a variety of ecological conditions will continue to provide lessons about the control of incubation behaviour.

Acknowledgements

Parts of this work were supported by NSF grants IBN-9211581 and OPP-93–17356.

6 *Behaviour patterns during incubation*

D. C. Deeming

Contact incubation is reliant upon the presence of a parent bird in order to provide sufficient heat energy to raise egg temperature to the appropriate level for embryonic development. This necessity has moulded the behaviour of breeding birds as well as nest location and construction (Chapter 2). The presence of the bird on top of the eggs during incubation limits the range of other behaviours that can be exhibited by the bird during the incubation period. There is certainly a potential conflict for the bird, which has to balance its own maintenance requirements with the needs of the developing embryos (Chapter 20). This chapter explores the patterns of behaviour during incubation in birds by describing how birds of both genders occupy their time during incubation. Rather than listing numerous examples in different species the intention here is to provide an overview of patterns of behaviour and the factors which affect the incubation phase of reproduction.

Patterns of behaviour

Skutch (1957) classified the pattern of incubation behaviour based on the presence and activity of the adult birds (Table 6.1). In its simplest form the main categories were: (I) both parents incubate (biparental continuous incubation; Williams 1996), and (II) only one parent, male, or female incubates the eggs (androparental and gyneparental incubation respectively; Williams 1996). At almost 50% of all bird Families, incubation shared by both parents is the commonest pattern (Figure 6.1). Female-only incubation is found in over 37% of bird families but male-only incubation occurs in only 6% of families (Figure 6.1). Distribution of these behaviour patterns between the orders of birds (Table 6.2) shows that most of the Palaeognathae exhibit male-only incubation with only a few other families showing this behaviour. Non-passerine families of birds have almost 80% of the families exhibiting shared incubation. By contrast, passerine families represent 62% of the Families with female-only incubation (Table 6.2).

Skutch (1957) went on to refine his classification further largely based on the length of the incubation sessions (the periods on which the bird sits on the eggs) and recesses (periods during which bird is absent from the nest) performed by the parent birds (Table 6.1). Lengths of both the incubation sessions and recesses are highly variable ranging from only a few minutes in female-only incubation of many passerines (Kendeigh 1952; Skutch 1962) through to many days in shared incubation patterns of penguins and procellariform seabirds (Warham 1990; Williams 1995). As a consequence, the number of sessions each bird undertakes during an

Table 6.1 Categories of incubation pattern and examples as defined by Skutch (1957).

Incubation category	Gender type	Behaviour pattern	Variation	Examples
I. Incubation by both parents	A. Both by sexes simultaneously at two nests			Red-legged partridge (*Alectoris rufa*)
	B. By both sexes, alternately	1. Either sex may cover the eggs by night	a. Change over at intervals of approximately 24 hours	Ringed kingfisher (*Ceryle torquata*), Diving petrel (*Pelecanoides urinatrix*)
			b. Change over at intervals of greater than 24 hours	Adelie penguin (*Pygoscelis adeliae*),
			c. Change over at intervals of less than 24 hours	Herring gull (*Larus argentatus*), Cape wagtail (*Motacilla capensis*)
		2. Female covers eggs by night	a. Male takes one long session per day	Columbiformes and Trogoniformes
			b. Sexes alternate on nest several times a day	Antbirds (Formicariidae), warblers (Sylviidae)
		3. Male covers eggs by night	a. Female takes one long session per day	Ostrich (*Struthio camelus*), Pale-billed woodpecker (*Phloeoceastes guatemalensis*)
			b. Sexes alternate on nest several times a day	Woodpeckers (Picidae), American coot (*Fulica americana*)
II. Incubation by one parent alone	A. Incubation by female alone	1. One long recess each day		Quail (Phasianidae), Guan (*Pauxi pauxi*)
		2. Several recesses during day and night		Trochilidae, most Passeriformes
		3. Female sits continuously for many days, fasting	a. Fasting during incubation	Phasianidae and Anatidae
			b. Male feeds female during incubation	Bucerotidae
	B. Incubation by male alone	1. One long recess each day		Tinamiformes
		2. Several recesses during day		Jacana (*Hydrophasianus chirurgus*), Tinamiformes
		3. Male sits continuously for many days, fasting		Emperor penguin (*Aptenodytes forsteri*), kiwis (Apterygidae), emu (*Dromaius novaehollandiae*)
III. Incubation by more than two birds at a single nest	A. Eggs laid by one female	(Several adults assist the female)		Bush-tit (*Psaltriparus minimus*)
	B. Eggs laid by more than one female	(Several birds of both genders participate in incubation)		Anis (Cuculidae), Acorn woodpecker (*Melanerpes formicivorus*)
IV. Eggs incubated by birds of other species				Cuculidae, cowbirds (Icteridae), honey guides (Indicatoridae), black-headed duck (*Heteronetta atricapilla*)
V. Eggs incubated without animal heat				Megapodiidae

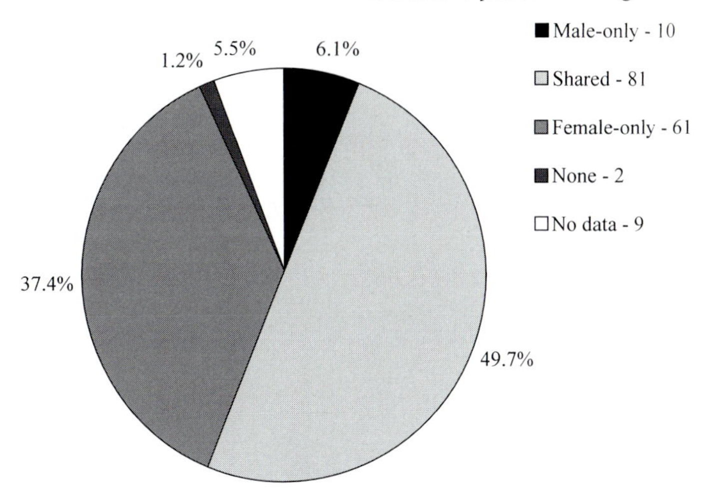

Fig. 6.1 Pie-chart showing the distribution of different patterns of incubation in the families of bird. The total number of families represented in each category is indicated in the legend.

incubation period can vary considerably. For some species, e.g. passerines and penguins (Skutch 1962; Williams 1995) the periodicity of the length of the sessions and recesses can be very regular but in many other cases there are many factors, which affect the constancy of incubation (see below).

When incubation is shared (type I) the two parents usually alternate on the one nest. The only example in this category seems to be the red-legged partridge (*Alectoris rufa*) in which the female lays eggs in two nests and relies on its mate to incubate the second nest whilst she incubates the first (Goodwin 1953). In other species that share incubation, Skutch (1957) made a distinction in terms of which gender covered the eggs at night. For the category in which either gender incubates during the night distinction was made between the length of the incubation sessions: nest relief can occur at (a) approximately 24 hour intervals, (b) intervals greater than 24 hours, and (c) at intervals less than 24 hours (Table 6.1). For the other two behaviour patterns (female-, or male-only incubates during the night), distinction was made between the length of time the mate spends on the nest during daylight. Hence, when the male incubates at night the female can incubate for one long session during the day, e.g. the ostrich (*Struthio camelus*) or the two genders can alternative on the nest several times a day, e.g. woodpeckers (Picidae).

Incubation by one gender alone (type II) is split into three behaviour patterns for (A) female and or (B) male birds: (1) one long recess during the day; (2) several recesses during the day and night (gyneparental intermittent incubation; Williams 1996); and (3) the bird sits continuously for many days (Table 6.1). A female incubating continuously may fast (e.g. waterfowl; Afton and Paulus 1992) or it is fed by the male bird during incubation (assisted gyneparental incubation; Williams 1996), e.g. hornbills (Bucerotidae; Kemp 1995). There are no recorded instances of females feeding incubating male birds. For those species where the male is on the nest continuously, e.g. the emu (*Dromaius novaehollandiae*), the bird is forced

Table 6.2 Distribution of basic patterns of incubation behaviour between the different orders of birds compiled from Perrins and Middleton (1985). Data are the numbers of families in each order exhibiting the incubation pattern. 'None' represent families of exclusive brood-parasites or the megapodes that bury their eggs.

Order	Male-only	Shared	Female-only	None	No data
Struthioniformes		1			
Rheiformes	1				
Casuariiformes	2				
Apterygiformes	1				
Tinamiformes	1				
Sphenisciformes		1			
Gaviiformes		1			
Podicipediformes		1			
Procellariiformes		4			
Pelecaniformes		6			
Ciconiformes		6			
Anseriformes		1	1		
Falconiformes		1	4		
Galliformes			5	1	
Gruiformes	2	8	2		
Charadriiformes	3	12	1		1
Pteroclidiformes		1			
Colombiformes		1			
Psittaciformes			1		
Cuculiformes		3			
Strigiformes			2		
Caprimulgiformes		5			
Apodiformes		2	1		
Trogoniformes		1			
Coliiformes		1			
Coraciiformes		3	6		
Piciformes		5		1	
Passeriformes		17	38	1	7
Total	10	81	61	3	8

to fast and can lose a lot of weight before the eggs hatch (Buttemer and Dawson 1989).

Skutch (1957) also distinguished between the source of the eggs where incubation is carried out by more than two birds at a single nest (type III). Type A, exemplified by the bush-tit (*Psaltriparus minimus*) involves a single female laying eggs in the nest with several adults assisting with incubation (Addicott 1938). Type B, exhibited by the acorn woodpecker (*Melanerpes formicivorus*), groove-billed ani (*Crotophaga sulcirostris*) and Pukeko (*Porphyrio porphyrio*), involves eggs being laid in a nest by several females and being incubated by several adults (Craig and Jamieson 1990; Koford *et al.* 1990; Stacey and Koenig 1991). Brood-parasitism (type IV) is dealt with in Chapter 17 whereas incubation by megapodes (type V) is described in Chapter 13.

Distinctions between incubation by one or both of the parents were based on logic rather than any phylogenetic relationships (Skutch 1957). It is difficult to recognise any relationship between phylogeny and incubation patterns. For instance, male-only incubation is found in families in seven different orders of birds. It is likely that the pattern of incubation behaviour adopted by different species of bird reflect the restrictions imposed upon them by their environment. Closely related species can adopt quite different incubation behaviour patterns. For instance, with one exception all penguin species exhibit shared incubation (Williams 1995). The emperor penguin (*Aptenodytes forsteri*), however, adopts male-only incubation because the extreme nesting environment on Antarctic ice and the prolonged incubation and rearing phases precludes this species from sharing incubation duties (Williams 1995). Similarly, within Coraciiformes, three of nine families exhibit shared incubation whilst the other families adopt female-only incubation with the hornbills adopting a unique incubation behaviour of being imprisoned in the incubation hole and completely fed by the males (Kemp 1995). Although many species of passerines exhibit female-only incubation characterised by several recesses per day, there are other species, which exhibit female-only incubation, combined with feeding by male birds, e.g. the yellow-billed magpie (*Pica nuttalli*; Verbeek 1973) and there are others that share incubation, e.g. the starling (*Sturnus vulgaris*; Drent *et al.* 1985). Possible factors affecting evolution of incubation behaviour are discussed below.

Activity of birds during incubation

A requirement to be in contact with the eggs for long periods of time to maintain the appropriate temperature for development restricts the behavioural repertoire of parent birds. Even in those species where contact incubation is suspended, i.e. megapodes, and for at least some of the time in the Egyptian plover (*Pluvianus aegyptius*; Howell 1979), the parent birds still spend considerable amounts of time tending the incubation site. There is considerable variation in the number and duration of both incubation sessions and recesses. Passerines exhibiting female-only incubation (Conway and Martin 2000b) have small eggs and average recesses of 10.2 min (SE = 0.6) taking a mean of 3.6 trips h^{-1} (SE = 0.2). By contrast, waterfowl (Afton and Paulus 1992) have larger eggs and take recesses averaging 61.2 min (SE = 8.2) with only 3.2 trips d^{-1} (SE = 0.4). During sessions, the incubating bird has a restricted behavioural repertoire imposed upon it by the need to maintain contact between the eggs and the body. On the whole the adult simply sits on the eggs and is only able to perform behaviours associated with the nest (Skutch 1976). Egg turning is a significant behaviour during incubation sessions that is described in Chapter 11. During recesses, the birds are free to carry out any behaviour they wish.

Behaviour during incubation sessions

A study of the time-budget of incubating Adélie penguins (*Pygoscelis adeliae*), which have incubation sessions of over 12 days, showed that at different times during the incubation period the birds spend most of their time in the prone incubation position with the remaining time spent upright (Derksen 1977). In the prone position (94% of observations), resting and sleeping predominate with the remaining time spent largely on agonistic and territorial defence behaviours with a minor amount of nest building. The time-lapse method of study did not always allow behaviour to be recognised so there is a relatively high proportion of unknown behaviours (Figure 6.2). Once the birds are upright (6.36% of observations) their behavioural repertoire increases although agonistic and territorial behaviours form over half of the observations in this position. Rotations, where the bird moves around the nest, and comfort and preening behaviours form the bulk of the remaining observations (Figure 6.2). Egg turning was only possible whilst the birds were upright and appears to be a minor behaviour during incubation (Figure 6.2). This is almost certainly an artefact because the time-lapse interval of one minute was unable to register all turning events by the birds.

Studies of incubating willow grouse (*Lagopus lagopus*) and capercaillie (*Tetrao urogallus*) show that incubating females spend most of their time sitting in one position with its eyes closed (Pulliainen 1971, 1978). Other recorded behaviours included rising from the nest to turn eggs (16 times per day for capercaillie), rising to re-settle and face in a new direction (3.3 times per day), a swaying movement to settle on the nest, preening (28.3 times per day), and nest tending (4.0 times per day). Female capercaillies were also absent from the nest 1.9 times per day, forming 3.9% of the day (Pulliainen 1971).

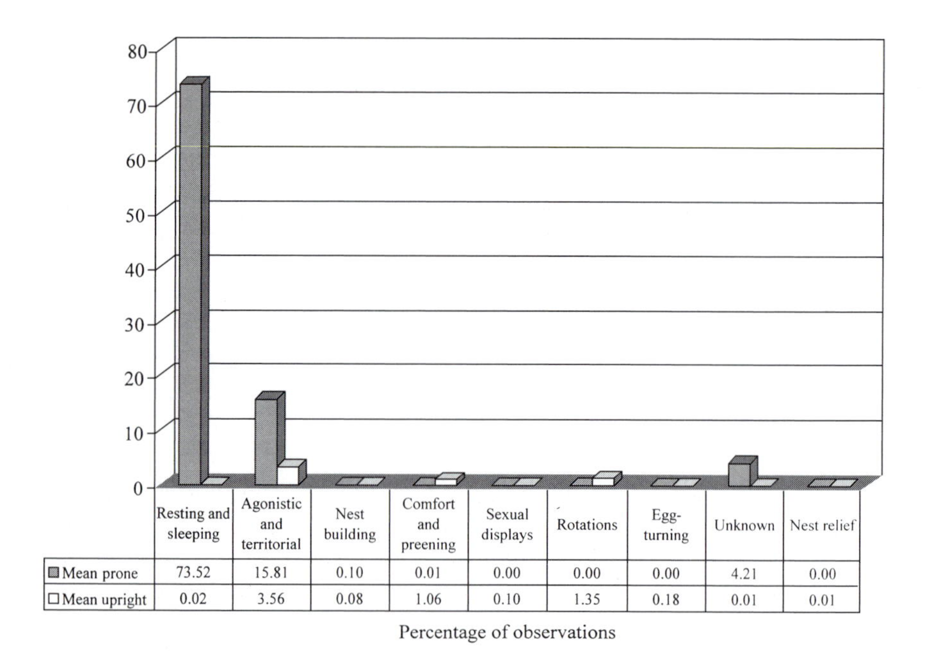

	Resting and sleeping	Agonistic and territorial	Nest building	Comfort and preening	Sexual displays	Rotations	Egg-turning	Unknown	Nest relief
■ Mean prone	73.52	15.81	0.10	0.01	0.00	0.00	0.00	4.21	0.00
□ Mean upright	0.02	3.56	0.08	1.06	0.10	1.35	0.18	0.01	0.01

Fig. 6.2 Time-budgets of Adélie penguins (*Pygoscelis adeliae*) during incubation. Data from Derksen (1977).

Gulls (*Larus* spp.) perform several complex behaviour patterns during incubation sessions (Beer 1961, 1965; Baerends 1970; Baerends *et al.* 1970; Impekoven 1973). These include looking down at the eggs whilst sitting on the eggs, and whilst standing above the nest. There is nest-building activity and the birds will vocalise when re-settling on their eggs or in response to external stimuli (Impekoven 1973). Re-settling after standing, or at nest-relief, consists of a fixed sequence of behaviour patterns: chest-dropping, waggling and quivering (Beer 1961). The mallard duck (*Anas platyrhynchos*) also has a waggling behaviour during re-settling (Caldwell and Cornwell 1975).

Many species of passerines exhibit tremble-thrusting, where the bird probes the sides and bottom of the nest (Haftorn 1994). This activity is carried out throughout incubation and is positively correlated with nest attentiveness. Tremble thrusting is considered to keep ectoparasites from the nest cup and to help maintain nest hygiene. It may also partially replace egg turning behaviour (Haftorn 1994) although Hartshorne (1962) considered movement of eggs to be one of its primary functions (Chapter 11).

Maintenance behaviours are important for the adult during incubation sessions. Incubating birds are often required to endure adverse weather conditions and so in hot conditions birds may pant or use gular flapping to increase evaporative cooling (Skutch 1976). On sunny days incubating common nighthawks (*Chordeiles minor*) constantly change position so as to face away from the sun, a behaviour absent during cloudy days (Weller 1958). The desert-nesting Egyptian plover routinely buries its eggs in sand whether the bird is present or not (Howell 1979). During the night, early morning and late evening the eggs are uncovered and contact incubated by the adult birds. During the mid-morning rising ambient temperature is sufficient to allow the birds to leave the eggs buried but during the hottest part of the day the adults are in constant attendance. To ensure that egg temperature does not rise above critical levels the adults regularly collect water in their breast feathers so as to soak the eggs. During cloudy days, when ambient temperature is lower, egg soaking behaviour is suspended. Wetting the eggs has also been observed in little terns (*Sterna albifrons*) and skimmers (*Rynchops flavirostris*) both of which nest in exposed sunny locations (Hardy 1957; Turner and Gerhart 1971). In other Charadriiformes, a cooling effect for egg covering is discounted because wetting of nest material can inhibit or prevent egg-covering behaviour (Maclean 1974). Incubation in extreme cold conditions is a less common problem. Male emperor penguins huddle together during incubation to minimise the effects of the Antarctic wind (Williams 1995).

Franks (1967) showed that incubating ringed turtle doves (*Streptopelia risoria*) reacted to artificial changes in egg temperature. Cool eggs (less than 30°C) produced shivering and feather raising by the birds whereas hot eggs (over 40°C) elicited gular flapping and preening. Similar behaviours were observed in zebra finches (*Poephila guttata*) under similar conditions (Vleck 1981a). Davis *et al.* (1984a) showed that patterns of attentiveness in the sparrow (*Passerculus sandwichensis beldingi*) were changed by manipulation of egg temperature but this was independent of prevailing ambient temperature.

In colony nesting species nest sites are at a premium and incubating birds often have to interact with con-specifics to maintain their position in the colony. Individual nesting sites in penguin colonies are often deliberately located to minimise contact with neighbouring birds although movement of birds through the colony elicits agonistic behaviour from incubating birds (Figure 6.2).

The value of the nest and its contents has led to the evolution of a variety of strategies to minimise loss of eggs. Location of nests and cryptic concealment of the sitting bird often leads to protection of the clutch and adult (Drent 1975). Simply sitting motionless can be sufficient to escape detection by a predator (Skutch 1976). Often an incubating bird has to respond to the immediate presence of a potential predator. Circumspect departure from, and arrival at, the nest can maintain the camouflage of the location of the eggs. Alternatively, departure from a nest is highly conspicuous with the intention of distracting the potential predator from the nest. This is seen in many ground nesting species, e.g. plovers (Drent 1975; Skutch 1976) and the ostrich (Bertram 1992), and characterised by the incubating bird leaving the nest and producing a distraction display often feigning injury (Skutch 1976). After the bird has drawn the threat away from the nest it spontaneously recovers and makes its own escape before returning to the nest once the threat has passed.

Other species use threats and attack to defend the nest. Raising the wings and throwing them forward can make a bird look larger than it is and the threat of attack is often sufficient to deter a predator. The hole-nesting tit (*Parus major*) reacts to nest disturbance by spreading its tail and flicking its wings against the sides of the nest cavity whilst hissing (Gosler 1993). Terns and skuas usually resort to physical attack of potential risks to nests (Drent 1975; Skutch 1976). Female eider (*Somateria mollissima*) will sit tight during the approach of a predator only leaving the eggs at the last moment but defecating on the eggs as she leaves. The noxious smell has been shown to deter predation of the eggs (Swennen 1968).

Nest relief

During shared incubation changeover between the birds on the nest can often involve an elaborate nest relief ceremony. This is very common in colony nesting seabirds and water birds (e.g. Ardeidae). A need to positively identify the correct nesting partner has meant that pair-bonding behaviours established during courtship are often simply maintained during incubation (Skutch 1976). In most other species nesting in isolation there is little or no ceremony at nest relief. Usually, the eggs are covered at all times in shared incubation but in some species the incubating bird will leave the nest before its mate has arrived (Skutch 1976). In many instances the drive to incubate is so strong that the bird has to be driven from the nest by its mate. Nest relief is often an inconspicuous behaviour in order to prevent unwanted attention from potential predators of eggs or in some cases the adults. Small shearwaters and petrels nest in burrows and nest relief only takes place during the night because of the threat of predation to the returning adults by gulls (Cramp 1977).

Maclean (1974) reviewed the behaviour of egg-covering found in Tinamiformes, Anseriiformes, Podicipediformes, and Charadriiformes. Eggs are covered with nest material or sand at times when the bird leaves the nest. In the black-necked grebe (*Podiceps nigricollis*) clutch size is correlated with the extent to which eggs are covered in response to disturbance (Broekhuysen and Frost 1968). Eggs were left uncovered in only 26% of instances of disturbance when there were 3 or more eggs in the nest compared with 62% of instances when there was only 1–2 eggs present. Although egg-covering is geographically widespread in grebes and waterfowl, in other orders it is restricted to hot arid environments. Maclean (1974) considered that for much of the time egg-covering was primarily concerned with concealment of eggs from potential predators although in some species there may be an important role in thermoregulation. Davis *et al.* (1984b) report that eggs of the pied-billed grebe (*Podilymbus podiceps*) can be left covered for the whole of the day and are only contact incubated at night.

Behaviour during recesses from incubation

Once a bird is relieved from the duty of sitting on eggs it is usually free to carry out any behaviour. In some species at least the bird assumes behaviour patterns typical of pre-breeding periods. The time budget of white-winged scoter (*Melanitta fusca deglandi*) ducks is little changed between periods before and during egg-laying and during incubation recesses; diving and dive pauses take around 60% of their time (Brown and Fredickson 1987). The only difference of note is a replacement of resting behaviours with maintenance behaviours (preening, bathing, comfort activities). The time-budget of female tundra swans (*Cygnus columbianus*) during pre-laying periods and during incubation recesses (Figure 6.3) showed similar rates of foraging, and a reduction in stationary behaviour (resting) matched by an increase in maintenance/comfort behaviours (Hawkins 1986). By contrast, male swans spent less time on the nest but during incubation recesses, stationary behaviour was replaced by more locomotion and foraging behaviours with the later rising to match the time commitment of females during recesses (Figure 6.3). High feeding levels of females during pre-laying periods may reflect the energy demands of egg formation. It is probably safe to assume that the primary activity of all birds during voluntary incubation recesses reflects a need to feed to restore body reserves depleted during the incubation sessions.

Egg recognition

The presence of eggs in a nest has important effects on the behaviour of parent birds. Visual stimulation by eggs in the nest cause male emus to go broody. It is unclear whether such stimulation is widespread in birds. Female ostriches are reported to recognise their own eggs and exclude those laid in the nest by other females (Bertram 1992) but it is unclear how this is possible given the lack of any distinguishing marks on the cream-white eggs. The ability to recognise their own eggs seems to be relatively rare in birds (Poulsen 1953) and many species seem unable to recognise the presence of eggs laid by brood parasites despite clear differences in appearance (see Chapters 17 and 19). Ability of birds to retrieve

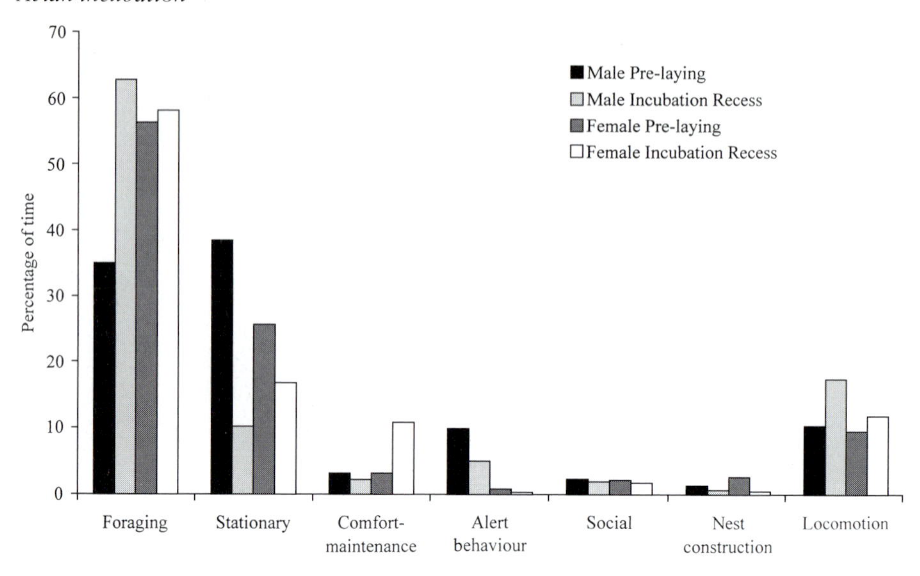

Fig. 6.3 Time-budgets of male and female whistling swans (*Cygnus columbianus*) during pre-laying periods and incubation recesses. Data derived from Hawkins (1986).

eggs lost from the nest is also variable with some species apparently ignoring eggs only a short distance from the nest (Poulsen 1953). Other ground-nesting species are adept at rolling eggs back into the nest (Poulsen 1953; Skutch 1976). The incubating bird usually ejects eggs that are cracked or broken (Poulsen 1953).

The drive to incubate has meant that birds will accept and sit upon a variety of objects irrespective of whether they resemble their eggs (Skutch 1976). Extensively studied in shore birds and gulls where many model 'eggs' painted in unnatural colours or of unnatural shapes were still sat upon. Often replacement objects were very large and served as super-stimuli despite being too large to sit upon. Marking eggs with coloured paint in order to study egg turning behaviour was shown to adversely affect the very behaviour the experiment was designed to study (Poulsen 1953; Holcomb 1969; Chapter 11). Fortunately, the willingness of birds to accept egg-shaped objects has meant that telemetric eggs can be used to monitor nest conditions (e.g. Howey *et al.* 1984; Chapter 11).

Activity of non-incubating parent birds

In many species where one parent assumes complete responsibility for incubation behaviour the mate is relieved of all obligations to the nest. Whether the mate remains within the vicinity of the nest depends on its duties post-hatching. In ratites exhibiting male-only incubation, the female departs after completion of the clutch because the male oversees the precocial hatchlings. Similarly when mate selection occurs at a lek the female does not form a pair-bond and is obliged to incubate and rear on her own. However, in many species the male bird will remain around the nest to help with nest defence or feeding of its mate. In some species nest defence includes attack and distraction behaviours as well as briefly occupying the nest when the female leaves to forage.

Constancy of incubation

Constancy of incubation, i.e. the time that the eggs are in contact with an adult bird usually expressed as a percentage of incubation time (Kendeigh 1952; Skutch 1962), is important in determining the pattern of incubation. In order to understand this relationship better, a database was established that recorded the %attentiveness of different birds from a total of 451 species from 24 orders of birds on the basis of three broad categories of incubation behaviour: shared, female-only and male-only. Nest attentiveness was high in most of the orders represented in the database with the mean value falling below 85% attentiveness in only two orders (Apodiformes and Passeriformes; Figure 6.4).

For shared incubation (N = 124 species) attentiveness ranged from 58–100% with a mean of 92.6% (SD = 9.4) and a median of 96.1% with a skew in the data towards higher rates of attentiveness (Figure 6.5). For those species (N = 46) with precocial and semi-precocial young (Apterygifromes, Podicipediformes, Gaviiformes, Procellariiformes, Anseriiformes, Gruiiformes, Charadriiformes and Caprimulgiformes) mean attentiveness was 95.2% (SD = 5.5) with two-thirds of the species recording attentiveness of 95–100% (Figure 6.6a). In those species (N = 77) with altricial and semi-altricial hatchlings (Sphenisciformes, Pelecaniformes, Ciconiformes, Falconiformes, Colombiformes, Strigiformes, Trogoniformes, Coraciiformes, Piciformes and Passeriformes) mean attentiveness

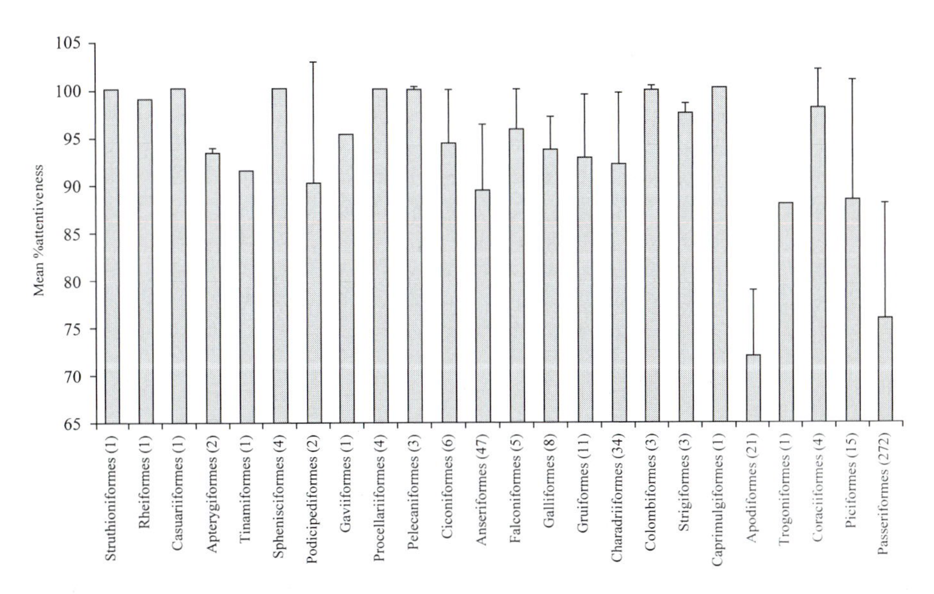

Fig. 6.4 Mean (+SD) percentage attentiveness in different orders of birds. Total number of species for each Order represented in brackets in x-axis.

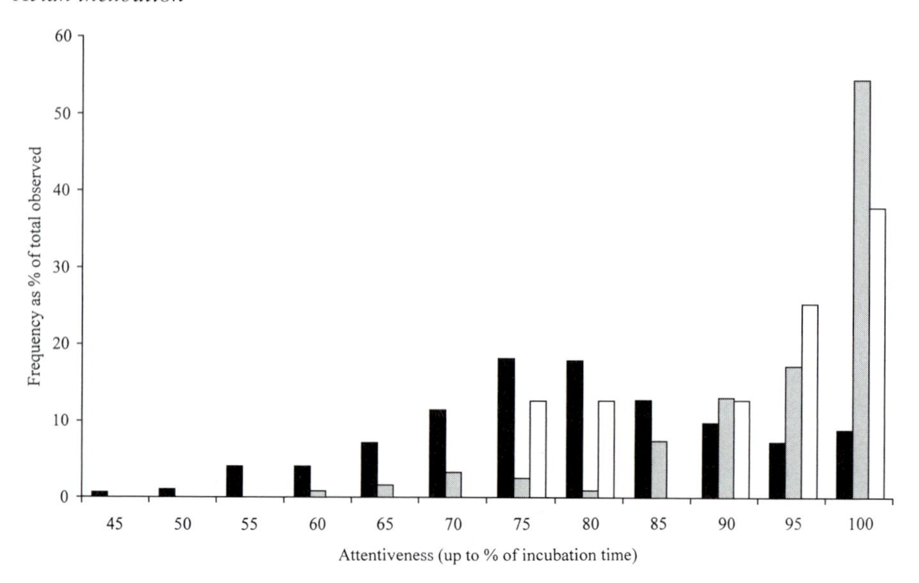

Fig. 6.5 Distribution of percentage attentiveness for shared incubation (grey columns), female-only (black columns) and male-only (white columns) incubation patterns. Data are expressed as a percentage of the total frequency in each 5% category.

was significantly lower (mean of 91.1%, SD = 10.8; one-way ANOVA, $F_{1,122} = 5.67$, $p < 0.05$) but the range of values was very much wider (Figure 6.6a).

For female-only incubation (N = 319 species) attentiveness ranged from 45 to 100% of the incubation period (Figure 6.5) and with a mean of 76.8% (SD = 12.0) and a median of 76.7%. At the extremes of the range there were more examples of high attentiveness than low attentiveness (Figure 6.5). Range and distribution of %attentiveness differed between groups based on hatchling maturity (Figure 6.6b). Those species (N = 61) with precocial young (Anseriiformes, Galliformes and Charadriiformes) exhibited significantly ($F_{1,317} = 111.54$, $p < 0.001$) higher attentiveness (89.4%, SD = 6.5) compared with those species (N = 258 from Apodiformes, Coraciiformes and Passeriformes) with altricial young (73.9%, SD = 11.1). For Passeriformes (N = 272 species) %attentiveness was significantly different between shared and female-only incubation patterns ($F_{1,270} = 70.66$, $p < 0.001$) with mean values of 89.0% (SD = 11.4, N = 41) and 73.4% (SD = 10.9, N = 231) for shared and female-only incubation respectively.

For the eight species showing male-only incubation, mean attentiveness was 89.9% (SD = 10.9) with a median of 92.2%. Three of these species, rhea (*Rhea americana*), emu and emperor penguin, were 99–100% attentive and all exhibited fasting during the long incubation period. The fourth, the North Island kiwi (*Apteryx australis mantelli*) has an attentiveness of 92.9% (McLennean 1988). The boucard tinamou (*Crypturellus boucardi*) was attentive for 91.4% of the recorded incubation period taking one long recess every second day (Lancaster 1964). The remaining species were phalaropes (*Phalaropus* spp.) and

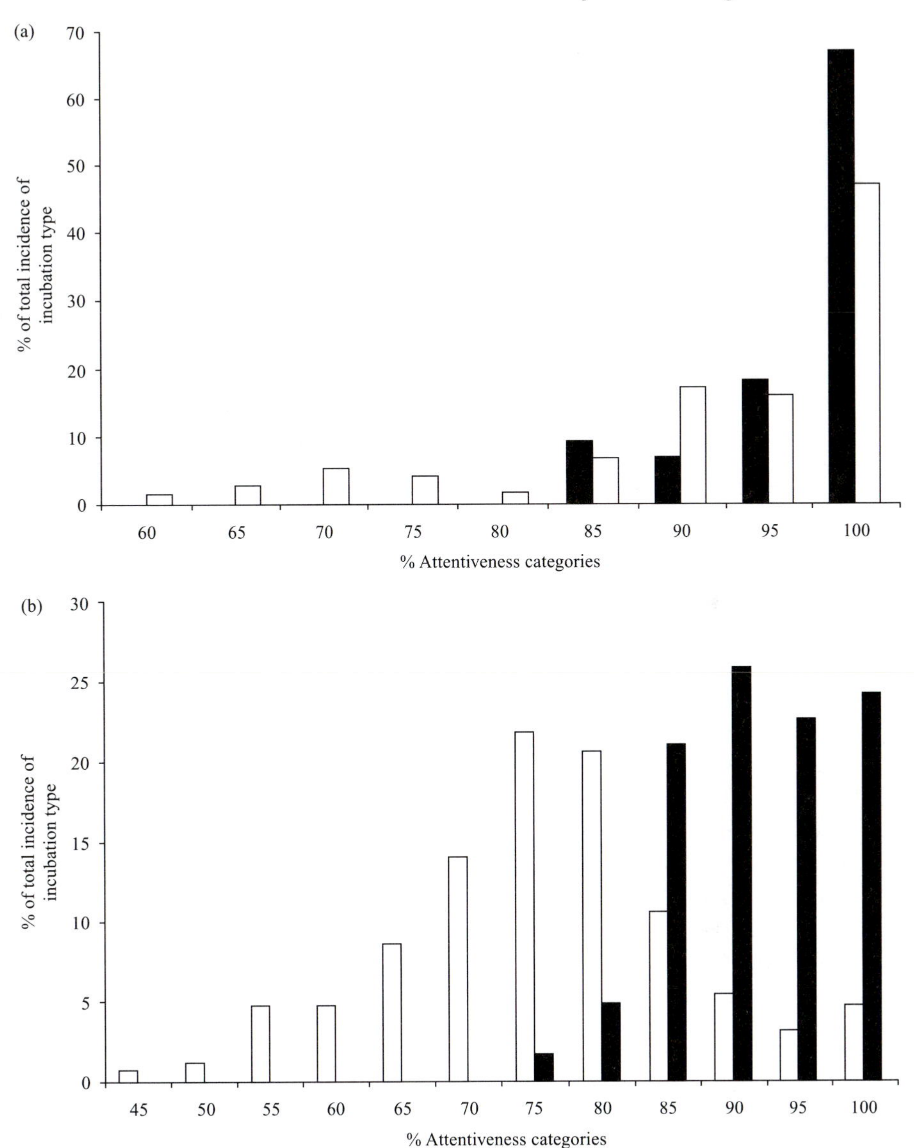

Fig. 6.6 Distribution of percentage attentiveness for species with precocial and semi-precocial, and altricial and semi-altricial, hatchlings (black and white columns respectively) for (a) shared incubation and (b) female-only incubation patterns. Data are expressed as a percentage of the total frequency in each category.

had attentiveness of less than 80%. With the exception of the penguin all of these species produce precocial hatchlings.

Many values for %attentiveness were determined from observations of bird activity at nests during daylight hours and often for short periods (see Skutch

1962). Often these data do not take into account the time spent on the nest overnight. Female tree sparrows (*Spizella arborea*) have a diurnal rhythm for nest attentiveness (Weeden 1966). Between 10.00 and 14.00 h attentiveness was only 58.5%, rising to 61.5% from 14.00 to 18.00 h. After 18.00 and up to the start of the night rest the attentiveness was 74%. During night rest on the nest attentiveness was 100%. As the bird became more active around dawn and the early morning attentiveness decreased (72% from end of night rest to 06.00, and 64% from 06.00 to 10.00 h). Nolan (1978) showed a similar pattern for the prairie warbler (*Dendroica discolor*) with %attentiveness high during the early morning and late evening and lowest during mid-day. %Attentiveness declined through the day in the barn swallow (*Hirundo rustica*; Smith and Montgomerie 1992). The average of 75.8% for attentiveness in the dusky flycatcher (*Emphidonax oberholseri*) is only for daylight hours and during the night attentiveness was 100% (Morton and Pereyra 1985). It is likely that many records of %attentiveness recorded for single gender incubation are only representative of the diurnal time-budget and will under-estimate the complete proportion of the 24-hour time budget for incubation. For shared incubation behaviour the activity of both parents during the day mean a higher %attentiveness during daylight hours and so the effects of recesses on the 24-hour incubation time-budget are much reduced.

Factors affecting nest attentiveness

Skutch (1962) recognised that incubation constancy was affected by a variety of factors influencing reproduction in birds. These included the participation of one or both parents, nutritional requirements, nest type, stage of incubation, and climatic conditions. Skutch (1962) also considered the bird's 'temperament' to be of importance but this seems to be a rather subjective view of behaviour.

Relative contribution of each parent to incubation is a critical factor in determining nest attentiveness. When both parents are incubating the constancy of incubation is high irrespective of the maturity of the offspring. Where one gender is left to incubate then there seems to be a dichotomy in the pattern of incubation dependent upon the maintenance requirements of the adult. Hence, attentiveness is high if the incubating bird is large and is able to survive off body reserves during the incubation phase. This pattern is seen in the female-only incubation pattern of galliformes and waterfowl (Pulliainen 1971, 1978; Hawkins 1986; Afton and Paulus 1992; Flint and Grand 1999). In these species the precocial young are able to feed themselves leaving the parent to quickly restore reserves. In those species which share incubation but have prolonged sessions, e.g. petrels and penguins, the incubating bird also loses a considerable amount of weight during the incubation session which is restored during the long recesses (Warham 1990; Williams 1995).

In many species with altricial young the female bird has to maintain body condition during incubation to ensure that she is fit to feed her offspring. Hence, compared with shared incubation, during female-only incubation attentiveness is relatively low (around 60–80% with an average of ~75%) because of the need of the bird to forage during the day. High attentiveness in species with altricial species is often possible because of the male bird feeding the female during incubation, e.g. Falconiformes and the yellow billed magpie (Verbeek 1973). In the meadow pipit (*Anthus pratensis*) female attentiveness was positively correlated with feeding by the male bird (Halupka 1994).

Nest type was seen by Skutch (1962) to influence %attentiveness during incubation with those species producing elaborate or large nests being more interested in building and maintaining the nest structure. As a result humming-birds (Trochilidae) and bush-tits (*Psaltriparus* spp.) were considered relatively inattentive parents (71% and 56% respectively) because their thick-walled or lichen-covered nests may well retain heat well during the parent's absence (see Chapter 2). Loss of heat from the eggs still means that nest-building activity is restricted to the warmest parts of the day (Skutch 1962).

Stage of incubation is also critical in determining nest attentiveness. In species with female-only incubation, true incubation may take a few days to become established but Skutch (1962) suggested that there was little evidence to support the idea that attentiveness increases as hatching approached. Rather it was maintained during the bulk of incubation and continued to the end of hatching. This view is supported by data for mallard (Caldwell and Cornwell 1975) but data for females of several species of waterfowl show that %attentiveness progressively decreases as incubation progresses (Afton 1980; Aldrich and Raveling 1983; Brown and Fredrickson 1987; Yerkes 1998). In those species with shared incubation %attentiveness does not change with incubation period but the relative proportions of incubation performed by the two parents may alter (Hawkins 1986; Brua personal communication).

Climatic conditions have a profound affect on the degree of incubation constancy. Environmental temperature is a major influence on how the birds incubate, particularly in those species with female-only behaviour patterns. In the dead sea sparrow (*Passer moabiticus*) there is a negative correlation between %attentiveness and lower ambient temperatures with the least time on the nest being spent at an air temperature of 35°C (Yom-Tov *et al.* 1978). Above this temperature %attentiveness increases dramatically as the bird seeks to maintain the egg temperature below that of the environment in order to maintain an appropriate temperature for development (Drent 1975). Afton (1980) showed that weather variables explained more of the components of incubation rhythm in small ducks than in larger geese and swans. Rain also affects incubation behaviour although it depends on the severity and extent of the rainfall (Skutch 1962). In those species exhibiting female-only incubation, rain during a recess usually elicits a return to the nest to shield the eggs. Rain during an incubation session appears to discourage the bird from leaving the eggs leading to longer than normal sessions.

Ambient temperature often affects the length of both sessions and recesses with the former showing a positive correlation, and the latter showing a negative correlation, with ambient temperature (e.g. Kluivjer 1950; Haftorn 1978, 1984). However, Conway and Martin (2000a) showed that the length of both sessions and recesses in the orange-crowned warbler (*Vermivora celata*) were positively correlated with environmental temperature but only in the range of 9–26°C; at higher and lower temperatures there were no correlations. In many other species there is no correlation between temperature and length of sessions and recesses.

The effect of temperature on sessions and recesses shows no simple pattern and factors such as female fat reserves, and the extent of feeding by the male during incubation, can affect how ambient conditions affect behaviour (Afton 1980; Halupka 1994; Conway and Martin 2000a).

Evolution of incubation behaviour

Incubation behaviour is very conservative in birds with almost all species relying on contact with the adult to raise egg temperature to a level suitable for embryonic development. The classification of incubation patterns (Table 6.1) provides little insight into the evolution of different patterns of incubation behaviour. It is tempting to assume that male-only incubation in the Palaeognathae is a primitive behaviour pattern but this group does not exclusively follow this trend. Widespread distribution of shared incubation between orders strongly suggests that this was the original type of contact incubation (Kendeigh 1952; Skutch 1957). Loss of the one or other gender in an incubation role appears to be related to the individual life-histories of different groups of birds. For instance, in Gruiformes the majority of species share incubation but bustards (Otididae) exhibit female-only incubation whereas hemipodes (Turnicidae) exhibit male-only incubation (Skutch 1957). Passeriformes exhibit both shared and female-only incubation (Table 6.2) and within individual genera different species can exhibit different incubation patterns (Skutch 1957).

It is questionable whether some of factors considered important by Skutch (1957) actually reflect driving factors in evolution. For instance, it is not clear whether the presence of brood-patches on male birds, the extent of sexual dimorphism in plumage and singing behaviour of males are characters that have shaped the evolution of incubation patterns or whether they are a consequence of different reproductive strategies.

To date an underlying basis that could explain the various patterns of behaviour in all birds has not been proposed. Drent (1975) suggested that the approach of comparing heat exchange of eggs could yield interesting results but to date the limited work on the relationship between attentiveness and egg temperature have provided equivocal results (Turner 1994a). Research has concentrated on the relationship between %attentiveness and egg and ambient temperatures in a small number of species often with small eggs. The prediction that attentiveness should decline as embryonic metabolic heat production increases as incubation proceeds has been borne out in a few species of waterfowl but not other smaller species (Turner 1994a). There are data, however, for eggs of the herring gull and blue tit (*Parus caeruleus*) which suggest that egg mass is related to %attentiveness and this idea is investigated further here.

Egg mass and incubation attentiveness

Drent (1975) suggested that rate of egg cooling may play a role in determining the length of incubation recesses. In the starling (*Sturnus vulgaris*) recovery time for egg (8 g) temperature is longer than exposure time for short exposure periods

(Drent *et al.* 1985). In the herring gull egg (95 g), there is a very unfavourable relationship between cooling and re-warming which suggests that recesses of longer than 1 minute should be avoided. Male and female herring gulls share incubation and so daytime attentiveness is 97.5% (Drent 1975). Observations confirmed that 75% of all incubation recesses were less than 1 minute and 93% were less than 2 minutes. By contrast, small (1 g) eggs of the blue tit are incubated in well insulated nests and rate of cooling is reduced and the rate of re-warming is fast suggesting that the incubation recesses could be much longer (Drent 1975). Attentiveness of 75% is exhibited by the female blue tit, which incubates alone (Kendeigh 1952), and observations showed that 50% of recesses for these small eggs were between 5 and 8 minutes and 75% were between 3 and 10 minutes in length.

This limited data implied that egg mass could be related to attentiveness in birds. To investigate this further, initial egg mass (IEM), available for 421 of the 451 species, was plotted against %attentiveness. As egg mass decreases then the minimum attentiveness decreases although maximum attentiveness is little affected by egg mass (Figure 6.7a). There was a significant positive correlation between IEM and %attentiveness for all species combined (Table 6.3). For eggs with a mass below 1 g then attentiveness ranged from 54.5 to 100% although the smallest eggs had attentiveness values between 70 and 80%. %Attentiveness varied between 45 and 100% for those species with egg mass between 1 and 10 g although most of the high attentiveness values were for species with shared incubation (Figure 6.7a). With a few exceptions (10 out of 98 species) %attentiveness was above 80% for species with egg mass between 10 and 100 g (Figure 6.7a). Above an egg mass of 100 g only three out of 38 species exhibit %attentiveness less than 90%. Above an egg mass of 500 g all species have over 99% attentiveness.

In order to better compare the effect of incubation type mean values for %attentiveness were calculated for egg mass rounded up to categories on a logarithmic scale for shared, female-only, and male-only incubation types (Figure 6.7b). For shared incubation, despite a significant positive correlation, egg mass had relatively little effect on %attentiveness (Table 6.3). For female-only incubation, there was a highly significant positive correlation (Table 6.3) and for egg mass below 10 g mean values for attentiveness were between 70 and 80%, some 20% less than shared incubation of eggs of similar mass. Between 10 and 100 g the mean values had increased above 80% (5–10% lower than in shared incubation). Above 100 g mean attentiveness was generally above 95% in all types of incubation. The few

Table 6.3 Spearman's Rank correlation coefficients (r) for the relationships between %attentiveness and initial egg mass and total clutch mass for all species combined, for species exhibiting shared, female-only and male-only incubation patterns.

	All		Shared		Female only		Male only	
	DF	r	DF	r	DF	r	DF	r
Initial eggs mass (g)	419	0.527***	114	0.262*	295	0.489***	6	0.866**
Total clutch mass (g)	352	0.514***	99	0.134^{NS}	243	0.541***	6	0.863*

DF = Degrees of freedom; NS = not significant; * = $p < 0.05$; ** = $p < 0.01$; *** = $p < 0.001$.

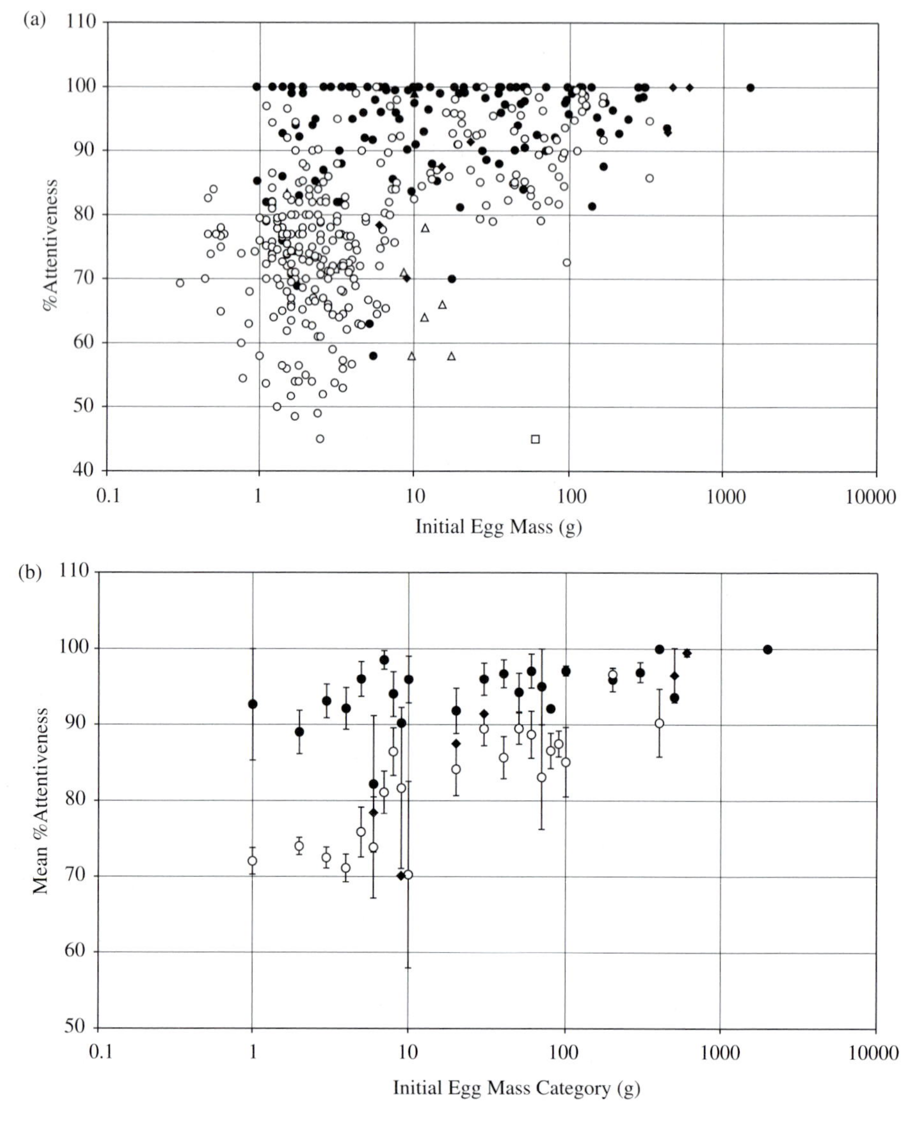

Fig. 6.7 (a) The relationship between initial egg mass and percentage attentiveness in birds. (b) Mean (±SE) percentage attentiveness for categories of initial egg mass based on a log scale. Black circles indicate species where incubation is shared, black diamonds have male-only incubation and white circles indicate female-only incubation. In (a) the square is the superb lyrebird (*Menura novaehollandiae*), the black triangle is the toucan (*Ramphastos sulfuratus*) and the white triangles are species of birds of paradise (Paradisaeidae).

examples of male-only incubation matched the pattern exhibited by female-only incubation (Figure 6.7b; Table 6.3). This pattern of distribution strongly suggests that in *Aves* that egg mass is a significant factor influencing minimum attentiveness when there is only one parent incubating.

Egg mass, thermal characteristics and incubation behaviour

It seems that mass is a significant factor in determining the thermal characteristics of eggs. Time constants (t) for cooling scale with egg mass (Turner 1985) and thus smaller eggs lose and gain heat faster than larger eggs (Tazawa *et al.* 1988a). Under contact incubation, small eggs are also less heterothermic than larger eggs, i.e. in a 2 g egg of a blue-gray gnatcatcher (*Polioptila caerulea*) there is almost no temperature gradient from the brood patch to the bottom of the egg (Turner 1987a). As egg mass increases significant temperature gradients develop and are most apparent in an ostrich egg. As development proceeds the embryonic circulation makes the eggs less heterothermal and so towards the end of incubation these temperature gradients are significantly reduced (Turner 1987a).

Turner (1994c) introduced the concept of thermal impedance in descriptions of the maintenance of temperature in contact incubated eggs which experience an intermittent pattern of sessions and recesses typical of single gender incubation (Chapter 9). Using infertile fowl eggs it was demonstrated that efficiency of heating an egg depended on the length of the incubation session with short sessions (<60 minutes) having a high impedance and inefficient transfer of heat. Low impedance heating associated with longer incubation sessions were more efficient and had much more influence on egg temperature (Turner 1994c; Chapter 9). Turner (1994c) also showed that the time constant of a contact incubated fowl egg is less than half the time constant of a convectively incubated egg of the same mass. Unfortunately, time constants for contact incubated eggs of different masses are unavailable at present and so the impedance values can only be estimated using the relationships between egg mass and time constants and thermal resistance (Turner 1985, 1994c). Nevertheless predicted thermal impedance values for eggs of different masses are revealing (Figure 6.8). As egg mass increases, the length of the incubation session associated with low thermal impedance, i.e. more efficient heating (impedance = resistance) gets progressively longer (Figure 6.8). Hence, short incubation sessions of less than 100 minutes will be energetically inefficient for eggs of 100 g but low impedance heating can be achieved during incubation sessions of around 10 minutes for 1 g eggs.

Convection cooling characteristics of the eggs after the bird has left are also important. Turner (1994a) showed that removal of the brood patch leads to loss of heat from the upper surface of the shell but also, via conduction, into the contents of the egg (Figure 9.10; Chapter 9). For small eggs (<10 g) there will be high rate of cooling from the shell surface but because temperature at the centre of the egg is close to that of the brood patch there will be a flow of heat from the core to the surface. As a result the temperature of the exposed surface will be maintained for some time after the bird departs. During the first half of development the embryo is located just under the surface and so the cooling effect experienced by the embryo during parental absence will be diminished. Furthermore, the egg surface temperature perceived by the returning bird will be relatively high. During an incubation session brood patch contact in small eggs involves high thermal resistance but low impedance heating leading to an investment in energy within the egg which can be released during the recess which follows. Of course, the rate

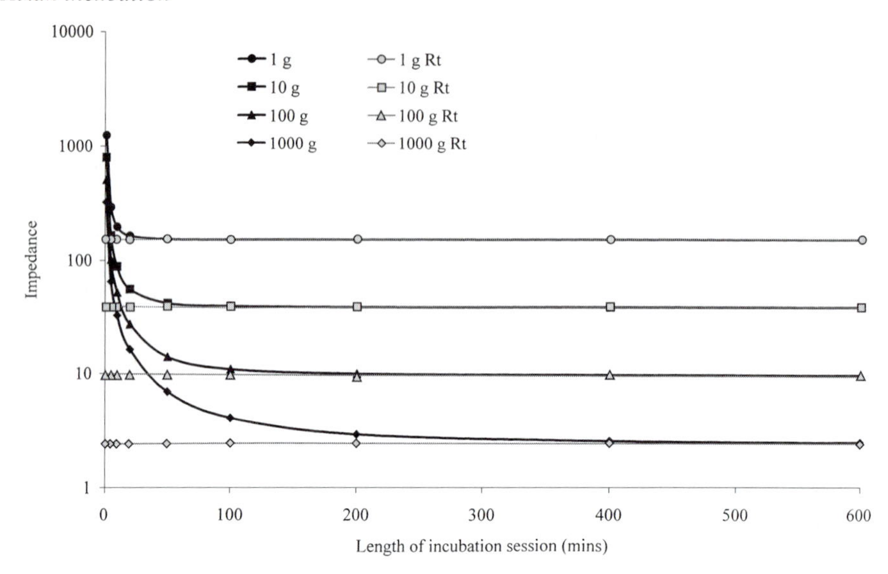

Fig. 6.8 Predicted values for thermal impedance (°C W^{-1}) and thermal resistance (R_t; °C W^{-1}) for different sized eggs plotted against the length of the incubation session (in minutes). Data were calculated from equations for time constants and thermal resistance from Turner (1985) and for phase delay and thermal reactance from Turner (1994c). Note the log scale on the Y axis.

of cooling will be related to the temperature gradient between the egg and ambient air and so environmental conditions affect the length of the recesses more than the length of the sessions (Conway and Martin 2000a). A bird incubating a small egg can afford to spend longer time from the nest because the small egg size, in combination with low impedance heating during short sessions and nest insulation characteristics, means that the embryo temperature is maintained relatively well and can be restored very quickly. These thermal characteristics allow low %attentiveness during incubation in small eggs. Development of the circulation has relatively little effect on the temperature gradients in the egg (Turner 1987a, 1991) and so %attentiveness will not change as incubation proceeds.

In larger eggs (10–100 g), once the parent leaves the nest a temperature gradient not only exists between the egg and the air but also between the surface and core of the egg (Turner 1987a). Conduction of heat from the eggshell surface to the cooler core will be more significant than convective loss to the air above the exposed surface (Turner 1994a; Chapter 9; Figure 9.10). Short periods of exposure lead to rapid cooling of the shell surface which is perceived by the incubating bird. This, and the longer session times required to achieve efficient low impedance heating, mean that attentiveness has to be high (over 80%; Figure 6.7) so as to maintain the appropriate eggshell temperature at the optimal thermal environment for the embryo.

Hence, the thermal characteristics of eggs weighing more than 10 g dictate that %attentiveness has to be high all of the time and often to achieve this both genders have to participate in contact incubation. In those species where shared incubation has been abandoned then the incubating bird has to adopt a different strategy to ensure the high %attentiveness required by the egg. Hence, in female-only incubation of galliforms and waterfowl, and male-only incubation of tinamous, the incubating bird has to fast for much of the incubation period and usually only takes infrequent, relatively long recesses during a day. Long periods of contact incubation lead to a high investment of heat energy in the egg contents that raises core temperature. During a long recess period, heat energy moves into the centre of the egg but with time heat loss from the egg surface draws heat energy towards the embryo at the surface. In this way heat input to the embryo is maintained during the time spent off the nest which is so essential for maintenance of the bird. Furthermore, in large eggs the developing embryonic circulation and metabolic heat production cause temperature gradients in the egg to progressively diminish (Turner 1991) and so the rate of eggshell cooling will diminish allowing attentiveness to drop as incubation proceeds.

Large thermal gradients develop away from the brood patch in eggs over 100 g. Absence from the nest leads to rapid cooling of the shell surface because of loss of heat to the nest air but more importantly conduction of heat into the egg contents (Chapter 9; Figure 9.10). Thermal impedance characteristics of these eggs mean that high efficiency heating can only occur during very long incubation sessions (Figure 6.8). Perception of the drop in surface egg temperature means that both parents are required in most species to ensure the appropriate egg temperature for embryonic development throughout incubation. Storage of heat in the egg may be insufficient to have any benefit during any recesses. Large adult body size allows for high attentiveness due to fasting in male-only incubation of smaller ratites.

Thermal characteristics of an ostrich egg may explain why this species is anomalous in ratites by not exhibiting male-only incubation. Timing of breeding in ostriches is relatively labile (Bertram 1992; Deeming and Ar 1999), which may prevent a seasonal build-up of reserves seen in other species. This would mean that the male bird is unable to successfully endure a fast during incubation and so the female is needed for a daytime session to ensure the high attentiveness of incubation demanded by the thermal characteristics of the eggs. The great spotted kiwi (*Apteryx haastii*) also departs from the typical ratite pattern with the female partially sharing incubation by relieving the male during the night allowing him to forage (McLennan and McCann 1989). This has been attributed to the cold climate during nesting which prevents males from maintaining high attentiveness. Both genders in this species lose mass during incubation. Nevertheless, attentiveness is over 92% in this species and comparable to the male-only incubation of the North Island brown kiwi (McLennan 1988).

Effect of clutch mass

The logic of the argument that the thermal characteristics of eggs are affecting nest attentiveness, could be undermined by the fact that many species of bird do not incubate single eggs but have clutches of many eggs. Biebach (1984) showed that the energy cost of incubation increased with larger clutch sizes in the starling although recess length and ambient temperature were more important in determining the energetics of warming a clutch (Biebach 1986). Blagosklononv (1977) showed in the pied flycatcher (*Ficedula hypoleuca*) that attentiveness was related to clutch size: large clutches required a higher %attentiveness and feeding by the male. Siegfried and Frost (1974) showed that the cooling rate of a clutch of moorhen (*Gallinula chloropus*) eggs was equivalent to the cooling rate of a single egg of the same mass.

To investigate this further, clutch number was determined for 354 species of known egg mass and %attentiveness was plotted against total clutch mass (Figure 6.9a). There was a significant positive correlation between total clutch mass and attentiveness (Table 6.3) with clutch masses below 40 g having a wide spread of %attentiveness (50–100%). Attentiveness for the large single eggs of the birds of paradise was now typical of multiple egg clutches of the same mass (Figure 6.9a). Between 40 and 1,000 g minimum attentiveness was 80% and above 1,000 g %attentiveness was above 95% in all but one species. For shared incubation, attentiveness averaged over 90% (Figure 6.9b) with no correlation with clutch mass (Table 6.3). For both female-, and male-only incubation, clutch mass was significantly correlated with attentiveness (Table 6.3). Below a clutch mass of less than 30 g then attentiveness was around 75% but as clutch mass increased then attentiveness progressively increased to above 90% at 100 g (Figure 6.9b). At high clutch masses, female-only incubation achieved averages of over 85% and as clutch mass increased then attentiveness was almost complete.

If the thermal characteristics, in particular thermal impedance values, of different sized eggs are important in determining minimum attentiveness then the length of incubation sessions should be more influenced by egg mass than the length of incubation recesses. Relationships between initial egg mass, and clutch mass, and the lengths of incubation sessions and recesses were investigated in 145 species exhibiting female-only incubation in Anseriformes, Galliformes, Apodiformes and Passeriformes. For both initial egg mass and clutch mass, length of the incubation session increased much faster than length of the incubation recess (Figure 6.10a). Hence, as egg mass gets larger incubation sessions increase in line with the prediction from impedance values: a clutch mass of 1 g in hummingbirds have incubation sessions around 10 minutes, and at a clutch mass of 100 g the incubation session is around 100 minutes (Figure 6.10b). The regression estimates in Figure 6.10 show that the increase in %attentiveness is more a function increasing session length than decreasing recess length.

Of course, there are many factors that can modify the incubation strategy of individual species. It would seem, therefore, that for species with small eggs, variation in attentiveness may not necessarily reflect the need of the bird to forage but rather indicates the ability of the bird to leave its egg for more time during incubation. This allows the minimum %attentiveness to be low. In larger eggs the birds are required to sit on the eggs for longer sessions, minimum attentiveness is higher and the females have to adopt a strategy of fasting. Highly attentive female-only incubation of large eggs in Falconiformes is only accommodated by the strategy of the male feeding the female during incubation.

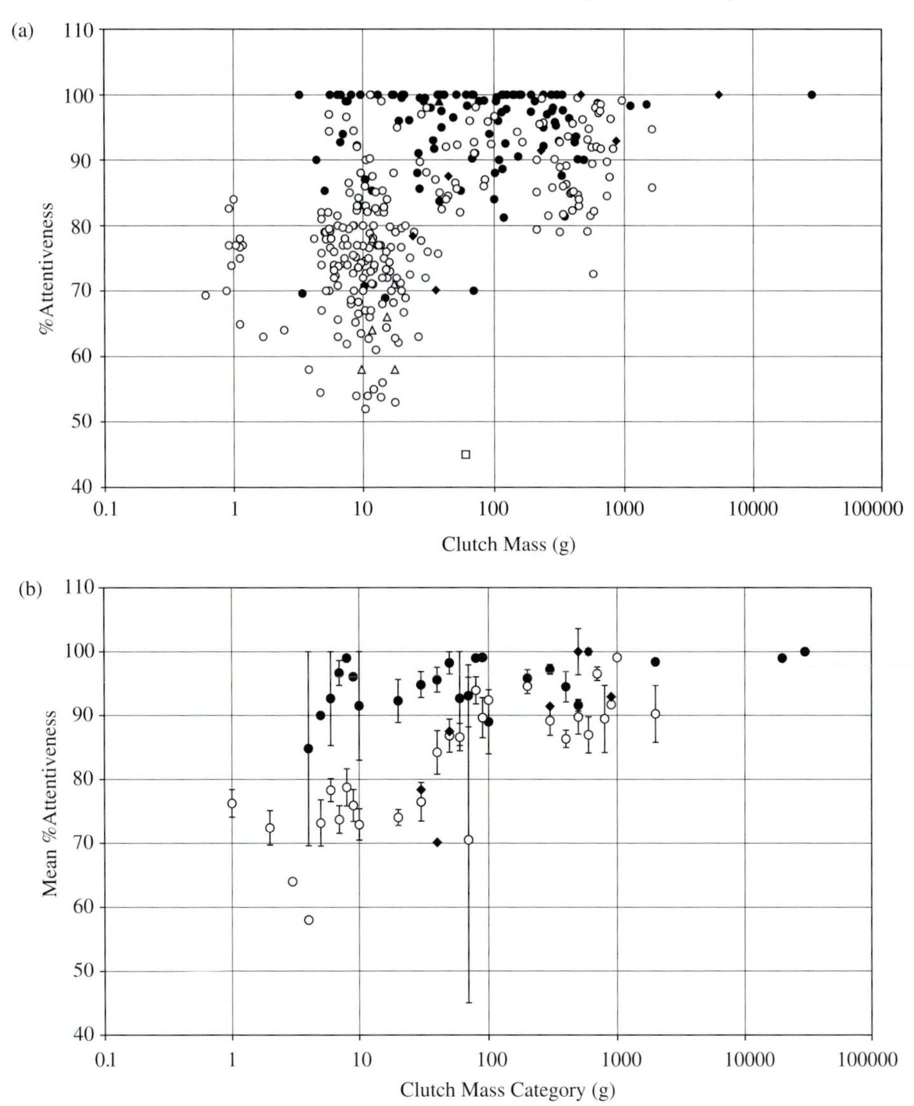

Fig. 6.9 (a) The relationship between total clutch mass (g) and percentage attentiveness in birds. (b) Mean (± SE) percentage attentiveness for categories of clutch mass based on a log scale. Black circles indicate species where incubation is shared, black triangles have male-only incubation and white circles indicate female-only incubation. In (a) the square is the superb lyrebird (*Menura novaehollandiae*) and the white triangles are species of birds of paradise (Paradisaeidae).

%Attentiveness and egg mass

It would seem that those studying the patterns of attentiveness during avian incubation may have been asking the wrong question. Rather than trying to explain why Passeriformes, with their small eggs, have low attentiveness, the more interesting

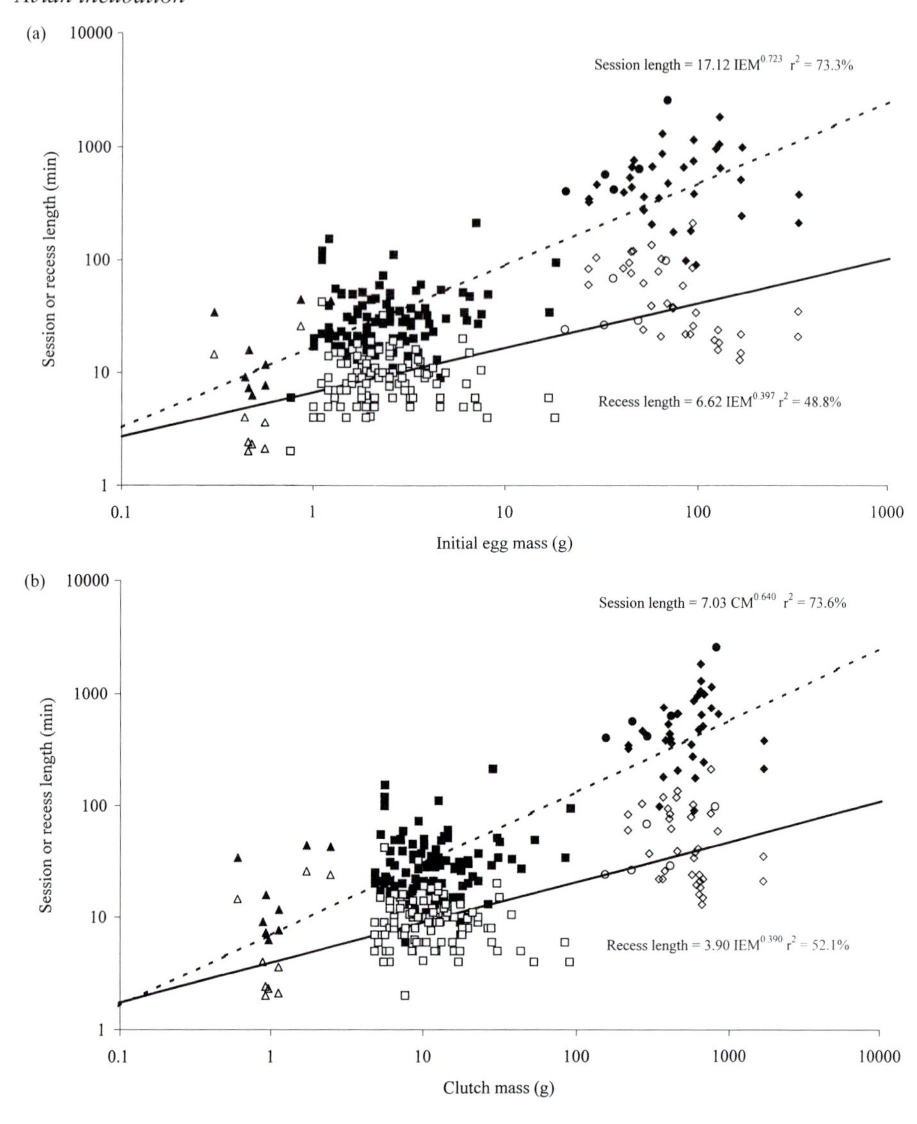

Fig. 6.10 Log–log relationships between (a) initial egg mass, and (b) clutch mass, and the length (in minutes) of the incubation session and recess (closed an open symbols respectively) for waterfowl (diamonds; data from Afton and Paulus 1992), galliformes (circles; Pullianen 1971, 1978; Cramp 1977; Naylor *et al.* 1988; Eaton 1992), passerines (squares; Courtney and Martin 2000b) and hummingbirds (triangles; Vleck 1981b). Regression estimates are as indicated.

question is why do larger birds need to sit on their eggs so much? Variation in %attentiveness exhibited by Passerines may reflect the reproductive strategy, nest morphology, prevailing climate, and food resources prior to and during incubation of individual species (Conway and Martin 2000b). However, low %attentiveness is made possible by the thermal characteristics of the small egg and clutch masses. As egg mass increases then the incubating bird is forced to maintain contact with the egg so as to maintain eggshell surface temperature and to efficiently introduce heat into the egg. This hypothesis predicts, therefore, that the cooling rate of the *upper surface* of contact incubated eggs (as opposed to the core temperature) will be negatively correlated with egg (and clutch) mass. Direct experimental evidence to support this idea is lacking at present but there is some evidence to support this concept.

Attentiveness in the village weaverbird (*Ploceus cuullatus*) was positively correlated with the cooling rate of the eggs within nests with different insulation properties (White and Kinney 1974). A correlation between cooling rate of nests and recess length was described for the spruce grouse (*Falconipennis canadensis*), a species with female-only incubation of 95.5% attentiveness where the bird takes long recesses 20–35 minutes (Naylor *et al.* 1988). Experimental manipulation of egg temperature is also very effective at altering the behaviour of incubating birds (Franks 1967; Drent *et al.* 1970; Vleck 1981a; Davis *et al.* 1984) and vasodilation of the brood patch (Mitgård *et al.* 1985). Moreover, de-sensitising the brood patch with anaesthetic led weaverbirds to incubate at higher temperatures and for longer sessions than control birds (White and Kinney 1974). These results suggests that the bird monitors egg temperature closely through its brood patch (chapter 7) and adjusts its behaviour accordingly.

The nest-bird incubation unit (Chapter 1) may have evolved to ensure that cooling rates of eggs of any given mass are matched with the attentiveness by the adult(s) mediated by monitoring of eggshell temperature by the brood patch. Although more research is needed to investigate this suggestion, it seems that the considerable variation in incubation behaviour patterns, and lack of phylogenetic relationships, can be explained to a great extent by the thermal characteristics of eggs laid by different species. As initial mass increases, the thermal characteristics of the egg mean that the bird perceives a lower eggshell temperature and has much to be more attentive to maintain a thermal environment suitable for embryo development. As has been shown to be the case in other aspects of the incubation environment, e.g. nest humidity, it seems that egg mass is a major underlying factor in determining attentiveness during incubation behaviour of birds.

Acknowledgements

Many thanks to Scott Turner, Graeme Ruxton and Bob Brua for their constructive comments on previous drafts of this chapter.

7 *Parent–embryo interactions*

R. B. Brua

During incubation, the embryo undergoes phenomenal transformations with specific senses developing in a well-defined progression. First, non-visual photic sensitivity appears followed by tactile, vestibular, proprioceptive, auditory, and finally, true visual capabilities develop (Gottlieb 1968). Eggs are also exposed to a variety of stimuli, including light, temperature changes, mechanical stimulation, and various auditory sounds. Thus, the ability of the embryo to react to certain stimuli is dependent on when the specific sensory system is functional. For example, an embryo is incapable of learning parental vocalisations if exposed to these stimuli prior to the development of the auditory sensory system. Lastly, the ability of an embryo to respond to certain stimuli may also be governed by its degree of maturity at hatching, i.e. altricial versus precocial.

All of these sensory systems have the potential to function in parent–embryo interactions, or possibly between embryos. These systems make ideal areas of research on factors affecting embryonic behaviour, hatching patterns, parental incubation behaviour, parental investment, parent–offspring conflict, and lastly, parent–embryo communication. Despite a long history of anecdotal and empirical interest dating back to the late 19th century, most research has been associated with imprinting studies. More recently, however, other aspects of parent–embryo interactions have begun to gain attention in the literature. Since a plethora of research has been devoted to imprinting, with several relatively recent reviews (Bolhuis 1991, 1999; Bolhuis and Van Kampen 1992; Van Kampen 1996) this aspect of parent–embryo interaction will not be discussed in this chapter. Instead, this chapter focuses on two areas of interactions, embryo–embryo and parent–embryo, along with hypotheses associated with each area of interaction. Most research has centred on clicking noises and embryonic vocalisations, and their implications for embryo–embryo and parent–embryo interactions.

Embryo–embryo interactions

Few interactions are possible among embryos during incubation, except during late stages of embryonic development, typically the period immediately after internal pipping until hatch when auditory stimuli are present. The most obvious auditory stimuli consist of embryonic clicking noises, vocalisations, bill tapping, breathing, and heartbeats, but limb movements, head-lifting, and beak-clapping are also audible (Vince 1969).

Clicking noises

Clicking, a sharp metallic sound, has been recorded in many altricial and precocial species (Oppenheim 1972), although it has not been observed in pied-billed grebes (*Podilymbus podiceps*; Driver 1967) or black-necked grebes (*Podiceps nigricollis*; Brua personal observations). Clicking is not produced by tapping of the bill against the shell but is produced by respiratory movements of the embryo shortly after internal pipping (Driver 1965, 1967). More specifically, clicking sounds are probably a result of air passing over the syrinx during the initial functioning of the respiratory system (Forsythe 1971). Across species, the frequencies associated with clicking sounds are in the range of 3–8 kHz (Vince 1969; Forsythe 1971). Rates of clicking vary among species and between precocial and altricial species (Table 7.1) although insufficient species have been studied to reveal any specific patterns.

Embryo–embryo interactions have been explained as relating to the synchronisation of hatching. In precocial species synchronising the hatch is proposed to be beneficial by co-ordinating departure from the nest (Lack 1968), reducing exposure to nest predation (Clark and Wilson 1981), or creating an earlier hatch date (Flint *et al.* 1994). Hatch synchronisation could occur in two possible ways: (1) Females could start complete incubation with laying of the last egg, or (2) by accelerating or retarding the development of sibling neonates, via embryo–embryo communication consisting of clicking sounds. Although initiation of incubation after the last egg may appear to be common, there is evidence of considerable amount of variation in the developmental rate of embryos in some precocial species (Afton 1980). More work is need to confirm that hatch synchrony is related to similar rates of development within a clutch.

Clicking sounds have been shown to synchronise the hatch of several precocial species, despite first laid eggs being up to 2 days ahead in development of

Table 7.1 Rates of clicking in precocial and altricial species. Data from Vince (1966a) except for that of *Numenius americanus* which is from Forsythe (1971).

	Clicking frequency (clicks s^{-1})
Precocial species	
Colinus virginianus	1.4–2.7
Excalfactoria chinesis	1.5–3.2
Coturnix coturnix	1–3.3
Phasianus colchicus	1.8
Perdix perdix	1.4–1.7
Gallus gallus	1.1–1.6
Numenius americanus	10–12
Altricial species	
Passer montanus	0.9–1
Turdus ericetorum	0.5–0.8

later laid eggs due to increasing nest attentiveness and incubation by the female during laying. However, it appears that hatch synchronisation can only occur if eggs are in contact with each other and during the last 2–3 days of incubation (Vince 1969; Woolf *et al.* 1976). In several laboratory experiments, bobwhite quail (*Colinus virginianus*) and Japanese quail (*Coturnix coturnix*) were shown to hatch synchronously due to accelerating (Vince 1966b) and retarding (Vince 1968) the development of other embryos in eggs touching each other. Bruning (1973) reported that hatching in the rhea (*Rhea americana*) could be accelerated by 9 days if there was egg–egg contact. Hatch dates in the domestic fowl embryo were also advanced in response to an artificial stimulus (White 1984). Similarly, in a field experiment, lesser snow geese (*Chen caerulescens*) embryos were able to synchronise hatch by accelerating the hatch time of later embryos and by extending the incubation period (Davies and Cooke 1983).

The stimulus used to accelerate the hatch time of delayed embryos is possibly related to breathing of the more advanced embryos. This stimulus not only increases hatch time but also advances other developmental processes, such as lung ventilation and yolk sac withdrawal (Vince 1972). Thus, stimulated embryos progress through normal developmental stages but in a shorter period of time. However, the stimulus used to slow development of advanced embryos is unknown, but Vince *et al.* (1984) suggest that slow breathing of delayed embryos may defer the onset of the fast click-breathing stage, which may retard development and hatching.

Embryonic vocalisations

Embryonic vocalisations appear to be common in precocial species but rare in altricial species (Oppenheim 1972). These vocalisations are distinct from clicking sounds and typically begin one to several days prior to hatching (Gottlieb and Vandenbergh 1968), and prior to external pipping as in several grebes (Driver 1967; Brua 1996). These vocalisations occur despite the many energetic and gas exchange difficulties faced by embryos in the pipped egg (Whittow and Tazawa 1991), which suggests that they may play an important functional role. Precocial duck (*Anas platyrhynchos*) and fowl embryos are capable of producing three distinct vocalisations within a frequency range of 2.5–5 kHz: distress, contentment, and brooding-like calls (Gottlieb and Vandenbergh 1968). Sonographs of contentment calls tend to be composed of a symmetrical distribution of ascending and descending sounds, whereas sonograms of distress vocalisations are comprised almost entirely of sharp descending frequencies, and were similar spectrographically between fowl and duck embryos (Gottlieb and Vandenbergh 1968). However, brooding-like calls tended to differ between fowl, a short ascending note, and duck embryos, which resembled contentment sounds (Gottlieb and Vandenbergh 1968). Calls of altricial budgerigar (*Melopsittacus undulatus*) embryos were composed of a single band within a frequency range of 4–6.6 kHz, and did not differ spectrographically among varying incubation temperatures (Berlin and Clark 1998). How frequent an embryo vocalises is dependent on embryo temperature (see below), but black-necked grebe embryos at normal incubation temperatures produced approximately 1.5 bouts of calls min^{-1}, consisting of 6 calls $bout^{-1}$, resulting in approximately 5 calls min^{-1} (Brua *et al.* 1996).

Although embryonic vocalisations were implicated as the only stimulus provided in a laboratory experiment investigating the hatching intervals of the asynchronously hatching glaucous-winged gull (*Larus glaucescens*), it is likely that clicking sounds were also present (Schwagmeyer *et al.* 1991). Nevertheless, hatching intervals of eggs in contact with each other were shortened between the second and third egg of a three egg clutch compared with clutches in which eggs were incubated 10 cm apart. However, hatching intervals were not reduced in a concomitant field experiment. Schwagmeyer *et al.* (1991) suggested that lowered parental attentiveness, or alternative auditory stimuli (parents or colony), possibly explain the difference between the laboratory and field experiments. However, more research is needed to ascertain what stimuli influence hatching intervals, if intervals can be shortened, and its implications in asynchronous avian species.

It seems apparent that in many species hatch synchrony is related to clicking sounds produced by embryos, which can accelerate the hatch of delayed neonates within the clutch. More research is needed to determine the factors responsible for retarding the hatch of more advanced embryos. Also, a fruitful avenue of research would be to determine how the embryo responds to acceleration or retardation. Metabolic and hormonal studies on embryos may help to determine the physiological response of an embryo to achieving hatch synchrony via acceleration or retardation. Basic information is needed in regards to fitness costs for parents and offspring stimulated to hatch sooner or individuals being delayed. Although Vince (1972) mentions that viability of a stimulated, i.e. accelerated, hatchling is impaired slightly, more studies are needed to assess viability beyond the period immediately following hatch to determine if subsequent survival or growth of stimulated or delayed individuals is impaired during the pre-fledging period.

Parent–embryo interactions

Avian parents affect the rate of embryonic development directly by determining the onset of incubation (Chapter 18), temperature during incubation (Chapter 9), nest microclimate (Chapter 10), and the rate of egg turning (Chapter 11). However, the parent(s) must strike a balance between feeding, predator avoidance, and egg attendance for proper embryonic development (Drent 1973, 1975; Chapter 6). How the embryo influences its own microenvironment by manipulating behaviour of the parent during incubation has received little attention.

Recent work has suggested that nitric oxide (NO) may play a role in communication between the embryo and the adult (Ar *et al.* 2000). NO is emitted at low concentrations (a few parts per billion) from a variety of altricial and precocial eggs although it can only be quantified during later stages of incubation. At 18 days of incubation a fowl egg had a NO emission of 123 pmol h^{-1}. Hypoxia caused an increase in NO emission whereas cold temperatures caused a reduction in NO release. NO synthase (NOS) has been isolated in the chorio-allantoic membrane of domestic fowl, quail and turkey (*Meleagris gallopavo*) embryos. In NOS-inhibited incubating turkey females the rate of egg re-warming was reduced. Further research is required but NO may serve to act in communication between the embryo and adult thereby reinforcing the idea of embryo–adult symbiosis during incubation (Chapter 9).

Parental manipulation by an embryo may occur during late incubation, when the embryo first begins to vocalise within the shell. Several hypotheses, although not mutually exclusive, have been developed about the relationship between parental responses and embryo activities or behaviour. One suggested hypothesis for these embryonic vocalisations is that they inform the parents of the impending hatch of the embryo and prepare them for parental care duties beyond incubation. Others have argued that vocalisations create an atmosphere in which parent–offspring recognition can occur thereby leading to parental acceptance of the hatchlings. Another explanation suggests that embryonic vocalisations act as an anti-abandonment strategy, thereby signalling its viability status to the parent and preventing its abandonment. A relatively recent hypothesis suggests that embryonic vocalisations may also act as a form of communication that signals the thermal status of the embryo to the parent. Parental care of recently hatched chicks may interfere with incubation of the remaining eggs leading some authors to suggest that the parent and the embryo are in conflict during incubation. However, in stark contrast, some birds have been witnessed assisting the hatch of their offspring. In this section, these hypotheses are discussed.

Parent–offspring recognition

In colony-nesting birds the ability to recognise your own offspring is critical so that crucial resources (food, parental defence, reduction in energy expenditure) are not wasted on unrelated offspring. Similarly, recognition of parents should be beneficial since it might reduce straying of chicks, which can lead to chick harassment or even mortality. Thus, the ability to accept offspring and parent–offspring recognition is paramount, and may be initiated with the onset of embryonic vocalisations.

Beer (1966) reported that black-headed gulls (*Larus ridibundus*) incubating vocalising, pipped eggs were likely to accept chicks added to the nest, whereas adults incubating non-pipped eggs were likely to be more aggressive. Laughing gulls (*Larus atricilla*) appear to respond in a similar manner (Impekoven 1976; but see Miller 1972). In the black-necked grebe, stray chicks that closely approach foreign nests, or actually attempt climbing onto these nests are usually rejected. However, if peeping eggs or newly hatched young are in their nest, recently hatched stray chicks are often accepted (Brua personal observations).

Tschanz (1968) and Impekoven (1970, 1973) noted that incubating parents would rise, poke at the eggs and call in response to embryonic vocalisations. Furthermore, embryos sometimes would vocalise in response to parental vocalisations. It was suggested that some form of parent–embryo communication was occurring. Using playback experiments, Tschanz (1968) demonstrated that common guillemot (*Uria aalge*) embryos were able to discriminate between parent and neighbour vocalisations. Gottlieb (1988) and Bailey and Ralph (1975) reported that mallard (*Anas platyrhynchos*) and ring-necked pheasant (*Phasianus colchicus*) chicks were capable of learning calls to which they were exposed to as embryos. For these synchronous hatching precocial species, which leave the nest shortly after hatching, the ability to recognise or learn the maternal call may be critical since the female leads them away from the nest shortly after hatch, and at times, the female is not visible to the following offspring (Bailey and Ralph 1975).

However, Impekoven and Gold (1973) reported that laughing gull chicks exposed to calls as embryos were unable to discriminate between familiar and novel vocalisations. An experiment with little terns (*Sterna albifrons*) revealed that parents unexposed to embryonic vocalisations were more aggressive, but accepted their own chicks (Saino and Fasola 1996). By contrast, those birds exposed to vocalisations were less aggressive and accepted newly hatched chicks, even if they were not related genetically. Since adults previously exposed to their own calling embryos readily accepted cross-fostered chicks, parental acceptance occurred but offspring recognition by adults did not occur. Saino and Fasola (1996) concluded that embryonic vocalisations function in the switching from incubation to parental care. However, most studies on birds have reported that parents do not recognise their offspring until there is some selective advantage to do so, such as when the broods or offspring begin to mix (Beer 1980). Obviously, more research is required to ascertain when or if parent–offspring recognition occurs and the implications of recognition or non-recognition of offspring or parent.

Signal impending hatch

The traditional explanation for embryonic vocalisations is that they signal the impending hatch of a chick, which cause parents to switch from incubation to begin brooding and feeding the hatchlings (Norton-Griffiths 1969; Impekoven 1973, 1976; Templeton 1983; Brua 1996; Saino and Fasola 1996). However, few studies have examined how embryonic vocalisations influence parental incubation behaviour, although some anecdotal reports exist. Results from these studies on a variety of species suggest that the frequency or duration of several aspects of incubation change after the embryo begins to vocalise.

For example, most studies have reported that the amount of egg turning typically increases with the onset of embryonic vocalisations for common guillemots (Tschanz 1968), oystercatcher (*Haematopus ostralegus*; Norton-Griffiths 1969), laughing gull (Impekoven 1973, 1976), American white pelican (*Pelecanus erythrorhynchos*; Evans 1989), black-necked grebe (Brua 1996), and budgerigar (Berlin and Clark 1998). By contrast, the rate of egg turning decreased in herring gulls (*Larus argentatus*; Lee *et al.* 1993). Similar results were found for the amount of rising and re-settling over the eggs, but Brua (1996) and Lee *et al.* (1993) reported that initiation of embryonic vocalisations caused a decrease in rising and re-settling. Also, nest material manipulations or additions increased (Chamberlain 1977; Evans 1989; Brua 1996), whereas Impekoven (1973) found no relationship between nest material manipulations and embryonic vocalisations. Impekoven (1973) and Chamberlain (1977) provided anecdotal information that the 'off-duty' parent tends to spend more time near the nest when embryonic vocalisations begin, and Brua (1996) reported that off-duty black-necked grebes spent more time near the nest after exposure to embryonic vocalisations. Lastly, playback experiments of embryonic vocalisations cause American white pelican adults to remove their foot webs from the eggs and push the eggs back between their legs to continue incubation (Evans 1988a).

Although a majority of studies have reported the impact of embryonic vocalisations on parental behaviour, some studies have found that embryos equally respond to parental vocalisations or certain aspects of incubation behaviour, such as egg turning. For example, Tschanz (1968) reported that adult vocalisations increased hatching activities in common guillemots, and embryonic motility tended to increase after hearing the 'crooning' call of laughing gulls (Impekoven and Gold 1973). Similarly, egg turning tended to increase hatching activities and calling of common guillemot embryos (Tschanz 1968). Domestic fowl embryos increased the frequency of backthrust movements after being turned than at rest, although the rate of bill clapping did not differ (Impekoven 1976). Black-necked grebe embryos emitted more vocalisations when turned than when at rest (Brua *et al.* 1996).

Taken together, a variety of evidence in relation to changes in parental incubation behaviour suggests that embryonic vocalisations appear to create a transitional period between incubation and chick-rearing that is beneficial to the embryo. For example, parents exposed to embryonic vocalisations were less aggressive to their offspring than parents unexposed to vocalisations. In a laboratory experiment, Impekoven (1976) reported that domestic fowl embryos turned at a higher frequency hatched earlier than eggs that were turned at a lower frequency, and this effect was more pronounced if the increase in egg turning rate occurred during the last two days of incubation.

The application of nest material may have some insulating capacity and/or may prevent eggs from rolling out of the nest cup, the region of optimal incubation temperature (Evans 1989). Newly hatched black-necked grebe chicks attempting to climb onto the parent's back sometimes fall backward onto the nest platform or occasionally into the water and may drown (Brua personal observations). Intensified nest building during the peeping egg period creates a larger nest rim around the incubating parent, and may prevent still poorly thermoregulating, wet black-necked grebe chicks from drowning if it falls off the parent's back at this critical early stage (Brua 1996). In a similar vein, spending an increased amount of time near the nest after embryos begin to vocalise, may be important if a recently hatched, wet chick falls into the water, thereby, the off-duty parent could respond by positioning itself near the chick so that it could climb onto its back to be brooded (Brua 1996). Also, remaining near the nest may expedite the change in incubation duties of adults and thereby reduce cooling, in accord with the care-soliciting hypothesis (see below), and support for this idea is reported in black-necked grebes (Brua 1996). Potentially, the ultimate response in creating a transition period from incubation to parental care in response to embryonic vocalisations is the presentation of food at the nest prior to hatching (Beer 1966; Emlen and Miller 1969; Norton-Griffiths 1969; Chamberlain 1977; Brua 1996; Berlin and Clark 1998).

Care-solicitation

Evans (1988b) proposed that embryonic vocalisations in American white pelicans act as 'care-soliciting signals' in response to parental neglect during hatching. Parental neglect occurs when parents shift primarily from incubation to brooding, which subjects the remaining eggs to lower and more variable incubation temperatures (Evans 1989, 1990a; Lee *et al.* 1993; Evans *et al.* 1994). Long term exposure to sub-optimal incubation temperatures can delay hatching (Lundy 1969; Mirosh and Becker 1974; Geers *et al.* 1983). However, domestic fowl, ring-billed gull (*Larus delawarensis*), and American white pelican embryos chilled only after pipping had significant delays in hatching time and an increase in hatching intervals among eggs within clutches (Evans 1990b, 1990c). Although hatchability and hatching intervals in herring gull nests were not affected by reduced incubation temperatures, but were reduced when temperatures dropped below 30°C in laboratory experiments (Lee *et al.* 1993). In contrast to sub-optimal incubation temperatures, domestic fowl embryos exposed to elevated incubation temperatures (44°C) for 1 hr, had similar hatching times as controls, but embryos left at 44°C for 2 hr died (Oppenheim and Levin 1975).

Beyond hatching delays or hatchability, other significant physiological or developmental affects to embryos can occur if exposed to sub-optimal incubation temperatures. Drent (1973) reports that incubating embryos at below normal temperatures caused disproportionate growth of the heart. Spiers and Baummer (1990) found that a 1°C decrease in temperature reduced the wet mass of a 5 g, 16 day-old quail embryo by 0.35 g (7%), and a drop of 2°C could result in a one-day delay in normal growth. Malleefowl (*Leipoa ocellata*) embryos used more energy and hatched with smaller energy reserves when incubated at low temperatures than when incubated at normal temperatures (Booth 1987a). Specific effects on reproductive fitness have also been postulated. Female Japanese quail embryos exposed to a slow decrease in temperature over 12 hr had reduced ovarian tissue and fewer ovarian follicles, which could negatively influence fertility and future reproductive potential (Callebaut 1991). However, much more research is needed to determine the effects of non-normal incubation temperatures on embryos and subsequently on future viability and fitness of these exposed embryos.

Neglected embryos must be able to inform their parent(s) of their thermal status, so that parents can respond appropriately to the needs of the embryo. Since embryos are unable to escape to warmer microhabitats or generate sufficient metabolic heat alone, the only remaining avenue available to embryos is to communicate their thermal status by vocalising (Evans 1988a; Evans *et al.* 1994). In fact, anecdotal evidence that embryos vocalise in response to below normal incubation temperatures has long been available (Gordon and Warren 1895). However, only relatively recent empirical studies of altricial and precocial species have reported that embryos respond to reduced or above average incubation temperatures by vocalising. Below normal incubation temperatures caused altricial American white pelican embryos (Figure 7.1), but not altricial budgerigars, to increase the rate of vocalisation (Evans 1988b, 1990d; Berlin and Clark 1998). However, for both species the rate of embryonic vocalisations increased

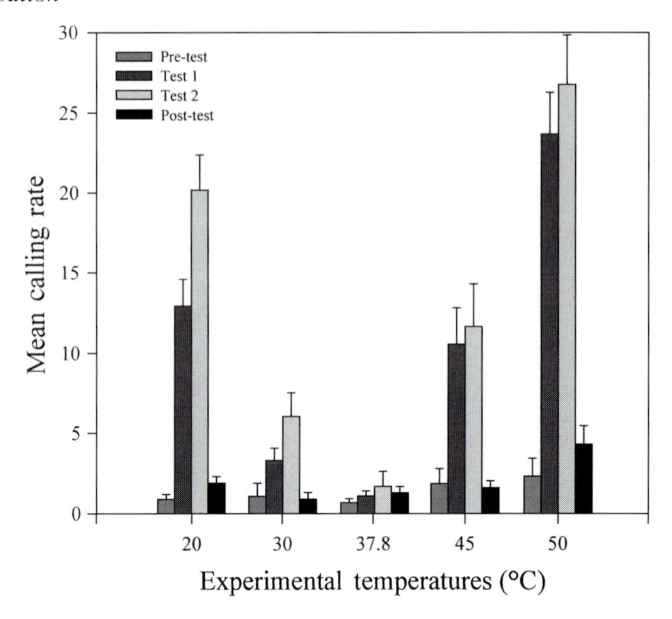

Fig. 7.1 Rate of vocalisation (mean + SE, calls min^{-1}) of American white pelican (*Pelecanus erythrorhynchos*) embryos in response to cold, normal, and warm incubation temperatures. Each testing period lasted 5 minutes with the pre- and post-test periods conducted at normal (37.8°C) incubation temperatures. Significant differences in vocalisation rate was detected among test periods and temperatures, however pre- and post-test periods did not differ from normal incubation temperature. Data from Evans (1990d).

when exposed to hyperthermic incubation conditions (Figure 7.1). Similar results in relation to lowered incubation temperatures have been found in several precocial species, such as American coot (*Fulica americana*; Bugden and Evans 1991), black-necked grebe (Brua *et al.* 1996), ring-billed gull (Evans *et al.* 1994), and herring gull (Evans *et al.* 1995). For example, during cold-water treatments the overall vocalisation rate of black-necked grebe embryos nearly doubled, bouts of calls became more disjointed, and chilled embryos produced significantly fewer calls per bout (Brua *et al.* 1996).

Several studies have reported that embryos can detect temperature changes rapidly, often within one minute, and subsequently respond to these changes by vocalising (Evans 1988b, 1990d; Bugden and Evans 1991; Brua *et al.* 1996). Similarly, once embryos detect the application of warmer temperatures the rate of vocalisation declines and returns to normal levels (Evans 1988b, 1990d; Bugden and Evans 1991; Evans *et al.* 1994, 1995; Brua *et al.* 1996). Thus, by vocalising, embryos may induce the incubating parent to take corrective measures to restore normal incubation temperatures. Indeed, two separate field investigations on American white pelicans have suggested that embryos convey, via vocalisations, their thermal status to their parents, and the parent often responds by improving the thermal microenvironment of the embryo (Evans 1990e, 1992). These results have lead to the suggestion that vocalisations act as honest signals of embryonic need, which in this case signals a need for warmth, and has been specifically tested and supported in ring-billed gulls (Evans *et al.* 1994), American white pelicans (Evans 1994), and herring gulls (Evans *et al.* 1995).

Anti-abandonment strategy

Occasionally viable, non-pipped eggs are abandoned, especially if only one egg is yet to hatch. This is possibly related to slow growth rates of late-developing embryos, which may be a result of exposure to lower than normal incubation temperatures (Evans 1990b). Often these abandoned eggs are viable and within several days of hatching. In response to being turned, black-necked grebe embryos emitted many more vocalisations than when at rest, and when turned back to their original up position there was a reduction in the number of vocalisations (Brua *et al.* 1996). Brua *et al.* (1996) suggested that black-necked grebe embryos may simply be signalling its parent that it is alive and about to hatch. Similarly, Prinzinger (1974, 1979) suggested that black-necked grebe embryos vocalised to prevent adults from abandoning the nest, and Simmons (1955) suggested a similar function for peeping vocalisations in great crested grebes (*Podiceps cristatus*). Pre-pipped embryos also commonly peep when simply touched or jiggled, and this is possibly as effective as turning the egg in eliciting embryonic vocalisations. Further laboratory experiments, as well as field observations of parental behaviour, are needed to distinguish the anti-abandonment and anti-turning hypotheses, both of which are special cases of the care-solicitation hypothesis. Also, the anti-abandonment hypothesis needs to be directly tested to determine if embryonic vocalisations can prevent or delay nest departure for species that raise young away from the nest.

Parent–offspring conflict (asynchronous hatching)

Onset of incubation before clutch completion results in hatching asynchrony and subsequent size hierarchies within the brood, which can lead to death of later-hatched offspring, and has been termed the 'paradox of hatching asynchrony' (Stoleson and Beissinger 1995). Asynchronous hatching has been argued to be adaptive since it allows for facultative brood reduction, and the resulting fitness benefits associated with hatching synchrony or asynchrony has influenced its evolution (Lack 1954; Clark and Wilson 1981). However, it seems intuitive that the benefits associated with synchronous hatching mentioned above would be advantageous to parents and all asynchronous hatching offspring, but especially for later-hatching offspring.

If individuals from last-laid eggs are at competitive disadvantage to older siblings, but can potentially influence hatching intervals by vocalising, one would predict that the selfish interests of late-hatching offspring would be selection for shorter hatching intervals or even synchrony, thereby, improving their fitness. By contrast, if hatching intervals facilitate brood reduction then hatching interval selection may have already been optimised, thereby maximising parental and brood-mate fitness, but to the detriment of young hatching last. Thus, one can envision the potential for parent–offspring conflict to arise.

Parent–offspring conflict is predicted to occur over optimal amounts of parental care, such that offspring favour greater amounts of care than parents are selected to provide (Trivers 1974). In relation to this chapter, an appropriate example would be incubation temperature of parentally neglected last-laid eggs. For asynchronous hatching birds, parental neglect in the form of egg cooling of last-laid eggs may extend hatching intervals, have substantial implications on embryonic growth and energy usage, or may cause death. In many species dominance hierarchies and competitive asymmetries are established soon after hatching, with older siblings dominant and more competitive than younger siblings (Stoleson and Beissinger 1995). Also, Forbes and Ankney (1987) found that chick survival decreased with increasing hatching intervals. As a consequence, parental neglect may have lethal or sub-lethal consequences for late-hatching offspring of asynchronous-hatching species, particularly in species with obligate or facultative brood reduction. However, Evans and Lee (1991) suggested that parental neglect is more an inevitable consequence of asynchronous hatching than a brood reduction strategy.

In a clever experiment, Evans *et al.* (1995) tested for parent–offspring conflict in relation to parental neglect of last-laid herring gull eggs. Temperature preference of vocalising embryonic herring gulls was 1°C below that of neglected non-vocalising eggs. Thus Evans *et al.* (1995) concluded that no conflict was evident, since incubation temperature preference of embryos was slightly below that provided by incubating parents. Also, embryonic temperatures of young hatching last have been observed at levels of neglect, but the level of parental neglect has not adversely affected hatching intervals (Evans 1990d; Lee *et al.* 1993; Evans *et al.* 1994), suggesting lack of parent–offspring conflict since embryonic temperatures were sufficient to maintain normal hatching intervals. Schwagmeyer *et al.* (1991) came to a similar conclusion for glaucous-winged gulls. Although information regarding conflict and hatching asynchrony is needed for hyperthermic incubation conditions, future studies should also emphasise the factors affecting the onset of incubation and subsequent hatching intervals in relation to hatching asynchrony (Stoleson and Beissinger 1995).

Parental hatching assistance

Hatching strategies of birds have been classified as symmetrical and asymmetrical (Chapter 4), whilst parental hatching assistance is considered a subsidiary hatching technique since assistance compliments but does not replace the actual hatching behaviour of the chick (Bond *et al.* 1988b). Parental assistance to hatching embryos has been documented in a variety of altricial and precocial species, and typically is comprised of enlarging the pip hole, extending the line around the egg, or dumping the chick out of the egg (Bond *et al.* 1988b). For example, Senar (1989) reported that siskin (*Carduelis spinus*) pecked at the pip-hole, making it larger, and eventually picked the shell up and dumped the chick into the nest. However, the importance of embryonic vocalisations in parental hatching assistance has not been studied adequately. The only study assessing the impact of vocalisations on parental hatching assistance was performed by Berlin and Clark (1998) on budgerigars. Playback experiments revealed no difference in the rate of nibbling on the egg between vocalising and silent periods. This response differed from actual observations of hatching, which suggested that egg-nibbling was a prominent behaviour during hatching, and parents were able to locate and assist vocalising embryos among older vocalising brood-mates (Berlin and Clark 1998). However, much more research is needed to assess the functional role of embryonic vocalisations in attaining parental hatching assistance.

Importance of parent–embryo interactions

This chapter reviewed several aspects of embryonic auditory noises that function in embryo–embryo and parent–embryo interactions. It is apparent that clicking noises are used by embryos to synchronise hatching by delaying or accelerating embryos. Although more research is needed to evaluate if embryonic vocalisations function among embryos, these vocalisations appear to be critical in parent–embryo interactions, and indeed are probably multi-functional. Embryonic vocalisations reliably signal embryonic thermal status and create a smooth transition from incubation to brood rearing by altering parental behaviour to ensure offspring acceptance and to begin foraging for offspring.

Acknowledgement

I thank Charles Deeming for comments on earlier versions of this chapter.

8 *The brood patch*

R. W. Lea and H. Klandorf

The vast majority of birds achieve successful incubation of eggs through the transfer of heat between parts of their body, usually but not always their ventral surface, and the clutch of eggs. That region of the incubating bird making contact with the egg is commonly termed the incubation or 'brood patch'.

Although the evolution of this structure is open to conjecture, an attractive possibility is that the appearance of a brood patch was a definitive step in the development of birds from reptiles and it has been thus suggested to share a common ancestry with the mammalian mammary gland (Long 1969). The nature, formation and structure of this patch vary enormously between species, reflecting the diverse mechanisms and methods adopted by different species in the incubation of eggs. In some birds, such as the domestic pigeon (*Columba livia*) and ring dove (*Streptopelia risoria*), there is a form of pocket or 'apterium' on the ventral surface of both sexes, which is devoid of feathers throughout adult life. In other species, patch formation occurs periodically and closely related to the period of egg laying and incubation. Furthermore, though both genders may have the potential to develop a brood patch, it is a general observation that it only occurs in the gender actually involved in incubation.

In the majority of species, brood patch development involves a number of morphological changes occurring to the ventral body area: de-feathering (mainly down feathers); a significant increase in the folding of the skin together with an infiltration of leukocytes (oedema formation); a thickening of the cornified layer of the ventral skin surface (epidermal hyperplasia); and an increase in both size and number of local blood vessels (vascularisation). The structure is surprisingly sophisticated in that the musculature of arterioles supplying blood to the patch also increases and thus can achieve a shut-down of blood flow to this region when the parent is off the nest (Peterson 1955). In some species, such as the house sparrow (*Passer domesticus*), there is also an increase in the amount of underlying defatted tissue (Selander and Yang 1966).

Taken together, these dramatic changes are superbly designed so as to facilitate a closer contact and more effective heat transfer between the parenting bird and the surface of the eggs, whilst minimising any possible damage caused to the skin by the sustained period of contact time. The function of the brood patch during the incubation period of the breeding cycle is complex and multi-factorial. Much still remains to be understood and one early suggestion for a possible purpose of the patch, that the bird finds relief from the peripheral irritation of the developing brood patch by sitting on eggs, remains plausible (Lehrman 1955). This chapter considers the principle morphology of the avian brood patch and the incredibly diverse variations that exist on this theme in birds. Factors known to be involved in the formation of the patch are then considered before finally the mechanisms of action and variety of functions are discussed.

Morphology

In species that develop brood patches, between one and three patches may be located in medial and/or lateral positions of the ventral apteria (Figure 8.1). The earliest histological studies of brood patch development were on domestic fowl (*Gallus gallus*) and the European coot (*Fulica atra*; Barkow 1830). Freund (1926), Koutnik (1927), and Lange (1928) continued these investigations in poultry and the morphology of the brood patch was described by Bailey (1952) in the white-crowned sparrow (*Zonotrichia leucophrys*), Oregon junco (*Junco*

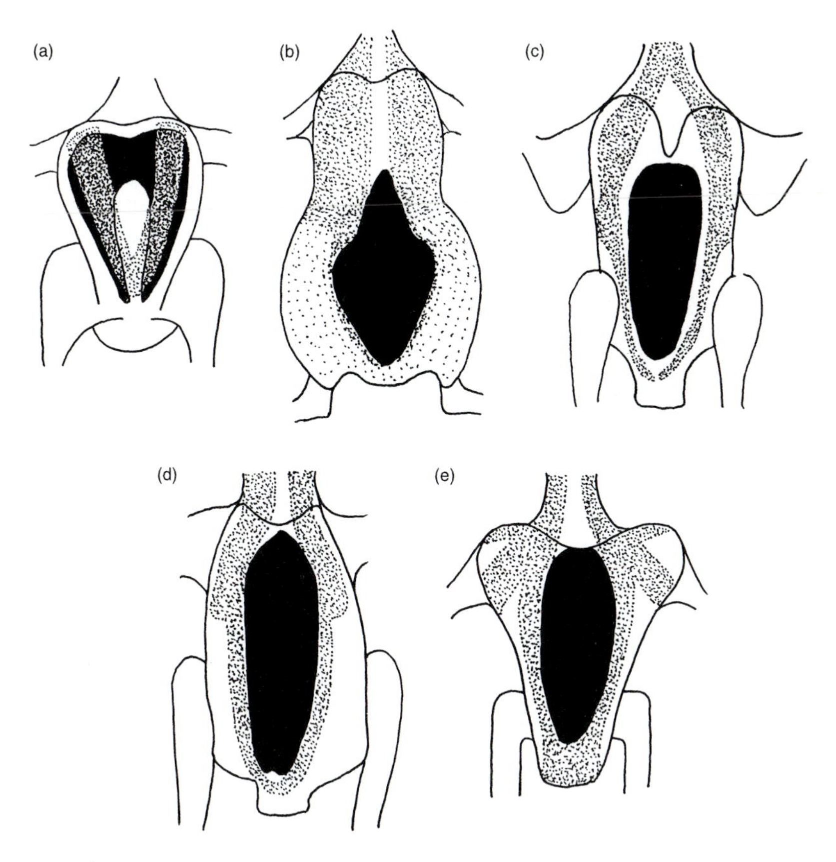

Fig. 8.1 Schematic representation of the ventral surface showing the relative size of the brood patch (black) and feather tracts (stippled) in (a) California quail (*Lophortyx californicus*); (b) red-necked grebe (*Podiceps grisegena*); (c) white-crowned sparrow (*Zonotrichia leucophrys*); (d) rook (*Corvus frugilegus*); and (e) marsh harrier (*Circus aeroginosus*). Re-drawn from Bailey (1952) and Drent (1975).

oreganos) and song sparrow (*Melospiza melodia*), by Selander and Kuich (1963) in the red-winged blackbird (*Agelaiius phoeniceus*) and the house sparrow (also Acharya and Menon 1984), by Durant *et al.* (2000) in the barn owl (*Tyto alba*), by Bailey (1955) in the tinamou (*Nothoprocta ornate*), by Peterson (1955) in the bank swallow (*Riparia riparia*), by Jones (1969a) in the California quail (*Lophortyx californicus*), and by Hinde (1962) in the domestic canary (*Serinus canaria*). Changes in the morphology of the ventral skin were also characterised in the domestic canary during the development of the brood patch (Hinde 1962) while Peterson (1955), Selander and Yang (1966) and Jones (1968) followed changes in the weights of various skin components during patch formation. It is from these studies that changes in the epidermis and dermis during development of the brood patch have been elucidated.

De-feathering

De-feathering is the loss of feathers from the ventral surface and is initiated during nest building such that in many passerines the skin is free of feathers by the time the first egg is laid (Jones 1971). There is some variation in the timing of this process because in the house sparrow it is not completed until the last egg (Selander and Yang 1966) whereas in the starling (*Sturnus vulgaris*; Lloyd 1965a), de-feathering is not complete until well into incubation. In other species de-feathering may not be complete until the end of incubation (Jones 1971). The length of time taken for completion of de-feathering is variable ranging from 24 hours in the white crowned sparrow (Bailey 1952) through to 25 days in the California quail (Jones 1969a). The loss of feathers is not due to plucking but is considered to be a kind of special moult (Bailey 1952). Anseriformes do not form a brood patch (Bailey 1952) but do pluck feathers from their breast for nest insulation resulting in a bare patch region of skin on the ventral surface (Weller 1959; Hanson 1962). As was suggested by Jones (1971) it would be interesting to investigate whether this behaviour is a precursor to a physiologically control moult in other orders.

Epidermal development

The epidermal layer of avian skin arises from embryonic ectoderm on the first day of incubation. By the second day, the single layer of original cells begins to divide and develops into a superficial layer of squamous cells, the epitrichium, and a deeper, compact layer of cylindrical cells. New cells are continually generated in this deeper germinative layer, the *stratum germinativum*.

Reproductively quiescent adult birds exhibit a thin *stratum germinativum* that consists of one to two layers of cylindrical cells with oval nuclei and a comparatively thin *stratum corneum*, which is the superficial corneous layer. The subsequent histological changes associated with patch development in passerines, for example, are initiated before the onset of egg laying and are associated with the increase in oestrogen concentrations (Bailey 1952). During the formation of the brood patch cells of the epidermal *stratum germinativum* undergo hyperplasia and hypertrophy, while the nuclei become more rounded. Associated with these changes the *stratum corneum* thickens with the result that it increases 2–5 fold as compared to non-breeding birds. Evidently this epidermal thickening helps to protect the dermis of the exposed skin from possible injury (Jones 1971).

Dermis and sub-dermis

The dermis originates from embryonic mesoderm with that portion of the dermis that covers the ventral and ventro-lateral regions of the body arising from the outer cells of the somatic layer. This layer of mesoderm is an unsegmented sheet that is originally lateral to the somites. On day 12 of incubation in the fowl, connective tissue and muscles of the dermis begin to differentiate in the mesenchyme (embryonic connective tissue that is composed of loosely associated stellate cells and a dispersed extracellular matrix) beneath the dermis while blood vessels and nerves extend into the germ layers.

Dermis of adult birds consists of a thick, compact zone of fibrous connective tissue, which is associated with an underlying zone of reticular connective tissue. However, the boundary between the dermis and the underlying sub-dermis is not readily distinguishable. Sensory nerve endings, blood vessels, smooth muscle and fat cells, components associated with skin physiology and feather follicles are also located in the dermis (Rawles 1960). During brood patch formation there is a marked reduction in dermal fat, muscle mass and feather papillae. Fat is an important part of the sub-dermis and is found both as a layer and as discrete bodies bound by fascia to the underlying muscles (Stettenheim 1972). An increase in subcutaneous adipose tissue has been measured in the brood patch of incubating female barn owls (Durant *et al.* 2000). The increase in the thickness of this layer was suggested to be due to the migration of lipids from other depot fat stores (e.g. medullar bones) in the body to the subcutaneous adipose tissue of the brood patch. It was speculated that accumulation of adipose tissue in this region limits heat loss and/or adjusts the transfer of heat to the clutch.

In female turkeys (*Meleagris gallopavo*), eight nerves, which arise from thoracic vertebra 3 to synsacrothoracic 1, innervate the brood patch (Book *et al.* 1991). Afferent sensory input arising from the brood patch was suggested to play a role in clutch termination, for although denervated hens continue to lay and incubate eggs, less time is spent on the nest and the normal rise in serum prolactin concentrations measured in incubating hens is inhibited (Book *et al.* 1991; Chapter 5). Concomitant with the development of the ovaries and egg laying in canaries, there is an increase in tactile sensitivity of the ventral skin. The changes in sensitivity of the skin have been postulated to influence nesting behaviour as well as the selection of nesting materials (Hinde *et al.* 1963).

Brood patch formation is also associated with an increase in the thickness of the dermis as a consequence of tissue hyperplasia and oedema in addition to the contribution provided by the hypervascularisation (Selander and Yang 1966). The dermal oedema is associated with leucocyte infiltration into the interstitial spaces surrounding the larger vessels (Lange 1928; Lloyd 1965), which leads to a general flabbiness of the superficial skin. These structural changes in the skin allow closer contact of the brood patch with the developing eggs. However, normal tissue aging of birds is characterised by an increase in the oxidative chemical modification of proteins as evidenced by an increase in the accumulation of glycoxidation products in the dermal connective tissue of birds (Klandorf *et al.* 1999). These advanced glycoxidation products, such as the cross-link pentosidine, are produced by secondary modification of proteins by products of carbohydrate oxidation. Long-lived proteins such as collagen (skin collagen has a half-life of approximately 15 years) accumulate these cross-links, which ultimately contributes to a loss in their elasticity and function. The connective tissue of skin is composed primarily of type I collagen, whose function is to provide resistance to force, tension and stretch. Brood patch formation may thus be associated with a specific reduction in existing cross-links in the collagen of ventral skin enabling the skin to adapt to the specific requirements associated with the incubation period.

Vascularisation

In association with the hormonal changes at the time of egg lay, the arterioles, capillaries, and venules in the brood patch increase in number. This hypervascularity is associated with an increase in the thickness of the muscular walls that line the arterioles, which led to the suggestion that the increase in the efficiency of controlling local blood flow enhances the ability of the skin surface to transfer heat to the eggs (Drent 1975). Development of the arteriolar musculature also permits a more efficient reduction in blood flow to the brood patch during periods when the parent bird is off the nest. Cardiac output of female broody bantam fowl was twice that of non-broody birds whereas their metabolic rate was increased by only one-third compared to that of non-broody hens (Brummermann and Reinertsen 1992). An increased rate of non-nutrient blood flow through arteriovenous anastomoses (AVAs) in the brood patch has also been established (Midtgård 1988). AVAs are medium sized blood vessels (low-resistance channels) through which blood can be shunted from arterioles to venules without passing through the capillaries and thus are believed to play a thermoregulatory role in most skin areas. In the domestic fowl brood patch AVAs are richly supplied with adrenergic, acetylcholinesterase-positive and vasoactive intestinal peptide (VIP)-immunoreactive nerve fibres (Midtgård 1988). Adrenalin and noradrenaline constrict the AVAs, whereas acetylcholine and VIP act as vasodilators (Hales *et al.* 1982; Midtgård 1988).

Species differences

This general pattern of brood patch formation differs considerably between species. Galliforms exhibit a relatively lower degree of vascularity and oedema compared with a typical passerine (Jones 1971). In passerines, brood patch development (epidermal hyperplasia, hypervascularisation and oedema) is generally concomitant with the initiation of de-feathering and continues to increase after de-feathering is complete, reaching maximal development during the middle to late incubation phase (Hinde 1962; Selander and Kuich 1963; Lloyd 1965; Selander and Yang 1966; Sykes 1993). In most species the increase in the number of blood vessels precedes the increase in size while oedema is maximal during incubation (Lloyd 1965; Selander and Yang 1966). By contrast, brood patch development (hyperplasia and hypervascularisation) in the California quail is initiated during early incubation and reaches a maximum during late incubation and early stages of brooding. Further, this increase in blood vessel size begins earlier and peaks after the increase in vessel number (Jones 1971) whereas leucocyte infiltration of the brood patch replaces the oedema (Jones 1968). A difference in the morphological expression of these features has been attributed to the differences in clutch size (Genelly 1955), passerines producing a comparatively smaller clutch (2–6 eggs) than galliforms (4–20 eggs). Patch formation is delayed in species with large clutches in order to synchronise hatching; brood patches formed before egg lay is complete would result in the premature incubation of eggs leading to asynchronous hatching and an increase in mortality (Jones 1971).

Occurrence of brood patches

The presence of a brood patch is ultimately linked with incubation practices characteristic for the species under consideration, although a variety of environmental and ecological factors can play a significant role in determining both the rate and extent of brood patch development. Many species, including galliforms, Strigiiformes, and some passerines, have large central brood patches (relative to body size; Figure 8.1), which can accommodate several eggs. In other species (e.g. Laridae, Gruiidae), each egg has its own brood patch (Figure 8.2; Skutch 1976). Crossbills (*Loxia leucoptera*) are considered flexible opportunistic breeders and are potentially able to initiate breeding throughout the year if adequate food supplies remain available (Deviche 1997). However, at high latitudes (Alaska) the number of breeding females with brood patches declines with advancing season, suggesting that the breeding season is constrained by the proximal environmental factors environmental temperatures and short photoperiod.

Both sexes sharing incubation and each developing a patch (Jones 1971) is very common and is found in the Podicipediformes, Procellariiformes, Columbiformes, Piciformes, most Charadriiformes and Gruiformes, and some Passeriformes and Falconiformes. Thus, both sexes of European starlings (Lloyd 1965), California quail (Jones 1969) and the laughing gull (*Larus atricilla*; Jones 1971) respond

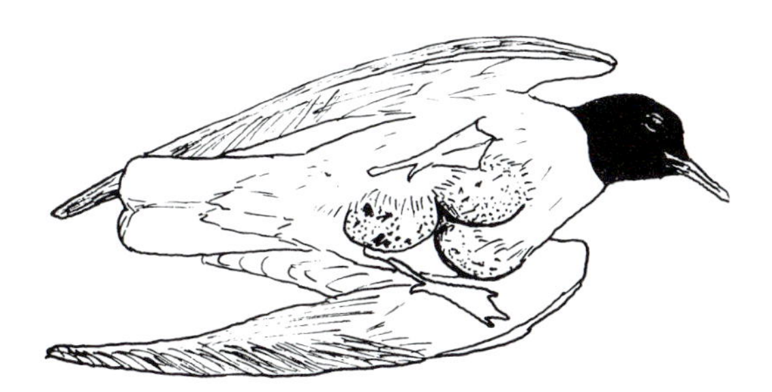

Fig. 8.2 The position of the three-egg clutch in the laughing gull (*Larus atricilla*) indicating the individual brood patch for each eggs (Re-drawn from Fisher and Peterson 1964).

similarly to the administration of exogenous hormone to induce brood patch formation although stimuli associated with nesting behaviour of the female are essential for brood patch development in male starlings (Lloyd 1965). Parental brood patches in Fiordland crested penguins (*Eudyptes pachyrhynchus*) develop after the laying of the first of a two-egg clutch (St. Clair 1992).

Interestingly, American kestrels (*Falco sparverius*) have difficulty in heating all of their eggs simultaneously for incubating birds develop three discrete brood patches which do not effectively correspond to the modal clutch size of five eggs (Wiebe and Bortollotti 1993). Although all incubating falcons develop brood patches, the size of the brood patch is correlated only with the body size of the female. Male American kestrels develop smaller patches than females, a finding consistent with their secondary role during the incubation phase. Eurasian male kestrels (*Falco tinnunculus*) rarely incubate the eggs and usually do not develop brood patches (Village 1990). Black-billed gulls (*Larus bulleri*) also develop three brood patches but incubate only two eggs (Beer 1965). Adult birds were observed to rise and resettle more often if provided a third egg. Most other species of gulls have three brood patches suggesting that black-billed gulls have comparatively recently undergone selection for a smaller clutch size while the number of brood patches has not yet adjusted to the tactile stimulation from the reduced clutch size. Brood patch size may thus be a developmentally plastic trait which is affected by environmental stimuli. Thus the tactile stimulus associated with a incubating a large clutch of eggs may further augment the release of hormones associated with feather loss in the ventral apteria (Wiebe and Bortolotti 1993).

Alternatively, only the female develops a brood patch in Galliformes, Strigiformes, most Apodiformes and some Falconiformes although in some species the male retains the potential to develop a brood patch (Jones 1969). In zebra finches (*Taeniopygia guttata*) only the female develops a brood patch but both sexes incubate the eggs (Zann and Rossetto 1991). The brood patch forms at the start of incubation and regresses soon after the eggs hatch. Females incubate at night whilst both sexes share incubation in the day. Even though the males fail to develop a brood patch there were no consistent differences in the temperature of the developing eggs heated by the two sexes nor did the sexes differ in their ability to re-warm cold eggs. However, male zebra finches maintain a region devoid of feathers throughout the year with overheating of the eggs a more likely problem due to the less temperate climate associated with its breeding locality. Administration of testosterone and prolactin in male Wilson's phalarope (*Phalaropus tricolor*) or red-necked phalarope (*Phalaropus lobatus*) has been shown to induce a brood patch regardless of time of year (Johns and Pfeiffer 1963). Only the female Harris' hawk (*Parabuteo unicinctus*), which carries out the bulk of the incubation, develops a brood patch (Vleck *et al.* 1991). The brood patch is normally fully functioning by the time the clutch of eggs is laid. The female rarely leaves the nest area during the nesting season and is provided food by her mate and the other birds in the group. In white-crowned sparrows, house finches, juncos (Bailey 1952), red-winged blackbirds (Selander and Kuich 1963), domestic canaries (Steel and Hinde 1964), fairy martins (*Hirundo ariel*; Magrath 1999) and house sparrows, incubation is carried out by the female and only this gender develops a brood patch. Male domestic canaries may develop a brood patch but do not share in incubation of the eggs. For most species studied, both genders respond identically to the administration of exogenous hormone even though only the female may develop a patch during the breeding season. It has thus been suggested that the failure to develop a patch in one gender is due to the lack of appropriate hormonal stimulation rather than a lack of sensitivity of the ventral skin to these hormones (Jones 1971).

Although unusual, there are also some species in which the male incubates the eggs and exclusively develops a brood patch; examples include the phalaropes, and some sandpipers (Scolopacidae) in the Charadriiformes, the Turmicidae in the Gruiformes, and the Tinamiformes. Amongst ratites, male-only incubation is typical of all species, except ostriches and some kiwis (*Apteryx* spp.). Male great spotted kiwis (*Apteryx haastii*) develop a brood patch and incubate what is recognised as a comparatively large egg (McLennan and McCann 1991). However, the breeding burrows of the kiwis are located in cold mountain environments and it has been suggested the energetic costs associated with incubation are substantially high enough that the male alone cannot meet them. For this reason females (who do not develop a brood patch) marginally contribute to incubation of the egg in order for the male to forage at night as well as limit the cost to the male of re-heating the eggs. Males maintain the core temperatures of the egg at about 28–31.8°C whilst females only achieve core temperatures of 28–28.5°C. Development of a brood patch in male Wilson's phalarope and red-necked phalarope has been shown to depend on both the photoperiodically induced testicular changes as well as feedback associated with nesting behaviour (Jones 1971). Alternatively, in breeding male California quail, a brood patch is formed only when the female is actually incubating or abandons the nest (Jones 1970). Paradoxically, Skutch (1957) was unable to establish a correlation between the presence of a brood patch in males and their incubation behaviour. In some species males may also incubate without having a brood patch and conversely, may not incubate the eggs despite developing a brood patch.

In addition to these generalisations, there are species, such as the Cassin's auklet (*Ptychoramphus aleuta*; Payne 1966; Manuwal 1974) and the ostrich (*Struthio camelus*) and other ratites (Skutch 1976), that are capable of incubating the clutch without evidence of developing a brood patch. Parasitic cowbirds (*Molothrus* spp.) are among a very few number of species which do not develop any form of brood patch (Höhn 1962; Selander and Kuich 1963).

Pelicans (Pelecanidae) and gannets and boobies (Sulidae) birds cradle their eggs in the webbing of their feet during incubation whilst resting in a prone position (Nelson 1978; Johnsgard 1993). The egg is only moved in between the legs during hatching (Evans 1988a; Johnsgard 1993). These species have a wide latitudinal distribution from the tropics to the sub-arctic. Utilising this incubation technique heat loss to the environment during periods of diving for food in the cold sea is minimised. For example, the red-footed booby (*Sula sula*) is a tropical seabird, whose large webbed feet are capable of maintaining relatively high egg temperatures (36.4°C) during incubation (Whittow *et al.* 1990). By comparison, the Australasian gannet (*Morus serrator*) in New Zealand maintains mean egg surface temperatures of 34.9°C during the first 4 days of incubation and 36.5°C thereafter (Evans 1995). The gannet (*Morus bassana*) has a foot web measuring 63 cm^2 compared with an egg surface area of 40–50 cm^2 (Nelson 1978). By contrast, in cormorants (Phalacrocoracidae) the eggs lie on top of the foot web where they are pressed against the ventral surface of the prone bird (Johnsgard 1993).

King penguins (*Aptenodytes patagonicus*) and emperor penguins (*Aptenodytes forsteri*), which do not rely on a nest (Chapter 2), incorporate both their feet and a skin flap to incubate a single egg (Handrich 1989a; Williams 1995). This brood pouch of the parent is maintained at a constant 38°C despite harsh environmental temperatures (Chapter 16).

The endocrine basis for brood patch development

The species specific development of a brood patch is dependent on a number of factors although birds can be divided into two broad groups. What is true for the vast majority of species is that the brood patch is not fully developed throughout the year but shows a seasonal development closely synchronised to the breeding cycle. Re-feathering of the brood patch occurs prior to the postnuptial moult in many species. For example, in canaries, re-feathering had begun just ten days after the removal of eggs (Hinde 1962). In the bantam fowl, feathers re-grew on the brood patch between 14 and 21 days after the chicks had hatched (Sharp *et al.* 1979).

Despite early scientific recognition of the structure (e.g. Freund 1926) it was not until the work of Bailey (1952) that brood patch formation was conclusively shown to be under hormonal control. In the white-crowned sparrow and the Oregon junco, de-feathering and vascularisation was produced with exogenous hormones, particularly oestrogen and prolactin. Subsequent studies involving administration of exogenous hormones in non-passerine species, such as the California quail and phalaropes, confirmed a hormonal involvement in brood patch formation (Jones 1969; Johns and Pfeiffer 1963).

Associated with morphological changes, sensitivity of the brood patch also changed significantly with respect to the breeding cycle. A progressive increase in the tactile sensitivity of the ventral skin of the canary occurring at the time of egg laying was reported by Hinde *et al.* (1963). Exogenous oestrogen was shown to cause such an increase in skin sensitivity in the ventral area of the canary (Hinde and Steel 1964). Interestingly, this effect was not so dramatic in ovariectomised females (Steel and Hinde 1964), which led to the suggestion that another ovarian hormone, probably progesterone, in addition to oestrogen was also involved in brood patch formation.

These pioneering studies stimulated numerous investigations of the effectiveness of administration of newly available purified preparations of both gonadal and pituitary hormones to a variety of species in inducing brood patch development. In a comprehensive review Jones (1971) analysed the data resulting from these studies with respect to whether the male or female was the natural incubator of eggs. It was apparent that, though there was a wide species variation, the ability to develop a brood patch was not confined to the sex that normally incubated the eggs, but was much more dependent on the endocrine environment experienced by the individual. In the house sparrow (Selander and Yang 1966) and red-winged blackbird (Selander and Kuich 1963), it is only the female that incubates and develops a brood patch in the wild. However, treatment with exogenous oestrogen, either alone or in combination with progesterone or prolactin, was effective in inducing the four main features of a brood patch (de-feathering, epidermial hyperplasia, increased vascularisation and oedema formation) in both genders. This general observation persisted even though the actual effective hormone combination did differ from species to species. For instance, in the male-only incubating phalaropes, administration of oestradiol benzoate, either alone or in combination with prolactin was ineffective (Johns and Pfeiffer 1963). However, testosterone propionate in combination with prolactin was extremely effective in producing a patch in both genders. In the California quail, in which the female predominantly incubates in the wild, it was diethylstilboestrol (rather than oestradiol) and prolactin, which proved to be the most effective combination in brood patch formation in both the male and the female (Jones 1968, 1969). Such studies consistently implied that it is the endocrine environment experienced by the particular bird rather than the actual sensitivity of the targeted tissue, which determines whether or not a particular sex develops a patch. Since the early study by Bailey (1952), a differential action of the hormones upon the morphological changes involved in brood patch formation have been indicated. In this respect oestrogen is particularly effective in increasing vascularisation, whilst prolactin was more concerned with de-feathering. In general, it is suggested that steroid hormones together with prolactin are responsible for de-feathering and vascularisation; whilst in combination with progesterone induced epidermal thickening and an increase in tactile sensitivity (Drent 1975).

However, brood parasites (Chapter 17) are most probably an exception to this. Administration of oestradiol, progesterone or prolactin, either alone or in combination with one another, failed to produce any brood patch development in the parasitic brown-headed cowbird (*Molothrus ater*). This would indicate that loss of tissue sensitivity has developed in avian species that have no requirement for heat transfer to a clutch of eggs (Selander 1960; Selander and Kuich 1963). Compared to species that develop a brood patch, the ventral skin of cowbirds is relatively unresponsive to the administration of exogenous hormone (Jones 1971). Interestingly, such hormone administration was successful in producing a brood patch in a near relative, the red-winged blackbird (Selander and Kuich 1963).

Furthermore, studies (see Jones 1971) also indicate that there is a relationship between the gender which incubates, and the form of steroid that is most effective with prolactin to produce a full brood patch. In those species in which the female alone incubates, it is oestrogen; whereas in those species in which the male incubates it is androgen. Consistent with this, species which might be considered as intermediate, such as the California quail and laughing gull, are sensitive, in terms of brood patch development, to both oestrogen plus prolactin and androgen plus prolactin (Jones 1968, 1971).

Hormonal effects, such as hypervascularisation and epidermal hyperplasia induced by the administration of oestrogen with prolactin in the domestic fowl, may also differ in their site specificity (Jones *et al.* 1970). This study used juveniles of both genders with grafts of ventral abdominal skin to the dorsum and *vice versa*. In the control birds, injections of oestrogen and prolactin achieved a full brood patch, i.e. de-feathering, dermal hypervascularisation and epidermal hyperplasia. However, ventral skin grafted to the birds back showed only epidermal hyperplasia while dorsal skin *in situ*, or on the bird's ventral surface, showed no response at all. Thus, epidermal hyperplasia is not a direct effect of hypervascularisation but is specifically hormonally stimulated, while dermal hypervascularity is both site specific and tissue specific. This study also indicates that the morphological changes involved in brood patch formation may occur independently of one another.

These early studies were consistent with the development of a brood patch in a particular species being dependent upon those changes which occur in circulating gonadal steroids and prolactin around the time of egg laying and onset of incubation. Generally, the brood patch of passerine species became noticeable before egg laying whilst in the Galliformes, patch development took place later during egg laying and becoming complete as incubation became fully established. If Galliformes developed a brood patch early on in the reproductive cycle, it would result in incubation of the eggs beginning well before the clutch had been completed (Jones 1971).

In early studies definitive conclusions were difficult because of the lack of direct evidence of circulating concentrations of these hormones in the different avian species. Attempting to accurately mimic the natural situation with administration of purified extracts and hormone preparations, often of pharmacological rather than physiological concentration, was a notoriously difficult process. Some studies reported hormone administration to produce complete patch development (e.g. Steel and Hinde 1963) whilst several others were only able to induce partial brood patch formation (e.g. Selander and Yang 1966). Since the 1970's, however, development of numerous specific radioimmunoassays has allowed accurate measurement of the plasma concentration of many circulating hormones, including gonadal steroids and prolactin. Such studies have shown that the actual endocrine profile exhibited by different species over a breeding cycle, though possessing general similarities, can differ markedly in profile. This is consistent with the earlier studies in which the effectiveness of exogenous hormones with respect to brood patch formation varied with species.

In a study on the female bantam fowl (Lea *et al.* 1981), daily blood samples were taken for hormone analysis together with the measurement of nesting frequency and brood patch development. The first day of incubation was taken as the first full day of nesting during which the female bantam failed to lay an egg and was on the nest for over 90% of the 24-hour period. An increase in brood patch score was first clearly seen around 5 days before the onset of incubation (Figure 8.3) and was the first overt signs that hens were about to become broody. While de-feathering was completed by the first day of incubation, oedema formation and increased vascularisation continued for several days thereafter. This observation is, therefore, entirely consistent with the view that the development of the brood patch is

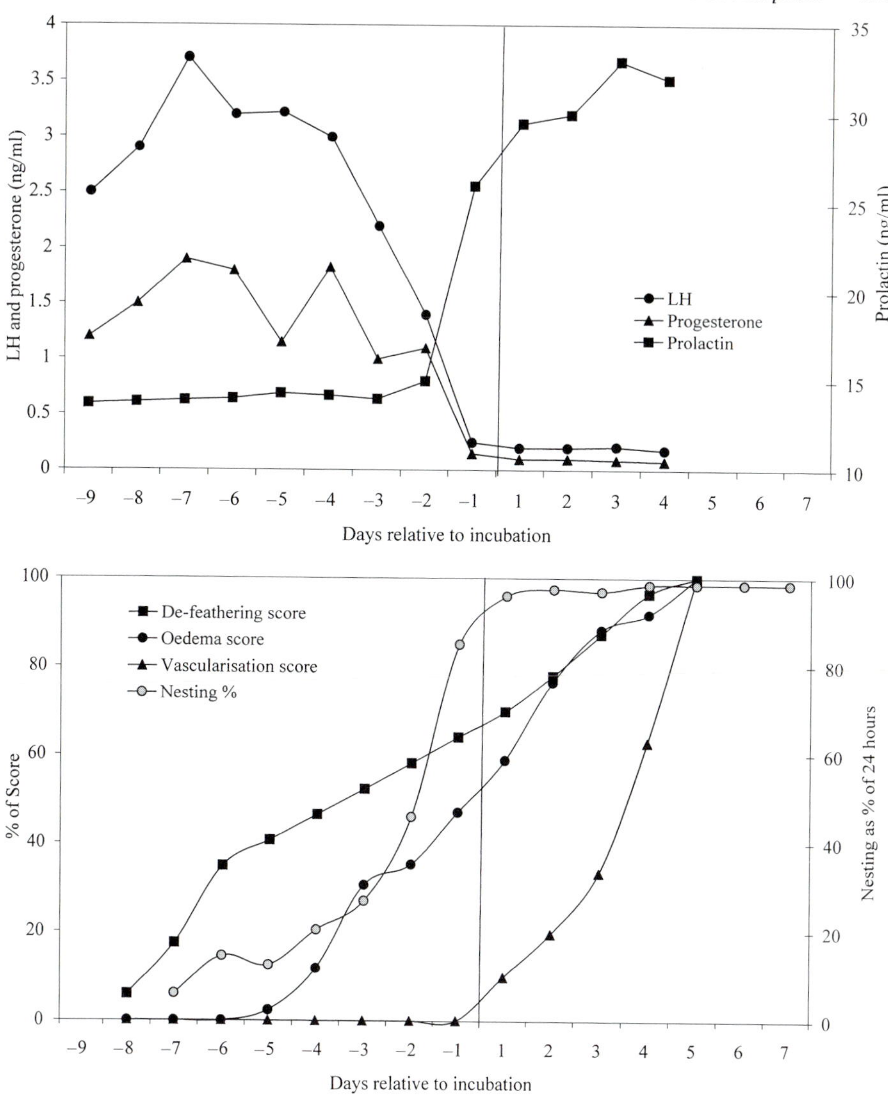

Fig. 8.3 Daily changes in plasma hormone concentrations, time spent nesting and development of brood patch in a group of female bantam fowl (*Gallus gallus*) around the onset of incubation. (Data from Lea *et al.* 1981, unpublished.)

stimulated by prolactin acting synergistically with oestrogen. Similarly consistent reports correlating changes in plasma hormone levels with brood patch formation have been shown in other species such as the ruffed grouse (*Bonasa umbellus*; Etches *et al.* 1979) and Harris' hawk (Vleck *et al.* 1991). Recently, it has been shown that the hormonal control of brood patch formation is, in turn, influenced by environmental factors. For instance, in the white-crowned sparrow, females kept on long days but low temperatures (5°C) failed to develop ovarian follicles or a brood patch as compared to those maintained at 30°C (Wingfield *et al.* 1997).

Mechanisms of action of the brood patch: evidence of importance of tactile sensitivity

Given the complex nature of the patch, the control of its development and the large species variation in form it is unsurprising that the structure is considered to have a number of functions and mechanisms of action. In this section the evidence that the brood patch is actually part of a neuroendocrine loop and the importance of tactile stimulation is considered.

Studies on the domestic canary (Hinde 1965) and European starling (Lloyd 1965b) confirmed that environmental stimuli affect the rate of brood patch formation. Presence of the non-incubating male partner was one significant factor. In the canary, presence of the male influences the onset of nest building as well as patch formation. It is probable that this effect is achieved through a stimulatory action of the mate's presence upon oestrogen secretion in the female via the visual pathway. However, it is well known that the avian dermis and basal layer of the epidermis is richly supplied with sensory receptors (Portman 1961) and provides an excellent mechanism for the transduction of tactile stimulation to also be involved. Such a system has been considered to form part of a 'neuroendocrine loop' (Patel 1926), involving the stimulation of the ventral surface, which, in turn, stimulates the hypothalamus–pituitary axis to effect the secretion of prolactin from the anterior pituitary gland. In this way, tactile stimulation from the brood patch is integral to both the hormonal completion of the developing patch itself and to facilitating those hormonal actions upon behaviour, and other physiological processes, which are manifest during this stage of the cycle (see Chapter 5).

The importance of tactile stimulation from the brood patch in the stimulation of hormone secretion varies with respect to the species studied and the form of incubation adopted by that particular species. For instance in those species with single gender incubation the bird is typically in direct contact with the eggs for 75% or more of daylight hours (Chapter 6), numerous studies have demonstrated a very clear link between the brood patch stimulation and release of prolactin. Simple removal of eggs from such incubating birds results in a sharp, dramatic fall in elevated prolactin levels in the domestic fowl (Sharp *et al.* 1988), turkey (*Meleagris gallopavo*; El Halawani *et al.* 1984), mallard (*Anas platyrhynchos*; Hall 1987) and canary (Goldsmith *et al.* 1984). In the turkey, forced nesting, with associated brood patch stimulation, results in a significant increase in the concentration of plasma prolactin. Subsequent studies were performed to ascertain clearly whether the actual tactile stimulation received from the patch or some other mechanism is responsible for the elevated levels of prolactin during incubation. Thus in the incubating mallard, anaesthetising the brood patch caused a significant depression in prolactin levels (Hall and Goldsmith 1983) and denervation of the brood patch in the turkey prevented any rise in prolactin usually associated with the onset of incubation (Book *et al.* 1991).

In other species in which the incubation pattern is more complex such a direct relationship is not necessarily the case. In the ring dove, the sexes share incubation with the male sitting during the day and the female during the night such that each sex is away from the nest for several hours each day. However, persistently elevated levels of plasma prolactin are observed in both sexes throughout the twenty-four hour period (Lea *et al.* 1986). Perhaps a more extreme example of factors other than brood patch stimulation being important in prolactin secretion during incubation is found in the king penguin where bouts off the nest can last for several weeks without any decline in elevated hormone levels (Lormée *et al.* 1999). It is probable additional factors other than brood patch stimulation, particularly the visual stimuli of a nest and eggs, are also involved in the neuroendocrine loop and the relative importance of each component varies with species. Visual stimuli from the nest and eggs may also be important even in those species whose incubation pattern involves virtual constant tactile contact by a single gender throughout the incubatory period. Denervation of the brood patch in the domestic fowl failed to inhibit a nesting induced increase in plasma prolactin (Richard-Yris and Lea unpublished observations) and visual stimuli significantly affected incubation in the red junglefowl (*Gallus gallus spadiceus*; Meijer 1995).

With this in mind the functions of the patch may be divided into two groups. Firstly, those which result from an input from the patch relative to the neuroendocrine system and secondly, those which are a direct consequence of the patch itself.

Indirect action of the brood patch

Incubation behaviour

Much evidence now exists that elevated levels of plasma prolactin are involved in the expression of incubation behaviour in several avian species (Buntin 1986; Chapter 5). If the brood patch were an integral part of a neuroendocrine loop, in which tactile stimulation sustains the elevated levels of plasma prolactin, then it would be expected to see an effect of brood patch stimulation upon incubation behaviour. Early studies investigating tactile stimulation of the brood patch in the expression of incubation behaviour, however, were not conclusive. In the domestic pigeon, denervation of the brood patch did not affect the ability of the bird to incubate (Medway 1961). In this particular species, however, perhaps this result was not particularly surprising. Both sexes share in incubation and are each off the nest for several hours a day and stimuli other than contact with eggs are thus considered important. Furthermore, Columbiformes appear different to many other species so far examined in that the level of plasma prolactin does not begin to increase until several days after incubation has been established (Lea *et al.* 1986). By contrast, in the mallard duck, in which the female incubates alone, application of local anaesthetics or denervation of the brood patch although significantly reducing the levels of plasma prolactin, failed to demonstrate a clear action upon incubation behaviour (Hall and Goldsmith 1983; Hall 1987). However, nest deprivation and return studies in Galliformes have demonstrated that the motivation to incubate persists for several days after the concentration of plasma prolactin has fallen (e.g. Sharp *et al.* 1980) and earlier studies, therefore, may not have studied the incubatory behaviour for a long enough period.

Brood patch denervation significantly reduces nesting behaviour and prevents the initiation of incubation in the turkey (Book *et al.* 1991). Furthermore, in this species the tactile stimulation the brood patch receives from the nest itself, and not necessarily from the eggs, is sufficient to maintain elevated concentrations of plasma prolactin and associated incubation behaviour (Book and Millam 1991). Consistent with these findings, canaries with denervated brood patches also showed decreased nesting (Kern and Bushra 1980). In conclusion, evidence suggests that in those species in which incubation of the eggs involves continuous tactile stimulation of the brood patch, then this form of stimuli has become pre-eminent in sustaining elevated prolactin levels associated with the expression of incubation behaviour. By contrast, in birds whose incubation patterns involve long periods away from the nest, tactile stimulation is less vital.

Termination of egg laying

Avian species can be grouped into either determinate layers or indeterminate layers. Whereas a determinate layer will produce a set number of eggs, an indeterminate layer may produce a range of egg numbers before deciding the clutch is complete and incubation begins. An exhaustive study of avian egg laying patterns (Frith 1959a), led Haywood (1993a) to classify clutch-size control mechanisms in indeterminate layers. In this system, 'semi-determinate layers' were classified as species in which the number of ovarian follicles is restricted by the females contact with the first egg of the clutch. The only example of this type was among the Adélie (*Pygoscelis adeliae*) and gentoo (*Pygoscelis papua*) penguins. By contrast, indeterminate layers were species where the number of follicles ovulated was entirely dependent on extrinsic factors. Three forms of indeterminate layer were identified: brood parasites who used visual cues; megapodes who use thermal information; and the vast majority of species who use tactile stimulation which develops between the brood patch and the eggs.

An early study by Hinde *et al.* (1963) claimed that in the canary increasing tactile sensitivity of the brood patch was responsible for termination of egg laying. In indeterminate layers, in addition to sustaining prolactin secretion during incubation, it is possible, therefore, that tactile stimulation of the brood patch may result in an antigonadal action. As an antigonadotrophic action of prolactin has been postulated in many species, whether this possible action of the stimulated brood patch is direct or indirect is unclear.

An antigonadotrophic action of incubation, which appears to be a direct effect, is reported in at least one determinate layer. In the ring dove, following a period of nest deprivation, when the nest and eggs are returned and the birds return to incubation there is an immediate and sharp decline in the concentration of plasma LH before any increase in basal levels of prolactin is observed (Lea and Sharp 1989).

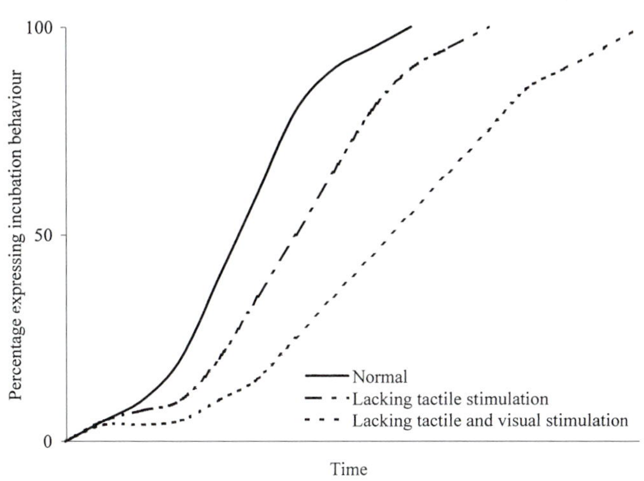

Fig. 8.4 Schematic diagram demonstrating summative nature factors involved in the development of incubation behaviour appropriate to an indeterminate layer such as the domestic fowl (adapted from Meier 1995).

It is also known that visual and tactile stimuli provide important information to the female bird about the number of eggs in the nest (Murton and Westwood 1977; Kennedy 1991; Haywood 1993a). However, there are few studies that have been specifically designed to discriminate between the sensory inputs derived visually from, that obtained from stimulation of the brood patch. In the bantam fowl, visual stimuli alone were sufficient to judge clutch size (Steen and Parker 1981). A closely controlled study in the indeterminate laying red junglefowl allowed Meijer (1995) to conclude that visual and tactile stimulation from the eggs were additive in their effect on termination of lay and onset of incubation (Figure 8.4). It was suggested that continuous egg removal will produce a situation where the tactile stimuli from the nest alone are eventually sufficient to terminate egg laying and initiate incubation. However, when the female can see an increasing number of eggs, the visual stimuli advances these events Meijer (1995). Such a situation is consistent with similar observations made in the denervated turkey hen (Book *et al.* 1991).

Functional activity of the brood patch

Development of the avian embryo is dependent on heat supplied from an external source. In the vast majority of avian species the source of this heat is through the brood patch, which has developed the necessary morphological changes to facilitate this transfer (Drent 1975). Several studies have demonstrated that heat transfer operates in such a way so as to keep egg temperature between narrow tolerance limits and very close to the optimal development temperature for the embryo (see review by Haftorn, 1988).

It is still not clear, however, how essential the full morphological development of the brood patch is to the successful incubation of eggs. There are several species in which both sexes actually actively share in the incubation of the clutch of eggs but in which brood patch development only occurs in the female, e.g. the red-billed firefinch (*Lagonostica senegala*; Payne 1980). The absence or presence of a brood patch in males is no clear indication of whether the male actually incubates the eggs (Skutch 1957).

The zebra finch is one species in which both sexes participate in egg incubation but only the female develops a patch. In this bird, however, no consistent differences were observed in egg temperature when incubated by either sex, or in the sexes ability to re-warm cold eggs. Fluctuations in incubation temperature also did not differ consistently between the sexes (Zann and Rossetto 1991). It is possible that the male overcomes the problem of ventral feathers effectively insulating the eggs from his skin, by having a higher basal metabolic rate than females (Welty 1975).

In addition to morphological changes observed in brood patch development, numerous studies have shown there are important physiological changes in the brood patch, which occur during the incubation period. Many avian species are 'intermittent incubators' in which the parent frequently leaves the nest to forage (Skutch 1957, 1962; Chapter 6; Chapter 15). The eggs of these species are subjected to repeated cycles of warming and cooling and the brood patch is vital in relaying to the parent the current state of egg temperature (Chapter 9). The heat to re-warm the eggs may be surplus heat accumulated by the parent during the preceding foraging bout or, more probably, it may come from increased metabolic heat (Biebach 1986). Thus artificial cooling of eggs under an incubating female bantam results in an increase in its breathing rate (Haftorn and Reinertsen 1982, 1985), oxygen consumption (Biebach 1979; Vleck 1981a) and heart rate (Gabrielsen and Steen 1979) as well as inducing shivering (Tøien *et al.* 1984). The heating phase continues until the egg temperature sensed by the brood patch reaches the 'exit temperature' (White and Kinney 1974; Zerba and Morton 1983a, 1983b; Chapter 15), whence the parent is then able to abandon the nest for its next foraging phase. The duration of the foraging bout may be limited to the time the egg takes to cool to some minimum temperature (Zerba and Morton 1983a, 1983b; Chapter 6). Experimental manipulation of egg temperature is very effective at altering the behaviour of incubating birds (Franks 1967; Drent *et al.* 1970; Vleck 1981a; Davis *et al.* 1984a). Attentiveness in the village weaverbird (*Ploceus cuullatus*) was positively correlated with the cooling rate of the eggs within nests with different insulation properties (White and Kinney 1974). De-sensitising the brood patch with anaesthetic led weaverbirds to incubate at higher temperatures and for longer sessions than control birds (White and Kinney 1974). These results suggest that the bird monitors egg temperature closely through its brood patch and adjusts its behaviour accordingly.

Artificial cooling of eggs results in heat production of the incubating hen, which was five times greater than the resting level (Tøien *et al.* 1986). One study in the incubating female bantam involved direct measurements of blood flow in the brood patch and revealed localised vasodilation due to a direct temperature influence on the smooth muscle cells of the brood patch vasculature. This contrasted with other peripheral vascular beds, which conserved heat by vasoconstriction in response to local cooling (Midtgård *et al.* 1985). Therefore, physiological responses are specifically related to increasing the heat transfer through the brood patch. Such mechanisms are important in increasing heat transfer to eggs at a low ambient temperature and heating up an egg, which had been left untended for a period, would be faster. In addition, local control would help diminish any temperature gradients between the eggs of a clutch. It has been shown in the incubating bantam fowl that the skin of the thoracic area of the brood patch demonstrates an increased cold sensitivity of their brood patch receptors, as compared to the same area in non-incubating birds (Brummermann and Reinertsen 1991). Furthermore, the thermal input is significantly increased compared to other skin areas and core temperature mediated via increased flow through arteriovenous anastomoses in this area (Brummermann and Reinertsen 1992). It has been suggested that this forms a reflex mechanism involving brood patch 'thermoreceptors' and that hormones such as oestrogen and prolactin increase the sensitivity of this response. Adrenergic innervation in these structures in the brood patch may well have evolved for this purpose (Midtgård 1988).

In addition, more intense vasodilation was observed in the brood patch area during moderate cooling of the eggs and more intense vasoconstriction during stronger cooling (Brummermann and Reinertsen 1992). The complexity of this system indicates that the heat transfer to the egg is not merely a local, temperature induced vasoactivity combined with increased metabolic rate but rather an extremely sophisticated adaptive central nervous thermoregulatory system utilising a network of feedback loops for different signal/effector systems (Brummermann and Reinertsen 1991).

Studies using the incubating bantam have shown that shivering induced in response to an experimentally cooled egg is a result of both peripheral and central thermoreceptors (Tøien 1993a). It is unclear exactly where the peripheral receptor input originates but possibly it may be in the brood patch, in tissue layers underneath the patch or on the venous side of the circulatory system before cold blood mixes with other blood in the right atrium (Tøien 1993a). Interestingly, comparative studies on the black grouse (*Lyrurus tetrix*), a species from the sub-arctic region, have demonstrated that the capacity of this bird during incubation to transfer heat to its eggs was even more effective than that observed in the tropically-derived bantam fowl (Tøien 1993b).

A further consideration, one that has received little attention, in having close contact between a naked part of the skin and the egg is the effect upon the water moisture of the eggshell. Indeed data on the coverage of eggs by a brood patch as a percentage of the egg surface area is sparse although it may be around 20% of the shell area (Kendeigh 1973; Handrich 1989a; Yom-Tov *et al.* 1986). Studies on the king penguin, in which a single egg is incubated by a skin flap derived from the feet, has demonstrated that this has important consequences upon such parameters as eggshell conductance, egg water loss, and the associated exchange of carbon dioxide and oxygen across the upper part of the shell (Handrich 1989a).

Finally, it should be noted that as the embryo grows, the developing circulation increasingly facilitates the transfer of heat into the egg. Such a development leads to the expectation that the incubation periods are initially long and then gradually shorten (Turner 1994a, 1997; Chapter 9). The engineering of the patch to match this development requires further study.

Further research into brood patch form and function

Future studies could aim to develop a more comprehensive understanding of how the micro-engineering of the brood patch changes during the incubation period and how these are designed to accommodate the demands of the developing embryos. In particular, it is far from clear how much of the surface of an egg is in contact with the brood patch or whether this varies between species. Other studies could investigate those specific areas of the brain receiving sensory information from the brood patch and how these regions are involved in the expression of incubation behaviour to achieve successful hatching. Changes in sensitivity of these neural loci to hormones during the breeding cycle could also be evaluated.

Acknowledgement

Many thanks to Charles Deeming for assistance with preparing the figures.

9 *Maintenance of egg temperature*

J. S. Turner

Contact incubation of eggs is one of the most conspicuous features of avian biology. Its purpose is the maintenance of a warm and steady egg temperature, and during incubation, the parent bird undergoes remarkable changes in its behaviour and physiology, all seemingly directed to meeting this need. For example, the pectoral skin of incubating birds, commonly the female, but in some instances the male as well, develops into a fleshy and well-vascularised brood patch, naked of feathers (White and Kinney 1974; Grant 1982; Tøien *et al.* 1986; Chapter 8). While incubating, the brood patch is pressed against one surface of the egg, warming it, a process known as contact incubation. The transfer of heat into the egg is regulated through adjustments to both blood flow through the brood patch and heat production by the parent, mediated through temperature sensors in the skin of the brood patch (Collias 1964; White and Kinney 1974; Midtgård *et al.* 1985; Chapter 8). Likewise, many birds construct nests to insulate the eggs against losses of heat from their exposed surfaces. In some instances, the nest completely encloses the egg in its protective environment, although many nests are cup-shaped, open at the top to accommodate the parent whilst it sits on the eggs (Collias 1964; Grant 1982; Skowron and Kern 1984; Chapter 2). Many birds are steady incubators, sitting on the eggs without interruption from the completion of the clutch to hatching. However, many are intermittent incubators, leaving the nest periodically during the day to feed or defend the territory around the nest. These absences seem timed to limit the extent to which the eggs cool during the absence: in colder conditions, the absences are shorter, while warmer conditions are correlated with longer absences (Turner 1994a; Chapters 6 and 15).

Incubation involves a transfer of heat between the parent and embryo, which in principle can be understood as a physical process mediated by the physiology of both parent and embryo. The intention of this chapter is to provide a brief overview of the physical and physiological principles underpinning the maintenance of egg temperature. These principles have potentially far-reaching implications for the ecology and life histories of birds. Thus, it is essential that they are properly understood from the outset. Many attempts to make this link rely on erroneous assumptions about the ways heat flows through eggs, usually from a misplaced desire for simplicity. The transfers of heat between parent and egg are marvellously subtle, however, and it is argued that these subtleties are important features in how parent and offspring manage the flows of heat between them. Understanding these subtleties is impossible if they are simplified away from the beginning.

Maintenance of egg temperature: steady state

At its simplest, the maintenance of egg temperature is a straightforward problem of energy balance:

$$Q_{in} + Q_{out} + Q_s = 0, \tag{9.1}$$

where Q_{in} = rate of heat flow into the egg, Q_{out} is rate of heat flow out of the egg and Q_s = rate of heat storage in the egg; all rate terms in units of watts (W; J s^{-1}). If egg temperature is steady, i.e. not varying with respect to time, the storage term Q_s is zero. In this section, this is assumed to be the case, and the matter of the unsteady state, i.e. where $Q_s \neq 0$, is described later.

When egg temperature, T_{egg}, is steady, equation 9.1 can be rewritten as follows:

$$Q_{emb} + Q_{bp} + K(T_{air} - T_{egg}) = 0, \tag{9.2a}$$

where Q_{emb} = rate of embryonic heat production (always a net addition of heat), Q_{bp} = heat transfer rate between the brood patch and egg (commonly an addition of heat, but sometimes a removal of heat if the egg is too warm), K = thermal conductance of the egg (W °C^{-1}), and T_{air} = air temperature (°C). Evaporation is a potential avenue of heat loss, but these are generally negligible compared to the other flows of heat (Turner 1985). The conductance term, K, is a proportionality that relates a rate of heat transfer to the temperature difference driving it. It encompasses a variety of factors, including surface area of contact between brood patch and egg, heat transfer by blood flow and conduction, and so forth. Based on this, the parental cost of incubation can be estimated simply:

$$Q_{bp} = K(T_{egg} - T_{air}) - Q_{emb}. \tag{9.2b}$$

This equation states that the parent's energy cost is determined by the heat lost from the egg to the environment [$K(T_{egg} - T_{air})$], offset by the addition of heat to the egg by the embryo [Q_{emb}]. Thus, the energy cost of maintaining an egg's temperature depends upon three things: the environmental temperature, the heat production by the embryo, and the thermal conductance of the egg. A number of simple predictions follow from this. For example, it predicts a linear dependence of parental incubation costs on environmental temperature. Also, it predicts that the parent's incubation costs should decline during the incubation period as the embryo's own production of heat increases.

Simple models are not subtle tools. They are useful mainly when the system to be analysed is itself simple. Unfortunately, much of the important biology that underpins the maintenance of egg temperature is subtle, most of it residing in the conductance term, K. As written, the conductance term is primarily a physical property of the egg and environment: traditionally there has been little appreciation of the extent to which this term is under physiological control, both by the parent and the embryo. The perils of ignoring these physiological factors are illustrated by the implied reduction of the parent's energy costs during the incubation period. As the embryo develops, its rate of heat production increases. According to equation 9.2a, this should lessen the subsidy of heat required from the parent.

In fact, the opposite is true: as the embryo matures, the parent's heat subsidy required to maintain egg temperature increases (Turner 1991).

The key to understanding this disparity involves understanding the distribution of temperatures in the egg, the so-called temperature field. In any body, heat flow is driven by differences in temperature. Where heat goes and how rapidly it flows there is governed by how the potential energy gradients driving its flow are distributed. In some instances, these temperature distributions are simple, and simple models of heat flow describe them accurately. In other instances, including heat flow through eggs, the temperature fields are more complicated.

The temperature field

The starting point for any analysis of heat flow is Fourier's Law of conduction, which is generally expressed as a differential equation (Thomas 1980):

$$dQ = -k\,dA\,dT/dx, \tag{9.3}$$

where k is the thermal conductivity of the material through which heat flows ($\mathrm{W\,m^{-1}\,{}^{\circ}C^{-1}}$). Heat flow through any object will depend not only upon the temperature differential (dT, °C), but also upon the object's shape, which is accounted for by the differential terms dx (m) and dA ($\mathrm{m^{-2}}$). The simplest such shape is a flat plate of thickness x and cross-section area A, for which case Fourier's Law is solved as:

$$Q = Ak(T_1 - T_2)/x. \tag{9.4}$$

Implicit in this equation is a temperature field which exists between both surfaces of the plate, and which can be expressed graphically as a series of isotherms within the object (Figure 9.1). The temperature field in a flat plate is uniform, its isotherms parallel to the plate surfaces and evenly spaced through the plate. This simply indicates that the temperature gradients driving heat flow are everywhere the same. As such, the vectors for heat flow are likewise uniform and parallel (Figure 9.1).

Fourier's Law is easily solved for other shapes (Thomas 1980), such as heat flow through a cylindrical shell of thickness t and outer radius r_o:

$$Q = 2\pi Lk(T_1 - T_2)/\ln(r_o/(r_o - t)), \tag{9.5}$$

or, more germane to the problem of eggs, through a spherical shell:

$$Q = 4\pi r_o(r_o - t)k(T_1 - T_2)/r_o. \tag{9.6}$$

The temperature fields and heat flows in such cases are somewhat more complicated than those in a flat plate. In a spherical shell, for example, the isotherms may not be uniformly spaced, and the vectors for heat flow will diverge from the inside surface of the shell to the outer. In both cases, however, the temperature field, and the heat flows associated with it, are functions of only one spatial dimension, specifically the distance x from one side of a flat plate to the other, or the radial distance r from the inside surface of a spherical shell to the outer (Figure 9.1). Such temperature fields are said to be one-dimensional.

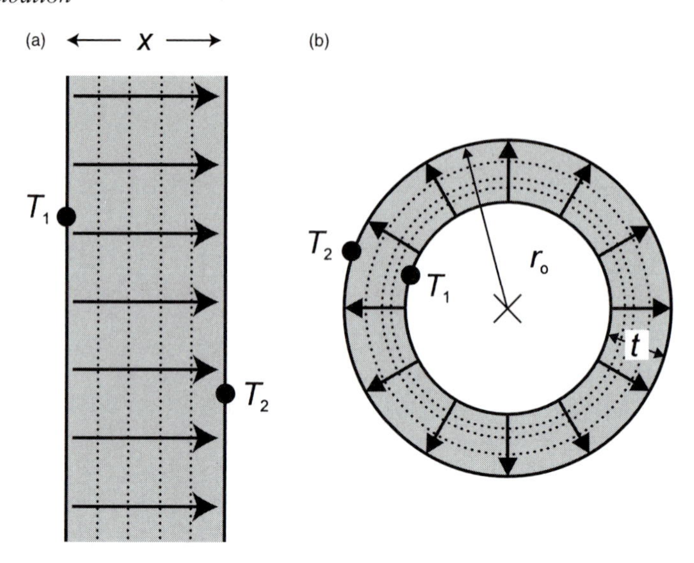

Fig. 9.1 Isotherms and heat flow vectors in two one-dimensional heat flow regimes. (a) One-dimensional heat flow through a flat plate of thickness x with temperatures T_1 and T_2 on either side of the plate. (b) One-dimensional heat flow through a circular tube or spherical shell of outer radius r_o and thickness t. Heavy arrows represent heat flow vectors. Dotted lines represent isotherms.

One-dimensional temperature fields and heat flows are a special case. More common are circumstances in which heat moves through objects in complex spatial patterns that cannot be reduced to a single dimension. One example of obvious relevance to eggs would be heat applied to a limited surface on a block of material, similar to the way a brood patch might warm an egg. As heat flows through the block, it spreads from its point of origin. The vectors for the heat flow can be resolved into three mutually perpendicular components, x, y and z (Figure 9.2). For such cases, Fourier's Law resolves the temperature field into three partial differential equations.

$$dQ_x = -k\,dA_x\,\partial T/\partial x, \tag{9.7a}$$

$$dQ_y = -k\,dA_y\,\partial T/\partial y, \tag{9.7b}$$

$$dQ_z = -k\,dA_z\,\partial T/\partial z. \tag{9.7c}$$

The temperature field is now three-dimensional, because the temperature at any locality in the block is a function of the three variables, x, y and z.

Treating heat flow through any object as one-dimensional is essentially a statement that only one of the partial differentials, say, $\partial T/\partial x$, matters. This, in turn, commits one to the converse assumption that the other partial differentials, in this case $\partial T/\partial y$ and $\partial T/\partial z$, and the heat flows that go with them are negligible. Similarly, two-dimensional heat flows are those in which only one of the partial differentials can be safely neglected. Three-dimensional heat flows, of course, are

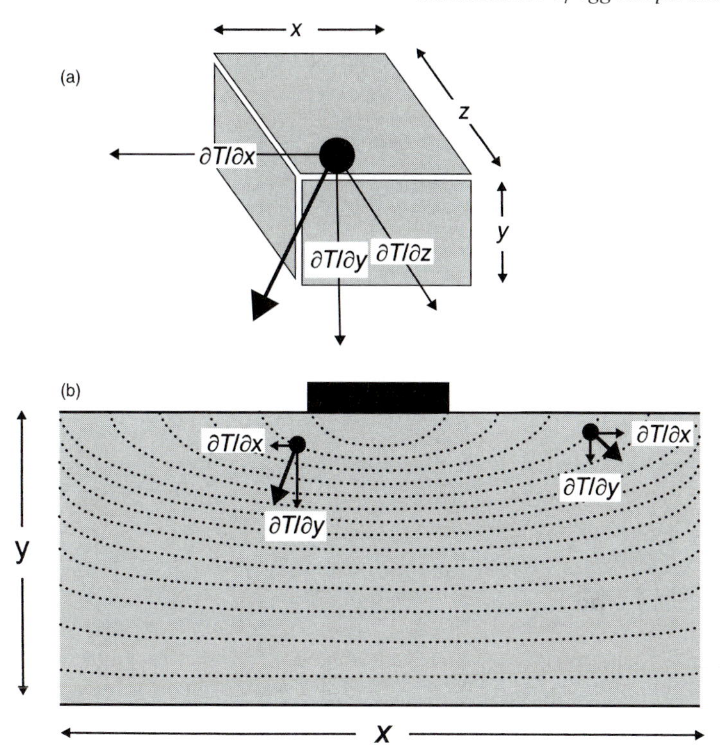

Fig. 9.2 Multidimensional heat transfer in solid bodies. (a) A heat flow vector (heavy arrow) that results from temperature gradients (light arrows), ∂T, in three dimensions, ∂x, ∂y and ∂z. (b) Hypothetical scenario for two-dimensional heat flow through a solid body (light grey) warmed at one locality by a heated strip (dark grey). In this circumstance, temperature gradients in the z dimension are *nil*, i.e. $\partial T/\partial z = 0$. Heat flow is governed by temperature gradients in the x and y dimensions, $\partial T/\partial x$ and $\partial T/\partial y$, which varies from place to place in the block.

those circumstances in which none of the partial differentials can be ignored. This underscores a general principle of heat exchange theory. One can always approximate multi-dimensional heat flows as one- or two-dimensional. However, the accuracy of such approximations will depend on the extent to which the ignored partial differentials are indeed negligible. If they are not, the simplified estimate will be seriously in error.

Temperature fields in eggs

It is noteworthy that much of the literature on the energetics of egg incubation presumes that heat flow through eggs is one-dimensional. This is only true in a few circumstances (Turner 1991). Of the likely circumstances eggs might be found in, there are four likely, but not exclusive, types, two of which result in a one-dimensional temperature field. (1) One-dimensional radial fields, in which

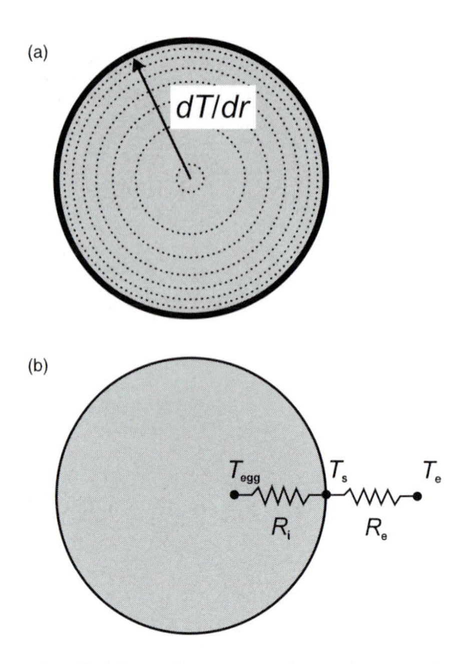

Fig. 9.3 One-dimensional radial heat flow in a spherical egg. (a) Isotherms (dotted lines) and heat flow vector (arrow) in the radial dimension r, down a temperature gradient dT/dr. (b) Electrical analogy for one-dimensional heat flow in a spherical egg, where heat flow is limited by an internal resistance, R_i, spanning an egg centre temperature T_{egg} and the egg surface temperature T_s, in series with an external resistance R_e spanning egg surface temperatures and the environmental temperature T_e.

the temperature variation is predominantly in the radial dimension (Figure 9.3). (2) One-dimensional axial fields, in which the temperature variation is along an axis that runs through the egg (Figure 9.4). (3) Two-dimensional fields, in which the temperature variation is differentiable into a combination of two axial or radial components (Figure 9.5). (4) Three-dimensional fields, in which the temperature field is a function of three radial or axial components.

One-dimensional radial temperature fields arise when the egg is surrounded by a uniform fluid medium, or by nest materials which incorporate insulating layers of still air. Such a circumstance could arise for eggs left for a time by an incubating parent. In a one-dimensional radial field, the egg's thermal conductance can be simplified to two conductances in series (Figure 9.3): an internal conductance, K_i, which governs the flow of heat from the interior of the egg to the eggshell surface; and an external conductance, K_e, which governs the flow of heat from the surface of the egg to the environment. The egg's thermal conductance, K_{egg}, is therefore:

$$1/K_{egg} = 1/K_e + 1/K_i, \tag{9.8a}$$

or in terms of the inverse of the conductance, the thermal resistance:

$$R_{egg} = R_e + R_i. \tag{9.8b}$$

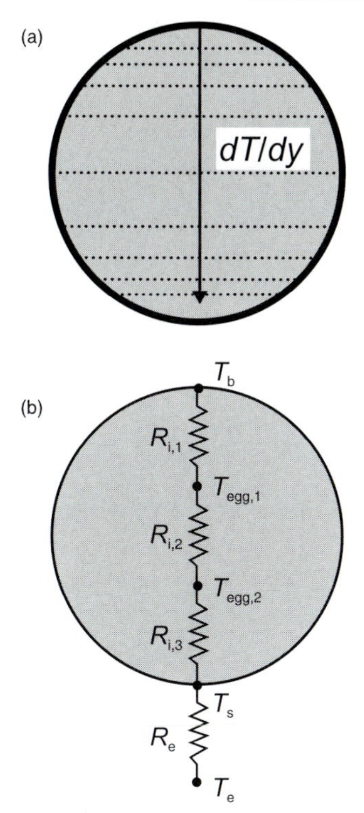

Fig. 9.4 One-dimensional axial heat flow in a spherical egg warmed at one surface by a brood patch. (a) Isotherms (dotted lines) and heat flow vector (arrow) in the axial dimension y, down a temperature gradient dT/dy. (b) Electrical analogy for one-dimensional heat flow in a spherical egg, where heat flow is limited by several series internal resistances, $R_{i,1} \ldots R_{i,3}$, spanning the brood patch temperature T_b, several internal egg temperatures $T_{egg,1} \ldots T_{egg,2}$, and the egg surface temperature T_s, in series with an external resistance R_e spanning egg surface temperatures and the environmental temperature T_e.

In air, the egg's external resistance is generally larger than its internal resistance (Turner 1985), with the result that $R_{egg} \approx R_e$. The extent to which the external resistance dominates the total conductance of the egg depends upon its size and the cardiac output by the embryo (Table 9.1), but it is unlikely that the external resistance will ever be less than 74% of the egg's total resistance. For most eggs, the external resistance accounts for more than 90% of the egg's total resistance (Table 9.1). The disproportionately high external resistance signifies two important features in the maintenance of egg temperature. First, any factor which influences R_e, such as protection from wind, reflection of solar radiation and so forth, will have a large effect on the egg's thermal resistance. Secondly, because the external resistance is so high, there is little potential for the embryo to have any physiological control of thermal conductance of its egg. The reductions in R_i which might result from the embryo's burgeoning circulation have relatively

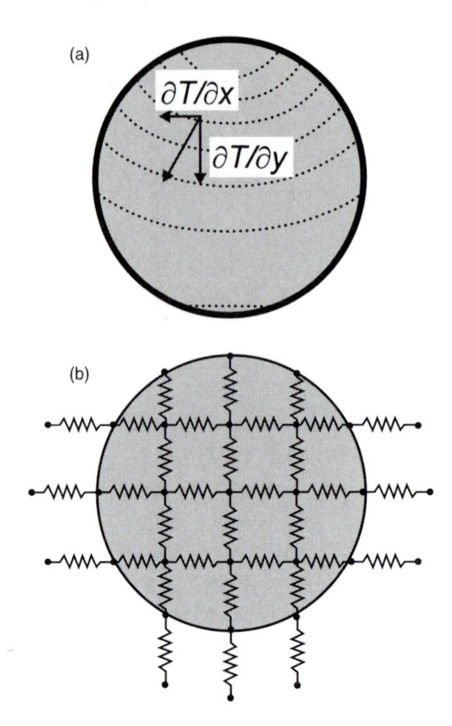

Fig. 9.5 Two-dimensional heat flows in a spherical egg warmed at one surface by a brood patch. (a) Isotherms (dotted lines) and heat flow vectors (arrow) in the horizontal (x) and vertical (y) dimensions, resolved into the temperature gradients $\partial T/\partial x$ and $\partial T/\partial y$. (b) Electrical analogy approximation for two-dimensional heat flow in a spherical egg warmed at one surface by a brood patch, where heat flow is limited by an array of internal resistances which feed heat to the egg surface where its flow is further limited by an array of external resistances.

Table 9.1 Estimated internal and external resistances of birds' eggs at start and end of incubation (Turner unpublished).

Egg mass (g)	R_e (°C W^{-1})	R_i (°C W^{-1})		R_{egg} (°C W^{-1})		R_e/R_{egg}	
		Start	End	Start	End	Start	End
2	100.7	6.0	5.1	106.7	105.8	0.94	0.95
10	38.4	3.5	2.6	41.9	41.0	0.92	0.94
60	13.4	2.0	1.1	15.4	14.5	0.87	0.92
150	7.6	1.4	0.7	9.0	8.3	0.84	0.92
1500	1.9	0.7	0.2	2.6	2.1	0.74	0.92

Where: R_e = External resistance; R_i = internal resistance; R_t = total resistance.

little effect on the egg's total thermal resistance, except in very large eggs, several hundred grams or more in mass (Table 9.1).

One-dimensional axial gradients are probably rare in naturally incubated eggs, but these are often invoked as an approximation of the two-dimensional gradients that commonly would arise in an egg warmed by a brood patch (Ackerman and Seagrave 1984; Turner 1991). In a contact-incubated egg, the normally expected two-dimensional gradient could approach a one-dimensional axial gradient when heat flows through the egg between a strong source at one surface, and a strong sink at the other. An egg warmed at one surface by a brood patch while its antipodal surface sat in water, on ice or on cold ground could qualify as a sufficiently strong source and sink. A one-dimensional axial gradient could also result when heat loss from the exposed sides of the egg is prevented. The natural situation that comes closest is probably found in the *Aptenodytes* penguins, where the egg is warmed on one side by a brood patch, cooled on the other by sitting on the penguin's cold feet, and is enveloped in a well insulated pouch (Burger and Williams 1979; Handrich 1989a).

In most contact-incubated eggs, a two-dimensional temperature field is more likely (Turner 1991). The two-dimensional fields in a contact-incubated egg show strong vectors for heat flow to the exposed surfaces of the egg adjacent to the brood patch, while heat flow to the antipodal surface of the egg is fairly weak. This is a crucial distinction: a one-dimensional axial gradient assumes, among other things, that heat from the brood patch (equivalent to an energy cost to the parent) ultimately leaves the egg uniformly over its surface. In a two-dimensional field, this is not the case: of the parent's heat imparted to the egg through the brood patch, more leaves the egg from surfaces near the brood patch, while relatively little leaves the egg from surfaces opposite the brood patch. This has some interesting consequences, both for the total energy cost for incubation, considered below, and for the role the embryo might play in the regulation of its own egg temperature. In the initial stages of incubation, when there is little or no embryonic circulation of blood, heat flows through the egg by conduction only. Consequently, most of the heat imparted to the egg leaves the egg at surfaces near the brood patch. The temperature of the egg's antipodal surface is very close to air temperature and little heat leaves the egg there. As the embryo's circulation develops, this promotes the axial transport of heat, warming the egg's antipodal surfaces, and increasing the loss of heat therefrom. Thus, the embryo's circulation affects the energy cost of incubation in a way that it could not if the egg was surrounded by air. For an egg of the domestic fowl (*Gallus gallus*), Turner (1991) estimated that the increase of embryonic circulation during incubation will roughly double the parent's energy costs of maintaining the egg temperature. This stands in marked contrast to the oft-stated presumption that the parent's energy costs will decline as the embryo's own heat production increases through the incubation period.

Three-dimensional temperature fields arise in circumstances where there are significant asymmetries in the sources or sinks for heat flow. For example, the air cell of the egg is usually located at one end of the egg's major axis, while the major axis, for its part, usually lies approximately parallel to the parent's brood patch. Thus, heat will flow through one of the egg's major poles differently than through the other. Another conceivable circumstance would be in a clutch of eggs where each egg has been warmed differently than its clutch-mates. Temperature fields in eggs in the centre of a clutch surrounded by warm eggs could have temperature fields that are approximately one-dimensional. Eggs at the periphery, for their part, might have within them substantial horizontal temperature gradients, warmed on one side by its clutch-mates and cooled on the other by the outside environment.

Common fallacies of steady-state incubation energetics

It would seem, therefore, that the energy cost of incubation depends crucially upon what types of temperature fields exist in the incubated eggs. There can be no presumption that one particular type of field will be common. It follows that any model of energy costs that assumes just the opposite, either explicitly or implicitly, will often be inaccurate. Consider, for example, the famous Kendeigh equation, which purports to estimate the energy cost of incubating a clutch of n eggs (Kendeigh 1963):

$$Q_{\mathrm{bp}} = nMc_{\mathrm{p}}(T_{\mathrm{egg}} - T_{\mathrm{air}})(1 - c)/\tau, \tag{9.9}$$

where M = egg mass (kg), c_{p} = egg specific heat (J kg^{-1}°C^{-1}), c = fraction of egg surface covered by brood patch and τ = time constant for cooling (s). In the Kendeigh equation, the term (Mc_{p}/τ) is equivalent to the egg's thermal conductance, K (W °C^{-1}). In Kendeigh's original formulation, the time constant τ was measured from eggs cooling in still air, conditions in which the temperature field in the egg will be one-dimensional in the radial dimension, and surface temperatures will be uniform. Kendeigh also assumes that heat loss from a contact-incubated egg is simply that lost from an egg cooling in air, corrected by the fraction of surface area left uncovered by the brood patch, $(1 - c)$. This presumes that the distribution of surface temperatures in a contact-incubated egg is uniform as they are likely to be in an egg cooling in air. This is incorrect because the temperature field in a contact-incubated egg is qualitatively different: it is a two-dimensional field, in which different rules for heat transfer apply.

This assumption introduces some significant errors in the estimates of energy cost calculated from Kendeigh's equation (Turner 1991). For an unembryonated fowl egg, Kendeigh's equation estimates an energy cost that is roughly double the measured cost when the egg is incubated by an artificial brood patch. The source of the error is clearly in the presumption that the temperature fields in eggs cooling in air describe the temperature fields in contact-incubated eggs. An egg cooling in air has a high external thermal resistance, and uniformly high surface temperatures. A contact-incubated egg has high surface temperatures only near the brood patch, and is significantly cooler at the egg's antipodal surfaces. Heat losses from the egg, and the energy costs of keeping the egg warm, are commensurably lower.

Consider also the conflicting roles of embryonic circulation and metabolism (Turner 1991). The thermal conductance of an egg cooling in still air is uninfluenced by the embryos' circulation, but its temperature is significantly influenced by the embryo's metabolism. For example, at the end of the incubation period of a fowl egg, the embryo dissipates heat at a rate sufficient to raise egg temperature 2–3°C above air temperature (e.g. Meir and Ar 1990). At the same time, the egg's thermal conductance in still air remains unchanged throughout the incubation period, despite the substantial increases in internal blood circulation. Equation 9.2a suggests that as the embryo matures, the parent's energy cost should decline as the embryo provides more of the heat required to keep the egg warm. For a fowl egg, Kendeigh's equation predicts the energy costs of maintaining egg temperature should decline by about 30% through the incubation period. However, the energy costs measured by an artificial brood patch actually *increase* by about 25%. Again, the error results from unrealistic assumptions about the temperature distributions inside contact-incubated eggs. One-dimensional radial fields where the external resistance is large compared to the egg's internal resistance limit the extent to which the embryo's blood circulation can affect the egg's thermal conductance. This leads to the erroneous implication that K in equation 9.1 is assumed by Kendeigh's equation to be constant, or at least controllable principally through externally-driven factors. However, the situation Kendeigh's equation is intended to simulate results in two-dimensional fields inside the egg. Based upon the evidence that embryonic circulation can influence energy costs so strongly, this signifies that the term K clearly is not constant. Furthermore, it opens the possibility of active control by the embryo of the parent's energy costs of incubation, particularly near the end of incubation, when the embryo has the neural machinery for temperature regulation, even if it lacks the thermogenic capacity for it.

Finally, consider how incubation energy costs should scale with egg size (Turner 1991). The thermal conductance of an egg cooling in air is a composite conductance for heat flow by convection and radiation from the surface of the egg. This composite resistance scales to roughly the 0.6 power of egg mass. By implication, the energy cost of incubation predicted by Kendeigh's equation should also scale to 0.6 power of egg mass. However, in a two-dimensional temperature field, the axial component of heat flow through the egg is important, and this should scale to the 0.33 power of egg mass. The thermal conductance of a contact-incubated egg is a composite conductance that melds these two, and should therefore scale to a power of egg mass intermediate between 0.6 and 0.33. Turner (1991) estimated the scaling exponent to be 0.45.

Thus, by conflating the temperature fields that characterise two very different thermal regimens, approaches like that adopted by Kendeigh (1963) are prone to serious errors in estimating the energy costs of maintaining egg temperature. Three obvious ones spring to mind. First, the energy cost of incubation is presumed to be more than it really is. Second, the energy cost of incubation is presumed to decrease through the incubation period, when in fact it increases. Third, the effect of egg size on energy cost of incubation is exaggerated.

Maintenance of egg temperature: unsteady state

In steady conditions, the net heat exchange of the egg with its surroundings (which include the parent) is *nil*, and the temperature of the egg does not change with respect to time. If the net heat flow is no longer nil, egg temperature changes at a rate proportional to the rate of heat storage, Q_s (equation 9.1). This can be either negative (with a decline of egg temperature) or positive (with an increase of egg temperature). Birds' eggs often experience this so-called unsteady condition.

For example, intermittent incubation involves a cycle of visits to, and absences from, the nest by the incubating parent, with the result that egg temperature cycles synchronously through a periodic excursion of temperature (Turner 1994a, 1997). The energy cost of maintaining the egg's temperature includes both the cost of keeping the egg warmer than the surroundings and the costs involved in changing the egg's temperature (Figure 9.6).

At its simplest (Turner 1997), the energy storage term, Q_s, is proportional to the egg's thermal capacity, C (J°C^{-1}), and the magnitude of the temperature change, dT (°C):

$$dQ_s = C\,dT/dt. \tag{9.10a}$$

The egg's thermal capacity is the product of its specific heat, c_p (J kg^{-1}°C^{-1}) and its mass, M (kg), so equation 9.10a can also be expressed:

$$dQ_s = Mc_p\,dT/dt. \tag{9.10b}$$

This simple description of the energy storage term could be used to estimate the energy costs of re-warming an egg following an absence from the nest. As in the simple application of Fourier's Law to the maintenance of steady temperature, this simple conception is only as good as the assumptions that underlie it. If these assumptions are not met, errors will arise in any estimates of the energy cost of re-warming derived therefrom. Also, the dimension of time is now involved and so these errors will pose a sort of double jeopardy, involving not only energy

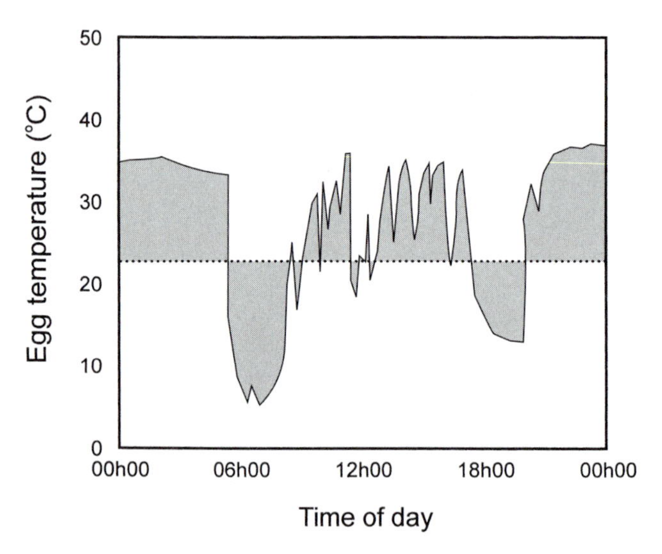

Fig. 9.6 Transient and steady components of egg temperature, illustrated using representative daily records of temperature (solid line) of the eggs of the dusky flycatcher (*Empidonax oberholseri*; after Morton and Pereyra 1985). The average egg temperature represents the steady component over the course of the day (dotted line) of egg temperature. The transient components, those involving a net storage or release of heat from the egg are represented by the shaded regions between the actual and average temperatures of the egg.

costs but time costs as well (Turner 1994a, 1997). For example, the time a bird can spend away from its nest, and hence the time that can be spent foraging, patrolling territories and defending the nest, will be influenced by how rapidly the eggs change temperature, both in how fast they cool when the parent is absent from the nest, in how fast they can be re-warmed when the parent returns, and by how many eggs are in the clutch. How differently sized clutches of eggs and eggs of different masses affect nest attentiveness is discussed in Chapter 6.

The unsteady temperature field

Unsteady flows of heat through eggs are also driven by temperature fields (Turner 1994b). An unsteady temperature field consists of two components: a steady component, which is equivalent to the temperature fields in steady conditions, and an unsteady component, expressed as a time-dependent change of temperature superimposed on the steady temperature field. The unsteady component is driven by a so-called unsteady forcing, in the case of eggs, a pulse of heat applied intermittently to the egg surface through a brood patch. This method of warming forces a periodic oscillation of temperature in the egg, the so-called unsteady component, about some average, T_0, the steady component.

The unsteady temperature field is most easily visualised by forcing the egg temperature with a sinusoidal wave of heating from an artificial brood patch (Figure 9.7). This forces a sinusoidal variation of temperature throughout the egg, the amplitude and phase of which varies with locality in the egg. In general, at greater distances from the brood patch, the amplitude of the unsteady temperature

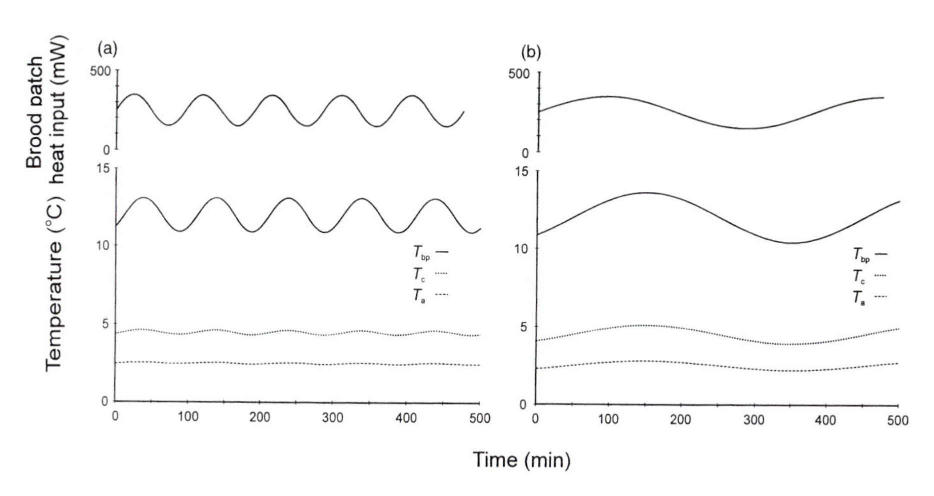

Fig. 9.7 Transient heat flows and temperatures in a fowl egg warmed by an artificial brood patch. Egg temperature was forced by a sinusoidal input of heat from the artificial brood patch which had an amplitude of 100 mW around a mean heat input of 250 mW (top panels). Egg temperatures were measured at the shell beneath the brood patch (T_{bp}, solid lines), the egg centre (T_c, dotted lines) and the eggshell antipodal to the brood patch (T_a, dashed lines). (a) Sinusoidal heat input with a period of 100 min. (b) Sinusoidal heat input with a period of 400 min. After Turner (1994c).

diminishes, and becomes increasingly out of phase with the heat wave forcing the temperature (Turner 1994a, 1994b).

The effective thermal capacity

This distribution of unsteady temperatures indicates that not all parts of a contact-incubated egg participate equally in the unsteady flows of heat through it. In particular, the unsteady heat flows are strongest near the brood patch, and diminish with distance from the brood patch (Turner 1997). Simple formulae (e.g. equation 9.10) presume that the temperature change is uniform throughout the egg. It demonstrably is not (Figure 9.7). This means that the actual cost of re-warming must be weighted by the distribution of temperature change within the egg. Those regions of the egg closest to the brood patch will experience large increases of temperature, and the costs of re-warming those parts of the egg will be high. By contrast, those regions of the egg which experience smaller changes of temperature will incur re-warming costs that are commensurably lower. Finally, those regions of the egg which experience no temperature change will incur no re-warming costs at all, as if this portion of the egg's mass was thermally not there. Thus, a contact-incubated egg will have an 'effective thermal capacity' which should be considerably less than the more commonly used 'gravimetric thermal capacity' implied by an equation like 9.10, i.e. the product of the egg's mass and its specific heat (Turner 1997). The effective thermal capacity is smaller and so the re-warming costs will therefore be less than might be predicted from the egg's gravimetric thermal capacity. Time spent re-warming is also a function of thermal capacity, meaning that time costs should also be considerably less. Turner (1997) has estimated the effective thermal capacity of a fowl egg might be as little as 15% of its gravimetric thermal capacity.

The concept of the thermal impedance

Effective thermal capacity of an egg is less than its gravimetric thermal capacity because the unsteady flows of heat penetrate unevenly into the egg. It follows that the effective thermal capacity will vary depending upon how readily the unsteady component of heat flow penetrates the egg. If the unsteady heat flow penetrates the egg poorly, only a small region of the egg will experience a temperature change, and the effective thermal capacity will be small. On the other hand, if the unsteady heat flow penetrates far into the egg, a larger portion will experience a temperature change, and the effective thermal capacity will be larger.

The temporal component of the unsteady heat flow is very important in determining how effectively it penetrates the egg (Turner 1994c). Again, this is easily demonstrated by forcing egg temperatures using a sinusoidally-varying heat input from an artificial brood patch. When the brood patch's heat input varies at high frequency (short period), the unsteady flows of heat are confined to a small region of the egg close to the brood patch. Temperatures in regions of the egg far from the brood patch oscillate only weakly (Figure 9.7). However, low frequency inputs of heat spread through the egg more broadly (Figure 9.7). As a result, temperatures of the egg far from the brood patch oscillate more strongly than they do under high frequency inputs (Figure 9.7).

These temporal effects further complicate the analysis of egg re-warming. Fortunately, this problem has much in common with unsteady energy flows through a variety of physical systems, such as are found in alternating current flow through an electronic circuit or acoustic energy flow through pipes. There is a substantial body of theory developed to deal with such flows, and this should, in principle, be applicable to the problem of unsteady energy flows through contact-incubated eggs. One of the more useful aspects of this theory is the notion of *impedance*, a sort of transient-state analogue to the egg's thermal resistance (Turner 1994c). Put simply, the thermal impedance, Z_t, accounts both for the steady and unsteady components of energy flow through the egg:

$$Z_t = (R_t^2 + X_c^2)^{0.5}, \tag{9.11}$$

where R_t = the steady state resistance through the egg, by definition insensitive to the temporal component of the heat input, and X_c = the capacitative reactance, which varies with the frequency of the heat input (all terms are expressed in units of $^\circ C\,W^{-1}$). Contact-warmed eggs behave like a low-pass thermal filter, so that their thermal impedance declines with the period of the unsteady heat flow driving the temperature. At very long periods (low frequencies), i.e. where the period approaches infinity, the thermal impedance is equivalent to the thermal resistance (Figure 9.8), and heat supplied from the brood patch penetrates deeply into the egg. For periods shorter than the egg's time constant, τ (roughly 20 min for fowl eggs), the impedance increases sharply, rising several orders of magnitude higher than the steady thermal resistance (Figure 9.8). Here, the unsteady component of heat supplied through the brood patch penetrates the egg poorly. Again, the embryo has some significant control over this. As the embryo's circulation develops, it distributes both the steady and unsteady components of heat flow more widely through the egg, resulting in a decline of the egg's thermal impedance over all frequencies.

The thermal impedance of an egg suggests some interesting possibilities for ways birds might manage the unsteady heat flows through their eggs (Turner 1994a). Most of these are unexplored, but could form the basis for future investigation. Imagine, for example, how an intermittently incubating bird might manage its incubation energy budget through the incubation period. Re-warming the egg should involve primarily warming of the embryo: it makes little sense to warm those parts of the egg not occupied by the embryo. If only those parts of the egg occupied by the embryo could be selectively re-warmed, energy savings could accrue to the parent. Early on in incubation, the embryo and yolk are free to rotate within the egg, such that the embryo is automatically positioned near the egg's upper surface. During this time, the incubating parent may need to warm only those regions near the egg's upper surface, i.e. that region occupied by the embryo (Rahn 1991). This the parent could do by adjusting the visitation schedule to favour high impedance warming. High frequency input of heat are now favoured, which are confined to those regions of the egg occupied by the embryo.

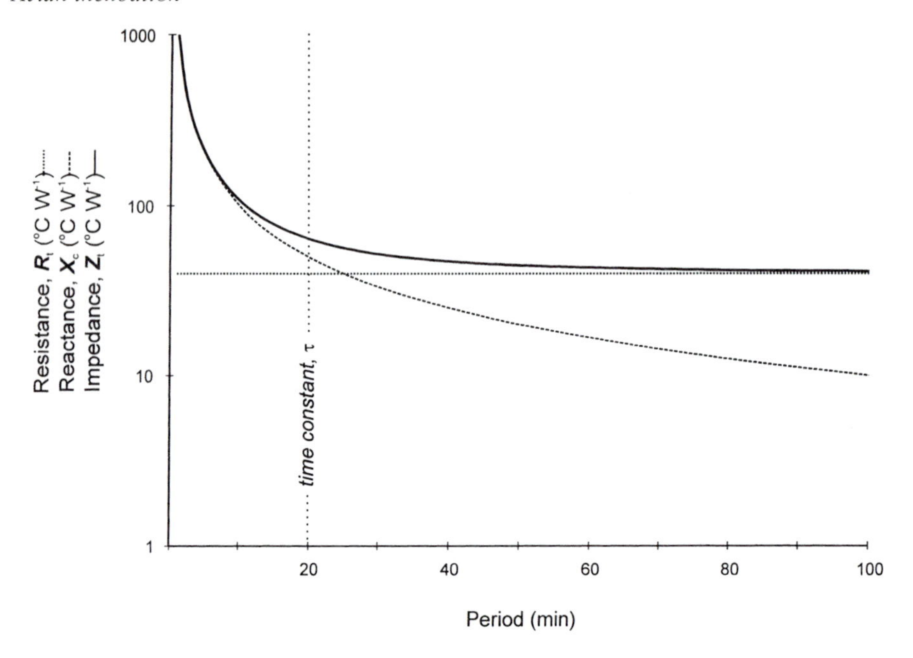

Fig. 9.8 The components of the thermal impedance and their variation with period of the transient component of the heat input from the brood patch. Thermal impedance of a domestic fowl egg, Z_t (heavy solid line), is the vector mean of the egg's steady thermal resistance, R_t (dotted line) and the egg's capacitative reactance, X_c (dashed line), all terms in units of $°CW^{-1}$. The time constant τ for a contact-warmed fowl egg is indicated by the vertical dotted line. After Turner (1994c).

As the embryo grows, however, it comes to fill more and more of the egg. High impedance warming, which warms only a narrowly circumscribed region of the egg near the brood patch, might no longer warm the whole embryo effectively. The parent might then shift its visitation schedule to one that favours low-impedance warming, which would penetrate further into the egg and warm all parts of the egg occupied by the embryo.

Thermal impedance and natural incubation rhythms

Subtle aspects of the maintenance of egg's unsteady temperatures, such as concepts of thermal impedance, or effective thermal capacity, emerge clearly when egg temperatures are forced by well-behaved sinusoidal heat inputs from an artificial brood patch. Whether such subtleties have any relevance to living eggs under actual brood patches is presently unknown. One obvious problem is whether the actual unsteady heat inputs to eggs by incubating parents bears any resemblance to the well-behaved sinusoidal heating one can impart to an egg using an artificial brood patch. Whatever the unsteady heat inputs to a naturally incubated egg might be, they are unlikely to be well behaved. Thus, the temptation might be strong to dismiss these biophysical subtleties as laboratory curiosities, with little relevance to the 'real world' situation faced by eggs and their parents in nature.

Fortunately, there is a way to evaluate such temptations critically, and these rely crucially on concepts like the thermal impedance. Any periodic phenomenon, whether it be sinusoidal or not, can be resolved into a series of sinusoidal components, each with a particular amplitude and phase (Trimmer 1950). This is the Fourier series:

$$f(t) = \sum_{n=0 \to k} (a_k \cos kt + b_k \sin kt), \tag{9.12}$$

where $f(t)$ is some function with respect to time, t. The function $f(t)$ can be resolved into a summed series of sinusoidal components of amplitudes a_k and b_k. Thus, any periodic forcing of egg temperature, whether it be sinusoidal or not, can, in principle, be resolved into a series of sinusoidal components, each governed by the thermal impedance appropriate to the frequency of that component (Trimmer 1950; Turner 1994b). The response, namely the egg's unsteady temperature, can be reconstructed from the sum of the unsteady responses to each of the sinusoidal components of the forcing. Consider, for example, a forcing of egg temperature by a square-wave input of heat. A square wave can be resolved into the following infinite series:

$$Q_s = Q_{bp} - Q_{bp,A} P(\sin t + \sin 3t/3 + \sin 5t/5 \ldots)/2\pi, \tag{9.13}$$

where Q_{bp} is the average heat input, equivalent to the steady component of heat from the brood patch, and $Q_{bp,A}$, is the amplitude of the heat input, and P is the period of the dominant frequency (Figure 9.9). The thermal impedance of the egg at that frequency weights the effect of each of the sinusoidal heat inputs on egg temperature. Thus, the high-frequency components will heat the eggs less than the low-frequency components. The actual change of egg temperature will be the sum of the forcing at each of the sinusoidal frequencies. Thus, the transient heat flow into the egg is derived from the sum of all the sinusoidal heat forcings that go into this series. The result is a damped oscillation of egg temperature that reflects the differential heating effects of the various components of the series.

One of the more interesting implications of this approach is the notion that not all joules imparted to the egg during an intermittent bout of incubation will be equal. Some (those in the high frequency components) will penetrate the egg only slightly, while others (those in the low frequency component) will spread more broadly through the egg. In short, to estimate the costs of re-warming accurately, it is not sufficient to simply measure the egg's temperature excursion. Rather, one must construct a power spectrum that accounts for the differential effects of the different frequency components of the warming.

Relaxation of the unsteady temperature field

There is a final complication in eggs driven by unsteady heating that needs to be considered. During visits to the nest, the eggs presumably are warmed by contact with a brood patch, and this will result in the eggs taking on the two-dimensional temperature fields characteristic of contact incubation. When the parent vacates

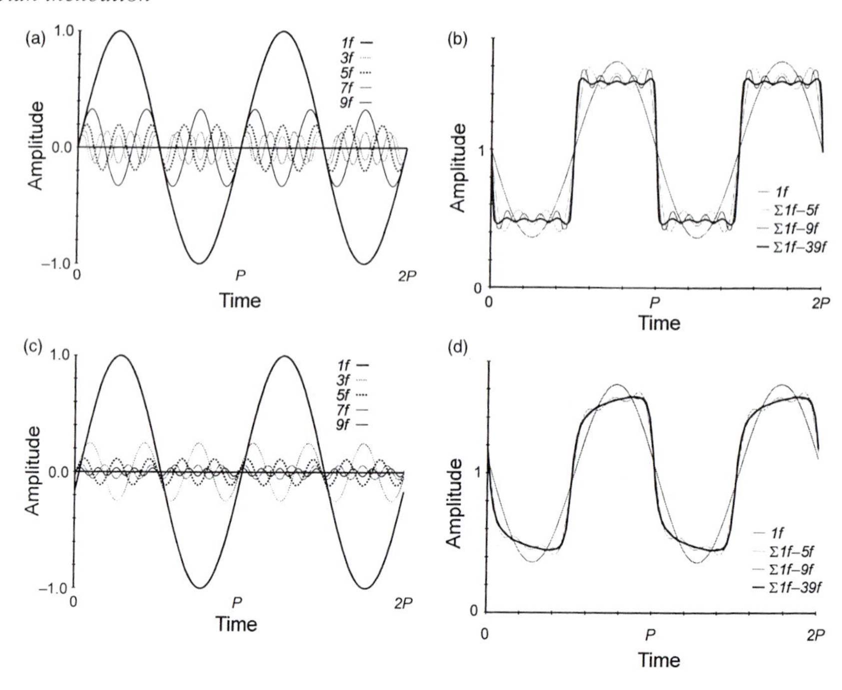

Fig. 9.9 Partial decomposition and reconstitution of a square-wave forcing of egg temperature and the temperature response of the egg. (a) The first five sinusoidal components of a square-wave forcing of heat input from a brood patch. (b) Reconstituted square wave response showing the effects of summing various of the sinusoidal components of the forcing, including the fundamental frequency (f), the sum of the components of the first three components ($\Sigma 1f - 5f$), the sum of the components of the first five components ($\Sigma 1f - 9f$), and the sum of the components of the first twenty components ($\Sigma 1f - 39f$). (c) The first five sinusoidal components of the response of egg temperature to the forcings of egg temperature by a sinusoidal wave of heat input from a brood patch in *a*. (d) Reconstituted temperature response of the egg forced by the components in *b* including the response at the fundamental frequency (f), the sum of the responses of the first three components ($\Sigma 1f - 5f$), the sum of the responses of the first five components ($\Sigma 1f - 9f$), and the sum of the responses of the first twenty components ($\Sigma 1f - 39f$).

the nest, however, the egg's temperatures will shift toward the one-dimensional radial temperature fields characteristic of eggs surrounded by still air. The process whereby a temperature field shifts from one configuration to another is known as relaxation. In a cooling egg, relaxation is evident as a two-phase pattern of temperature change (Turner 1987b; Figure 9.10). Early in the cooling phase, the egg cools at a rate dominated by the redistribution of heat from near the brood patch to the centre. As the egg assumes the one-dimensional temperature field typical of cooling in air, the egg's cooling rate changes because heat is now flowing uniformly out of the egg across its surface (Figure 9.10). Thus, the cooling of the egg is a two-phase process, governed by two time constants, one early in the transient for the relaxation time, τ_1, and the other, τ_2, later in the transient for

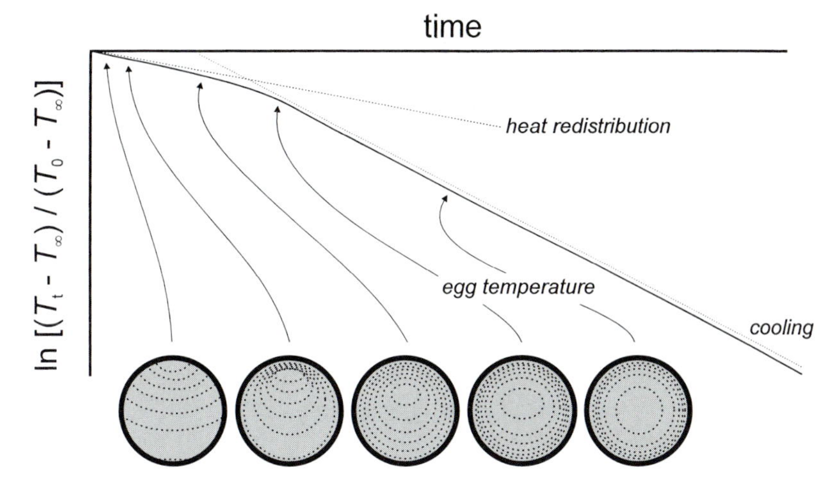

Fig. 9.10 Hypothetical course of temperature change as egg temperatures relax from a distribution of temperatures characteristic of contact-warming, to the temperature distribution characteristic of an egg cooling in air. Temperatures are expressed as the natural log of a dimensionless ratio of temperature at any time t, T_t, cooling toward an equilibrium temperature of T_∞, starting at an initial temperature T_0. Cartoons show likely temperature distributions, indicated by isotherms (dotted lines) at various times through the transient.

normal cooling of the egg:

$$T_{\text{egg}} = T_\infty + (T_0 - T_\infty)(\alpha e_{+}^{-t/\tau_1} \beta e^{-t/\tau_2}), \tag{9.14}$$

where α and β are coefficients representing the ratios of the time constants (Turner 1987b). The reverse should happen when the parent returns to the nest. The returning parent first encounters an egg containing a one-dimensional radial temperature field, and imposes upon it the two-dimensional field characteristic of contact incubation. The egg will similarly change temperature in two phases.

The implications of this type of temperature change are largely unknown, except for the work of Hainsworth and Voss (Chapter 15). However, it may be substantial, particularly in those circumstances where the period of the incubation rhythm is shorter than the egg's time constant for cooling. For example, imagine an intermittently incubating bird which, during its visit to the nest, engages in high-impedance heating of the egg, so that the added heat is confined to a small region close to the brood patch, this region surrounded by a sharply graded temperature field into the egg. This will enable the bird to deposit a load of heat, and leave the nest quickly. Once the bird leaves, this sharply graded temperature field will shift within the egg as the heat deposited there is redistributed (Figure 9.10). The parent's deposit of heat is therefore retained in the egg, rather than being lost, because it migrates preferentially into the egg rather than out through the egg's warmed surface. This could be a useful heat conservation measure, because heat that is retained in the egg is heat that will not have to be replaced on a subsequent visit.

Limitations of the biophysical approach to maintenance of egg temperature

The biophysical approach to egg temperature outlined above assumes that temperature maintenance, i.e. incubation, is largely a matter of how incubating birds and embryos manage the flows of heat into and out of the egg. This approach follows ultimately from the notion that natural selection is the result of a sort of cost–benefit analysis. Differential reproduction results when one phenotype is able to mobilise more energy to do reproductive work than another. Phenotypes can be favoured as a result of increases of physiological efficiency, so that a greater proportion of an animal's total energy budget is diverted to reproduction. However, it can also result from seemingly costly behaviours and investments: as long as more reproductive work accrues from a costly investment, it will be favoured. Contained within this energetic definition of selection is a seductive promise: understand the energetics of an organism and you open a window on all features of an animal's biology, such as its behaviour, its physiology, its life history, any aspect that is subject to evolution by natural selection. The biophysical approach to the maintenance of egg temperature embodies this promise: understand the energy flows between parents and eggs and you understand all features of avian biology that follow from it, such as attentive behaviour, parental care, and so forth. Realising this promise was an important motivation for the pioneers of the field such Charles Kendeigh and his contemporaries.

As has been shown, the flows of heat through eggs are far more complex and subtle than these pioneers ever imagined it could be. This rich complexity presents a three-sided challenge to the promise just outlined. One the one hand, understanding these biophysical subtleties can reveal tremendous possibilities for ways incubating birds might manipulate their incubation energy budgets. For example, applying impedance concepts to the heating of eggs suggests that costs and benefits accrue differently to behaviour patterns that result in 'high-impedance' heating of the egg compared to 'low-impedance' heating (Turner 1997). On the other hand, the complexity makes a rigorous experimental approach to the problem of incubation energetics more difficult. The difficulty arises for the same reasons economics or sociology are experimentally difficult sciences. The systems under study are themselves adaptable which means that it is sometimes uncertain whether a result is due to the experimental treatment, or due to an adaptive response of the system to the novel and artificial environment posed by the experiment. Finally, these complex and subtle features of egg incubation operate in a natural environment that is itself highly variable and unpredictable. Turbulent winds, for example, can introduce such high variation in heat exchange rates that other, more subtle sources of variation, such as egg colour, nest insulation, time spent away from the nest, etc., are swamped. So, while theoretical tools are available for quantifying the thermal interaction between parent, egg and embryo in great detail, applying them to the 'real world' might be problematic.

Given these problems, it may be worthwhile at this stage to step back and ask: what is to be gained by pursuing an ever more sophisticated physical understanding of the maintenance of egg temperature? If a perfect understanding of the biophysics of heat transfer between parent and egg is attained, could we then say that we understand incubation?

The biophysical approach to egg temperature reflects the reductionist faith that a system can be understood through knowledge of the behaviour of its parts. While very powerful, inherent in reductionism is the tendency to alienate the parts of the system from the system itself. For example, an enzyme can be analysed as an entity in itself, which can illuminate such enzyme-level questions as how the active site operates, or how substrate is turned into product. This approach, however, divorces the enzyme from the biochemical, cellular, organic and organismic *milieu* in which it normally operates, such that a thorough understanding of active site kinetics may be of only limited value in understanding how the enzyme works in the system of which it is a part.

The initial biophysical studies of incubation energetics, exemplified by Kendeigh (1963), similarly offered a useful reductionist framework for answering the seemingly simple question: how much energy is required from the parent to keep its eggs warm? However, these initial approaches alienated the egg profoundly: from the parent, treated now as a disembodied regulated heat source; from the embryo it contained, denying the embryo any physiological control over its own temperature; and even from the environment in which it lived. As we have come to understand the biophysics of incubation more fully, the alienation between parent and embryo has eased to a degree. For example, we now know that the embryo can, in some contexts, affect its own temperature through physiological distribution of heat from the parent. The important question now is not whether or not the embryo can control its own temperature, but whether, and how often the embryo is found in contexts that enable it to control its own temperature. Such questions might be more profitably approached by examining how the egg, embryo, parent, and environment function together as a system, identifying and studying the mutually interacting feedbacks and physiological systems which govern the behaviour of the whole.

The egg as part of a thermal mutualism

Treating the egg, embryo, parent and environment as a unified system (i.e. the bird-nest incubation unit described in Chapter 1) is tantamount to describing egg temperature as the outcome of a type of symbiosis, which includes the components (i.e. the embryo, the parent and so on) and the feedbacks which regulate the flows of energy between them (Figure 9.11). Symbiosis can take the form of a mutualism, a physiological phenomenon whereby two disparate organisms associate in some mutually beneficial way. A mutualism usually involves partners that have complementary physiologies, mechanisms for one partner adjusting the physiology of the other, and some means of controlling the common environment shared by the partners. Consider, for example, the quintessential mutualism, the lichen (Ahmadjian 1993). The heterotrophic and autotrophic physiologies of the fungal and algal partners complement one another, each partner providing nutrients that the other partner could not synthesise on their own. At the same time,

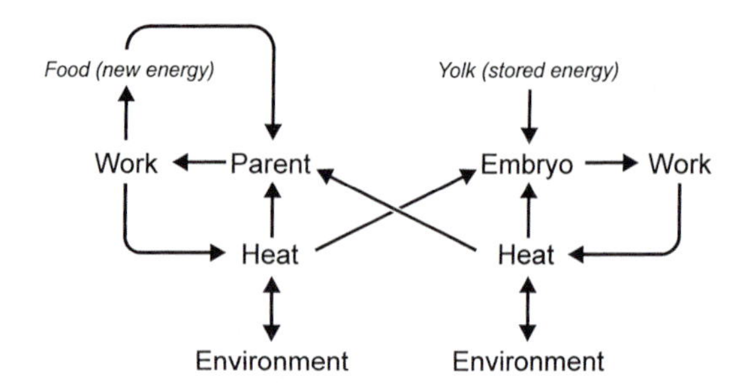

Fig. 9.11 Sketch of the elements of a thermal mutualism between parent and embryo. The control of this mutualism resides in the control of the crossed arrows representing heat transfers between parent and embryo.

the physiology of the autotroph is regulated by a feedback from the heterotroph, e.g. the algae's rate of photosynthetic carbon fixation is regulated by the production rate of urea in the fungus. Finally, the intimate association between the alga and fungus ensures that environmental perturbations do not disrupt the orderly exchange of matter and energy between the partners.

An incubated egg meets some of the criteria for being a mutualism. For example, there is complimentarity of metabolisms: the embryo requires a high temperature to grow, but is not itself capable of providing the heat required to elevate the egg temperature more than a degree or so. The heat required by the embryo must come mostly from the parents, supplied through the brood patch. There is also an intimate association between parent and egg. The egg is usually contained within a nest of some sort, and the transfer of heat is usually through the intimate contact between the brood patch and the egg. Finally, there are feedbacks whereby this transfer of heat is governed. For example, heat flow between the brood patch and egg appears to be regulated by adjustments to blood flow through the brood patch and heat production by the parent (White and Kinney 1974; Vleck 1981a, 1981b; Tøien *et al.* 1986). The insulation of nests is adjusted to some degree to the local conditions (Ponomareva 1971; Schaefer 1976; Webb and King 1983a, 1983b; Kern 1984; Kern and van Riper 1984; Skowron and Kern 1984). Still largely unknown, though, is the extent or mechanisms of feedbacks within the embryo, or between the embryo to the parent, but there is tantalising evidence these exist. For example, fowl embryos appear to have the neural regulatory machinery to regulate egg temperature several days prior to hatching, although the ability to actually do so is limited by the resistance of the eggshell to exchange of respiratory gases, which in turn limits the thermogenic capabilities of the embryo *in ovo* (Tazawa *et al.* 1988b). Also, the embryo might, through some pheromonal or other mechanism, signal to the parent that it requires more heat. Indeed, recent work has shown that nitric oxide appears to play a role in embryo–parent interactions during incubation (Ar *et al.* 2000). It is even conceivable that the embryo, through altering the building behaviour of its parents, could actually control the structure of its own nest.

What is 'egg temperature'?

If the egg is part of a thermal mutualism, this puts the question 'what is egg temperature?' in a particularly critical light. Consider, for example, the title of this chapter. A contribution on the 'maintenance of egg temperature' implies some knowledge of just what egg temperature is. Biophysically, though, the question is nonsensical. Eggs do not have a temperature, they have temperature fields (Table 9.2). which can behave in some complicated ways. In this light, the term 'egg temperature' has to be qualified somehow. Is it temperature of the embryo? Is it temperature of the egg's exposed surface? Is it temperature of the shell underneath the brood patch? All these temperatures co-vary (Rahn 1991), and simple measurements of either egg temperature or brood patch temperature offer little insight into which is the important temperature that is maintained.

Certainly, a case can be made for designating any of the temperatures in an egg as the most important temperature. For example, one could credibly argue that the embryo's temperature is the most critical, on the grounds that it must be kept within the limits of the embryo's thermal tolerance. When the egg is treated as an autonomous individual, separate from the parent that incubates it, such embryo-centred biases colour the types of questions asked. For example, what is the nature of the 'physiological zero', the low temperature at which embryonic development suspends? Similarly, what is the nature of the embryo's tolerance to either high or low temperatures? These are clearly important questions, but coming as they do from an essentially reductionist perspective, they imply that the embryo is essentially a 'physiological problem' presented to the parent, which the parent then 'solves' through behaviour and adaptation, similar in kind to other 'physiological problems' posed to the parent by cold temperatures, high insolation, water scarcity and so forth.

If the embryo is thought of as an active partner of a thermal mutualism, however, the focus shifts to the critical feedbacks and regulators that drive the behaviour of the system as a whole. This poses a whole new set of criteria for deciding which, of all the temperatures in the egg, is the important one. For example, an important component of a thermal mutualism is the temperatures which are sensed by both partners, and which form crucial control points for the system itself. In this perspective, the embryo's temperature is of little relevance, because only one of the partners in the mutualism can be aware of it. The other partner,

Table 9.2 Body, brood patch and egg temperatures in various bird species, based on Rahn (1991).

	Average	SD	*N*
Body	39.53	1.36	23
Brood patch	38.39	1.61	27
Egg centre	35.85	1.80	103

namely the parent, cannot directly sense the temperature of an embryo deep within the egg. A more relevant temperature might be one that both partners can sense, and perhaps control. The best candidate for the regulated temperature in an egg–parent mutualism would be the egg surface contacting the brood patch. Incubation behaviour, life history, thermal tolerance and so forth will centre around and adapt to the maintenance of the egg's surface temperature, rather than the maintenance of the embryo's temperature *per se*.

10 *Nest microclimate during incubation*

A. Ar and Y. Sidis

The vast amount of written information about bird nests and their significance to the eggs, parents and nestlings, calls for a revision with the hope that significantly new insights to this subject will emerge. Bird nests are so highly variable in their appearance, size, building material, structure, location, temporal appearance and species specificity, that more then one kind of classification is possible. In fact, many reviewers over-emphasise rare, curious and unusual features of nests. Hanging nests, communal nests, odd building-material nests, soil and foliage covered nests, floating nests, nests in tunnels, on cliffs, etc., seem to catch the eye of the keen observer (e.g. Collias 1964) and mask the fact that most nests are 'ordinary'. Even so, nests are so species-specific that an expert can easily recognise a bird species by its nest (e.g. Welty 1982; Hansell 2000). The variability of nest forms among species, in particular in warm climates, can only partly explain the microclimate they provide. It may be related more to niche competition, protection and camouflaging against enemies during this critical period, whereas factors such as shading or entrance orientation may contribute more to the microclimate of the nest (e.g. Facemire *et al.* 1990; Williams 1993a; Sidis *et al.* 1994).

Reptiles, which usually lay their eggs in substrates of high thermal inertia, high thermal conductance and high humidity, such as soil, almost always abandon their eggs to the mercy of nature. Hence, eggs are relatively sheltered from daily temperature and humidity fluctuations and mainly encounter the mean seasonal soil temperature. Soil temperature around reptile nests, although fairly stable, tends to be low compared to bird nest temperatures. This seems to determine the relatively slow embryonic development of the poikilothermic reptiles (Seymour and Ackerman 1980; Ackerman 1994). Birds, which are homeotherms, embark on a different strategy: they keep their eggs warm, close to their own body temperature (Huggins 1941), thus enabling the embryos to develop relatively quickly. This strategy requires that parents regulate egg temperature (Drent 1975). It also requires, if energy is to be saved, development in a medium of low heat conductance (= high insulation), such as that provided by still air, and thus, birds usually brood their eggs in air above ground. However, this raises other problems: (1) eggs are exposed to the desiccating action of the unsaturated air (e.g. Rahn *et al.* 1977a; Meir *et al.* 1984); and (2) may be exposed to the heat dissipating action of wind (e.g. Walsberg and King 1978a, 1978b; Webb and King 1983a; Walsberg 1985). Both these problems are partially solved by shielding the eggs with a nest.

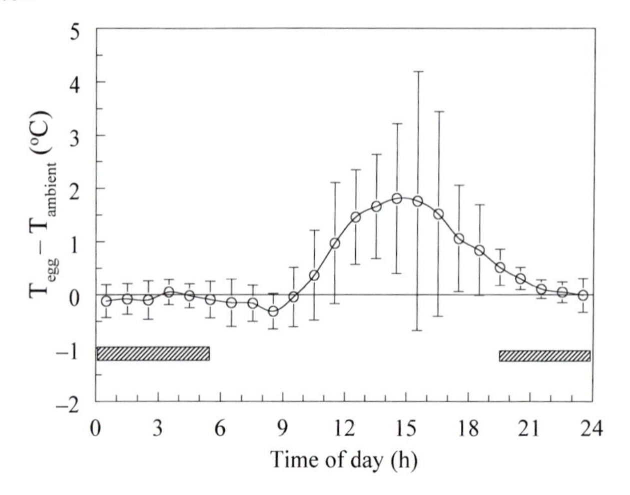

Fig. 10.1 Average temperature differences between the unattended interior of a Palestine sunbird (*Nectarinia osea*) nests (N = 12) and ambient air, during a 24 h cycle in a summer day. Dark bars represent night-time.

Data show that the trends of temperature and humidity in the interior of abandoned, unattended or empty nests, conform to, or are only slightly different from, those of the free air around them. This occurs if the nest is protected (e.g. by foliage) from direct sunlight (Yom-Tov *et al.* 1978; Ar and Sidis 1991; Sidis *et al.* 1994). Some rise in temperature may occur during daylight due to indirect radiation (Figure 10.1). Amplitude dampening or temporal delay may occur due to nest mass inertia and its wind shielding effect. However, this is mainly true in very specific cases such as in tunnel and hole nesters (e.g. White *et al.* 1978; Birchard *et al.* 1984, Ar and Piontkewitz 1992). It is the presence of the incubating bird in the nest, which, together with the nest structure, creates a local environment different from that of the air 'outside'.

It is important to note that, in spite of large differences in nest size, shape, and location, the parameters of microclimate inside the nest relevant to all its inhabitants (parent and eggs or hatchlings) are rather similar in most cases. For example, mean egg temperature in the nest varies among birds at a very narrow range (Huggins 1941). This is because: (1) most birds have chosen their breeding season and nest location such that climatic conditions would be both as suited and as predictable as possible; and because (2) they care for their eggs and young. Thus, incubation and brooding behaviour are very significant in determining the nest microclimate.

Beyond the obvious requirements for substrate, camouflage, shelter, and defence for the eggs, the hatchlings and the sedentary incubating parent, there are four important factors that have to be satisfied in the nest during incubation in nature. These are: (1) appropriate temperature; (2) appropriate humidity; (3) appropriate respiratory gas composition; and (4) egg turning. Certainly, when all four are provided correctly in an incubator, eggs can be successfully hatched without the presence of the parents. Hence, ideally the incubating parents should fully provide these microclimate elements. The problem is that such provisions may be in conflict with the normal life demands of the parent birds – their need to forage and to conserve heat (Ricklefs 1974; Carey 1980; Drent *et al.* 1985). How birds either optimise or compromise their own or their offspring demands, while trying to maintain 'ideal' nest microclimate will be part of the following discussion.

'Nest' has different meanings for altricial hatchlings, which remain within it after hatching, and for precocial hatchlings, which leave it after hatching (see also Chapter 2), and so it is preferred here to address nest microclimate in relation to eggs/embryos and incubating bird only. It is assumed that, in principle, the nest microclimate plays essentially the same role for the incubating bird and the hatchlings. Whereas the last two have the behavioural and physiological capacities to cope with unfavourable situations, the embryo inside the egg does not have such capabilities. As a result, embryonic development may be easily affected by climatic stress.

Definition of nest microclimate

The nest microclimate must be subdivided into two components: (1) the microclimate of the eggs; and (2) the microclimate of the incubating bird (or nestlings). These components may differ and sometimes conflict. The four essential conditions for normal development of eggs, which are fulfilled by the attending parent in the nest, differ considerably from the optimal microclimate of the incubating parent. The optimal microclimate for the incubating parent is closer to the ambient climate near the nest. Often, the incubation season is the favourable season of the year but in other cases, the nest provides a climatic shelter from adverse conditions. Figure 10.2 summarises and groups most of the direct and indirect factors involved in the determination of the nest microclimate.

Temperature regulation of the eggs

Avian egg temperature

The temperature in the centre of the egg in natural incubation ranges from 32 to 38°C (Huggins 1941; Drent 1975; Webb 1987). The common use of artificial eggs for such measurements introduces errors since foreign materials may have different heat capacities and different heat conductances and do not take into account both embryonic heat production and cooling by water evaporation (Ar 1991a). Moreover, the temperature at the centre of the egg does not represent an accurate embryonic temperature for the early embryo, which is not at the egg centre but floats at the top of the egg, adjacent to the shell and the brood patch. Egg temperature under the incubating parent also varies. It changes vertically across the egg, from the brood patch contact area on the upper side of the shell to the area in contact with the nest substrate on the lower side (e.g. Drent 1975; Vleck *et al.* 1983; Swart and Rahn 1988; Groscolas *et al.* 2000; Figure 10.3). This temperature gradient is more pronounced during the first part of the incubation. Thus, although

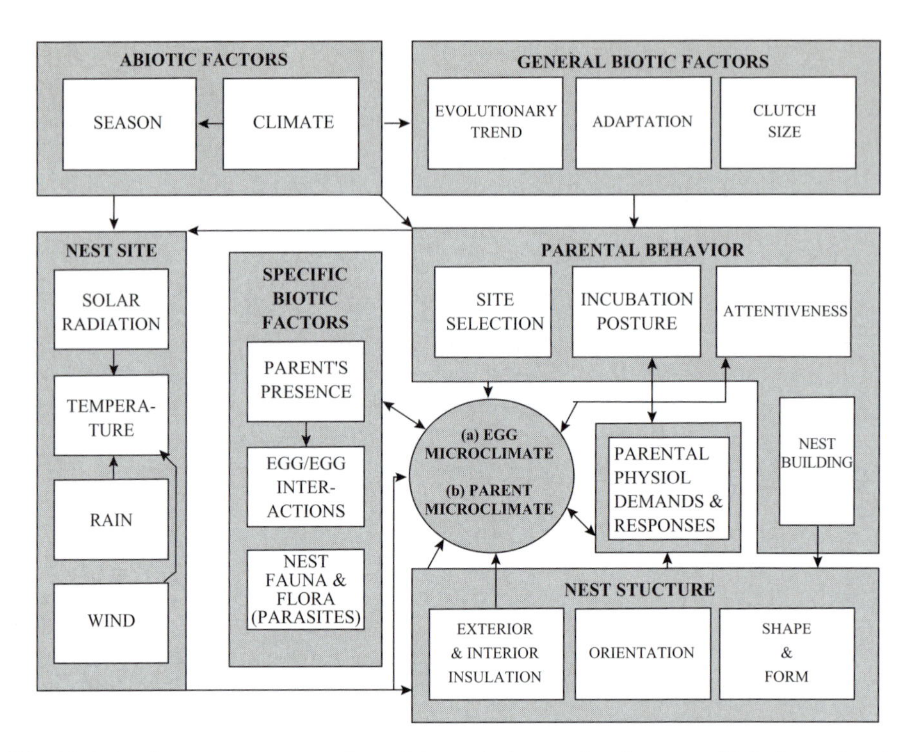

Fig. 10.2 A schematic representation of the main interacting factors, which determine the nest microclimate before and during incubation, for embryos and/or adults.

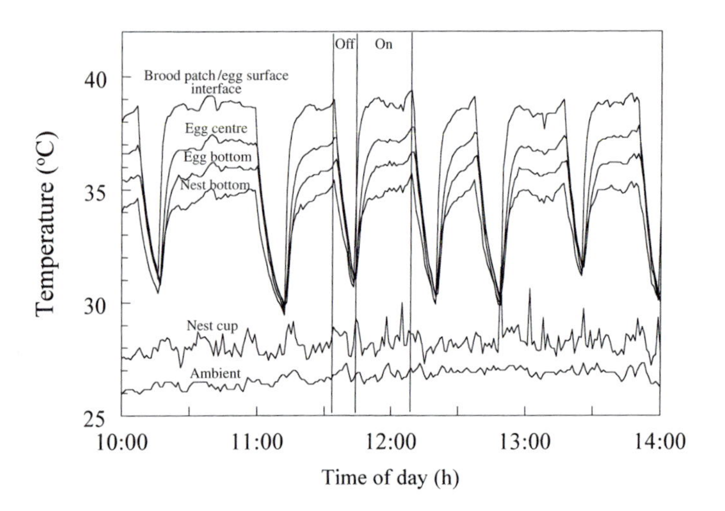

Fig. 10.3 Temperature changes in the nest and the egg of the Palestine sunbird (*Nectarinia osea*). Time sequences of temperatures for ambient, external side of nest cup, nest bottom, egg bottom, egg centre and egg surface in contact with the brood patch, are shown during a relatively cool day. Off-nest times are marked by a temperature decline towards ambient temperature. On-nest times are characterised by a rise in temperature, followed by a plateau. The plateau value is highest for the brood patch/egg surface interface and lowest at the egg bottom.

the egg-centre temperature averages the top and bottom egg temperatures, the embryo itself benefits from the local warmth at the top of the egg. At this time, the embryonic circulation is not yet fully developed and most heat is transferred by conduction across the egg. Later in incubation, after the chorio-allantoic membrane (CAM) has developed, heat convection by blood becomes important and heat is distributed in the egg by the embryonic blood circulation. This causes an increase in heat loss, which is only partly compensated by the metabolic heat production of the embryo, and may require a larger parental heat investment in order to keep egg temperature constant (Turner 1991; see Chapter 9).

The role of the parent

Either one or both parents usually regulate the temperature of bird eggs in the nest. They first pick a suitable nest site and nesting materials, build the nest, and then closely attend the eggs laid therein (e.g. Collias 1964; Collias and Collias 1984; Walsberg 1985). Heat exchange between the bird and embryo, and the close regulation of egg temperature, have been a subject of intense research. Both active heating and cooling are involved (e.g. Russell 1969; Drent 1970; Yom-Tov *et al.* 1978; Howell 1979; Grant 1982; Bergstrom 1989; Brummermann and Reinertsen 1992; Figure 10.4).

It is commonly accepted that avian egg temperature is kept high and relatively constant during incubation, but actually, there is a large variation in the 'normal' temperature of bird eggs, mainly because of parents' off-nest periods (Webb 1987). Many researchers observing bird behaviour, have noted off- and on-nest periods

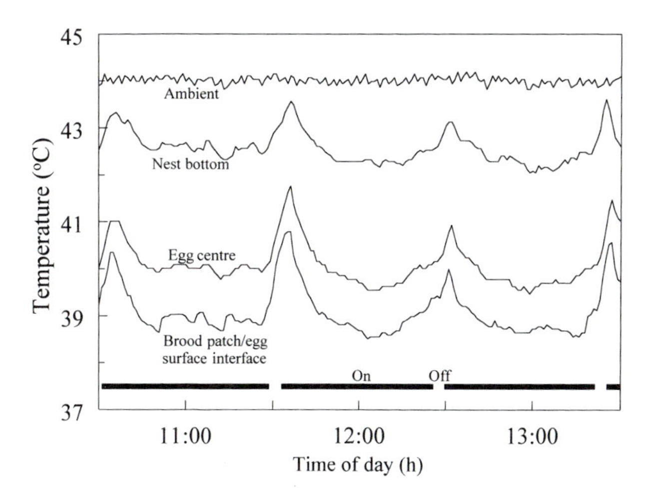

Fig. 10.4 Temperature changes with time in the nest and egg of the Palestine sunbird (*Nectarinia osea*) during exposure to heat. Short off-nest times (probably for liquid imbibition) are marked by a temperature increase towards ambient temperature. Long on-nest times (characterised by tight sit and panting) exhibit a decline in temperature followed by a mild increase towards the next nest departure (probably because of severe body water loss). Egg temperature values are lowest for the brood patch/egg surface interface (inverse temperature gradient) and lower than under mild conditions (compare with Figure 10.3).

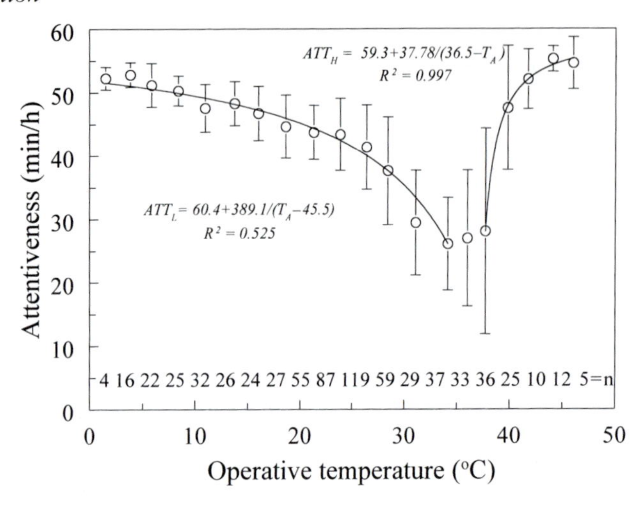

Fig. 10.5 Time in nest of the Palestine sunbird (*Nectarinia osea*) as a function of operative temperature (ambient temperature as recorded and averaged from a black body at the nest site). Observations (mean ± SD) are shown with the corresponding n values. The equations describe the curves passing through the points below (ATT_L) and above (ATT_H). Minimal attentiveness (*circa* 30 min h^{-1}) occurs at an ambient temperature close to body temperature. Maximal observed attentiveness was higher than 50 min h^{-1} at both extreme cold and hot operative temperatures.

during daytime and their relation to ambient temperature. The attentiveness of the incubating bird increases at both colder and warmer temperatures from certain minima at ambient optimal temperatures (Yom-Tov *et al.* 1978; Figure 10.5). At night-time, however, attentiveness is close to 100% and variation of egg temperature around the mean is smaller (Siegfried and Frost 1974). Data for the Palestine sunbird (*Nectarinia osea*), which exhibits female-only intermittent incubation, show that, although the mean daytime egg centre temperature (34.8°C, mean of 6 nests, total of 144 h recording) is not significantly different from night temperature (34.6°C, 113 h), the coefficient of variation at night is half the daytime value (4.0% *versus* 7.9%).

Birds actively regulate the temperature of their eggs during incubation. They closely attend the eggs throughout the entire embryonic development (Carey 1980; Grant 1982). Birds respond to temperature change cues coming from that part of the egg surface in contact with their brood patch during periods of nest attentiveness, and exchange heat between their body and the eggs over this contact area (Jones 1971). The sensory inputs to the parent are known to affect behavioural patterns such as the duration of incubation sessions and recesses and tightness of sit on the eggs (Drent *et al.* 1970; White and Kinney 1974; Drent 1975; Tøien 1989). Cold eggs evoke changes in parent body temperature and/or heat production, general energy metabolism, shivering, non-shivering thermogenesis, changes in cardiac output and changes in blood flow in the brood patch (see below; Chapter 8).

In addition, birds of many species which nest in exposed areas, shade their eggs by standing above them to prevent overheating (Drent 1970, 1972; Grant 1982;

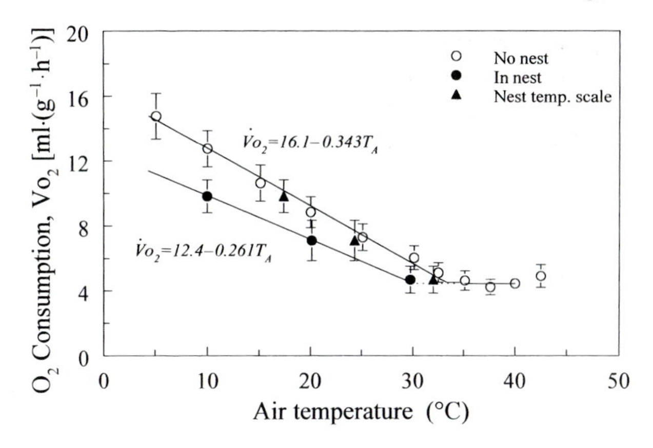

Fig. 10.6 Rate of oxygen consumption at rest (± SD), inside (closed circles) and outside the nest (open circles), of the Palestine sunbird (*Nectarinia osea*) as a function of ambient temperature. Triangles represent inside-nest values when the temperature scale (x axis) represents nest temperature (instead of ambient).

Howey *et al.* 1984; Ward 1990). Overheating is also prevented by wetting the eggs (Schardien and Jackson 1979; Grant 1982), or by covering them with sand and wetting it (Howell 1979). The temperature at the egg surface in contact with the parent determines the duration and efficiency of heat transfer (see also Chapter 6). Overheating of egg surface temperature may evoke physiological responses in the parent such as pressing the eggs to their bellies while panting, performing gular flattering or evaporating water from their skin (Drent 1972; Howell 1979; Marder and Gavrielli-Levin 1986). These responses may reduce body temperature to 2°C below normal (Russell 1969; Grant 1982; Walsberg and Voss-Roberts 1983; Marder and Gavrielli-Levin 1986; Figure 10.4). It is at this point where nest insulation becomes even more important for the parent in saving heat production, and for maintaining a favourable temperature for the embryo (Figure 10.6).

It appears that the egg mass is directly proportional to adult metabolic rate (Rahn *et al.* 1975). From the point of view of parental energy expenditure, this may indicate that on the average, all birds invest the same proportion of their metabolism in both producing an egg and in heating it.

The brood patch

Most birds develop a brood patch during the breeding season. It is a specialised area of the ventral skin of the adult that is applied to the eggs or young hatchlings (Drent 1975; Gabrielsen and Steen 1979; Kern and Coruzzi 1979; Haftorn and Reinertsen 1982; Midtgård *et al.* 1985; Aulie and Tøien 1988; Brummermann and Reinertsen 1991, 1992; Tøien 1993b; Chapter 8). Characterised by absence of the feathers, skin oedema, skin thickening and an increased skin vascularity, the brood patch exhibits increased blood flow, increasing the direct tactile contact between the adult and eggs, and thus facilitating efficient heat transfer to and from the incubating adult and the eggs (reviewed by Drent 1975; Chapter 8).

The degree of the brood patch development differs amongst species and sexes. Size and number of brood patches is correlated with clutch and egg size (Skutch 1957; Jones 1971; Drent 1975; Hawkins 1986; Wiebe and Bortolotti 1993). Usually the size and shape of the brood patch is such that it covers about 20% of the egg surface during actual incubation (Kendeigh 1973; Yom-Tov *et al.* 1986; Handrich 1989b). It is accepted that the brood patch senses and regulates egg temperature by varying its blood flow and heat output in comparison to other skin areas of the bird and in comparison to the brood patch in a non-attentive bird (Brummermann and Reinertsen 1991). This is in addition to the regulation of the tightness of contact by the incubation parent (Tøien *et al.* 1986). However, beyond the presumably reflexive responses to the sensory input from the brood patch, the mechanisms directly responsible for the actual changes in its heat transfer and blood flow are not known. Midtgård *et al.* (1985) showed a vasodilatation response to cold in the brood patch even after apparent nerve blockage, indicating that the response is local, and birds still exhibited resettling movements. It was recently found that intra-muscular administration of nitric oxide synthase inhibitor reduces the rate of re-warming and the final brood patch temperature in incubating barn owls (*Tyto alba*) and turkeys (*Meleagris gallopavo*; Ar *et al.* 2000).

The role of the embryo

Birds will incubate both non-fertilised and artificial eggs. This could indicate that no cues coming from the embryo are needed for normal natural incubation. Recently it has been found that, at least during the second half of incubation, embryos of several species emit nitric oxide through the shell (Ar *et al.* 2000). If this emission could augment blood circulation in the brood patch, then it may represent an additional communication avenue between the embryo and the parent, affecting in turn, the embryonic microclimate.

The heat production of the embryo is very small. For example, a domestic fowl (*Gallus gallus*) embryo just before pipping produces about 0.35 mW. As small as this might be, embryonic heat production nevertheless adds to the overall temperature balance in the nest, and may delay the cooling of the eggs to some degree. Heat loss from the egg increases with age as circulation develops. This is offset in part by the increase in embryonic heat production (Ackerman and Seagrave 1984; Tazawa *et al.* 1988b; Turner 1990, 1991), which is linked to increased metabolism with development, and to increased CAM circulation (Tazawa, 1980a; Ar *et al.* 1991). The relative contributions of brood patch and embryonic heat productions to the maintenance of egg temperature may differ in eggs of dissimilar masses (altered surface/volume ratio) which have different metabolic rates (Ar and Rahn 1985).

In addition, mature embryos of some species, although considered stenothermic (Freeman 1964; Aulie and Moen 1975; Nair and Dawes 1980), are capable of temporarily resisting the effect of cooling on their energy metabolism. This resistance was named 'incipient thermoregulation' (Whittow and Tazawa 1991).

So far it is not known whether the blood flow of the CAM responds to cues from the brood patch above it. It is known however that blood flow distribution in the CAM may differ from area to area (Paganelli *et al.* 1988). An intimate match between the circulations of the brood patch and the embryonic blood may increase heat exchange efficiency, similarly to the way embryonic and maternal circulations of the mammalian placenta optimise embryonic gas exchange (Longo *et al.* 1972).

Role of the nest

It is interesting to note that (with some exceptions) there is a general trend for nest structure to become more and more 'open' for larger and larger birds. Closed nests are found mainly among small species, while almost flat and open nests typify large avian species (Hansell 2000). While evolutionary and adaptive explanations may vary, the end result is that during incubation recesses, eggs tend to be much more exposed in the latter nests. To put it in a different way, eggs of small birds tend to be more insulated, protected from heating from direct sun radiation, and cooling by wind and rain. This may offset the need of small birds, with high specific metabolic demands, to be so attentive to eggs and so frequently leave the nest in order to forage for longer periods (Ricklefs 1974). The 'closed' nest shape also compensates for the higher surface-to-volume ratio of their small eggs (Turner 1985; Figure 10.5). In addition, many such species choose the location and orientation of the nest so that it is relatively protected from direct sun radiation, rain and is on the lee side of the wind (e.g. Facemire *et al.* 1990; Williams 1993a; Sidis *et al.* 1994). These characteristics may help to explain the wide variation in %attentiveness in small eggs (Chapter 6). These considerations are less important (from the point of view of the egg) when the parent is present in the nest.

The combined effect of the incubating parent and nest insulation on egg temperature

If it is assumed that the average temperature of eggs is kept constant when the parent is attentive in the nest, then their heat content is kept in balance. On this basis, several models for heat balance of eggs have been proposed, in which the parental contribution plays a major role (e.g. Calder 1973a; White and Kinney 1974; Walsberg and King 1978a, 1978b; Webb and King 1983a; Ackerman and Seagrave 1984). However, questions of which type of nest is better and why (from the point of view of insulation) have rarely been quantitatively answered (Whittow and Berger 1977; Walsberg and King 1978a, 1978b; Skowron and Kern 1980; Kern and Van Riper 1984; Kern 1984). According to Webb and King (1983a) the most important avenues for heat exchange of the eggs in the absence of the parent are solar radiation and the insulation stripping action of the wind, while the effect of nest wall insulation was relatively unimportant. Although this is true for open cup shaped nests, is it better to have a deep and narrow *versus* a shallow and wide nest? Does a nest canopy have a thermal function? Such questions seem to have relevance of equal importance to the energy budget of the incubating bird as well as to that of the eggs during an incubation recess. In contrast to Webb and King (1983a), Figure 10.7 demonstrates that the presence of the blackbird (*Turdus merula*) nest impedes egg cooling in still air by about 25–30%, but at wind velocities above $0.75\,ms^{-1}$, egg cooling times are halved.

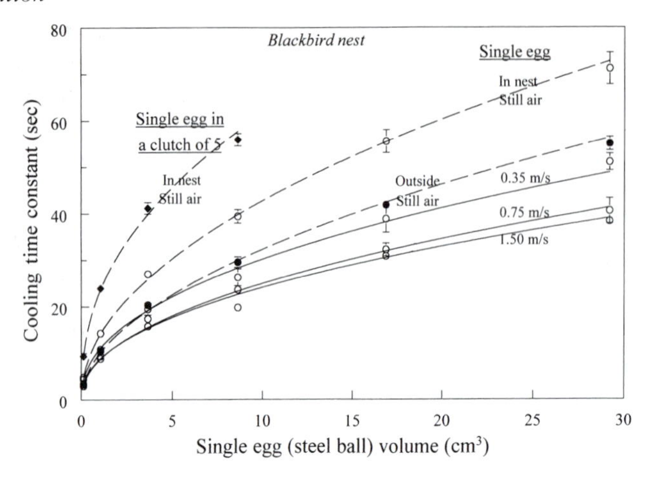

Fig. 10.7 The influence of wind of various speeds (full line curves) on cooling time constant (inverse of cooling rate) of single 'eggs' in the blackbird (*Turdus merula*) nest (n = 6) with reference to a single egg in still air (broken line curves) inside (open circles) and outside (closed circles) the nest. The influence of a five-egg clutch on the cooling time of a single egg among them in still air is also given (closed diamonds, broken line curve). The following regression equations were all highly significant, with r values higher than 0.99. For single eggs: In nest: still air: $Y = 13.641\ X^{0.496}$. In nest: wind at $0.35\ m\,s^{-1}$: $Y = 10.750\ X^{0.449}$. In nest: wind at $0.75\ m\,s^{-1}$: $Y = 8.544\ X^{0.451}$. In nest: wind at $1.50\ m\,s^{-1}$: $Y = 8.449\ X^{0.471}$. Outside: still air: $Y = 9.796\ X^{0.519}$. Y is in seconds and X is in cm^3. In a clutch of 5 eggs: In nest: still air: $Y = 22.875\ X^{0.429}$. In order to standardise the eggs, to avoid cooling by evaporation in wind and to avoid temperature gradients, they were replaced by steel balls of several sizes. Temperatures were measured in their centres. The product of specific heat and specific gravity of steel is about the same as that of eggs ($0.460\ J\,°C^{-1} \times 7.719\ g\,cm^{-1} = 3.844\ J\,°C^{-1}$ *versus* $3.313 \times 1.033 = 3.321$ respectively). These properties make steel balls an ideal model for testing factors of cooling and warming in nests. It is evident that, in contrast to large volume eggs, small eggs cool faster and more than in direct proportion to their size in a given nest. Note that the blackbird egg (*circa* $6.9\ cm^3$) would be on these curves at the point where further increase in volume would not change the cooling time constant by much.

Figure 10.7 shows that the presence of a clutch of eggs impedes cooling even further. This was also reported by Frost and Siegfried (1974) for eggs of the moorhen (*Gallinula chloropus*). It seems that at least part of the role of the parent in maintaining egg temperature is to shelter eggs from wind.

Thermal considerations of the incubating adult

The role of nest insulation

The role of the nest in creating a comfortable thermal environment for the incubating bird is so well understood that in the past it was almost unquestioned in a critical way (Drent 1975). The insulation provided by the nest and its contribution to the incubating bird has been quantified in physical terms by only a few investigators (Whittow and Berger 1977; Walsberg and King 1978a, 1978b; Skowron and Kern 1980; Webb and King 1983a; Kern 1984; Prinzinger 1992). However, many indirect observations indicate the importance of the nest to the energy balance of the parent (and nestlings). Most research was performed in cold or temperate climates, hence the scientific literature is biased towards effects of cold, and there are very little data on the effects of hot climates. The rate of cooling of a warm water flask inside nests was used by Palmgren and Palmgren (1939) to demonstrate change in insulation with the latitude of the nests. Similar correlations of insulation with altitude has also been demonstrated (Wagner 1955; Corley-Smith 1969; Kern and Van Riper 1984). Parental energetic savings from 15 to 65% in comparison to an exposed bird were calculated (Calder 1973b; White and Kinney 1974; Webb and King 1983a; Ackerman and Seagrave 1984; Prinzinger 1992).

An example of the difference in parental oxygen consumption (heat production), outside and inside the closed nest of the 7 g Palestine sunbird is given in Figure 10.6. The ratio of heat production outside the nest to that inside the nest, at ambient temperatures below the thermoneutral zone, was 1.48. The dry thermal resistance ratio (without evaporation) calculated for the same conditions (bird + nest *versus* bird alone), was 1.37. The energy saving was 32.4% ([0.48/1.48]100). The corresponding increase in thermal insulation due to nest presence was 27% ([0.37/1.37]100) and the nest-to-bird thermal resistance (insulation) ratio was 0.37.

Parental energy-saving and egg temperature regulation may act in opposition to each other, particularly in cold ambient temperatures (Vleck 1981a). The conflicting demands must be optimised or compromised. In fact, most birds choose moderate incubation seasons, sheltered nest sites and insulating materials for their nests. As a result, no consistent change in parental energy expenditure is evident during incubation compared with a non-incubating bird under the same measuring conditions. Prinzinger (1992), who reviewed 30 references, lists 19 species where energy expenditure was increased and 18 species where it decreased during incubation. The correlations found between nest attentiveness and either cold or hot environmental temperatures, indicate that at least small birds tend to minimise lengths of incubation sessions (Kendeigh 1963; Walsberg and King 1978a; Carey 1980; Biebach 1981, 1986; Haftorn and Reinertsen 1982; Williams 1993b; Figure 10.5). This may indicate that the conflict birds face is, in fact, the competition between time investment for parental care and for self maintenance (Drent 1972).

A simplified physical model for required nest insulation

Let us assume the following (Ar and Yom-Tov 1985): (1) the nest is built mainly to serve as an isolating 'cloth' around the incubating bird and eggs; (2) in closely fit nests, the temperature of the contact area between the bird and the inner side of the nest represents nest temperature. However, at the bottom of the nest, the eggs are 'stuck' in between. Therefore, the mean surface temperature of eggs in contact with the parent may be regarded as representing nest temperature (T_n); (3) the 'dry' heat ($\dot{H}$) produced by the bird (rate of metabolic heat production minus latent heat of evaporation) is dissipated through the nest material (at least in closed nests); (4) at night, ambient temperature (T_a) is below the thermoneutral zone of the incubating parent; and (5) attentiveness is 100%. Then, for a bird with body temperature T_b, the following simplified equation may be written for a steady state situation:

$$\dot{H} = 1/R_b(T_b - T_n) = 1/R_n(T_n - T_a) \tag{10.1}$$

where $1/R_b$ and $1/R_n$ are minimal body heat conductance and nest heat conductance respectively. R_b and R_n are the corresponding heat resistances (or insulations). It follows that:

$$(T_b - T_n)/(T_n - T_a) = R_n/R_b \tag{10.2}$$

If averaged values are used for the temperatures involved, such as $T_b = 40°C$, $T_n = 36°C$ and $T_a = 24°C$, then $R_n/R_b = 0.33$. This value is similar to the one found for the Palestine sunbird (see above). Moreover, since the temperatures used may be considered almost universal then the R_n/R_b ratio must be universal too, provided that the nest is built to provide insulation. Since R_b is different among species, as heat conductance of birds scales with body mass (Schmidt-Nielsen 1984), it appears that nest resistance (or its heat conductance) scales with bird mass in the same way as bird resistance (or its conductance). In any event, it seems that nest insulation is of the same order of magnitude as body insulation. Indeed Prinzinger (1992) has obtained figures for the blackbird showing that specific nest conductance and bird conductance are very similar. This should be also true for the area-specific insulation if birds are geometrically similar to their nests (simplified to a sphere or hemisphere).

The exposed area of the brood patch, where local R_b is minimal, is protected by the local R_n, where the nest wall is sometimes lined with feathers of the moulted brood patch. Hence the eggs are located between these two resistances. This may explain the similar temperature gradient across eggs early in incubation. A gradient of 2–4°C is found in small and large eggs alike, due to the fact that brood patch temperature and ambient temperatures during the incubation season are similar among species (Drent 1975; Swart and Rahn 1988; Groscolas *et al.* 2000; Figure 10.3).

Nest humidity

Absolute nest humidity is one of the factors determining water loss from eggs. Others are egg temperature (which in turn determines the saturated water vapour pressure under the shell), and eggshell water vapour conductance (Ar 1991a, 1991b). It is by now common knowledge that excessive or insufficient water loss both interfere with normal embryonic development (Robertson 1961; Ar and Rahn 1980; Meir and Ar 1986). Nest humidity, although important for successful development and hatching (e.g. Ar and Rahn 1980; Visschedijk *et al.* 1980), is not usually regulated actively and nest humidity parallels ambient humidity (Walsberg 1980, 1983a; Andersen and Steen 1986; Ar 1991a; Kern and Cowie 2000). Only a few cases have been recorded where birds actively wet their nest or eggs (Howell 1979; Schardien and Jackson 1979; Grant 1982). In an empty nest, very little humidity difference, if at all, exists between the interior and the exterior of the nest (Brown 1994). It was suggested that birds regulate nest humidity (Rahn *et al.* 1976, 1977a; Morgan *et al.* 1978). In fact, birds regulate humidity in the nest only by 'choosing' the 'right' nesting season and location, material and construction and simply by being part of the system. 'Being part of the system', or 'the bird-nest incubation unit' (Chapter 1), requires some explanation. The incubating parent heats the eggs and thus increases the water vapour gradient between the eggs and nest atmosphere, thereby augmenting egg water loss. At the same time, however, the parent covers a fraction of the egg with its brood patch, consequently preventing part of the egg from 'participating' in losing water vapour into the nest. For the covered fraction of the egg, humidity is at saturation (Ar and Yom-Tov 1985). The parent adds humidity to the nest by its own respiratory water loss and skin evaporation (Brown 1994) again demonstrating a built-in conflict in attentiveness behaviour. The concept of 'nest humidity' must include all these factors.

One explanation for the lack of 'sensitivity' to humidity of the parent is the fact that, at a relatively high egg temperature, the water vapour pressure head inside the egg is high. In cold environments, it is significantly higher than that of even fully saturated cold air, and even higher at less than 100% saturation (Ar 1991a). Thus, in order to prevent excess water loss, the eggshell itself is adapted by having a relatively low water vapour conductance as compared to reptiles and is adjusted to climatic conditions (Ackerman *et al.* 1985; Ar and Rahn 1985; Ar 1993; Chapter 16). Water loss and resulting water deficiency are determined purely by physical factors, over which the developing embryo has no control. This is in contrast to oxygen requirements for which embryonic physiological adaptations can occur. These facts may explain why the eggshell is mainly adapted to prevent excessive water loss, for instance, at altitudes where air is relatively dry and vapour diffusivity is high (Rahn *et al.* 1977b; Ar 1993; Chapter 16).

Best estimates of nest humidities, averaged over time, have been obtained gravimetrically by recording water loss from eggs in the nest, whose water vapour conductance had been determined beforehand in the laboratory. These eggs are either intact or emptied and filled with a desiccant (e.g. Rahn *et al.* 1977a; Rahn and Hammel 1982; Vleck *et al.* 1983; Swart and Rahn 1988). More sophisticated methods employing 'electronic eggs' have also been used (Howey *et al.* 1984). Long-term measurements with 'egg hygrometers' average all the above mentioned influences on the eggs. Using such measurements, it was found that the 'effective' (for the eggs) nest humidity is higher than that of the ambient air (Rahn *et al.* 1976; Rahn and Dawson 1979; Vleck *et al.* 1983). Rahn *et al.* (1977a) who used correctly absolute humidity values (water vapour pressures) to point out the direction of the gradients from the egg to the ambient, give a range of calculated nest humidities of 9–25 Torr.

When the rate of water loss in the nest (assuming that all water loss comes from the eggs) and the partial pressure difference of water vapour between the nest and the ambient air are measured, a value for 'nest ventilation' can be calculated (Rahn *et al.* 1976; Rahn and Dawson 1979; Vleck *et al.* 1983). Actually, it is not clear how much of this 'ventilation' is really convective and how much of the water vapour is dispersed by diffusion (Howe and Kilgore, 1987; see below). Here too, it is logical to assume that the presence or absence of the incubating bird in the nest and its incubation behaviour influence the effective humidity and thus a critical parameter determining water loss from the eggs. Averaged results show that the water vapour gradient between the eggs and the free atmosphere is divided such that about two-thirds of it is between the eggs and the nest, and one-third between the nest and the ambient air (Rahn *et al.* 1976; Vleck *et al.* 1983; Swart and Rahn 1988). However, this is based on egg water loss measured in the nest and water vapour conductance measured in the laboratory. As a result, by not taking into account that part of the shell that is covered by the brood patch, the water vapour conductance and the water vapour pressure in the nest are overestimated. For a 20% eggshell coverage, the gradient is recalculated to be divided equally between egg/nest and nest/ambient air. This leads to an interesting conclusion that on the average, eggshell water vapour conductance is similar to nest water vapour conductance.

Nest gas composition

The conclusion that eggshell and nest conductances are similar can be easily applied to respiratory gas cascades between the eggs and the ambient air through the nest. Towards the end of incubation, the carbon dioxide pressure (P_{CO_2}) under the eggshell is *circa* 40 Torr (Hoyt and Rahn 1980; Ar and Rahn 1985). This is also the total pressure difference to the ambient air. If this gradient is equally divided between the egg/nest and the nest/ambient air gradients, as is the case for water vapour, then one would expect a P_{CO_2} of 20 Torr for carbon dioxide in the nest at the end of the incubation. For a 'nest' which includes the incubating parent, P_{CO_2} should be sub-divided again between 40 Torr under the brood patch in contact with 20% of the shell and 16 Torr in the free atmosphere around the rest of the egg. The latter value is similar to the highest values found in the nest of bank swallow (*Riparia riparia*; Birchard *et al.* 1984), in the nest chamber of the European bee-eater (*Merops apiaster*; Ar and Piontkewitz 1992), in the nesting hole of the Syrian woodpecker (*Dendrocopus syriacus*; Mersten-Katz 1997), and in the nest of the domestic hen (Burke 1925, cited by Walsberg 1985). However, higher concentrations of carbon dioxide have been found within the cavity nest of a roosting pygmy nuthatch (*Sitta pygmaea*; Hay 1983, cited by Howe *et al.* 1987). The P_{CO_2} of 16 Torr is below the critical value affecting embryonic mortality shown in a series of reports in artificial incubation of fowl eggs (Taylor *et al.* 1956, 1971; Taylor and Kreutziger 1965, 1966, 1969). Lower maximal concentrations were found in rhinoceros auklet (*Cerorhinca monocerata*) and burrowing owl (*Athene cunicularia*) burrows (Birchard *et al.* 1984). The large variation in carbon dioxide concentrations found within nests and between species can be explained by the degree of exposure to wind and the differences in nest attentiveness.

Importance to eggs

Experimental manipulations of gas composition around incubated eggs, either acute or chronic, have demonstrated that an atmosphere as close to free air composition at sea level as possible is compatible with successful development and hatchability (Smith *et al.* 1969; Visschedijk 1980; Girard and Visschedijk 1987). As is the case with the level of water vapour in the nest, gas composition in the nest is partly influenced by the inhabitants of the nest. Eggs, incubating parent, hatchlings, commensals and parasites, all contribute to the reduction in oxygen and the increase in carbon dioxide in the nest. While the incubating parent is capable of increasing its lung ventilation, the eggs, which rely on gas diffusion through the shell, are thus dependent on partial pressures of respiratory gases in the nest. Among all nest inhabitants, the adult bird has the most significant influence on nest atmosphere (e.g. Howe and Kilgore 1987; Ar and Piontkewitz 1992), except for the nests of megapodes where microbial activity in nest material changes gas composition (Chapter 13). The absence of the parent, free convection and the action of wind, aerate the nest mainly during off-nest periods.

Role of parent and nest ventilation

The frequency of nest ventilation in black-footed and Laysan albatrosses (*Diomedea nigripes* and *Diomedea immutabilis*) was 0.77 and 1.30 times h^{-1} respectively (Grant *et al.* 1982). Several cases of behavioural nest ventilation have been observed in hole and tunnel nesters (Wickler and Marsh 1981; Howe and Kilgore 1987; Ar and Piontkewitz 1992). Ar and Piontkewitz (1992) were able to show that excursions through the long tunnel leading to the nesting chamber of European bee-eaters in and out of the nest, force-ventilate the chamber actively and maintain much higher oxygen and lower carbon dioxide partial pressures in it, as compared to the calculated values in the absence of ventilation. As the hatchlings grow and consume more oxygen, the frequency of feeding trips of the parents increases, augmenting active nest ventilation. The Syrian woodpecker (*Dendrocopos syriacus*), which nests and roosts in tree cavities, was observed to move up and down the cavity several hundred times each night, actively ventilating the cavity. In still air, and without these up-and-down ventilatory movements acting as a piston, a single roosting male would reduce the oxygen in this confined atmosphere to 18.4% (Mersten-Katz, Barnea, Yom-Tov and Ar unpublished).

As mentioned above, wind has a strong influence on nest ventilation. The orientation of the nest, or, in closed nests, of its opening, in relation to wind direction, are of significance (e.g. Ricklefs and Hainsworth 1969a; Facemire *et al.* 1990; Sidis *et al.* 1994; Chapter 2). Wind seems to improve internal gas conditions in the nesting chamber at the end of the tunnel of the European bee-eater (White *et al.* 1978).

Brood patch coverage conflict

Most research conducted on eggs in the laboratory has disregarded the fact that a significant area of the egg surface is covered during natural incubation by the brood patch (Rahn 1984; Yom-Tov *et al.* 1986; Handrich 1989b). However, as is the case with water loss from the eggs (see above), the parent which covers a fraction of the egg with its brood patch prevents part of the eggshell from exchanging gas with the nest atmosphere. This must reduce the 'effective' gas conductance of the shell.

Possible advantages of covering parts of the shell have not been investigated. It is speculated that reducing the 'activity' of oxygen in the young embryo adjacent to the covered shell may prevent uncontrolled oxidation and the activity of oxygen free radicals in the embryo (Ar and Mover 1994). Experimental eggshell covering has imposed water and gas exchange imbalance (Wakayama and Tazawa 1988; Turner 1990, 1991). Application of the brood patch to the egg may induce hypoxic hypercapnia. Ar and Girard (1989) have shown that late in incubation, lateral diffusion of oxygen under an area covering the shell is limited and embryonic gas exchange is altogether abolished from the centre of a covered area of about the size of a brood patch. Embryonic oxygen uptake is reduced even when the covered area is small; Burton and Tullett (1985) showed that taping eggshells of domestic poultry species reduced oxygen consumption and embryonic growth. It is also reduced in experimentally low oxygen atmospheres at altitude (Visschedijk *et al.* 1988; Ar *et al.* 1991). On the other hand, even under normal conditions, local oxygen pressures in the early embryo approach zero, due to diffusive uptake in the absence of circulation (Lomholt 1984). Later in development, after circulatory vessels have formed, and presumably mechanisms to offset oxygen free radical damages have developed, gas permeability of the shell above the embryo increases by an order of magnitude (Tullett and Board 1976; Paganelli *et al.* 1978; Lomholt 1984) and gas pressures in embryonic tissues increase somewhat (Tazawa and Mochizuki 1977).

Parent–embryo communication

It is not known whether there are avenues of communication other than heat exchange between the embryo and its incubating parent and foetal vocalisation before hatching (Vince 1966a; Chapter 7). Ar *et al.* (2000) showed that embryos of several species emit the gas nitric oxide (NO), which is known to regulate capillary blood flow by its vasodilator effect (e.g. Ignarro 1981). Since communication between bird and embryo can occur only through gas filled pores in the shell, other vascular tone regulators are excluded. Moreover, NO emission from eggs was found to be augmented in hypoxia (Figure 10.8) and reduced by hypothermia. The blood flow through the brood patch is far beyond the local respiratory needs of the tissue and may be considered close to arterial blood in its gas composition. It can provide a relatively high head pressure for oxygen, which presumably may help to maintain gas exchange across the shell under the brood patch. This idea merits further research.

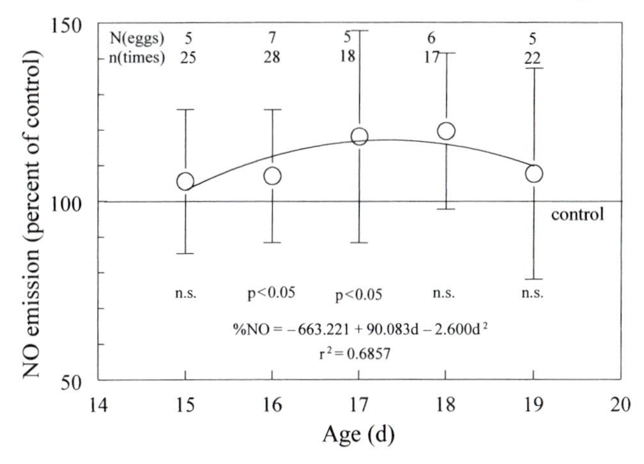

Fig. 10.8 Relative nitric oxide (NO) emission (± SD) from intact chicken embryos (days 15–19 of incubation) after 10 min exposure to hypoxia (14.1% oxygen). 100% represents emission of *circa* 0.01 nmol h^{-1} of NO. n(times) is the number of repetitions for N eggs. Data from Ar *et al.* (2000).

Egg turning

Deeming (1991) pointed out a fundamental difference between reptilian and avian eggs during incubation: While reptile embryos can die if eggs are turned, turning is a necessity for successful hatching of avian embryos. Thus, one of the important roles of the incubating parent is to turn the eggs in the nest, and this has become an essential part of the incubation behaviour (Drent 1975). It is interesting to note that the critical period (for turning in birds and lack of turning in reptiles) is the first part of the incubation. It is not the scope of this chapter to review the anatomical changes and the physiological consequences for the embryo in unturned eggs (see Chapter 11). However, there might be a link of the latter to nest microclimate: Wittman and Kaltner (1988) reported a reduced amount of sub-embryonic fluid in unturned Japanese quail (*Coturnix c. japonica*) eggs. Tazawa (1980b) found that albumen absorption was retarded in unturned chicken eggs. Deeming (1989c) shows that the reduction in sub-embryonic fluid (Deeming *et al.* 1987; Babiker and Baggott 1992) is carried over to the allantoic and amniotic fluids as well being expressed in the smaller than usual embryo.

All of these observations indicate a common source, the accumulation of the sub-embryonic fluid from the albumen by the embryonic disc during the first days of the development. This active accumulation requires active transport of ions from the albumen through the embryonic disc to the sub-embryonic space (Ar 1991b). The amount of ions in the albumen adjacent to the upper surface of the embryonic disc available for the absorption process at this stage may limit water transport, because the young embryo floats to the surface of the egg and is separated from the shell membranes only by a thin layer of albumen. In addition, due to the egg-to-nest humidity difference, this albumen continues to lose water vapour, which can cause a local shortage of available water if the egg is not turned. In reptile and megapode nests, where evaporation is limited, eggs do not have to be turned.

Recently, it has been found (Ar and Reizes unpublished) that the application of an artificial brood patch to fowl eggs for 12 and 24 h in the second half of the incubation causes a local reduction of *circa* 20% and 40% respectively in the number of blood vessels per unit area of CAM. This too may aid in explaining the role of egg turning behaviour as a requisite for normal embryonic development.

The nest microclimate: further research

There are four requirements needed for the completion of successful embryonic development in the avian nest. The appropriate temperature, humidity, respiratory gas composition and egg turning, are either provided, or modulated, by the nesting and incubation behaviour of the parents. This behaviour has evolved such that the parents choose the time and place to build an appropriate nest suited to the climatic conditions around it. The behaviour of the incubating parent(s) is either compromised or optimised, to confront the conflicts between their own needs and those of reproduction.

An important principle, which species and individuals seem to obey for long term survival through evolutionary processes, is the optimisation of a positive and efficient energetic cost–benefit ratio. Optimisation rules, if applied, narrow the number of solutions that birds can use for successful reproduction. Appropriate nest microclimate is one of the factors that have to be optimised at a rather narrow range. Understanding the relevant common features of nesting, beyond a simple description of the nesting events, through experimentation and research, may help creating appropriate management models, which would help both wildlife management and efficiency in the poultry industry.

Acknowledgements

Parts of the work were supported in part by grant 460/97 of the Israel Science Foundation – The Israel Academy of Sciences and Humanities to A. Ar; by Grant 88–00113 of the United States-Israel Binational Science Foundation to A. Ar and R. A. Ackerman; by a grant from the Inter-University JNF to C. Mersten-Katz; and by grants from the Ministry of Agriculture and Israel Poultry Council to A. Ar and M. Meir.

The skilled technical assistance of Ms S. Greenberg is acknowledged. We are in debt to Ms A. Belinsky for helping throughout the work and the preparation of the manuscript.

11 *Patterns and significance of egg turning*

D. C. Deeming

Turning, shifting or jabbing, is the process in which the egg is moved around within the nest changing both its location and, through rotation, its orientation. It is considered to be an integral part of successful incubation in almost all species of bird (Poulsen 1953) although as will be shown it is not a universal behaviour. This chapter reviews the process of egg turning during incubation using qualitative descriptions of the behaviour together with quantitative records of rates and angles of turning. Theories explaining the need for turning are discussed and a start is made in examining the evolutionary significance of this behaviour in avian incubation. This chapter does not deal with other behaviours during incubation which involve movement of eggs, e.g. retrieval of an egg that has rolled from a nest. Such activities are described in Chapter 6.

There are two main techniques employed to study egg turning during incubation. First, there is direct observation of a bird (or birds) sitting on the nest (e.g. Beer 1961; Caldwell and Cornwell 1975) although video recording of the incubation period (Pulliainen 1971, 1978) can make analysis easier. Second, many studies employ electronic or telemetric eggs (Howey *et al.* 1984; Boone and Mesecar 1989; Stetten *et al.* 1990; Gee *et al.* 1995) which employ position sensors or tilt switches in an artificial egg to record the time and angle of position changes. Restrictions on the smallest size of these 'eggs' tend to limit the species in which they can be employed. Temperature sensors have been used to record nest temperatures indicate when the incubating bird rises from the nest, often to leave to forage (Haftorn 1966, 1978). Although these events can often correspond to turning events (Howey *et al.* 1984) the latter do not always occur at every re-settling on the nest (Beer 1961; Drent 1970). Similarly, egg turning events can occur during incubation bouts without the bird leaving the nest (Epple and Bühler 1981). Although it would seem to some (Freeman and Vince 1974) that monitoring temperature can give an indication of rates of egg turning, it has to be considered as unreliable in this respect.

The process of egg turning

Egg turning prior to incubation

It is well known that egg turning occurs after the onset of incubation but there are reports of turning during egg laying. Derksen (1977) reported that eggs of the

Adélie penguin (*Pygoscelis adelie*) are turned prior to incubation. Potter (1989) showed that the first egg laid in a cardinal (*Cardinalis cardinalis*) nest was deliberately turned by the female before incubation was initiated. Sustained turning behaviour before completion of the clutch has also been observed in the blue tit (*Parus caeruleus*) where the rate was 8 turns h^{-1} (Jefferies personal communication). There is a need for more research to investigate the incidence of turning during clutch formation to clarify whether these reports are typical of passerines or of all birds.

Behaviour patterns

Brua (1996) defined turning as 'moving of the eggs by reaching beneath them with the bill' and this seems to be a useful working description of the process. The many descriptions of the actual process of egg turning largely follow the same pattern. In the mallard (*Anas platyrhynchos*) for example, the duck stands up from the nest and straddling the eggs or standing on the side of the nest, she pokes her bill between the eggs (McKinney 1951). Using the ventral aspect of her bill she pulls one egg towards the middle of the clutch thereby shifting the position of all of the eggs relative to each other (Caldwell and Cornwell 1975). The process of turning certainly moves eggs from one side of a clutch to the other (Stewart 1971). Willow grouse (*Lagopus lagopus*) and capercaillie (*Tetrao urogallus*) hens also use the ventral aspect of the beak and the front of the neck to move one or two eggs from the periphery to the centre of the clutch (Pulliainen 1971, 1978). The beak and lower facial disc is used to turn eggs in the owl (*Tyto alba*; Epple and Bühler 1981). In the black-billed gull (*Larus bulleri*) the egg is rolled against the feet by the ventral aspect of the bill and the egg usually turns about its short axis to lie with its long axis parallel with the long axis of the bird (Beer 1965). Similarly the gannet (*Morus bassanus*) uses the lower edge of its closed bill pointing backwards to move the egg between the webs of its feet (Nelson 1965). In the least bittern (*Ixobrychus exilis*) turning is achieved by simply 'jabbing the bill' into the nest cup (Weller 1961). 'Tremble-thrusting' is a common behaviour in titmice (*Parus* spp.) and the bluebird (*Sialia sialis*) where the female moves the eggs by pushing her beak with a trembling motion into the nest cup (Hartshorne 1962; Haftorn 1994). Haftorn (1994) reported that egg turning, where the bill is ploughed through the eggs from the periphery to the centre, was an independent, but relatively rare behaviour.

Gibson (1971) reported that turning of eggs by the American avocet (*Recurvirostra americana*) was achieved by the feet but did not describe the behaviour in detail. Handrich (1989a) suggests that erosion of the organic cuticle on the eggshells of king penguin (*Aptenodytes patagonica*) eggs was due to egg rotation between the feet and body as the incubating bird moves around. Egg turning with the bill was not observed in this species. In the Adélie penguin turning is achieved by the feet and by the bill (Derksen 1977). In black vultures (*Coragyps atratus*) the eggs are incubated on the toes and are frequently moved around the incubation site (Stewart 1974). No turning behaviour with the bill was observed and Stewart (1974) assumed that turning was achieved as the eggs were rolled around. Other reports suggest that turning is possibly achieved by the feet, e.g. Weaver and West (1943) for the siskin (*Spinus pinus*) and Skutch (1976) for a hummingbird, but the nature of incubation does make it hard to confirm whether this is true in many species. Shuffling of the eggs whilst re-settling on the nest was considered by Caldwell and Cornwell (1975) to have little effect on egg position in the mallard and did not contribute to turning. By contrast, swivelling on the nest by an incubating ostrich (*Stuthio camelus*) was considered by Sauer and Sauer (1966) to turn eggs and use of the bill to move eggs was limited to times when the male and female birds changed incubation duties. Considerable investigation is required to determine the role that movement of the feet plays in egg turning. The possibility that turning is achieved by the feet without rising from the nest could mean that in many species turning frequency could be much higher than has been reported from observations of behaviour.

Tinbergen (1953) found that the timing of turning was usually at the arrival at the nest although the incubating herring gull (*Larus argentatus*) would spontaneously rise and turn the eggs without any apparent external stimulus. Many reports of incubation behaviour report a similar situation and it is likely that this pattern is common in most species of bird but further research is required in order to confirm this.

Egg position during incubation

At oviposition the egg exhibits a neutral orientation because that the two ends of the egg contain albumen. Some studies of egg turning in the nest have revealed that egg orientation is affected by egg movement. Lind (1961) showed in the black-tailed godwit (*Limosa limosa*) that egg orientation was affected by the time of incubation. Turning appeared to release the egg from contact with adjacent eggs allowing the egg to adopt a position with the broad-end (air space) uppermost. As incubation proceeds the increasing weight asymmetry associated with the development of the air space means that the egg is naturally tilted. This effect was confirmed by Drent (1970) in his study of herring gull eggs. Using a telemetric egg in a greater sandhill crane (*Grus canadensis tabida*) nest showed that the air cell was rarely downwards and most of the time the egg normally had one of its four sides or the air cell end upwards (Gee *et al.* 1995).

The movement of the egg during turning has been reported to be via rotation around the long axis in waterfowl (Howey *et al.* 1984) but via rotation around the short axis in black-billed gulls (Beer 1965). It is not clear which direction prevails or whether this is related to egg size or shape.

Caldwell and Cornwell (1975) found that the average angle of turn for the mallard egg was 61.2° about the long axis. However, the length of incubation affected the average angle of turn through which an egg was moved. Towards the start of development the average angle was 66.2° (day 5) but declined steadily to 54.1° at day 23 leading Caldwell and Cornwell (1975) to suggest that there was an increasing weight asymmetry in the egg. A telemetric egg showed that, in the domestic bantam fowl (*Gallus gallus*), the mean degree of turning was 52.7° although the range of angles varied from 5–175° (Smith and Fox personal communication). By contrast, Boone and Mesecar (1989) showed that in the domestic fowl (*Gallus gallus*) the mean hourly angle of turn was 62.4°. In captive and wild buzzard (*Buteo buteo*) nests the mean angles of turn were 64.1 and 50.0° although the ranges were 5–175° (Smith and Fox personal communication). Bassett *et al.* (1996) found that 50% of the turns of emu (*Dromaius novaehollandiae*) eggs measured 0–45°, 30% measured between 46–90° with the remaining turns were up to 135°. In the black-necked grebe (*Podiceps nigricollis*) the angle of turn can be up to 180° (Brua 1996). Mountain plover (*Charadius montanus*) eggs have a minimal angle of turn averaging around 85–90° but this varied between 25–125° but without any discernible pattern of change throughout incubation (Graul 1975).

Turning frequency during natural incubation

Reports describing egg turning behaviour are relatively common (e.g. Poulsen 1953; Beer 1961; Drent 1970, 1975; Epple and Bühler 1981) but quantitative data are not commonly reported. For example, Zerba and Morton (1983b) recorded egg turning in mountain white-crowned sparrows (*Zonotrichia leucophrys oriantha*) but they only reported that 'egg positions changed frequently and randomly'. Other reports are simply imprecise. Lancaster (1964) reported that male boucard tinamou (*Crypturellus boucardi*) turned their eggs 5–10 times a day whereas Van Ee (1966) reported turning events of 9–21 times per day for the blue crane (*Tetrapteryx paradisea*). Contrary to the impression made by reviews of incubation (Drent 1970, 1975; Skutch 1976) numerous records of the frequency of egg turning are to be found in the literature.

Eighty two reports of the average frequency of turning (turns h^{-1} during daylight hours) for sixty one species from fourteen orders of birds are shown in Table 11.1. Rates of turning vary considerably between the different species ranging from no turning in megapodes through to a turning rate of 12 turns h^{-1} for the cedar waxwing (*Bombycilla cedrorum*). With the exception of the Galliformes, there is usually a fair degree of similarity between birds within the same order of birds (Table 11.1). It is encouraging to see that studies of the same species of bird yield similar rates of turning particularly where in some species, buzzard and whistling swan (*Cygnus columbianus*), the research was done in wild and captive conditions (Smith and Fox personal communication; Hawkins 1986).

Comparison of turning rates in male and female birds shows little difference in some species, e.g. black-necked grebe, moorhen (*Gallinula chloropus)* and Clark's nutcracker (*Nucifraga columbiana*). In other species, it can be either the male (least bittern and hooded crane [*Grus monacha*]) or the female (white-naped crane [*Grus vipio*]) which turns the eggs more frequently (Table 11.1). Reasons for these differences in rates of turning are not immediately clear. More data is required for those species where incubation is shared so as to study on the effects of gender on turning in greater detail.

Egg turning is not a behaviour restricted to daylight hours and in a few species at least turning is known to continue throughout the night although precise rates of turning are not often available. Marks *et al.* (1994) report that long-eared owl

Table 11.1 Frequency of egg turning events, initial egg mass, incubation period, and albumen content for bird species organised on the basis of different classes of hatchling maturity (as defined by Nice 1962).

Species	Turning frequency (turns h^{-1})	Initial egg mass (g)	Incubation period (days)	Albumen (% of egg contents)	References#
Precocial 1*					
Apterygiformes					
Apteryx australis	0	350	75	31.0	1, 2
Galliformes					
Megapodius pritchardii	0	75	49	35.0	3
Megapodius freycinet	0	108	42	33.0	3, 4
Megapodius cumingii	0	112	63	65.0	3
Alectura lathami	0	169	49	49.9	5
Leipoa ocellata	0	179	62	47.4	5
Macrocephalon maleo	0	231	62	37.5	3
Precocial 2*					
Anseriformes					
Anas platyrhynchos	1.11	51	28	59.0	6, 4
Anas platyrhynchos	1.20	51	28	59.0	7, 4
Melanitta fusca	1.80	92	27.5	56.0	72, 4
Branta leucopsis	0.72	103	24	56$	8, 4
Anser anser	0.78	150	26	55.0	8, 4
Branta canadensis	1.20	220	29	54.5	9, 4
Cygnus atratus	0.75	235	36	43.0	8, 4
Cygnus columbianus (captive)	1.30	280	32	52$	10, 4, 11
Cygnus columbianus (wild)	1.50	280	32	52$	10, 4, 11
Cygnus columbianus bewickii	1.67	290	29	52$	12, 4
Cygnus cygnus	0.60	330	35	52$	8, 4
Cygnus buccinator	1.04	330	35	52$	70, 4
Galliformes					
Bonasa bonasia	1.38	19	25	55.6$	13, 4
Lagopus lagopus	1.30	20.1	21	53.0	14, 4
Lagopus lagopus	1.25	20.1	21	53.0	15, 4
Lagopus lagopus	1.11	20.1	21	53.0	16, 4
Lyrurus tetrix	1.04	36	26	55.6$	13, 4,
Tetrao urogallus	0.71	48	28	58.0	14, 4, 17
Tetrao urogallus	0.60	48	28	58.0	18, 4, 17
Charadiiformes					
Recurvirostra americana (female)	1.58	29	24	50$	19, 4
Recurvirostra americana (male)	1.15	29	24	50$	19, 4
Precocial 3*					
Galliformes					
Phasianus colchicus	1.00	27	25	61.0	20, 4, 21
Gallus gallus	1.33	60	21	65.0	22, 4
Gallus gallus	1.33	60	21	65.0	23, 4
Gallus gallus	2.88	60	21	65.0	24, 4
Precocial 4*					
Podicipediformes					
Podiceps nigricollis (female)	0.60	21	21	76.0	25, 4
Podiceps nigricollis (male)	0.82	21	21	76.0	25, 4

Table 11.1 *Continued.*

Species	Turning frequency (turns h^{-1})	Initial egg mass (g)	Incubation period (days)	Albumen (% of egg contents)	References#
Gruiformes					
Gallinula chloropus (female)	5.00	25	21	68.0	26, 4
Gallinula chloropus (male)	6.00	25	21	68.0	26, 4
Chlamydotis undulata macqueenii	0.82	53	21	69.0	73, 74
Chlamydotis undulata macqueenii	0.68	53	21	69.0	75, 74
Antropoides virgo	1.10	128	28	65.5$	67
Grus monacha (female)	0.49	166	29	65.5$	65,
Grus monacha (male)	0.80	166	29	65.5$	65,
Grus monacha (female)	0.16	166	29	65.5$	66,
Grus monacha (male)	0.88	166	29	65.5$	66,
Grus monacha	1.38	166	29	65.5$	67
Grus vipio (female)	2.14	190	30	65.5$	27, 4
Grus vipio (male)	0.75	190	30	65.5$	27, 4
Grus canadensis tabida	1.71	195	35	65.5$	28, 4
Grus grus	1.15	195	29	65.5$	67
Grus leucogeranus	0.50	197	27	65.5$	29, 4
Grus japonensis	1.13	217	33	65.5$	67
Semi-precocial*					
Charadiiformes					
Larus bulleri	1.88	38	20	70.7$	30, 4
Larus ridibundus	1.76	38	23	71.0	31, 4, 32
Larus atricilla	1.19	42.1	26	63	71, 4
Larus argentatus	0.50	93	30	73.5	33, 34, 32
Larus marinus (female)	0.30	116	29	72.0	35, 4
Larus marinus (male)	0.34	116	29	72.0	35, 4
Semi-altricial 1*					
Falconiformes					
Accipiter nisus	3.00	23	35	78$	20, 4
Falco peregrinus	1.29	43	31	76$	36, 4
Falco peregrinus	1.76	43	31	76$	64, 4
Buteo buteo (captive)	3.59	51	36	76$	24, 4
Buteo buteo (wild)	3.87	51	36	76$	24, 4
Ciconiiformes					
Ixobrychus exilis (female)	1.60	10.5	19	76.2$	37, 4
Ixobrychus exilis (male)	3.55	10.5	19	76.2$	37, 4
Ardea herodias	0.50	70	27	68$	38, 4
Semi-altricial 2*					
Sphenisciformes					
Spheniscus demersus	6.00	103	40	79.0	39, 34
Strigiformes					
Tyto alba	4.48	24	33	76.0	40, 34
Altricial*					
Pelecaniformes					
Morus bassanus	2.00	105	44	83.0	41, 34
Pelecanus erythrorhynchos	0.80	154	30	83$	42, 34, 43

Table 11.1 *Continued.*

Species	Turning frequency (turns h^{-1})	Initial egg mass (g)	Incubation period (days)	Albumen (% of egg contents)	References#
Apodiformes					
Cypsiurus parvus	0	1.2	19	N/A∇	44, 45
Passeriformes					
Setophaga ruticilla	7.50	1.2	11	77$	46, 4, 47
Parus caeruleus	7.07	1.2	14	75.2	68, 69
Dendroica discolor	6.10	1.3	12	N/A∇	48
Peucedramus taeniatus	4.00	1.5	11	78$	49, 34
Carduelis carduelis	6.20	1.5	12	69	50, 51, 52
Emphidonax traillii	5.00	1.7	12	N/A∇	53, 54, 55
Motacilla lugens	3.41	3	14	N/A∇	56, 57
Bombycilla cedrorum	12.00	3.2	12	N/A∇	58, 59
Habia gutturalis	2.08	3.8	13	N/A∇	60
Habia rubica	2.73	4.3	13	N/A∇	60
Turdus migratorius	1.57	6.3	14	74	61, 62, 34
Nucifraga columbiana (female)	3.33	9.1	18	74.5$	63, 34
Nucifraga columbiana (male)	3.53	9.1	18	74.5$	63, 34

*Hatchling maturity class as determined by Nice (1962). $Value used from closely related species within the same genus or family. ∇N/A – data not available. #References: 1 – Rowe (1978); 2 – Calder *et al.* (1978); 3 – Jones *et al.* (1995); 4 – Carey *et al.* (1980); 5 – Vleck *et al.* (1984); 6 – McKinney (1952); 7 – Caldwell and Cornwell (1975); 8 – Howey *et al.* (1984); 9 – Kossack (1947); 10 – Hawkins (1986); 11 – Limpert and Earnst (1994); 12 – Evans (1975); 13 – Skutch (1976); 14 – Valanne (1966); 15 – Andersen and Steen (1986); 16 – Pulliainen (1978); 17 – Cramp (1980); 18 – Pulliainen (1971); 19 – Gibson (1971); 20 – Welty and Baptista (1988); 21 – Campbell (1974); 22 – Chattock (1925); 23 – Kuiper and Ubbels (1951); 24 – Smith and Fox (personal communication); 25 – Brua (1996); 26 – Siegfried and Frost (1975); 27 – Walkinshaw (1951); 28 – Gee *et al.* (1995); 29 – Johnsgard (1983); 30 – Beer (1965); 31 – Beer (1961); 32 – Cramp (1983); 33 – Drent (1970); 34 –Sotherland and Rahn (1987); 35 – Butler and Janes-Butler (1983); 36 – Herbert and Herbert (1965); 37 – Weller (1961); 38 – Pratt (1970); 39 – Duncan (personal communication); 40 – Epple and Bühler (1981); 41 – Nelson (1965); 42 – Evans (1989); 43 – Evans and Knopf (1993); 44 – Lack (1956); 45 – Moreau (1941); 46 – Sturm (1945); 47 – Sherry and Holmes (1997); 48 – Nolan (1978); 49 – Lowther and Nocedal (1997); 50 – Conder (1948); 51 – Cramp and Perrins (1994); 52 – Arieli (1983); 53 – Holcomb (1969); 54 – Baicich and Harrison, (1997); 55 – Skutch (1997); 56 – Nakamura *et al.* (1984); 57 – Badyaev *et al.* (1996); 58 – Putnam (1949); 59 – Witmer *et al.* (1997); 60 – Willis (1961); 61 – Schantz (1944); 62 – Howell (1942); 63 – Mewaldt (1956); 64 – Enderson *et al.* (1972); 65 – Winter *et al.* (1996); 66 – Fujimaki *et al.* (1989); 67 – Winter *et al.* (1999); 68 – Jefferies (personal communication); 69 – Ramsay (1997); 70 – Cooper (1977); 71 – Impekoven, (1973); 72 – Afton (1977); 73 – Schulz, Schulz, Paillat, Gaucher, and Eichaker (unpublished); 74 – Anderson and Deeming (2001); 75 – Deeming, Paillat and Hémon (unpublished).

(*Asio otus*) eggs are turned 5–12 times per night but did not report the incidence of daytime turning. In willow grouse and capercaillie turning was evenly distributed throughout day and night (Pulliainen 1971, 1978). By contrast, in greater sandhill crane eggs a telemetric egg showed that turning occurred at a rate of 1.72 turns h^{-1} during daylight but dropped to 0.96 turns h^{-1} during the night hours (Gee *et al.*

1995). Similarly, in three species of waterfowl turning rate during the night was only around half of that recorded during the day (Howey *et al.* 1984). In the moorhen egg turning by the male during the night was only 2.7 turns h^{-1} compared to 6 turns h^{-1} during the day (Siegfried and Frost 1975). Further work is needed to determine the diurnal patterns in rates of egg turning in other species of bird.

There are a few reports that follow the rate of egg turning during the entire length of the incubation period. In the willow grouse the rate of egg turning was 0.77 turns h^{-1} during the first half incubation compared with 1.20 turns h^{-1} during the second half of incubation (Pulliainen 1978). By contrast, Valanne (1966) reported that egg turning in this species was more frequent during the first half of incubation (1.43 versus 1.12 turns h^{-1}). A similar pattern was observed during incubation of the capercaillie where turning rates was 0.88 turns h^{-1} during the first half of incubation but this fell to 0.52 turns h^{-1} during the second half of incubation (Valanne 1966). Brua (personal communication) observed that during the first week of incubation the turning rates of black-necked grebe eggs were 0.56 and 0.47 turns h^{-1} for male and female birds respectively. During the second week turning rates increased to 0.80 and 0.64 turns h^{-1} and by the final week of incubation had reached 1.00 turns h^{-1} for both genders.

Photographic records of pied flycatcher (*Ficedula hypoleuca*) eggs every hour showed that day of incubation affected the distance travelled by eggs suggesting that egg turning activity is greater during late incubation. Eggs travelled on average 10 mm h^{-1} on days 4 and 8 of incubation but increased to over 13 mm by day 12 out of an incubation period of 14 days (Figure 11.1; Kern unpublished data). Weaver and West (1943) did not report exact rates of turning in the siskin but stated that turning frequency increased during hatching. In the least bittern turning rates during incubation were 1.7–3.5 turns h^{-1} but this rose to 7.1–7.5 turns h^{-1} during hatching (Weller 1961). Telemetric eggs showed that turning rates are higher during hatching, which Howey *et al.* (1984) attributed to the movement of hatchlings within the nest. In the black swan (*Cygnus atratus*) eggs were turned between 20 and 30 times per day up to day 15 of incubation but there was a decline to around 10–15 times per day by day 40 before rising again during hatching (Howey *et al.* 1984). Turning during the hatching process increases embryo vocalisations (Chapter 7). Turning rate is also affected by vocalisation by the chicks during hatching (Chapter 7). A peeping black-necked grebe egg was turned 5–6 times more frequently than was observed in silent eggs (Brua *et al.* 1996).

The effects of the weather on egg turning has been studied by Winter *et al.* (1999). In cranes the rate of egg turning was reduced by a half to two-thirds during periods of rain compared with dry weather (Table 11.2). It would be very interesting to see whether hot or cold weather or rain affect rates of turning in other species.

Beer (1965) showed that clutch size affected turning rates in the black-billed gull. The typical clutch size of three eggs had turning rates of 0.5 turns h^{-1} but this increased to 1.02 and 1.34 turns h^{-1} in clutches of two eggs or one egg respectively. In four-egg clutches the rate of turning increased to 4.66 turns h^{-1}. It is not known whether this situation is common in gulls or in birds as a whole.

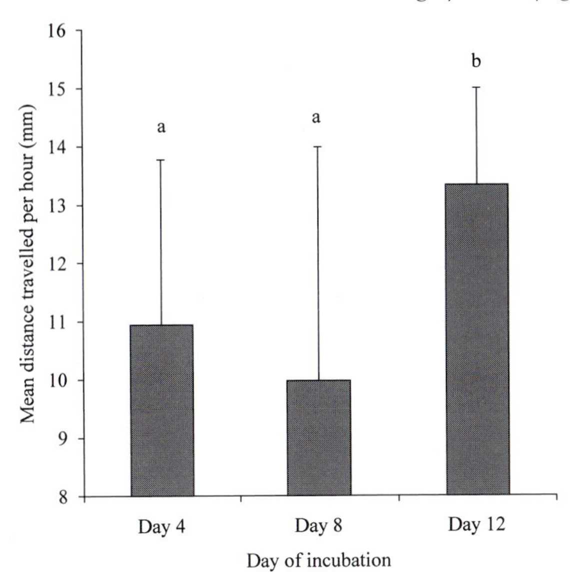

Fig. 11.1 Mean distance travelled per hour by pied flycatcher (*Ficedula hypoleuca*) eggs on different days of incubation (Kern unpublished data). Columns with different superscripts are significantly different at $p < 0.05$.

Table 11.2 Effects of rain upon the turning rates (turns h^{-1}) of four species of crane (Gruidae). Data from Winter *et al.* (1999).

Species	No rain	Rain
Antropoides virgo	1.10	0.32
Grus monacha	1.59	0.60
Grus grus	1.30	0.58
Grus japonensis	1.23	0.48

Birds which do not turn their eggs

With a few exceptions it is assumed that the vast majority of birds turn their eggs turning incubation. Megapodes are the most obvious example of birds that do not turn eggs, because the adult birds have lost direct contact with their buried eggs (Chapter 13). An incubating kiwi (*Apteryx australis*) cannot turn its egg due to the physical restrictions of incubating a large egg within a small burrow (Rowe 1978). Skutch (1976) also suggests that cramped incubation conditions in the nest of hermit hummingbirds (*Phaethornis* spp.) also precludes egg turning although it may be achieved by the feet. Palm-swifts (*Cypsiurus parvus*) lay a single egg which they glue to the nest which is fastened to the underside of a hanging palm frond (Moreau 1941; Lack 1956). Crested swifts (Hemiprocnidae) also attach their eggs to their nests with saliva and so the eggs remain unturned during incubation (Lack 1956).

Although not reported there are other birds which may not turn their eggs. For instance the fairy tern (*Gygis alba*) lays its single egg at the end of a tree branch. The precarious nature of this location appears to prevent much movement by the incubating adult (Dorward 1963; Niethammer and Patrick-Castilaw 1998) thereby possibly preventing turning. It is interesting that the incubation period in this species is uncharacteristically long for the mass of the eggs concerned. Other species of birds, e.g. the superb lyrebird (*Menura novahollandiae*; Lill 1986), and offshore and pelagic seabirds (Whittow 1980; Sotherland and Rahn 1987), have prolonged incubation periods during which eggs are often left unattended, and unturned, for long periods. Unfortunately there are no records of the rate of turning during periods of incubation by any of these birds although Grant *et al.* (1982) did record the frequency of nest ventilation in black-footed and Laysan albatrosses (*Diomedea nigripes* and *Diomedea immutabilis*) 0.77 and 1.30 times h^{-1} respectively. It would be interesting to see whether these rates always correlate with turning events.

Problems associated with studies of egg turning

Many ornithologists studying incubation find it difficult to record rates of egg turning because they have problems with interpretation of behaviours within nests and their role in turning of eggs. It is also unclear what role the feet play in egg turning. Egg turning has been studied by marking of eggs (Tinbergen 1953; Lind 1961; Drent 1970) but this has been seen to have adverse effects on turning behaviour and frequency. Holcomb (1969) demonstrated that passerines were able to distinguish a mark on the eggshell and turned the egg so that the mark was covered up for most of the time.

Artificial changes in clutch sizes in the black-billed gull affected the rates of turning (Beer 1965). An increase in clutch size from 2 to 3 eggs decreased turning frequency from 1.60 to 0.80 turns h^{-1}. Turning frequency in a four-egg clutch was 4.6 turns h^{-1} but this reduced to 0.6 turns h^{-1} once one egg was removed. By contrast, decreasing clutch size from 3 to 2 eggs increases rates of egg turning (0.4–3.2 turns h^{-1}). These experiments confirmed that egg turning was lowest in 3-egg clutches.

The data for turning frequency shown in Table 11.1 are derived from a variety of sources which vary in the rigour of investigation. Some studies involved observation of birds over long periods of time whereas other were for much shorter time intervals and the number of nests studied also varied from only one through to several. Some nests were studied through incubation whereas other studies were more restrictive. Only scientific studies are reported and every effort has been made to exclude others where the data may not be representative. For instance, one report of egg turning in the ostrich as excluded because it was based on only one pair of birds observed for two 10-hour sessions (Siegfried and Frost, 1975). This variability in the source of data cannot be helped and data are shown to indicate the values obtained in various studies in the hope that there will be more research into the rate of egg turning in birds.

Reasons for egg turning during incubation

Regular turning of eggs is known to be critical for normal development but it has received relatively little attention from researchers studying incubation in both nests and incubators. The bulk of the work on the turning requirements of eggs has been carried out on the domestic fowl with the aim of determining the optimal angle and frequency of turn so as to maximise hatchability in commercial operations (reviewed by Lundy 1969). As a result incubators employ turning systems which move eggs through 90° (45° either side of the vertical) every hour through to transfer into the hatcher (Deeming 1991; Wilson 1991).

An early explanation for egg turning involved re-distribution of the heat from the brood patch (de Reamur 1791; Caldwell and Cornwell 1975) but this has to be discounted because turning is still required in force-draught incubators where temperature gradients do not normally exist in eggs (Drent 1975). The usual explanation for egg turning has long been a need to prevent the embryo from adhering from the inner shell membrane (Dareste 1891; Eycleshymer 1906; Chattock 1925; New 1957). This explanation has been long accepted (Drent, 1975; Freeman and Vince 1974; Skutch 1976; Wilson 1991).

Whilst embryos in unturned eggs do adhere to the inner shell membrane, various studies into the effects of failure to turn eggs have shown significant effects on the physiology of domestic fowl and quail (*Coturnix coturnix*) embryos. Embryos in unturned eggs have reduced levels of sub-embryonic fluid (Deeming *et al.* 1987; Wittman and Kaltner 1988; Deeming 1989a; Babiker and Baggott 1992) due to diffusion gradients for sodium ions which develop within the albumen (Deeming *et al.* 1987; Latter and Baggott 1996, 2000b). Formation of sub-embryonic fluid is also being suggested as a reason for egg turning (Chapter 10). The rate of growth of the *area vasculosa* of the yolk sac membrane is slower in unturned eggs (Deeming 1989b) and the pattern of growth of the chorio-allantoic membrane is disrupted (Tullett and Deeming 1987). Failure to turn eggs also affects the volume of the allantois and amnion and slows the rate of embryonic growth (Tazawa 1980b; Tullett and Deeming 1987; Deeming 1989a, 1989b, 1989c, 1991). The oxygen consumption and heart rate of embryos in unturned eggs is also depressed compared with turned controls (Tazawa 1980b; Pearson *et al.* 1996). Furthermore, the levels of albumen proteins are reduced in the amniotic fluid after day 12 of incubation in the fowl in unturned eggs (Deeming 1991).

These results led to a suggestion that the physiological basis of egg turning lay in promoting the utilisation of albumen (Deeming 1991). It was argued that access to the water and protein in the albumen is critical for normal development. Indeed removal of albumen from turned fowl eggs at 3 days of development can simulate many of the symptoms of failure to turn eggs (Deeming 1989c). The prediction from this hypothesis was that eggs with differing amounts of albumen would require different rates of turning. Therefore, eggs producing altricial young, which have a high albumen content (relative to egg size), would require more turning than eggs from precocial species where relative albumen content is lower.

Deeming (1991) supported this hypothesis by pointing out that the eggs of megapodes and kiwis have low albumen contents, relative to egg size (Sotherland and Rahn 1987), and these eggs are not turned during development (Jones *et al.* 1995; Rowe 1978). Deeming (1991) also reports preliminary work which indicated that under artificial incubation parrot (altricial) eggs required more turning than guinea fowl (precocial) eggs. Harvey (1993) showed that during artificial incubation turning eggs of the night heron (*Nycticorax nycticorax*) once an hour appeared to prevent normal development of the yolk sac membrane and chorio-allantoic membrane (termed 'vein growth' as observed by candling) and no eggs hatched. Only when artificial turning was supplemented by turning by hand seven times per day was there sufficient membrane growth to support development through to hatching. This finding fitted the Deeming hypothesis because night heron eggs have an albumen content between 77 and 81% of the contents (Carey *et al.* 1980).

The data for turning frequency shown in Table 11.1 allow for a more rigorous testing of Deeming's (1991) hypothesis. Therefore, the reports of egg turning in nests have been supplemented with data on egg mass, incubation period and egg composition (Table 11.1) in order to test whether the frequency of turning during incubation is related to hatchling maturity and relative albumen content of bird eggs.

Factors affecting frequency of turning

For each species where turning frequency was recorded the following data were also collected (Table 11.1): (1) the category of hatchling maturity (Nice 1962), (2) initial egg mass (IEM, g), (3) the length of incubation period (I_p, days), and (4) where possible, the percentage of albumen in the contents of the fresh eggs (%Alb). In some instances it was not possible to find values for %Alb and data for closely related species (averaged from data collected for other species within the same genus or family but not the same order) were shown for comparison in Table 11.1. For some species, where egg composition data were unavailable even at the family level, %Alb data are not shown.

Prior to all analyses where more than one record for turning frequency was found for one species an average was taken. One-way analysis of variance on 60 species showed that class of hatchling maturity had significant effects on the frequency of egg turning during incubation ($F_{7,52} = 6.44$, $p < 0.001$; Figure 11.2a), IEM ($F_{7,52} = 4.94$, $p < 0.001$; Figure 11.2b) and I_p ($F_{7,52} = 23.11$, $p < 0.001$; Figure 11.2c). Analysis on %Alb was limited to those species where egg contents were known (28 species). Hatchling maturity also had a significant effect on %Alb ($F_{6,21} = 26.99$, $p < 0.001$; Figure 11.2d). Spearman's Rank correlations were employed to compare the relationships across those species where %Alb values were known. There were significant negative correlations between the frequency of turning and IEM, and I_p (Table 11.3). By contrast, turning frequency showed a significant positive correlation with %Alb (Table 11.3). IEM was significantly positively correlated with I_p, and significantly negatively correlated with %Alb (Table 11.3). There was a significant relationship between I_p and %Alb (Table 11.3).

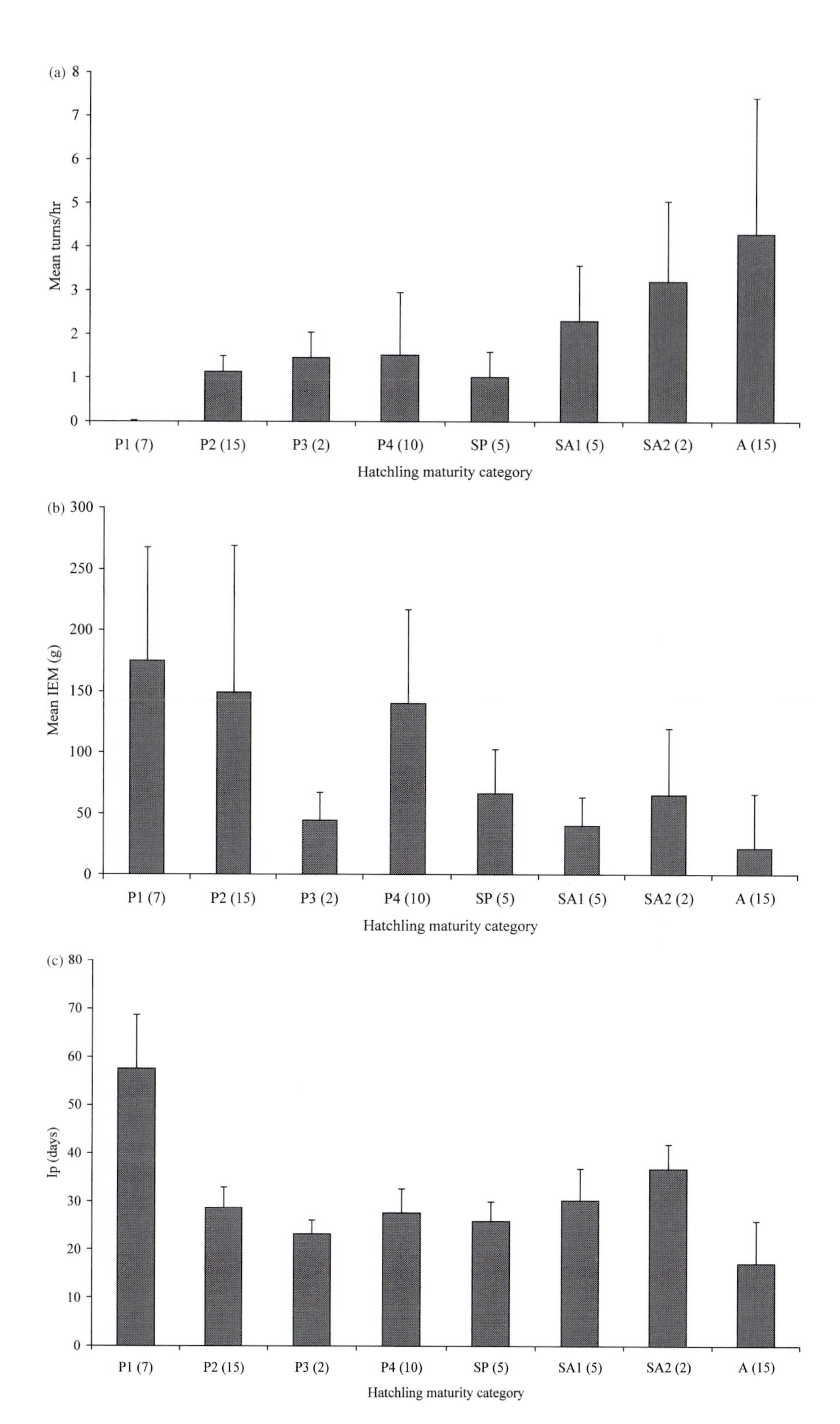

Fig. 11.2 *Continued.*

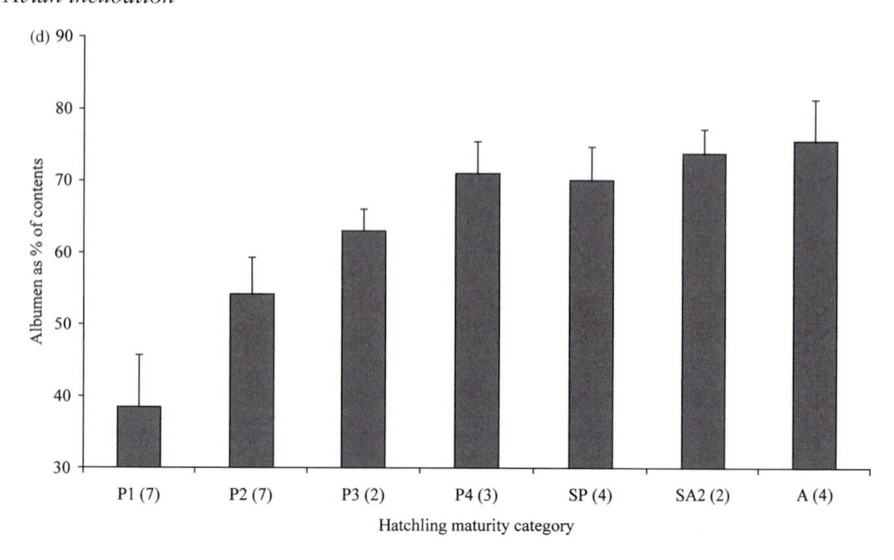

Fig. 11.2 (a) Turning frequency, (b) initial egg mass (IEM), (c) incubation period (I_p) and (d) albumen as % of contents, for different states of hatchling maturity (mean values + SD). Different symbols indicate hatchling maturity classes (P = precocial categories 1–4; SP = semi-precocial; SA = semi-altricial categories 1–2; A = altricial). Note that no data were available for % albumen content of eggs of the SA1 species. Numbers in brackets indicate number of species in group.

Table 11.3 Spearman's Rank correlation coefficients (r) for the relationships between frequency of turning, initial egg mass (IEM), incubation period (I_p) and relative albumen content (%Alb) for 28 species of bird. (DF = 26 and $p < 0.01$ in each case.)

	Frequency of turning	IEM	I_p
IEM	−0.583		
I_p	−0.544	0.793	
%Alb	0.622	−0.616	−0.532

It is interesting to note that some species in this sample showed a divergence from the relationship between IEM and I_p (Figure 11.3) as determined by Rahn and Ar (1974). Many species have shorter incubation periods than expected from the regression estimate. A few species have longer incubation periods than predicted but those with the largest residual from the regression model were the kiwi (*Apteryx australis*) and those species from the Megapodidae (Figure 11.3). Similarly the palm-swift has an incubation period of 19 days compared with 12.5 days predicted from its IEM.

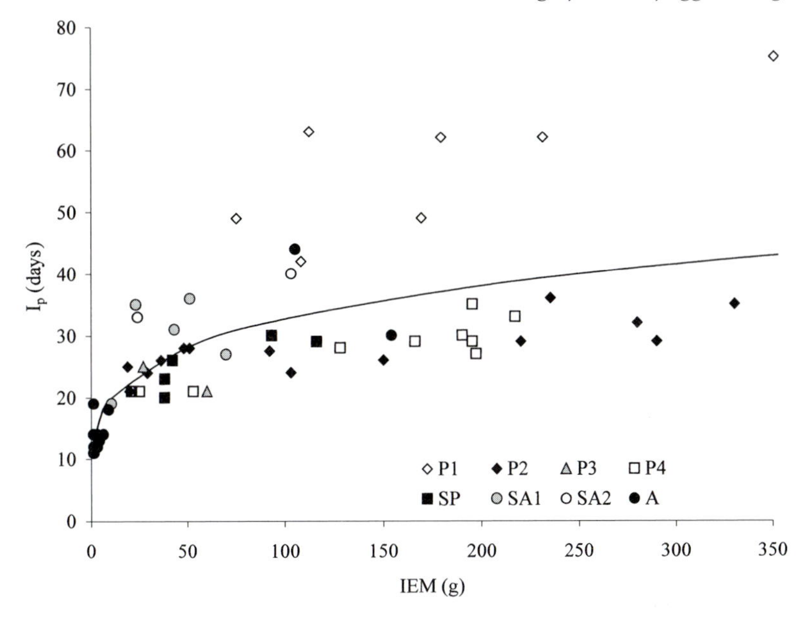

Fig. 11.3 The relationship between initial egg mass (IEM) and length of incubation period (I_p) for the species listed in Table 11.1. Different symbols indicate hatchling maturity classes (P = precocial categories 1–4; SP = semi-precocial; SA = semi-altricial categories 1–2; A = altricial). The line is derived from $I_p = 12.03 \text{ IEM}^{0.217}$ (Rahn and Ar 1974).

Phylogenetic relationships between the mean frequency of egg turning for each order of birds, based on the phylogeny of Cracraft (1988), are shown in Figure 11.4. Although it may seem that the Palaeognathe resemble reptiles in not turning their eggs, it is known that other ratites do turn their eggs (Lancaster 1964; Bassett *et al.* 1996) although turning frequency in the ostrich could be as low as 2–3 times per day (Siegfried and Frost 1974). It is fully accepted that the data presented here does not represent the broad range of avian orders but it is hoped that the research presented here will stimulate further research which will eventually complete the picture with regard to egg turning.

Evolutionary aspects of egg turning

The data presented in Table 11.1 broadly support the hypothesis that the rate of egg turning during natural incubation is correlated with the albumen content of the egg (Deeming 1991). Turning frequency increased with increasing albumen content of the egg and increasing immaturity of the hatchling. Given what is known about the morphological and physiological effects of failure to turn fowl eggs (see Deeming 1991), it is likely that albumen-rich eggs of semi-altricial and altricial species require high rates of turning in order to optimise membrane growth and fluid dynamics although experimental work is required to confirm this. Turning frequency is also correlated with IEM and I_p although this in part is probably due

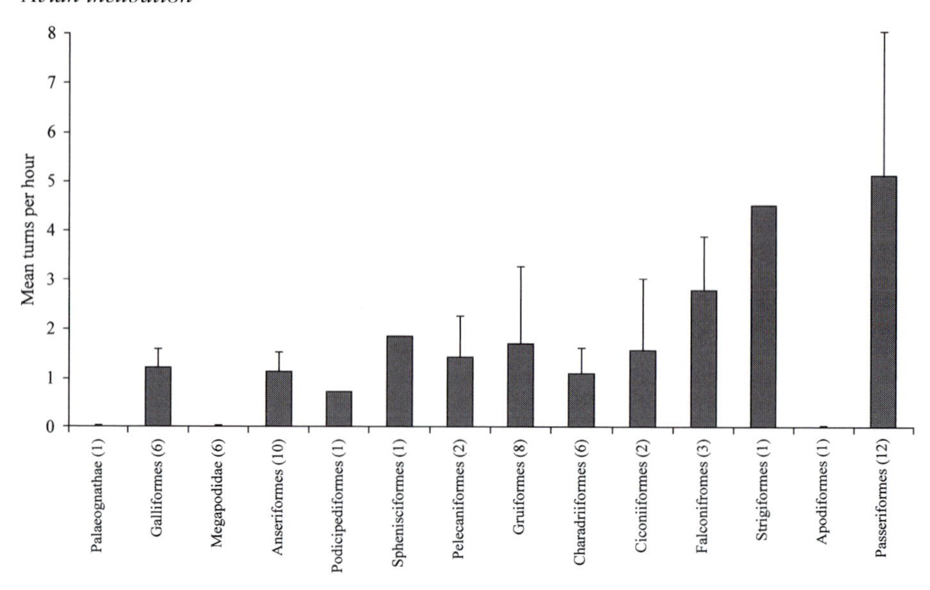

Fig. 11.4 Mean rates (+SD) of egg turning for different orders and families of birds based on the phylogeny of Cracraft (1988). Numbers in brackets indicate number of species in group.

to altricial eggs being smaller than precocial eggs (Vleck and Vleck 1987) and that I_p is highly correlated with IEM (Rahn *et al.* 1979). It is possible, however, that turning frequency is inter-related with egg mass and composition, and incubation period and it is interesting to speculate why egg turning evolved to be an integral part of avian incubation.

Extant crocodilians do not turn their eggs (Deeming 1991) and the embryo actually sticks to the inner shell membrane as a normal part of development (Deeming and Ferguson 1991c). It has to be assumed that the Archosaur ancestors of birds also did not turn their eggs during incubation because they were buried (Carpenter 1999; Chapter 1). The advent of contact between the adult animal and the eggs during incubation presumably started to increase the extent to which the eggs were moved as the animal moved around on the nest. Retrieval of eggs lost from the nest site as the adult settled down to incubate may have increased the movement of eggs. Such movement would have encouraged membrane growth and enhanced fluid metabolism would have optimised the rate of embryonic development and slowly begun to reduce the incubation period. One disadvantage of incubation is that it exposes the sitting adult to predation so any process which can speed up development will minimise this risk. As the role of albumen in avian development became more important (higher protein contents and greater volume) turning gave more of selective advantage by maximising the rate of embryonic development and so it became integrated in normal incubation behaviour.

Factors affecting developmental rate would include the maintenance of a high temperature for development. Contact incubation behaviour by the parents itself promotes maintenance of a stable and high thermal environment for development and enhances the survival of the eggs by offering some protection from predation. Evolution of altricial hatchlings would also serve to shorten the incubation period further because the birds are able to hatch at an earlier stage of development thereby releasing the parents from the constraints of contact incubation. Furthermore, the lower energy content of individual altricial eggs means that the investment in a clutch of eggs can be reduced and parental efforts can be concentrated in feeding hatchlings. The evolution of altricial young meant that the water content of the eggs had to increase and the role of turning in metabolism of this additional fluid became increasingly important. The high albumen content of altricial eggs meant that high rates of turning by individual birds would have a selective advantage because eggs would have hatched in the shortest possible time.

Lower rates of egg turning during incubation are possible in those species which have energy-rich eggs with relatively large yolks and small albumen fractions, long incubation periods and relatively safe nest sites. Hence, the buried, energy-rich eggs of Megapodes can cope without turning. Hole nesting birds like kiwis have highly energy-rich eggs with a long incubation period due to a relatively low incubation temperature (Rowe 1978). Many seabirds of the Procellariformes have safe nest sites in holes and their eggs are able to survive extended periods of chilling thereby extending their incubation periods. This has reflected on the composition of their eggs which have relatively high water contents (Sotherland and Rahn 1987). Even eggs of the altricial lyrebird can survive extended periods of chilling which increases their incubation period by around 20–50 days. Altricial pelecaniform birds may have relatively low rates of turning because their large eggs have long incubation periods and these species usually have relatively safe colonial nest sites. It follows that it can be predicted that the eggs of the fairy tern are not turned during incubation.

Areas for further research into turning during incubation

Further research is required in order to investigate the relationships between egg composition, nest site and egg turning. Given the considerable interest in natural incubation in birds it is regrettable that there appear to be so few reports of turning frequency in nests. It would be interesting to study how rates of egg turning vary according to diurnal cycle and to examine how turning rates are affected by the developmental age of the embryos. Furthermore, is the angle of turning affected by the composition of the eggs? It would also be extremely interesting to examine the morphological and physiological effects of different rates of turning in altricial species. It is hoped that this review will stimulate an interest in quantifying egg turning behaviour in birds from a wider variety of species.

In conclusion, the relationship between egg turning and albumen content of eggs has implications for our understanding of artificial incubation. Lack of investigation of turning rates in non-poultry species has meant that most incubation equipment has been based upon the principle of turning eggs once an hour. This rate would appear to be inappropriate for many species of birds. Therefore, aviculturists dealing with threatened or economically important species which produce semi-altricial or altricial hatchlings (e.g. parrots or raptors) have been using equipment which has probably not provided an optimal incubation environment. It is likely that machine design will have to be modified in order to improve the rates of egg turning for non-poultry eggs and thereby help in optimising utilisation of the albumen within the egg.

Acknowledgements

Many thanks to Glenn Baggott and Nick French for their constructive comments on previous drafts of this chapter and to Mike Kern, Nick Fox and David Jefferies for sharing unpublished data.

12 *Microbiology of natural incubation*

G. K. Baggott and K. Graeme-Cook

The interaction between the microbial flora of the environment and the avian egg has been investigated extensively for poultry species (Bruce and Drysdale 1991). Primarily, research has focussed on the spoilage of eggs during incubation and the contamination of the domestic fowl (*Gallus gallus*) egg when used as food, with much emphasis on the routes of infection, either transmission across the eggshell after lay or vertical transmission through the laying female. To combat microbial invasion of the egg, antimicrobial defences consist of two main components: a physical defence provided by the shell and a chemical defence primarily provided by the albumen (Board and Fuller 1974). The physical defences against microbial transmission arise from the eggshell acting as a resistance network comprised of cuticle, the pore canal and the shell membranes (Board 1980). Of these three components the cuticle is the most important in reducing the possibility of liquid water entering the pore canal, and so movement of bacteria into the egg (Sparks 1994). Eggs lacking cuticle readily absorbed liquid water (Sparks and Board 1984), and newly laid fowl eggs, where the cuticle is not complete, were less able to resist bacterial penetration of the shell (Sparks and Board 1985). However, fungal hyphae can easily penetrate this barrier if conditions are suitable for their growth, e.g. humidity (Board *et al.* 1964, 1994). The chemical defences of the egg consist primarily of the alkaline pH of the albumen and the presence of ovotransferrin, which reduces the availability of iron ions (Fe^{3+}) that micro-organisms need for their growth (Board and Fuller 1974).

Nevertheless, in poultry eggs bacteria can penetrate the shell and it is thought that three main factors determine whether this occurs: the temperature differential between egg and environment; the presence of liquid water on the egg surface, usually brought about by temperature fluctuations; and presence of contamination on the egg and in the environment (Bruce and Drysdale 1994). All of these factors could be present during natural incubation. In domestic species microbial contamination not infrequently leads to poor hatchability, although some pathogens, like *Staphylococcus aureus* and *Streptococcus* spp. were more effective in reducing hatching than, for example, *Enterobacter aerogenes* and *Micrococcus* spp. (Bruce and Drysdale 1991). Information on the types of micro-organisms that penetrate the shell and produce embryonic mortality is entirely restricted to commercial poultry and waterfowl in captivity (Bruce and Drysdale 1994). Eggs from the latter group mainly contained *Escherichia coli* and *Staphylococcus* spp. with smaller numbers of *Micrococcus* and *Pseuodomonas*. By contrast, 92% of isolates from fowl eggs from a commercial hatchery contained Gram positive cocci, mainly *Micrococcus* spp. and *Staphylococcus* spp., with very few Gram negative bacteria (Seviour *et al.* 1972). Potential human pathogens, such as Gram negative *Salmonella* spp., are known contaminants of the eggshell (Bruce and Drysdale 1994); they can penetrate the shell but are also transmitted vertically through the ovary of the laying bird (Humphrey 1994). Other pathogens known to be present in eggs, like *Campylobacter jejuni*, *Listeria mononcytogenes* and *Yersinia enterocolitica*, are thought to be transmitted only across the shell (Humphrey 1994).

For natural incubation there is very little information on the microbial flora of the eggshell or on the penetration of the shell by micro-organisms. Primarily, the focus of attention has been on fungal associations of nest material, mainly with regard to keratin-digesting (keratinolytic) fungi as these are known to be present on plumage (Pugh 1972), and on the presence of pathogens within nest materials (Hubalek *et al.* 1995). In the latter case the assumption is that these will lead, if not to a failure of hatching, then to nestling mortality but this has yet to be demonstrated. Furthermore, the relationship between the incubating bird and the microbial flora has only been explored for the starling (*Sturnus vulgaris*; Clark 1990), and, as is reported below, for some species of waterfowl.

Fungal associations of bird nests

The material from which nests are constructed often provide a suitable environment for the growth of fungi, especially as the adult plumage itself also provides a suitable substrate for keratinolytic fungi of the genera *Arthroderma*, *Chrysosporium* and *Ctenomyces* (Pugh 1972). Pugh and Evans (1970a) found that, in nests of 7 species of UK birds, about 27% of isolates were conidial *Arthroderma quadrifidum* and *Arthroderma uncinatum* and that these species were also present in half the nests examined. In addition, *Chrysosporium* spp. constituted 32% of isolates from 59% of nests. All of these fungal species were found at a higher frequency in nests than in the surrounding soils. By contrast, although *Arthroderma curreyi* was the most common fungus on adult plumage, and found quite frequently in nests, about 10% of nests testing positive, it was no more common in nests than in the surrounding soils.

Within the nest structure, the grass, twigs and moss of the outside of the nest of blackbird (*Turdus merula*), song thrush (*Turdus philomelos*) and hedge sparrow (*Prunella modularis*) had a lower pH and less fungi than the inner lining, where conidial *A. quadrifidum* was especially frequent. Keratinolytic fungi were found to be more abundant in nest materials of low water content, with 68% of isolates from nests where water content was <10%, and 15% from nests with water contents of 10–20%. Overall, *A. quadrifidum* was more abundant in nests than on bird plumage, whereas *A. curreyi* was more abundant on adult birds, especially blackbirds (and found in 14% of blackbird nests).

Pugh and Evans (1970b) found that fungi isolated from adult birds were found only on the outer feathers where the temperature was lower than at the skin surface. It is not surprising, therefore, that the nest material, often containing feathers at lower environmental temperatures, provides a suitable habitat for these keratinolytic fungi. Additionally, feather fats can augment the growth of these fungi. For example, *A. curreyi* growth was stimulated by feather fats of the blackbird and ring-necked pheasant (*Phasianus colchicus*). Interestingly, feather fats from the starling inhibited growth of this fungal species, as well as that of *Ctenomyces serratus* and *Chrysosporium keratinophilum*. Consequently, starlings yielded only 31 isolates of keratinolytic fungi from the plumage of 126 birds. When present, the keratinolytic fungi found on starling plumage are predominantly *Chrysosporium* spp., but *Scopulariopsis brevicaulis* and *Trichophyton terrestre* have also been isolated (Camin *et al.* 1998).

Nest boxes appear to supply a particularly favourable environment for fungi. Hubalek *et al.* (1973) examined 57 nests from 8 bird species in Czechoslovakia and keratinolytic fungi were found in 92% of nests (2.3 species per nest) with *Arthroderma ciferri* (61% of nests) and *Anixiopsis stercoraria* (40% of nests) as the predominant species. Cellulose digesting (cellulolytic) fungi were found in 88% of nests (mean of 2.8 species per nest) but only *Fusarium* spp. were frequent (40% of nests). Potentially pathogenic fungi were isolated from 74% of nests (mean of 1.2 species per nest) with *Scopulariopsis brevicaulis* present in more than half the nests examined. Again the water content of substrate had substantial influence on the mycoflora. Whilst *Ctenomyces keratinophilum* occurred in nests with a wide range of water contents (9–62%) *A. quadrifidum* and *Arthroderma cuniculi* were isolated only from nests material with maximum water content of 14%, a figure similar to that reported for these fungi in open nests (Pugh and Evans 1970a). Nest pH also was found to be important. *Ctenomyces keratinophilum* was isolated from nests with broadest pH range 4.5–7.9, and the greatest number of fungal species per nest were found in the range of pH 6.0–6.2; this was also the peak pH range for keratinolytic fungi. In the tree sparrow (*Passer montanus*) the number of fungal species and number of keratinolytic species was lower when nests contained eggs or young, possibly due to the higher nest temperatures (Hubalek *et al.* 1973). Repeated nesting of tree sparrows in the same box increased the number of fungal species and keratinolytic species. For example *Ctenomyces keratinophilum* and *Ctenomyces serratus* were found in 7% of nests on first nesting but had increased to 27% of nests when boxes were used a second time.

Nest boxes can provide a favourable environment for overwintering fungi. In nest boxes used by tree sparrows, Hubalek and Balat (1974) compared the fungal communities in 11 nest boxes in July, when young or eggs were in the nest, with the same boxes in April before breeding commenced. They found an increase in cellulolytic fungi (*Penicillium, Scopulariopsis)* but no change in the abundance and species numbers of keratinolytic fungi (mainly *A. ciferri*, *A. quadrifidum* and *Aphanoascus fulvescens*). Also overwintering were potentially pathogenic fungi (*Aspergillus fumigatus, Candida albicans, Rhizomucor pusillus*). This may have been due to the greater amounts of excreta in boxes by April, as they were used by roosting birds during the winter. It is noteworthy that nests can provide reservoirs of infection of avian pathogenic fungi (Table 12.1), for example *A. fumigatus* that causes avian aspergillosis, an important mycoses of birds, and *Aspergillus flavus* the cause of aspergillosis or aflatoxicosis (Hubalek *et al.* 1995). The potentially pathogenic, cellulolytic, *Fusarium oxysporum*, as well

Table 12.1 Potential pathogenic fungi of wild birds found in nests.

Fungal species	Host nest	Known disease and effect[A]	Reference[#]
Absidia corymbifera	*Turdus philomelos, Parus major, Hirundo rustica*	Zygomycosis: pulmonary disease	1
Aspergillus fumigatus	13 species; *Passer montanus*	Aspergillosis; respiratory disease	1, 2
Aspergillus flavus	5 species; *Passer domesticus*	Aspergillosis; respiratory disease	1, 3
Aspergillus nidulans	9 species	Aspergillosis; respiratory disease	1
Aspergillus niger	*Parus nigricollis, Passer montanus; Passer domesticus*		1, 3
Candida albicans	*Streptopelia decaocto, Passer montanus*	Candidiasis; alimentary tract infection	1, 2
Candida tropicalis	*Passer montanus*		1
Curvularia lunata	*Passer domesticus*		1
Emmensia parva	*Passer domesticus*		1
Fusarium oxysporum	*Passer domesticus*	Mycotoxicosis; toxin poisoning	3
Geotrichum candidum	*Streptopelia decaocto, Turdus philomelos, Carduelis cannabina*		1
Microsporum ripariae	*Riparia riparia*		1
Rhizomucor pusillus	8 species; *Passer montanus*		1, 2
Rhizopus arrhizus	*Falco tinnunculus*		1
Scopulariopsis brevicaulis	*Passer montanus, Parus major, Ficedulla albicollis, Sitta europaea, Sturnus vulgaris*; 29% of nests (total 270)		1, 4

[A]Nuttall (1997). [#]References: 1 – Hubalek *et al.* (1995); 2 – Hubalek and Balat (1974); 3 – Mazen *et al.* (1994); Hubalek *et al.* (1973).

as *Aspergillus* species were isolated by Mazen *et al.* (1994) from house sparrow (*Passer domesticus*) nests in Egypt (Table 12.1). These nests also contained non-pathogenic *Chrysosporium* species, with *Chrysosporium keratinophilum* (40% of nests) and *Chrysosporium tropicum* (39% of nests) being the most prevalent.

Hubalek (1978) using cluster analysis confirmed the existence of fungal communities associated with bird nests for over 283 nests of 90 bird species. The main fungal communities were classified as: the keratin decomposers, *A. quadrifidum*, *Chrysosporium serratus*, *Chrysosporium tropicum*; the psychrophilic fungi of frequent occurrence in nests, *Arthroderma versicolor*, *Chrysosporium pannorum*, *A. ciferrii*, *Arthroderma tuberculatum*; and the hygrophilic fungi that occur in old bird nests *Arthroderma fulvescens* and *Chrysosporium keratinophilum*.

Fungi can also form an integral part of the nest construction (Hansell 2000). Ten species of North American birds breeding in subalpine forests were found to incorporate the Horsehair Fungus (*Marasmius androsaceus*) into their nests (McFarland and Rimmer 1996). Whilst Melin *et al.* (1947) have shown that three species of this genus exhibit antibiotic activity against *Staphylococcus*, it was not the case for *M. androsaceus* with their assay conditions.

The nest bacterial community

The realisation that parasitic organisms are found in birds' nests has focussed attention on the ways in which birds respond to these nest cohabitors. Bird–mite relationships, for example, have provided a rich area of research on sexual selection and immunocompetence (Proctor and Owens 2000). However, far less attention has been paid to bacterial pathogens, despite the occurrence of pathogenic fungi in nests of a wide variety of avian species (Table 12.1). In a study of 43 nests of cavity-nesting titmice (*Parus major*, *Parus caeruleus* and *Parus ater*) nearly all of the nests contained Gram positive bacteria, including *Bacillus* spp., Micrococcaceae, *Streptococcus* spp., and a wide range of Gram negative bacteria, mainly from the Enterobacteriaceae including *E. coli*, in smaller numbers (Mehmke *et al.* 1992). Of the species isolated, only *Streptococcus* and *E. coli* have been shown to be pathogenic in birds (Nuttall 1997). However, although potentially pathogenic Gram negative bacteria were present in a third of nests, broods from only three of these nests did not fledge (Mehmke *et al.* 1992). The reuse of old nests by cavity nesters can, however, mean that new nest material is exposed to established bacterial populations. For example, the house wren (*Troglodytes aedon*) commonly uses old nests which were found to contain Gram positive *Bacillus* and *Staphylococcus* spp., together with potentially pathogenic *Pseudomonas* spp. and Enterobacteriaceae such as *Salmonella* spp. (Singleton and Harper 1998).

In a study of nestling tree swallows (*Tachycineta bicolor*) cloacal microbial populations were found to consist of a wide range of microbes: Gram negative enteric bacteria, Lactobacilli, with rather fewer *Staphylococcus* spp., *Campylobacter* spp., *Salmonella* spp., *Shigella* spp. and *Yersinia* spp., as well as fungi (Lombardo *et al.* 1996; Stewart and Rambo 2000). These microbial populations appear in the cloaca by 2–3 days after hatching and increase in numbers with nestling age (Mills *et al.* 1999). It is not known how nestlings acquire these microbial populations, although it has been suggested that they are transferred from the adult in regurgitated food (Lombardo *et al.* 1996). In the chimney swift (*Chaetura pelagica*) bacterial populations of hatchling saliva closely resemble that of the adult in the types of Gram positive and Gram negative bacteria present, and the adult saliva is, apparently, essential for nestling survival (Kyle and Kyle 1993). However, as adult saliva is an important part of the nest construction (Kyle and Kyle 1993), hatchling salivary bacterial populations could conceivably have originated from the nest materials rather than direct transfer from the adult. Certainly, in tree swallows the composition of the bacterial communities of the cloaca was a better predictor of the nest where the nestling was reared than nestling size (Lombardo *et al.* 1996). However, despite the avian nest providing, apparently, a suitable environment for a wide range of bacteria there are only a few studies exploring the means whereby breeding adults might control populations of potentially pathogenic organisms.

Behavioural control of potential pathogens

That birds often incorporate green plant material into their nest constructions is long established (Wimberger 1984; Hansell 2000). Of 137 passerines breeding in eastern North America that were surveyed by Clark and Mason (1985), those nesting in enclosed spaces (20% of total), and so usually using old nests, were more likely to incorporate green plants into their nests than passerines with open cup nests, who reused nests infrequently. Similarly, Wimberger (1984) found that, of 49 species of North American Falconiformes surveyed, 28 species used greenery and 22 of these species reused nests; whereas of the 21 species that used no greenery, only 8 reused nests.

A number of hypotheses have been proposed to explain the use of green plant material in nests. It has been suggested that it provides camouflage, although cavity nesters would hardly need this; that it retards water loss or provides shading, though again cavity nests would be expected to be a more humid environment; or that it provides insulation, although dry grasses and moss are better insulators than wet vegetation (Clark 1990). An alternative hypothesis is that green material provides protection against pathogenic organisms. Using the starling as a model, Clark and Mason (1985) investigated whether the plants used as nest material in North America were more effective in controlling parasites and pathogenic bacteria than other vegetation available to these birds. Using the green plants found in the nests, the ability of leaf disks to inhibit bacterial growth in nutrient agar was assessed both for potentially pathogenic bacteria, *Streptococcus aurealis*, *Staphylococcus epidermidis*, *Pseudomonas aeruginosa* and *E. coli*, as well as for mixed populations of bacteria swabbed from the nest material and nest cavity.

Of the five plant species preferred by starlings as nest material, *Agrimonia paraflora* was highly effective in inhibiting the growth of *Staphylococcus epidermidis, Streptococcus aurealis*, and the native bacterial populations. *Solidago rugosa* was moderately effective against *Streptococcus aurealis* and native bacteria, and *Daucus carota* moderately effective against *Pseudomonas aeruginosa* and *Streptococcus aurealis*. Of the two other plant species preferred by starlings, *Lamium purpureum* had no effect on bacterial growth in this test system and *Achillea millefolium* was moderately effective against native bacteria only (Clark and Mason 1985). Overall they found that plants effective at inhibiting bacterial growth were utilised more by starlings in their nest construction and attributed this anti-bacterial action to volatile compounds released from the plant material. Clark and Mason (1985, 1988) also found green plant material in nests retarded the emergence of lice *Menacanthus* spp. and the mite *Ornithonysus sylviarum*. However in European populations, the presence of green plant material in starling nests did not decrease nestling ectoparasite load, although these nestlings had increased body mass at fledging and increased immunocompetence (Gwinner *et al.* 2000).

In those plant species preferred by starlings as nest materials, more volatile compounds were found, and at higher concentrations, than in other vegetation also available to these birds (Clark and Mason 1985). It has been suggested, therefore, that starlings use odour rather than visual cues in selection of green plant material for nest construction. In starlings, volatiles from six plant species elicited strong electrophysiological responses from the olfactory nerves (Clark and Mason 1986, 1987). In conditioned avoidance behaviour tests these birds were able to discriminate between pairs of plant volatiles, from *Erigeron-Potentilla recta* and *Lamium purpureum-Alliaria officinalis*, as assessed by their avoidance behaviour. Nerve section abolished the ability to acquire this avoidance behaviour. Starlings also exhibited a seasonality in their ability to respond to odour cues. Male starlings, trained to respond to an artificial odour (cyclohexanone), had a lower threshold in April when they were sexually mature with enlarged testes and yellow bills (Clark and Smeraski 1990). For both males and females the frequency of a response to this odour peaked in April and ebbed in July through November, increasing in December–March. During the breeding season both male and females responded more to the natural odour of *Alliaria officinalis* during April, and when presented with both *Daucus carota* and *Lamium purpureum* they could discriminate between these plant odours only in April, not September (Clark and Smeraski 1990). Starlings, it would seem, possess the sensory capabilities to select green plant material on the basis of its potential to act as an anti-pathogenic agent.

It has been suggested that birds also use feather waxes as a fumigant (Jacob and Ziswiler 1982). Although, the specialised sebaceous gland, the uropygium, produces substantial amounts of waxes and fatty acids (Jacob 1978), it is now clear that these secretions have little to do with waterproofing the plumage, as thought previously, although preen gland secretion does make the plumage more flexible (Jacob and Ziswiler 1982). However, these secretions have been postulated to regulate the microflora of the plumage (Jacob and Ziswiler 1982), and perhaps participate in nest hygiene (Jacob *et al.* 1979). Jacob and Ziswiler (1982) report that the β-hydroxy fatty acids found in the uropygial gland secretions of the wood-pigeon (*Columba palumbus*), the red-backed shrike (*Lanius collurio*), the mallard (*Anas platyrhynchos*), and the Apterygiformes have fungicidal activity. Although these hydroxy fatty acids are more prevalent in ducks during the pre-incubation period (Jacob *et al.* 1979), and the antibacterial and antimycotic properties of the free fatty acids are widely asserted, there is relatively little evidence of their efficacy in this regard, either in the adult or during incubation.

Alkyl-substituted wax acids and alcohols are present in uropygial gland secretions of many bird species, and in culture 3-methyl-nonanoic, 4-methyl-heptanoic acid and 3,7-dimethyl octanol showed activity against *Microsporum canis*, *Microsporum gypseum*, *Scopulariopsis brevicaulis*, *Aspergillus niger* and *Candida* species, whereas 2-ethyl- and 2-butyl-substituted acids were active against *Staphylococcus* and *Streptomycetes* species (Jacob 1978). In the gannet (*Morus bassana*) a great variety of waxes are present in the uropygial gland secretions and one of these 3,7-dimethyl-octan-1-ol has antimicrobial activity (Jacob *et al.* 1997). Antibacterial activity was found at final concentrations of 3,7-dimethyl-octan-1-ol of $>0.1\%$, mainly for Gram-positive bacteria,

Staphylococcus aureus, *Staphylococcus epidermidis*, *Propionibacterium acnes*, but there was poor or no activity against the Gram-negative species, *Pseudomonas aeruginosa*, *Pseudomonas putida* and *Escherichia coli*. Fungicidal effects were demonstrable against yeasts (*Candida* spp., *Torulopsis*), moulds (*Aspergillus*, *Penicillium*) and there was substantial activity (at concentrations $<0.1\%$) against dermatophytes (*Trichophyton* spp., *Microsporum gypseum*).

In summary, chemical defence against pathogenic microbes during incubation has been inferred in the few bird species investigated. However, whilst the evidence is convincing for the starling that incubating birds select plant materials with antimicrobial activity to add to the nest, a role for secretions from the uropygial gland in controlling nest microbes remains unproven.

Microbes on the egg surface

Many of the studies cited above have focussed on nests built in cavities (natural or artificial) where nest materials may be reused. Keratinolytic fungi, for example, are frequent associates of these nests, which presumably reflects the extent to which nesting materials incorporate feathers. A survey of nests of the Tyrannidae, Corvidae and Fringillidae revealed that feathers were the second most frequent lining material (Hansell 2000). These nests may provide, therefore, an environment for maintaining specialist microbial populations and it is possible that the microbial communities of the eggshell surface may be determined by these populations.

The mandarin duck (*Aix galericulata*) is a cavity nester, introduced to the UK from Asia in the early years of the 20th century, that breeds widely in the southern UK (Davies and Baggott 1989a). In a nest box population in Berkshire, UK, the period for the completion of a clutch prior to the start of incubation was 15 days, the incubation period was about 33 days, and many females incubated large clutches; not infrequently in excess of 30 eggs (Davies and Baggott 1989a). About a quarter of eggs failed to hatch due to embryonic death or infertility (Davies and Baggott 1989b). The total bacterial counts per egg from the surfaces of the eggs of the mandarin duck from this site were highest for eggs prior to incubation when measured at two culture temperatures, 25 and 38°C (Figure 12.1). The numbers of bacteria able to grow at 38°C, which approximates to the temperature of incubation, showed a marked decline by the first week of incubation when compared with samples obtained before incubation. This decline persisted into the following weeks of incubation (Figure 12.1a, solid squares) with far less variability in the bacterial numbers. The numbers of bacteria able to grow at 25°C showed a similar, though less marked decline (Figure 12.1a, solid triangles). This change in total bacterial counts per egg was associated with an increase in the proportion of *Bacillus lichenformis* in these samples (Figure 12.1b). Evidently, during incubation *B. lichenformis* is able to survive/prosper rather better than other members of the shell surface bacterial flora, including other *Bacillus* spp.

Strains of *B. lichenformis*, were isolated at 38°C, but not at 25°C, from aquatic sediments adjacent to this site, as well as from incubating birds and the eggshell

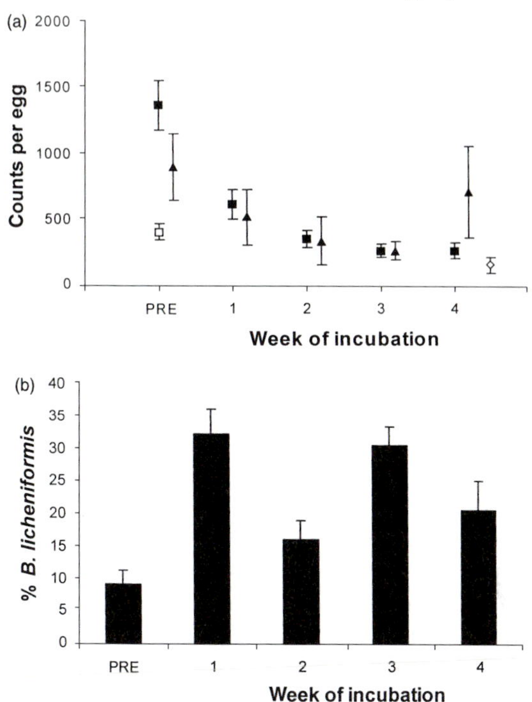

Fig. 12.1 (a) Total bacterial counts (c.f.u.), when cultured at 38°C (solid squares) and at 25°C (solid triangles), of samples obtained from eggs swabbed prior to (PRE) and during the four weeks of incubation in the mandarin duck (*Aix galericulata*). Mean fungal/actinomycetes counts per egg in 25°C cultures from unincubated eggs – open square; mean fungal/actinomycetes counts per egg cultured at 25°C for all weeks of incubation – open diamond. Fungal/actinomycetes counts for samples from incubated eggs cultured at 38°C were too small to be presented on this scale. Symbols represent mean values and standard error bars. Each egg was swabbed in the nest with sterile water and the swab suspended in 1 ml of this water; 50 ml aliquots were plated, in triplicate, onto nutrient agar (bacterial counts) or potato dextrose agar (fungal/actinomycetes counts) at and grown at 25 or 38°C for 3 days. Eleven clutches were sampled before and through the period of incubation: samples obtained from 12–40 eggs at each time point were cultured at 38°C, and samples from 3–4 eggs at each time point were cultured at 25°C. (b) For the same eggs as in A, the mean proportion (+SE) of the total bacterial counts per egg that were identified as *Bacillus licheniformis* for samples were cultured at 38°C. *B. licheniformis* was absent from cultures at 25°C.

surface. These strains exhibited negligible growth at 25°C, growing only at 38°C (Reed, unpublished observations). *B. licheniformis* was also isolated at 38°C from samples taken from plumage of incubating females and egg surfaces of shelduck (*Tadorna tadorna*), widgeon (*Anas penelope*), tufted duck (*Aythya fuligula*) and goosander (*Mergus merganser*), and was also present in the plumage of immature birds of these species, as well as shoveler (*Anas clypeata*), pintail (*Anas acuta*), mallard and mandarin duck. Strains of *B. licheniformis* have also been isolated from the plumage of a further 32 species of North American birds (Burtt and Ichida 1999). Thus, although the numbers of bacteria on the shell surface are considerably reduced during natural incubation, *B. licheniformis* may be least affected because of its ability to grow at incubation temperatures, as well as the continual addition of this bacterium to the egg surface from the plumage of the incubating bird.

In the mandarin duck, fungal/actinomycetes counts were low in culture at 25°C, the highest counts per egg being found for unincubated eggs (Figure 12.1a), and may be indicative of organisms imported into the nest by the incubating bird. When cultured at 38°C, very few samples from egg surfaces yielded any counts of fungi, which was not surprising as fungi present in the nest or on plumage grow best at temperatures lower than homeothermic body temperatures (Pugh 1972). In part, however, the low fungal/actinomycetes counts and the decrease during incubation of the numbers of bacteria able to grow at 38°C, including perhaps potential pathogens, could have been due to interactions between the microbial flora of the shell surface.

In three waterfowl species, *B. licheniformis* strains, isolated at 38°C from both the incubating female and the eggs of the same clutch, inhibited the growth of both Gram positive and Gram negative bacteria as well as the fungus *Pythium ultimum*, although there was no effect on the growth of the yeast *Saccharomyces cerevisiae* (Table 12.2). In addition, all of these strains hydrolysed casein as well as starch (Table 12.2; Heslegrave and Graeme-Cook unpublished observations). These strains, as well as additional ones from incubating mandarin duck and shelduck, when cultured in feather media (Williams *et al.* 1990), fully digested feathers (Heslegrave and Graeme-Cook, unpublished observations), and at rates comparable to those reported from other bird species (Burtt and Ichida 1999). Such feather-degrading activity might be expected as a keratinase has been characterised from strain PWD-1 of *B. licheniformis* (Lin *et al.* 1992). So how might these properties of *B. licheniformis* be important for natural incubation?

During natural incubation of the mandarin duck there is an increase in eggshell conductance to water vapour (see also Chapter 3). For example, when unincubated eggs of mandarin duck were added to incubated clutches for 7 days the conductance of these eggs increased by 72% (Baggott and Graeme-Cook 1997). Also, when *B. licheniformis*, isolated from the shell surface of eggs of this species, was inoculated on the cuticular surface of shell pieces and incubated at 38°C for 7 days at 40% relative humidity, there was visual evidence of both cuticle degradation and the growth of the bacterium (Figure 12.2). As the cuticle of the fowl egg consists of about 85% protein with carbohydrate, lipid and minerals (Wedral *et al.* 1974; Sparks 1994), the protease and glycolytic activity exhibited by *B. licheniformis* might well account for the degradation observed. Previously, only a pseudomonad has been reported to digest the cuticle of the fowl eggshell using a protease (Board *et al.* 1979). As it is known that the removal of the cuticle of eggs by chemical means can increase the egg conductance substantially (Deeming 1987; Thompson and Goldie 1990), it is conceivable that bacterial degradation of the cuticle could increase egg conductance by facilitating access of gases to the shell pores. This may be the basis for the increase in egg conductance to water vapour observed in

Table 12.2 Inhibition of microbial growth and casein hydrolysis by strains of *Bacillus licheniformis* isolated from incubating females and eggs of the incubated clutch for three waterfowl species.

Species	*Bacillus licheniformis* strain[a] from..	Gram positive *Micrococcus luteus*	Gram negative *Escherichia coli*	Fungus *Pythium ultimum* % growth	Yeast *Saccharomyces cerevisiae*	Casein hydrolysis
Anas penelope	Incubating female	++[b]	++	45[c]	–[b]	++[d]
	Eggs	++	+	36	–	+
Mergus merganser	Incubating female	++	++	51	–	++
	Eggs	++	+	42	–	+
Aythya fuligula	Incubating female	+++	–	40	–	+++
	Eggs	+++	–	93	–	+

[a]Bacteria were isolated from platings of eggs and bird swabs taken during the second week of incubation as described in legend to Figure 12.1. Identification of *B. licheniformis* was confirmed by API strips (BioMerieux). All tests were conducted in triplicate. [b]*B. licheniformis* was streaked down the centre of nutrient agar plates and incubated for 3 days at 38°C. Both *Micrococcus luteus* and *Escherichia coli* were then streaked at right angles and inhibition assessed after 24 hours. Distance of inhibition : + = 0–5 mm; ++ = 5–10 mm; +++ = 10–15 mm; – = no inhibition. [c]*B. licheniformis* streak was cultured for 3 days on potato dextrose agar and then a 10 mm plug of *Pythium ultimum* culture was placed 2.5 cm away from streak. % growth after 24 hours was calculated as – (the distance grown towards *B. lichenformis*)/(the distance grown in the opposite direction) * 100. [d]*B. licheniformis* was streaked onto 5% skimmed milk agar and incubated for 7 days at 38°C; + = zone of hydrolysis visible; ++ = zone 1–2 mm; +++ = zone > 2 mm.

the mandarin duck during natural incubation. The ability of *B. licheniformis* to survive at the temperature of incubation and its expression of antibiotic activity would assist in this process, by ensuring that this microbe thrives on the eggshell surface during natural incubation.

Improving our understanding of the microbiology of natural incubation

It seems, therefore, that the microbial flora of the eggshell surface during natural incubation is not simply an extension of the microbial community of the nest materials. The low fungal counts and the decrease in total bacterial counts observed during natural incubation may be due to both the increased egg temperatures and the antibiotic activity of the microbial flora. It is also possible that the shell surface is exposed to antimicrobial preen gland secretions from the incubating bird. In addition, although *Bacillus* species have been found to be the commonest bacteria in the nest materials of the few birds investigated, predominately titmice (Mehmke *et al.* 1992), the presence of bacilli, notably *B. licheniformis*, on the eggshell surface of waterfowl may be reinforced by a constant exposure to this

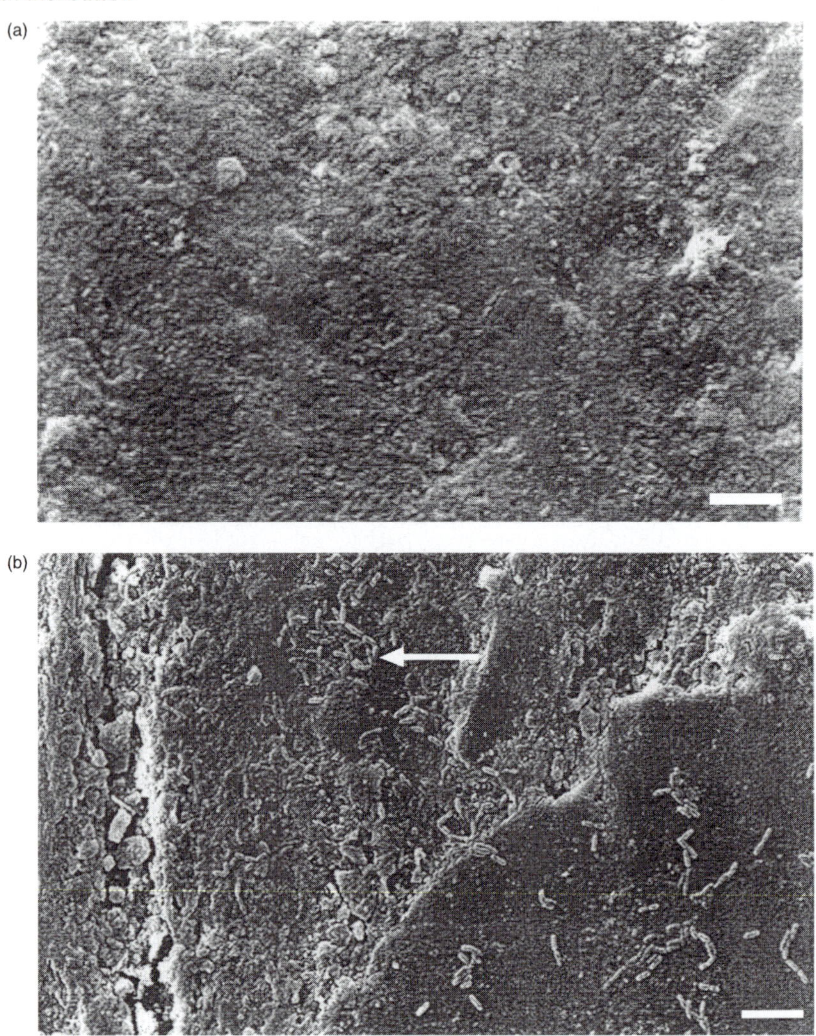

Fig. 12.2 Scanning electron micrographs of the cuticular surface the eggshell of the mandarin duck (*Aix galericulata*). (a) Representative shell piece from the control showing the lack of visible pore openings due to the cuticle overlying the shell. (b) A representative piece of shell 7 days after adding to the surface 1×10^6 c.f.u of *Bacillus licheniformis* in 10 μl of sterile water and incubating the piece at 38°C and 40% relative humidity. The bacteria exhibit signs of grwoth, forming rods clustered at the sites of degradation of the cuticle (arrow). Shell pieces were fixed in cacodylate buffered (pH 7) glutaraldehyde, and gold coated. Scale bars represent 10 μm.

bacillus on the plumage of incubating birds. Furthermore, a combination of the microbe's tolerance of incubation temperatures and the ability to degrade the shell cuticle may provide for *B. licheniformis* an environment sufficient for its growth on the eggshell surface and a consequent modification of the cuticle structure. There are a small number of bird species where it is reported that the shell permeability to water vapour increases early in natural incubation (Carey 1979; Hanka *et al.* 1979; Rahn *et al.* 1983; Booth and Rahn 1990; Kern *et al.* 1992; see Chapter 3), yet the means by which this occurs remains unknown. The hypothesis that shell surface bacteria and the incubating bird participate in a symbiosis leading to a modification of shell permeability by an alteration of cuticle structure would seem worth pursuing.

Currently, our knowledge of the microbial flora during natural incubation is rudimentary. It is clear that nest materials and contents provide a suitable environment for a range of specialised fungi and bacteria. In addition, the evidence is convincing that one species, the starling, modifies its nest construction in response to the presence of parasites and pathogens. There is virtually no information for natural incubation, in contrast to the artificial incubation of eggs of domestic species, as to whether, and how, micro-organisms penetrate the shell, or the responses of the incubating female, if any, to such a microbial challenge. The characteristics and numbers of microbes on the eggshell surface of waterfowl are relevant to this and would repay further investigation.

Acknowledgements

We wish to thank the Crown Estates for access to and permission to erect nest boxes for mandarin ducks. Thanks are also due to Dr Ruth Cromie of The Wildfowl and Wetlands Trust for collection of material from waterfowl species other than mandarin duck.

13 *Underground nesting in the megapodes*

D. T. Booth and D. N. Jones

The Family Megapodiidae is distributed throughout the islands of the Indo-Australian region east of Wallace's Line including New Guinea, Australia and the Philippines, as well as on numerous islands in the southern Pacific as far east as Tonga (Jones *et al.* 1995). Megapodes appear never to have been present on the mainland or larger islands of south-east Asia (such as Java, Sumatra and most of Borneo) but are present on the Nicobar and Andaman Islands in the Bay of Bengal (see Dekker 1989). At present 22 species are recognised (Jones *et al.* 1995). Recent paleontological data suggests, however, that an additional 23–33 species were exterminated soon after the arrival of humans in the island chains of southern Oceania over the last few millennia (Steadman 1999).

Megapodes are small to moderately sized birds, morphologically very similar to the galliforms (Clarke 1964). While sharing many physical and behavioural characteristics with other gamebirds, megapodes are currently regarded as a sister group of the Galliformes (Jones *et al.* 1995). The most remarkable characteristic of all megapode species is that they do not use body heat for incubating eggs but utilise external sources of environmental heat for this purpose. Indeed, the defining characteristic of this group is their use of a remarkable variety of heat sources for the incubation of their eggs. In this chapter, the significant aspects of incubation by megapodes are described, with particular emphasis on variability among the species.

Sources of incubation heat

The origins of megapode incubation?

Megapodes almost certainly evolved from a galliform ancestor, as early as the Pliocene (Boles and Ivison 1999), and this probably had a ground nest with large-clutches of synchronously hatching, highly precocial chicks commonly seen in extant galliforms. Although the origins of the exploitation of external heat for incubation remain unknown, the most plausible suggestion is that the first step away from body-heat incubation involved the covering of the eggs with organic matter immediately available near the nest, as grebes (Podicipedidae) do today (Clark 1964). These proto-megapodes would almost certainly have occurred in hot and moist equatorial rainforests where the high ambient temperatures enabled them to leave their eggs unattended without the eggs becoming chilled. Covering the eggs in a pile of leaf litter may initially have been employed by the parent to hide the clutch from potential predators. In the tropical rainforest environment, of course, any organic matter on the forest floor would have been prone to rapid decomposition; piling the material would have enhanced the natural processes of breakdown by micro-organisms, and some microbial heat production would certainly have resulted. This scenario suggests that the adults, having found that the eggs remained warm during their absence, may have left them unattended for increasingly longer periods. Thus, the first incubation mounds, now employed by the majority of extant megapode species, would have developed from typical galliform nesting behaviour (Clark 1964; Jones *et al.* 1995).

Heat released through the decomposition of organic material is a product of respiration of the enormous number of micro-organisms present in litter material. This process does not involve fermentation, as has been often stated (e.g. Frith 1956a, 1959b), and no part of the incubation site is anoxic (Seymour and Ackerman 1980; Seymour *et al.* 1986). Virtually nothing is known about the micro-organisms involved in this process although it is likely that a high diversity of bacteria and large amounts of heat-adapted fungi are present (Seymour 1985). A detailed investigation of both the biodiversity and heat-producing mechanisms of these organisms would be of great value to improving our understanding of a natural process that is central to megapode ecology (see also Jones 1999).

Incubation mounds

Regardless of the origins of megapode incubation, their use of the heat produced by microbial respiration is a remarkable exploitation of a readily available natural heat source. The incubation mounds constructed by modern-day megapode species are essentially structures that enhance the productivity of these decomposers. This is achieved, in many mound-building species (Jones *et al.* 1995), by the regular addition of fresh organic material, the thorough mixing of new materials into the mound, and the maintenance of moisture levels within through appropriate alteration of mound shape (Seymour and Ackerman 1980; Seymour and Bradford 1992). At a microclimatic scale, the placement of the mound may also have a significant influence on the subsequent performance of the micro-organisms. Most species appear to select locations dominated by plants with leaves suitable for rapid decomposition, and to avoid exposure to direct sunlight (Jones 1988b; Sinclair *et al.* 1999; Sinclair 2000).

Each of these features (and others; Seymour and Bradford 1992; Sinclair 2000) may be regarded as indirect means by which megapodes enhance the natural activities of decomposers. More direct actions, such as the direct manipulation of the mound to alter temperature in a particular direction, appear to be limited to a single species, the malleefowl (*Leipoa ocellata*; Frith 1956a, 1957). This species is the only megapode found in an arid environment and the maintenance of conditions suitable for the micro-organisms is dependant upon the careful conservation of moisture levels within the mound. Frith's (1957) experiments demonstrated clearly that the malleefowl was capable of remarkably sensitive alterations of temperatures. Other megapodes, almost all of which occur in sub-tropical or tropical regions, appear not to engage in such direct manipulation of internal mound temperatures (Jones *et al.* 1995). This is probably because the moister environmental conditions enjoyed by most species have not forced the extreme adaptations of the malleefowl. Indeed, Seymour and Bradford (1992) have demonstrated that a typical incubation mound, in this case, that of the Australian brush-turkey (*Alectura lathami*) will naturally achieve homeothermy provided it is of an appropriate size and is regularly supplied with suitable organic matter, conditions seemingly typical of most mound-building species (e.g. Jones 1988b; Palmer *et al.* 2000).

While the mound-maintenance activities of most megapodes may be less sophisticated than that of the malleefowl, all species seem capable of monitoring the temperatures of their incubation sites to a fine level. The physiological mechanism behind this capacity has not yet been investigated in megapodes. However, in chickens, which are capable of detecting temperature changes of 0.5°C (Sturkie 1976), the sensitive site appears to reside in the palate. Megapodes of both sexes often 'taste' the substrate into which eggs are to be laid, a common behaviour almost certainly associated with temperature monitoring (Frith 1956a, 1957; Jones *et al.* 1995; Sinclair 2000). In several species, females have been observed to reject a potential oviposition site within an incubation mound apparently because of inappropriate temperatures (Frith 1962; Jones 1988a; Birks 1996).

Other incubation heat sources

The ability to assess and monitor the temperatures of a warm substrate was obviously a crucial evolutionary adaptation for megapodes as they became increasingly dependent upon mounds of organic material for the incubation of their eggs. It was also an ability that enabled some species to explore other sources of external environmental heat as potential incubation sites.

Megapodes utilise two other sources of heat for incubation: geothermal activity, where the substrate is warmed through proximity to hot springs or gases associated with volcanic areas; and solar radiation, where substrates are warmed by exposure to direct sunlight. Table 13.1 lists the heat source used by each species known at the present time. Note that while most species use microbial heat (by building mounds), at least five species also (all but one a *Megapodius* species) use more than one source. Although these heat sources are also naturally occurring in the environment, its utilisation for incubation requires quite different procedures. Rather than manipulating a natural process (such as microbial decomposition) to enhance and exploit it for incubation, the use of these alternative sources is much more passive. The differing approaches by megapodes to the use of different heat sources is indicated in the terms 'mound builders' and 'burrow nesters' as a main distinction among the species.

Incubation technique

Mound builders versus burrow nesters

Most megapodes are 'mound builders'. They construct incubation sites by raking together large piles of damp organic matter into which, after becoming warmed

Table 13.1 Incubation type, heat sources and incubation temperatures reported for megapode species. Mound or burrow nesting and heat source from Jones *et al.* (1995), except for Sankaran and Sivakumar (1999) and Palmer *et al.* (2000). ? = suspected but unconfirmed.

Species	Nest type	Heat source	Mean temperature (°C)	Temperature range (°C)	Reference#
Alectura lathami	Mound	Microbial	31.4		1
			33.3	31–36	2
			32.9	30–35	3
Aepypodius arfakianus	Mound	Microbial	38		4
				29–34	5
Aepypodius bruijni	Mound?	Microbial			
Talegalla cuveri	Mound	Microbial			
Talegalla fuscirostris	Mound	Microbial			
Talegalla jobiensis	Mound	Microbial			
Leipoa ocellata	Mound	Microbial/Solar	34.0	29–35	6
			34.1	27–38	7
			33.9		1
Macrocephalon maleo	Burrow	Solar/Geothermal		32–36	8
				31–38	9
Eulipoa wallacei	Burrow	Solar/Geothermal?	32.0	31–33	10
Megapodius pritchardii	Burrow	Geothermal	36.0		4
				32–38	11
			32	28–37	12
Megapodius laperouse	Mound Burrow	Microbial/ Geothermal	31.3	30–33	13
Megapodius nicobariensis	Mound	Microbial/Solar[1]	32.1	29–35	14
Megapodius cumingii	Mound	Microbial			15
	Burrow	Solar			
Megapodius bernsteinii	Mound	Microbial			
Megapodius tenimberensis	Mound	Microbial			
Megapodius freycinet	Mound	Microbial	34.4	33–37	16
Megapodius geelvinkianus	Mound?	Microbial			
Megapodius forstenii	Mound	Microbial			
Megapodius eremita	Mound	Microbial		35–39	4
	Burrow	Geothermal/Solar		31–33	17
Megapodius layardi	Mound	Microbial			
	Burrow	Geothermal?			
Megapodius decollatus	Mound	Microbial			
Megapodius reinwardt	Mound	Microbial/Solar[2]		35–39	4
			30.5	29–38	18
			32.5	29–35	19
All mounds and burrows			33.1	27–39	
Mounds only			33.2	27–39	
Burrows only			32.7	28–39	

[1]Sankaran and Sivakumar (1999). [2]Palmer *et al.* (2000). #References: 1 – Seymour *et al.* (1986); 2 – Jones (1988b); 3 – Seymour and Bradford (1992); 4 – Frith (1956b); 5 – Kloska and Nicolai (1988); 6 – Frith (1956a); 7 – Booth (1987a); 8 – MacKinnon (1978); 9 – Dekker (1988); 10 – Baker (1999); 11 – Todd (1983); 12 – Göth (1995); 13 – Wiles and Conry (2001); 14 – Sankaran and Sivakumar (1999); 15 – Sinclair *et al.* (1999); 16 – Lincoln (1974); 17 – Roper (1983); 18 – Crome and Brown (1979); 19 – Palmer *et al.* (2000).

through the processes of decomposition, eggs are laid. These constructions are often massive; even the relatively modest mounds of the Australian brush-turkey (typically 0.8–1 m high and about 4 m in diameter) may weigh 2–4 tonnes (Jones 1988b). Mounds also vary greatly even within a species, both temporally and between individuals (Wiles and Conry 2001). Detailed measurements of the mounds of the Nicobar megapode (*Megapodius nicobariensis*) showed that mounds ranged in volume from 0.1 to 28 m^3, with some increasing by 25 times in size during one year (Sankaran & Sivakumar 1999).

Mound-builders tend to construct new mounds each breeding season, although they may site this on the base of the previous year's mound (Frith 1962; Jones 1988b; Jones *et al.* 1995; Sinclair 2000). Some, however, continue to add material to an existing mound and in some species, this can result in an extremely large construction. By far the largest are those of certain populations of the orange-footed megapode (*Megapodius reinwardt*) the most widely distributed of all megapode species. Whilst this species' mounds are often modest, e.g. 1–2 m high in the Northern Territory of Australia (Palmer *et al.* 2000), they can also attain very large dimensions. Mounds of 4.5 m height have been recorded in northern Queensland, Australia, and 3.5 in height in parts of Indonesia (Jones *et al.* 1995). Such mounds appear to grow to such dimensions through the accumulation of material over many years; continuous use for over 40 years has been reported (Jones *et al.* 1995). The results of such efforts can be enormous hillocks that may remain as prominent landforms for thousands of years (Bowman *et al.* 1999)

The initial step towards successful incubation in all megapodes requires depositing the eggs, which are laid at intervals averaging 2–13 days for individual females, (Jones *et al.* 1995; Baker and Dekker 2000) into substrate suitable for incubation. For mound building species this requires that the mound be excavated to enable access to material in the depths of the mound. This may be a simple (if laborious) process of digging a blunt cone-shaped depression (which typically take less than 30 minutes in species with modest sizes mounds) or may take much of the day in the case of the malleefowl (Frith 1962) because of the relatively loose material used in their mounds.

For species that do not construct mounds, egg-laying necessitates the female visiting an incubation site, which is often situated in a different location to that in which the birds live the rest of their lives. The primary exception to this are those species or individuals that lay their eggs in the mounds of conspecifics or even other megapode species, which are available locally (Jones *et al.* 1995; Sinclair 2000). Such sites may include beaches, exposed areas of friable land within the rainforest, or the lower slopes of volcanoes. Suitable sites are usually geographically widely dispersed and large numbers of birds may visit these sites during each breeding season and may have done so for very long periods of time. For example, the Moluccan megapode (*Eulipoa wallacei*) lays its eggs in black volcanic soil adjacent to beaches in Indonesia that are heated by the sun (Baker and Dekker 2000). Visiting this site requires a journey of considerable distance and is repeated for each egg. Females in this species simply dig a small depression of 50–100 cm depth, lay and roughly cover the egg with soil and leave the site immediately. By contrast, Melanesian megapodes (*Megapodius eremita*) lay their eggs in loose material at the base of pre-existing burrows which honey-comb their communal egg-laying grounds at the base of dormant volcanoes. These effectively permanent tunnels, which may be several metres in length, appear to have been formed by thousands of birds over unknown periods of time, each attempting to reach the soil level where temperatures are appropriate for incubation (Jones *et al.* 1995). These tunnels remain open because of the nature of the soil in the area and the structural support provided by tree roots and rocks in the area (Jones unpublished data). In other situations, these 'access burrows' often fill with friable substrate material after each visit and will need to be re-opened by each female seeking to use the site.

Mounds versus burrows: dichotomy or continuum?

Although convenient, the mound-building/burrow-nesting dichotomy may be less clear-cut than initially thought. Jones *et al.* (1995) showed that at least four megapode species employ both mounds and burrows in different parts of their range so the relationship between species and incubation site used may more usefully be regarded as a continuum. This applies to both the method of accessing the egg laying substrate, and the source of incubation heat being utilised. For example, several mound-building species appear to construct tunnels into the depths of their mounds, rather than excavating a conical depression from the top (Jones *et al.* 1995). This is clearly illustrated with respect to the mound of the orange-footed megapode (Figure 13.1; see caption note). In this particular case, the mound material is sufficiently compacted that these tunnels of friable material remain evident from year to year. Although the source of heat is that of microbial decomposition of the organic matter introduced into the tunnels by the birds, in physical structure this incubation site is very similar to that of the Melanesian megapodes nesting in volcanic areas described above.

There is also an increasing realisation that numerous megapode species utilise more than one heat source. Until the 1970s, it was generally believed that only malleefowl mounds were heated by two heat sources – solar and microbial respiration (Frith 1956b). However, the findings of more recent studies of other mound-building megapodes suggest that other species may also be reliant on solar heat in addition to that produced by organic decomposition (e.g. Sankaran and Sivakumar 1999; Sinclair *et al.* 1999; Palmer *et al.* 2000). To further indicate the use of multiple heat sources, it is now recognised that numerous burrow nesters apparently reliant upon solar radiation, typically incorporate organic material into the substrate of the incubation site (Jones *et al.* 1995).

Nest incubation conditions

Gaseous conditions

Compared with the open nesting environment of most bird species, the nest conditions of megapodes are higher in humidity and have high levels of carbon dioxide but low levels of oxygen (Seymour and Ackerman 1980; Seymour *et al.* 1986;

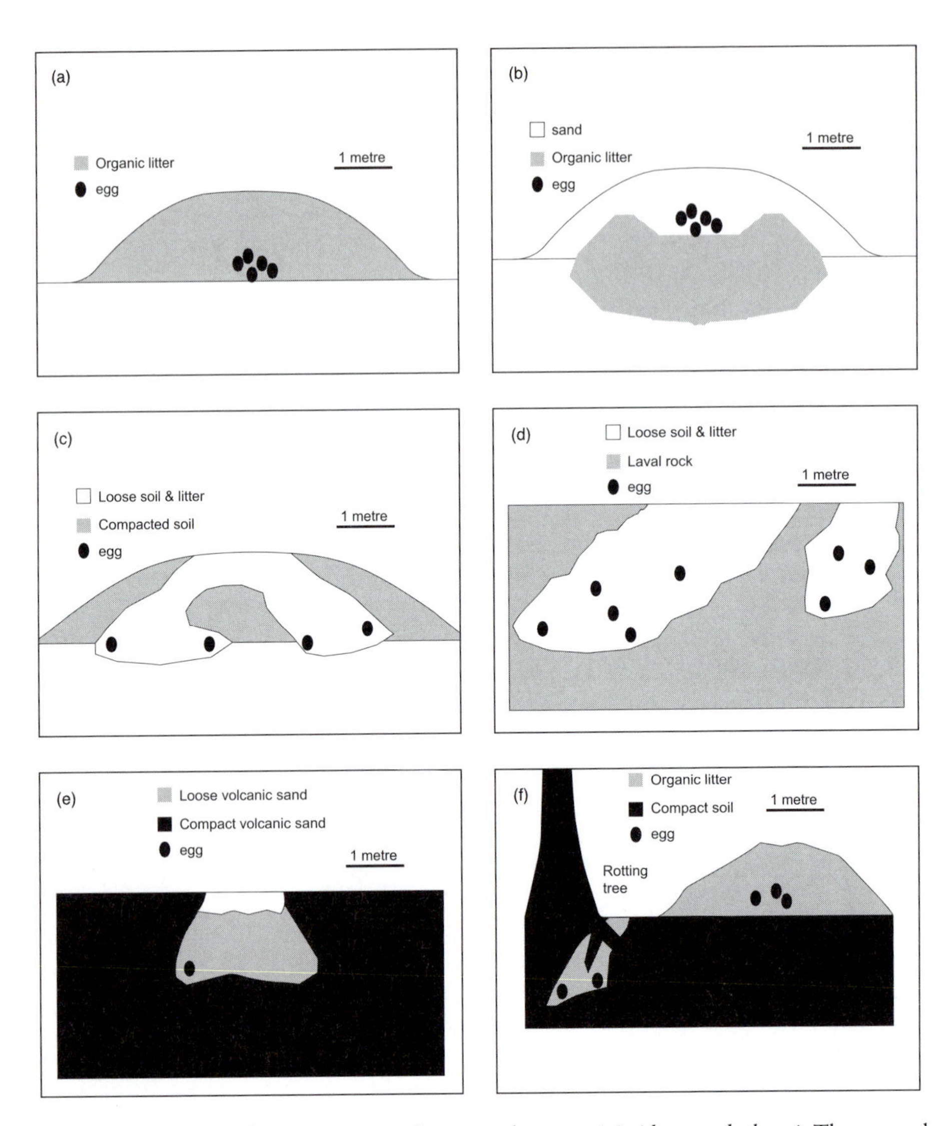

Fig. 13.1 Structure of various types of megapode nests. (a) *Alectura lathami*. The mound is constructed entirely of organic litter material raked together by parent birds. This is the most wide spread method of incubation amongst megapodes. After Booth (unpublished). (b) *Leipoa ocellata*. The mound is arguably the most sophisticated and consists of a bed of organic material overlaid by a sand blanket. After Booth (unpublished). (c) *Megapodius freycinet*. Egg laying sites consist of tunnels containing loose soil and litter dug into a larger mound of compacted soil. Adapted from Lincoln (1974). (d) *Megapodius pritchardii*. Eggs are laid in tunnels within solid geothermally heated larval rock formations. The tunnels are filled with loose soil and organic litter material. From descriptions in Göth and Vogel (1997). (e) *Macrocephalon maleo*. Eggs are laid in simple pits located in geothermally heated sand. This is the simplest form of incubation. Adapted from Dekker (1998). (f) *Megapodius cumingii*. Within a single population slightly different methods of incubation are used. In one method eggs are laid in tunnels constructed between rotting tree roots. Eggs are located at the bottom of the tunnel and the tunnel is filled with loose soil and organic litter material. In a second method, eggs are incubated in mounds constructed from organic litter material raked together from the ground. From descriptions in Sinclair *et al.* (1999).

Booth and Thompson 1991). In mound building species the altered respiratory gas tensions are the result of microbial respiration, respiration of the eggs themselves and gas exchange with the atmosphere which occurs by diffusion (Seymour and Ackerman 1980; Seymour *et al.* 1986; Seymour and Bradford 1992). The gaseous conditions experienced by eggs of two mound building species, the Australian brush-turkey and the malleefowl have been examined in detail (Seymour and Ackerman 1980; Seymour *et al.* 1986). In both cases the nest chambers were found to be moderately hypoxic ($P_{O_2} \sim 17.5$ kPa) and hypercapnic ($P_{CO_2} \sim 3.0$ kPa) compared to open nests ($P_{O_2} \sim 20$ kPa; $P_{CO_2} \sim 0.1$ kPa). Despite these unusual condition, the respiratory gas tensions experienced by embryos immediately before hatching (air space $P_{O_2} \sim 14.4$ kPa; $P_{CO_2} \sim 6.3$ kPa) were similar to that recorded for other bird species (Seymour *et al.* 1986, 1987), a result of adaptations of the eggshell pore system. More extreme respiratory gas tensions would occur if the nests became very wet as this would decrease the gaseous porosity of the nest and thus increase the resistance to diffusive gas exchange within the nest and also increase the rate of microbial respiration (Seymour *et al.* 1986). Such situations rarely occur in nature because birds of both species build up their mounds into dome-shaped structures during heavy rains and this prevents large volumes of water entering mounds. Mounds are also relatively well drained. However, the occasional bird is caught by surprise by torrential rain and in such cases embryo suffocation may occur (Brickhill 1987; Jones 1988a). Mound humidity is always high, usually greater than 95% RH (Seymour *et al.* 1987). However, during late summer malleefowl mounds may dry considerably so that relative humidity may drop as low as 77% (Seymour *et al.* 1987).

No information is available about the gaseous nesting environment of megapode species that utilise solar and geothermally heated soils for incubation. However if the soil has a low organic content, then only relatively small deviations from atmospheric conditions are predicted, especially as eggs are usually laid individually in different burrows. More extreme conditions may occur temporarily immediately following heavy rain as the soil's increased water content will decrease its gaseous porosity and increase resistance to diffusive gas exchange. In cases where several eggs are laid by birds in a single burrow such as the Polynesian megapode (*Megapodius pritchardii*) (Göth 1995; Göth and Vogel 1997), burrow gas tensions may also become more extreme.

Temperature conditions

Megapodes have less immediate control over incubation temperature compared to other bird species. Mound nesting species regulate mound temperature within limits by adjusting their mound tending behaviour (Frith 1956a, 1957, 1962; Crome and Brown 1979; Seymour and Ackerman 1980; Jones 1988b, 1989; Seymour 1991; Seymour and Bradford 1992; Palmer *et al.* 2000). Species that utilise solar and geothermally heated soils are very selective of the thermal environment where they bury their eggs (Weir 1973; Bishop 1980; Todd 1983; Jones 1989; Göth and Vogel 1997). The preferred incubation temperature of megapodes is 32–34°C (Table 13.1), but there is a wide range in reported incubation temperatures

Table 13.2 Variation in the incubation periods recorded in various species of megapode.

Species	Range in incubation period (days)	Initial egg mass (g)	Reference#
Alectura lathami	42–54	185	1
Leipoa ocellata	44–99	175	2
Macocephalon maleo	62–85	220	3
Eulipoa wallacei	49–99	100	4
Megapodius pritchardii	50–80	80	5
Megapodius nicobariensis	72–85	110	6

#References: 1 – Baltin (1969); 2 – Booth (1987a); 3 – Dekker (1988); 4 – Heij *et al.* (1997); 5 – Göth (1995) and Göth and Vogel (1997, 1999); 6 – Sankaran and Sivakumar (1999).

(27–39°C; Table 13.1). There is no obvious difference in temperatures recorded for mound and burrow nesting species (Table 13.1). During incubation the range in temperature individual eggs experience is usually quite narrow (2–3°C) but occasional larger ranges in temperature (5–10°C) are experienced (Crome and Brown 1979; Booth 1987a; Jones 1988b; Booth and Thompson 1991). As incubation proceeds the embryo grows in size and its metabolic heat production increases such that the egg temperature rises to 2–4°C above that of the surrounding soil (Booth 1987a; Seymour *et al.* 1987; Seymour 1991), so that in some cases egg temperature may approach 40°C. Within a species, temporal variation of temperature within a single nest throughout a nesting season, or between different nests, results in a wide variation in incubation period (Table 13.2).

Physiology of embryonic development

Egg characteristics and adaptations for gas exchange

Megapode eggs are 2–3 times larger in size compared to most bird of similar body size (Jones *et al.* 1995), and the eggs contain a large proportion of yolk (>50%) resulting in high energy contents (Vleck *et al.* 1984; Dekker and Brom 1990). Burrow nesting species have slightly larger yolk content than mound nesting species (Dekker and Brom 1990). These characteristics reflect the extremely precocial nature of megapode chicks (Sotherland and Rahn 1987). Perhaps the most specialised feature of megapode eggs is the relatively thin eggshell, only 50–80% of thickness of contact incubating species thickness, (Seymour and Ackerman 1980; Booth 1988), which has unusually shaped pores (Booth 1988; Booth and Seymour 1987; Booth and Thompson 1991). The thin eggshell is not a liability because once eggs are buried they are not subjected to mechanical damage, as is the case in open nesting birds. The unusual shape of the pores is an adaptation to development in high humidity, high carbon dioxide and low oxygen conditions (Booth and Seymour 1987; Seymour *et al.* 1986).

Megapode eggshell pores typically consist of a large diameter opening on the outside surface of the eggshell that is partially plugged with inorganic material

(Board *et al.* 1982; Booth and Seymour 1987). The pore tapers to a very small diameter approximately two thirds of the way through the shell, and persists at this small diameter until opening at the inside surface of the shell (Booth and Seymour 1987; Booth and Thompson 1991). This inner section of the pore offers the greatest resistance to diffusive gas exchange. When it is lost, due to calcium mobilisation from the inner eggshell surface during the latter third of incubation (Booth and Seymour 1987; Booth and Thompson 1991) eggshell gas conductance increases 1.5 to 3-fold. This facilitates the exchange of oxygen, carbon dioxide and water vapour between the embryo and nest environment at a time when metabolic demands are greatest (Seymour *et al.* 1986; Booth and Seymour 1987; Booth and Thompson 1991). By this mechanism megapode embryos experience similar respiratory gas tensions and water loss as open nesting avian species despite the challenging conditions of the underground nesting environment (Seymour *et al.* 1986; Booth and Thompson 1991).

Tolerance of variation in incubation temperature

Most bird embryos are very tolerant of relatively large (>10°C) decreases in incubation temperature for periods of time from a few hours to a few days (Webb 1987). During these cool periods embryos cease development but are able to recommence development once the temperature has been returned to a normal level. However, avian embryos are usually intolerant to chronic exposure to sub-optimal temperatures (2–5°C below optimal) because embryonic development continues but lethal teratogenic abnormalities can occur (Webb 1987).

Megapode embryos are remarkably tolerant to sub-optimal incubation temperatures. Presumably the inability of parent birds to regulate incubation temperature as precisely as brood incubators has lead to increased tolerance to long-term temperature fluctuations. Optimal incubation temperature for malleefowl eggs is 34°C with hatchability decreasing at higher and lower constant temperatures (Booth 1987a). Incubation of malleefowl and Australian brush-turkey eggs at low temperature results in different patterns of embryonic oxygen consumption and total energy expenditure (Figure 13.2). Incubation at higher temperatures results in a sigmoidal pattern of oxygen consumption (Vleck *et al.* 1984; Booth 1987a). During the plateau phase much energy is transferred from the yolk to internal fat bodies and the remaining yolk becomes internalised as residual yolk attached to the small intestine (Vleck *et al.* 1984; Booth unpublished). Embryonic tissue growth slows remarkably during the plateau phase of oxygen consumption as tissues mature in preparation for hatching (Vleck *et al.* 1984). Incubation at low temperature results in a prolonged plateau phase of oxygen consumption that is followed by a decrease in oxygen consumption before hatching (Booth 1987a; Figure 13.2), a pattern similar to ratite (Vleck *et al.* 1980), and most reptile eggs (Thompson 1989). This prolonged plateau phase before hatching is the primary reason for greater embryonic energy consumption during incubation at lower temperatures. However it is not known whether chicks from eggs incubated at lower temperature hatch with a larger body mass and smaller residual yolk than chicks incubated at higher temperatures as occurs in some chelonian and crocodilian

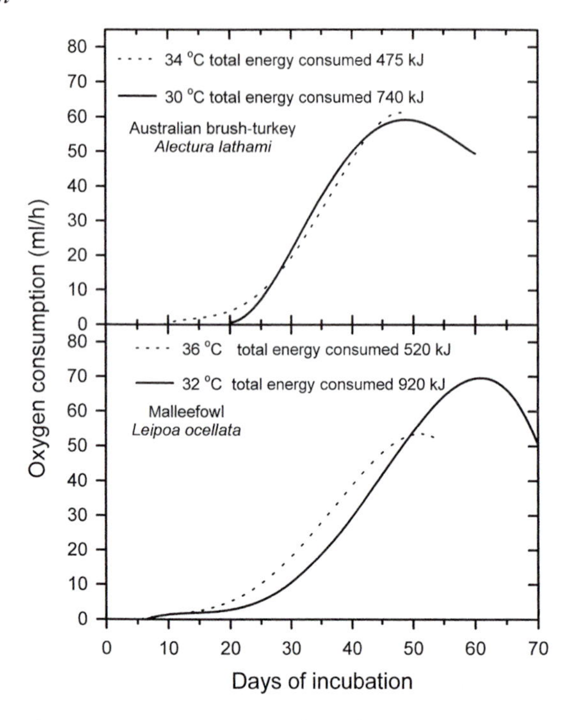

Fig. 13.2 Pattern of oxygen consumption and total energy expended during embryonic development in malleefowl and Australian brush-turkey embryos. Data from Vleck *et al.* (1984) and Booth (1987a and unpublished).

hatchlings (Booth and Thompson 1991; Booth 2000). It is also unclear whether incubation at different temperature influences fitness of the hatchling.

Hatching and escaping the nest

Megapode chicks do no use an egg-tooth on the end of the beak and neck muscles to break out of the shell as is the case in most birds (Jones *et al.* 1995). Instead they use their feet and movement of the shoulders to shatter the shell and break out (Frith 1959a, 1962; Seymour 1984b, 1991; Vleck *et al.* 1984; Göth 2000). Most avian species require a period of 24–48 hours to make the transition from chorio-allantoic respiration to lung respiration. During this time fluid in the bronchi and parabronchi is absorbed into the circulatory system and the lung grows in size as new parabronchi are formed. This allows respiratory gas exchange to be slowly switched from the chorio-allantois to the lungs (Seymour 1984b, 1991). However megapode chicks must make this switch within minutes (Frith 1959a, 1962; Seymour 1984b, 1991; Vleck *et al.* 1984; Seymour *et al.* 1987). The thin fragile nature of megapode eggshell makes the hatching process relatively easy, and sufficient exchange via the lungs occurs to keep the chick alive during this process. When the shell is broken and the chorio-allantoic membrane ripped between 5 and 25 ml of fluid (presumably allantoic fluid) drains from the egg in both Australian brush-turkey and malleefowl eggs (Seymour *et al.* 1987). This excess fluid is a result of lower than normal water loss during incubation and metabolic water production during embryonic development (Seymour *et al.* 1987).

After the egg is broken, chicks rest for long periods of time, e.g. for Australian brush-turkeys a mean of 16 hours and range 0.5–36 hours (Göth 2000) while their lungs absorb the excess fluid form parabronchi before they begin to dig their way out of the nest (Seymour 1984b, 1991; Göth 2000). Once hatched, malleefowl and brush-turkey chicks take between 24 and 55 hours to dig their way out of the mound (Frith 1959a, 1962; Seymour 1984b, 1991; Vleck *et al.* 1984; Göth 2000). During this time 5–10 minute busts of active digging are inter-spaced by 10–80 minute periods of rest (Vleck *et al.* 1984; Göth 2000). The burrow nesting Maleo (*Macrocephalon maleo*) chick may take several days to dig its way to the surface once hatched and also rests for long periods between active digging bouts (Jones *et al.* 1995). Digging out of the nest is an energetically demanding process and a considerable amount of energy in the residual yolk at hatching is used during this process (Vleck *et al.* 1984).

The digging out process has been described in detail for the Australian brush-turkey (Seymour 1991; Göth 2000). Chicks first form a small air-filled chamber about them. Next they scratch material from the top of the chamber and compress this freed material below them. Seymour (1991) described how chicks lie on their back and use their feet to displace mound material above them and then use their back and shoulders to compress the freed material beneath them. By contrast, Göth (2000) found chicks remained up-right during the digging process, lifting one foot at a time above the head to dislodge mound material and then compressing this material beneath them by treading on it with their feet. The digging out period is crucial in allowing the plumage to dry (Booth 1984, 1985; Göth 1995, 2000), for preening of plumage, and development of neuromuscular co-ordination and reaction to environmental stimuli (Göth 1995, 2000). On reaching the surface megapode chicks (Figure 13.3) are often physically exhausted and unstable on their legs for several hours which makes them particularly vulnerable to predation at this stage (Booth 1987c; Jones 1988a; Jones *et al.* 1995; Göth 2000).

Ontogeny of thermoregulation

Megapodes produce the most precocial hatchlings of any group of birds (Nice 1962), but see Göth (2000) for an alternate point of view. There is absolutely no parental care so brooding of chicks by parents to prevent hypothermia is not possible. Thus hatchlings need to be competent homeotherms on emergence from their nest environment. The first signs of homeothermic thermoregulation appear during the last week of incubation while the chick is still inside the egg when there is a transient increase in oxygen consumption in response to chilling of the egg (Booth 1987b). Chicks hatch with well-developed primary feathers and a thick layer of pennaceous feathers (Clark 1960) and this plumage provides excellent insulation once dried. As soon as the chick's plumage has dried (typically 3–6 hours after hatching) chicks are competent homeotherms being able to increase their metabolic heat production at least 3-fold by shivering thermogenesis

(a)

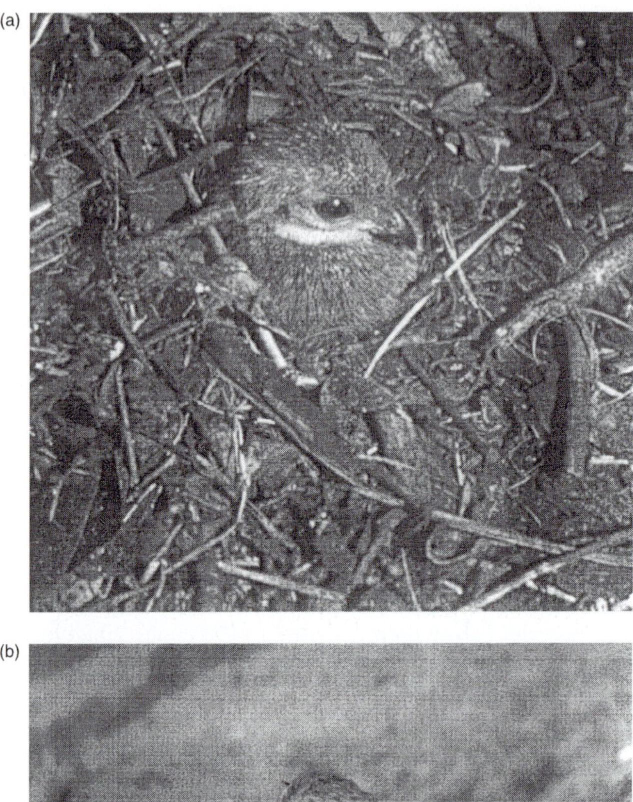

(b)

Fig. 13.3 Hatchlings of (a) the Australian brush-turkey (*Alectura lathami*) and (b) a malleefowl (*Leipoa ocellata*) emerging from their nests. Photographs by Darryl Jones.

in the face of a cold challenge (Booth 1984, 1985). The well-developed plumage of hatchlings is crucial for cold defence, and if the plumage becomes wet the ability to remain homeothermic is severely compromised. Indeed wet inclement weather after hatching has been documented as the cause of death in free-ranging malleefowl hatchlings (Priddel and Wheeler 1990).

Hatching success

Quantitative data for hatching success in natural mounds is only available for three mound building species the malleefowl, Australian brush-turkey and Nicobar megapode. For malleefowl in different locations hatching success of 52% (Frith 1959a), 84% (Booth 1987c), 64% (Brickhill 1987) and 92% (Benshemesh 1992) have been reported. Egg mortality was due to infertility, egg breakage by parents, failure to develop successfully, or predation by foxes. Predation by foxes was by far the largest cause of egg loss in the two areas where fox predation occurred (Frith 1959a; Brickhill 1987). For Australian brush-turkeys field hatching success of 87% (Jones 1988a), 72% (Jones and Everding 1991) and 77% (Birks 1996), have been reported with little evidence of predation of eggs. Likewise the hatching success of the Nicobar megapode was also high averaging 80% with no obvious signs of egg predation (Sankaran and Sivakumar 1999). These values for hatching success are generally higher than the mean of 69% for 158 species of contact incubating birds (Owens personal communication), suggesting that mound nesting is at least as good if not a better method of incubating avian eggs. Indeed, it can be argued that successful hatching in megapodes is equivalent to fledging in brood incubating birds, and fledging success averages only 48% in 192 species of brood incubating birds (Owens personal communication).

Areas for future research in megapode incubation

Several areas remain unexplored in megapode incubation. The gaseous nest conditions in burrow nesting species that use solar or geothermally heated soils for incubation need to be examined. If the respiratory gas tensions are not greatly different from the atmosphere, then it would be interesting to see if eggshells of these species have the same unusual pore structure seen in mound building species and which has been interpreted as an adaptation to hypoxic and hypercapnic conditions.

The question of whether or not incubation temperature affects the quality of hatchlings needs to be addressed. It would appear that embryonic energy expenditure is influenced by incubation temperature in malleefowl and the Australian brush-turkey (Figure 13.2), but this needs to be verified in other megapode species. In reptiles, incubation temperature can influence hatchling attributes such as amount of residual yolk, body morphology, and post-hatch behaviour and growth, and sex determination (Deeming and Ferguson 1991b) and thus incubation temperature may be an important determinant of hatchling survival and therefore reproductive fitness (Booth 2000). Given this situation in reptiles, the hypothesis that incubation temperature influences post-hatch fitness of megapode chicks needs to tested.

Amongst mound building species of megapode, malleefowl are the only species thought to be sophisticated active regulators of mound temperature (Seymour and Bradford 1992; Jones *et al.* 1995) as demonstrated by the manipulative experiments carried out by Frith (1957). Incubation mounds of other species are thought to largely self-regulate temperature without a great deal of input from the attending birds (Seymour and Bradford 1992; Jones *et al.* 1995). If this is so, then do attending birds have the ability to deliberately manipulate mound-tending activities in order to regulate mound temperature, or are they completely reliant on mound temperature self regulation for incubation? These ideas could be tested on mound building species by manipulative experiments in which mounds are artificially heated and cooled as in Frith's (1957) study to see if birds change their mound tending behaviour in response to artificial temperature manipulation.

Acknowledgements

We thank Ann Göth, Ross Sinclair, Sharon Birks and Sharon Wong for many discussions on megapode incubation techniques, and members of the megapode specialist group for suggestions and ideas.

14 *Characteristics and constraints of incubation in hummingbirds*

W. A. Calder III

Reproductive fitness is the crucial link by which lineages persist and evolve. The fundamentals of every field of study are manifested in this linkage: genetics, embryology, development, physiological regulation, and the behaviour of nest siting and construction, incubation, and acquisition and management of energy and food matter – all quantities of which are size-dependent. Hummingbirds (Order Apodiformes, Family Trochilidae) are some of the smallest birds and they lay some of the smallest eggs, in small clutches (Table 14.1). Mass-specific metabolic rates ($kJ\,d^{-1}\,g^{-1}$ body mass), like egg mass g^{-1} body mass, are inversely proportional to body size and are most intense in the size range of hummingbirds (Calder 1996; Brown and West 2000). In order to collect floral nectar, its main energy source, a hummingbird usually relies upon hovering flight which is energetically very expensive. In fact, when hovering, the rufous hummingbird (*Selasphorus rufus*) has the highest metabolic rate recorded for animals (Suarez 1992). In order to support this activity, the hummingbird needs a high daily nectar intake.

When energy reserves, derived from nectar intake through the day, are nearly depleted before there is enough light to resume feeding, the bird can conserve energy by entry into hypothermic torpor, i.e. 'a regulated state of physiological dormancy in which body temperatures may drop by 20°C or more from normothermic levels and metabolic rate may be reduced to as little as 5% of its normothermic value...' (Hiebert 1992). Torpor is a last resort for surviving a crisis when stored energy proves inadequate for maintaining a body temperature of 39–40°C through the nocturnal fast. Torpor is a good strategy for the survival of the female, but how does the accompanying compromise in regulation of body temperature affect the development, and indeed the survival, of hummingbird embryos during incubation? This chapter examines the incubation behaviour of hummingbirds to determine how the energetic demands of self-maintenance are successfully reconciled with the needs of their offspring.

Incubation in hummingbirds has received little attention. This neglect is especially surprising, when one considers that they constitute one in 29 of all bird species on the planet and include the smallest birds and the smallest pollinators outside of the Insecta. The sparse information available (Table 14.1) is from only a few species of the Trochilinae that breed at mid-to-high latitudes (30–60° N). Hence this review of hummingbird incubation is biased towards situations where steeper gradients for heat loss pose a greater challenge to incubation and fledgling success. However, the northern species illustrate the basic processes of keeping

Table 14.1 Comparison of female body mass, egg dimensions and mass, and incubation period for hummingbirds, a tody and small passerines. Data for incubation period predicted from egg mass, and egg mass predicted from (1) egg dimensions and (2) adult body mass, are also provided.

Species	Female mass (g)	Egg L (mm)	Egg B (mm)	Egg mass (g)	Incubation period (d)	Predicted incubation period**	Predicted egg mass (g) 1*	Predicted egg mass (g) 2$	Predicted egg mass 1/ egg mass	Predicted egg mass 2/ egg mass	References#
Apodiformes											
Stellula calliope	2.85	12.1	8.3	?	15–16		0.46	0.50			1
Cynanthus latirostris	3.01	12.5	8.2	?	15		0.46	0.53			2
Selasphorus sasin	3.25	12.7	8.6	0.39	12–22	9.2	0.51	0.56	1.31	1.44	3
Calypte costae	3.25	12.4	8.2	?	15–18		0.46	0.56			4
Archilochus alexandri	3.47	12.7	8.3	?	12–16		0.48	0.59			5
Selasphorus rufus	3.58	13.1	8.8	?	15–17		0.56	0.60			6
Archilochus colubris	3.61	13.0	8.6	?	12–14		0.53	0.61			7
Selasphorus platycercus	3.67	13.0	8.8	0.50	16–19	9.7	0.55	0.61	1.10	1.22	8
Calypte anna	4.01	12.7	8.5	0.50	14–19	9.7	0.50	0.66	1.00	1.32	9
Eugenes fulgens	6.72	15.4	10.0	0.97	?	11.4	0.84	0.99	0.87	1.02	10
Lampornis clemenciae	6.80	15.3	10.1	0.74	17–19	10.6	0.86	1.00	1.16	1.35	11
Mellisuga helenae	1.9	?	?	?	21–22						12
Mellisuga minima	2.4	?	?	?	12–13						12
Patagona gigas	19.3	?	?	?	12						12
Coraciiformes											
Todus mexicanus	6.5	16.0	13.5	1.44	21–22	12.5	1.60	0.97	1.11	0.67	13
Passeriformes											
Nectarina jugularis	8.1	17.0	12.0	?	13–14		1.34	1.15			14
Nectarina osea	7.9	15.4	11.2	0.99	13–14	11.5	1.06	1.13	1.07	1.14	15
Anthreptes metallicus	7.0	16.9	11.5	1.13	?	11.8	1.22	1.02	1.08	0.90	15
Polioptila caerulea	5.8	14.6	11.5	0.98	13	11.4	1.06	0.88	1.08	0.90	16
Auriparus flaviceps	6.3	15.3	11.0	0.96	14–18	11.4	1.01	0.94	1.06	0.98	17
Regulus regulus	5.5	13.6	10.4	0.78	15–17	10.8	0.81	0.85	1.04	1.09	18
Regulus ignicapillus	5.2	13.6	10.6	0.69	14–16	10.5	0.84	0.81	1.22	1.17	18
Regulus satrapa	6.1	13.3	10.4	9.77	~15	10.8	0.79	0.92	1.02	1.19	19

L = length; B = breadth; **Rahn and Ar (1974): Days = $11.5\, m_{egg}^{0.24}$; *Hoyt (1979): Mass = $5.48 \times 10^{-4}\, LB^2$; $Grant (1983) and Scott and Ankney (1983): $m_{egg} = 0.22\, m_{ad}^{0.79}$; References: 1 – Calder and Calder (1994); 2 – Mitchell (2000); 3 – Baltosser (1996); 4 – Powers and Wethington (1999); 5 – Baltosser and Russell (2000); 6 – Robinson *et al.* (1996); 7 – Calder (1993); 8 – Calder and Calder (1992); 9 – Russell (1996); 10 – Powers (1996); 11 – Williamson (2000); 12 – Schuchmann (1999); 13 – Kepler (1977); 14 – Frith (1982); 15 – Cramp and Perrins (1993); 16 – Ellison (1992); 17 – Webster (1999); 18 – Cramp (1992); 19 – Ingold and Galati (1997).

eggs warm, faced by all hummingbirds, conveniently exaggerated to make them more obvious.

Basic breeding biology

Hummingbird mating systems are characterised by polygyny, usually based around male territoriality centred around rich, but ephemeral floral resources, some in sequential seasonal phenologies (Waser and Real 1979). These territories attract females to the scenes of courtship displays and mating. At lower latitudes, however, some hummingbird species are 'trap-line feeders', i.e. they feed along regular routes rather than spending the day defending 'central patch' territories.

The use of a lek arena for communal courtship displays has been observed mostly in the Phaethornithinae and in a few species of the Trochilinae (Schuchmann 1999). Use of a lek by the broad-tailed hummingbird (*Selasphorus platycercus*) at temperate latitudes is questionable (Calder and Calder 1992). The breeding seasons of many species nesting at higher latitudes and elevation are brief insertions between two long migrations, limiting many hummingbirds to a single annual clutch (Calder and Calder 1992).

All nesting duties are performed entirely by females. Rare observations of male help at hummingbird nests seem to be just aberrant exceptions of questionable significance (Johnsgard 1997a; Schuchmann 1999). Clutch size is two eggs and chicks are altricial. Table 14.1 shows data for reproductive biology of 11 species of North American hummingbirds and for comparisons, similar relationships are shown for four small neotemperate passerines, and a tody (Coraciiformes; Todidae). Incubation times reported for the smallest and largest hummingbirds are also shown in Table 14.1.

The use of allometric predictions based upon female body mass and/or fresh egg mass leads to some generalizations. Firstly, although incubation period has been determined to be scaled to body mass (see below), empirical determinations do not consistently conform to the general trend. Incubation periods of hummingbirds are generally longer by about 40–60% than size-predictions, except for the bee hummingbird (*Melisuga helenae*) which looks more like a clutch of infertile eggs or a re-laying after predation. In particular, the wide range of incubation times could be explained by assuming that longer incubation reflects slowing in development due to food shortage which necessitates the female to resort more frequently to hypothermia. However, in the outdoor hummingbird aviary at the Arizona Sonora Desert Museum, where food availability was unlimited, hummingbirds all had incubation periods of 15–17 days. In order of both incubation period and, approximately in body size: 15 days for the calliope hummingbird (*Stellula calliope*), Costa's hummingbird (*Calypte costae*), broad-billed hummingbird (*Cynanthus latirostris*); 16 days for the black-chinned hummingbird (*Archilochus alexandrii*); and 17 days for the Anna's hummingbird (*Calypte anna*). These periods were lengthened if temperatures were cooler and females spent more time off the nest (Krebbs 1999, personal communication). In cases of incubation periods reported as 12 days (Table 14.1) actual onset of incubation may have been missed and cannot be taken seriously when they are not consistent with other species, often congeners, of approximately the same body size. Otherwise, incubation periods of hummingbirds and the Puerto Rican tody (*Todus mexicanus*) are generally longer than for the passerines.

Thirdly, there is considerable variation in reported egg masses and incubation periods, which must be considered suspect. Eggs lose water by evaporation from the time they are deposited in a nest and observations of 'fresh' eggs may have been a day or more after laying. Thus egg masses which are not consistent between species or in a size progression, e.g. blue-throated hummingbird (*Lampornis clemenciae*) and Allen hummingbird (*Selasphorus sasin*) which are consistent in egg dimensions were probably not weighed when fresh. By contrast, for three of four passerines and the Puerto Rican tody, actual egg sizes are larger than predicted. The relatively small eggs of hummingbirds may reflect constraints of the high exertion and metabolic demands of hovering for the females. If these apparent irregularities are excluded, it seems true that hummingbirds lay relatively small eggs, the ratios of allometric predictions to actual egg masses being considerably larger than 1, as are several of the comparisons with other kinds of small birds. Finally, readily available natural history documentation on hummingbird reproduction is meagre and clearly, more old-fashioned, but more accurate, natural history investigations are warranted!

Hummingbirds and incubation: a scaling perspective

In birds in general, incubation periods scale positively with egg size ($m^{0.22}$; Rahn and Ar 1974) and adult body size ($m^{0.17}$; calculated from: Rahn and Ar 1974; Scott and Avery 1983; Grant 1983). However, incubation of hummingbird eggs is up to 60% longer than predicted from their small mass (Calder 2000). Prolonged incubation periods increase the vulnerability of both adults and eggs to predation, desiccation, and stochastic events which affect nectar and small insect food resources. Slower incubation may be a direct consequence of resource shortage, which could also be related to lost development time when the bird-nest incubation unit and eggs are in hypothermia. After natural selection for subsistence on nectar and very small insects, do the quantitative details of other life history traits of hummingbirds fit what are predicted according to the scaling rules for larger birds? Have reproductive mechanics dictated the lower limit for avian body size? These questions are investigated by considering egg mass and surface area, and shell characteristics in order to gain some insight into why clutch size should be limited in hummingbirds to two eggs.

Mass and surface area

Avian egg size scales at less than linearity, so that the physically large eggs of large birds are smaller in proportion to their body size (Scott and Ankney 1983; Grant 1983). Therefore, the 1,692 g egg of a 120 kg ostrich (*Struthio camelus;* Hoyt 1979) amounts to 1.41% of the mass of the female. By contrast, a 3.5 g female broad-tailed hummingbird lays a 0.5 g egg that is 14.3% of female mass. Have hummingbirds hit an upper limit for provisioning energy into egg formation and are unable to synthesise more than two eggs? This seems not to be the case, because the Puerto Rican tody has a body mass of 6.5 g, which is larger than *Selasphorus* hummingbirds (but at the middle of the size range for other hummingbirds), yet egg mass is 22.2% of body mass and clutch mass can reach four eggs (Kepler 1977). In temperate Europe, the female goldcrest (*Regulus regulus*) has a body mass of 5.5 g and its egg mass is 0.78 g which is 14.2% of body mass (Cramp 1992). Although comparable to the broad-tailed hummingbird in relative egg size, clutch size in the goldcrest averages 10.4 eggs (Cramp 1992). Similarly, the 5.2 g firecrest (*Regulus ignicapillus*) has eggs with a mass of 0.69 g (13.2%) and an average clutch size of 8.8 eggs (Cramp 1992).

Perhaps it is the diet of nectar which limits synthesis of such relatively large eggs and determines how many eggs can be included in a clutch? Sunbirds (Nectariniadae, Passeriformes) are the nectarivore equivalents of hummingbirds in the eastern hemisphere and with a size range of 4–20 g the smaller species are equivalent to hummingbirds. A 7.9 g Palestine sunbird (*Nectarinia osea*) lays three 0.99 g eggs each 12.5% of female mass (Cramp and Perrins 1993) whereas other sunbirds regularly lay up to three eggs per clutch (Austin and Singer 1961; Prinzinger *et al.* 1979). The Nile Valley sunbird (*Anthreptes metallicus*) has a 7 g body mass, 1.13 g egg mass (16.1%) and is reported to incubate up to 4 eggs (Cramp and Perrins 1993).

Perhaps the surface area available for contact incubation is limited with relatively large eggs restricts the ability of the female to contact incubate a clutch of eggs? Walsberg and King (1978a) found that surface area of the adult bird (A_{body}) scaled with adult mass (m_{ad}):

$$A_{body} = 8.11\, m_{ad}^{0.67}. \tag{14.1}$$

Paganelli *et al.* (1974) found a similar scaling exponent for surface area of avian eggs (A_{egg}) with egg mass (m_{egg}):

$$A_{egg} = 4.84\, m_{egg}^{0.66}. \tag{14.2}$$

Egg mass scales with adult mass (Grant 1983; Scott and Ankney 1983):

$$m_{egg} = 0.22\, m_{ad}^{0.79}. \tag{14.3}$$

For successful incubation, female-to-egg surface contact should be sufficient for heat transfer. Here the area of the brood patch is assumed to be 20% (Turner 1991; Chapter 10) and so the area of the egg as a function of body mass can be calculated:

$$A_{egg} = 4.84 \times 0.20(0.22\, m_{ad}^{0.79})^{0.66} \Rightarrow 0.35 m_{ad}^{0.52}. \tag{14.4}$$

Combination of equations 14.4 and 14.1 allows the fraction of available incubation surface area (A_{inc}) of the adult necessary for female-egg contact to be determined:

$$\%A_{inc} = 100 \times [0.35\, m_{ad}^{0.52}/8.11\, m_{ad}^{0.67}] \Rightarrow 4.32\, m_{ad}^{-0.15}. \tag{14.5}$$

Therefore, as body mass increases then a proportionately smaller area of the body is in contact with each egg. Hence, for a 120 kg ostrich each egg is predicted to contact 0.75% of the adult body surface area (a total of 14.2% for a clutch of 19 eggs). By contrast, for a 3.5 g broad-tailed hummingbird, $\%A_{inc}$ is predicted to be 3.6% of the adult body surface area per egg (a total of 7.2% for the clutch of two eggs). However, for a 5.5 g female goldcrest, each egg would occupy 3.3% of the adult body area which equates to 33.5% of the whole body area for a clutch of 10 eggs.

With regard to mass and surface area, comparison of hummingbirds with other small birds suggests that there is no absolute physical or energetic constraint *per se* to laying more than a 2-egg clutch in hummingbirds. It is possible that the constraints on reproductive performance are not associated with incubation but may relate to the ability of the lone female hummingbird to provision more than two offspring.

Eggshell characteristics

Eggshell size limits the amount energy reserves that could be contained within the egg prior to laying. Scaling for strength of rigid eggshells, as exoskeletons, insures adequate protection for the delicate embryo from mechanical trauma and evaporation, yet cannot be excessive when it comes to hatchability (Ar *et al.* 1979). The eggshell is also an effective mediating barrier and factors such as gas conductance, functional pore area and pore length, all scale to initial egg mass (Rahn and Paganelli 1990).

Egg mass is scaled to adult body mass with approximately the same fractional exponent ($m^{0.79}$) as the adult standard metabolic rate ($m^{0.73}$; SMR = 'basal'), so it is reasonable to expect a proportional continuity between these two functions. Actual egg mass of a broad-tailed hummingbird is 87% of the allometric prediction. Lacking SMR data from the broad-tailed hummingbird, Lasiewski's (1963) SMR for the rufous hummingbird is substituted to find that it is 91% of the active phase (daytime) allometric prediction, but 116% of the resting phase (nocturnal) prediction. Lasiewski measured metabolic rates when birds were 'resting in the dark' (of the metabolic chamber), but not specifically in their night (resting phase) of the daily cycle, when metabolism is minimal (Aschoff and Pohl 1970). Assuming that daytime allometry, equivalent to the Lasiewski and Dawson (1967) equation, is the appropriate basis for this comparison, it appears as if hummingbird egg size and SMR are similarly lower than size alone would predict, although not in the same proportions. The incubation period is 60% slower, but the resting metabolism is only 10% slower. Comparable figures for the Peurto Rican tody are a 71% slower incubation and a 33% higher metabolism (Kepler 1977; Merola-Zwartjes and Ligon 2000). Available information suggests that sunbirds do not depart significantly from allometric predictions in either incubation time or metabolic rate. The only consistent correlation seems to be that hummingbirds lack the combined participation in time and effort investment by both parents. Thus if the female must not only lay and incubate the eggs, but also feed the hatchlings she is limited to a clutch of two relatively small eggs.

Eggs decrease in mass during incubation, due only to diffusive loss of water vapour, the mass of carbon dioxide release balancing the oxygen intake (Tullett

1984). Bird species whose eggs require longer than average incubation periods also have reduced porosity which lowers evaporative water loss rates in compensation (Chapter 3). This means that water loss throughout incubation comes very close to a typical loss of 18%, independent of egg size and adaptive modifications (Chapter 3). Limited records for broad-tailed hummingbird show that the longer incubation (160% of the scaling prediction) is compensated for by a reduction in rate of daily evaporation to 61% of the scaling prediction (Ar and Rahn 1980; Calder 2000). Therefore, as is seen in other species, e.g. the wedge-tailed shearwater (*Puffinus pacificus chlororhynchus*; Ackerman *et al.* 1980), the longer incubation period for a hummingbird is compensated for by a lower water vapour conductance ($1.6 \times 0.61 = 0.98$).

Incubation periods of hummingbirds are both relatively longer than scaling to size predicts (Table 14.1; Calder 2000, unpublished). In cooler climates at higher elevations or latitudes, nest insulation is thicker, so nest-building probably takes more time, further prolonging the nesting cycle. Furthermore, at higher elevations, lower barometric pressure increases the diffusivity of gases, including water vapour (Walsberg 1985).

The very low reproductive potential of hummingbirds probably is due to the combined effects of some (or all) of these factors: brief seasons, prolonged nesting cycle, higher rates of water loss, and the vulnerability of unguarded nests during maternal absence for foraging. Demographic constraints and slow recovery following stochastic events are inherent in the low reproductive potentials. Hence, there would have been good reason to expect natural selection of faster incubation, but that did not occur. Perhaps, therefore, hummingbird lineage has radiated to limits that are set for body size in reproduction.

Nest site selection

For hummingbirds, reproductive fitness is not a product of prolific reproductive output, which has been ruled out by small clutch size, short breeding seasons in some habitats, and intense mass-specific metabolic demands of self-maintenance by the female. Quality of care must compensate for this low reproductive potential. This quality of care begins with selection of nest location and its construction. Nest sites are probably located to optimise logistical considerations of distance to flowers, heat conservation during the hen's feeding recesses, and concealment by vegetation to reduce predation when the nest is unattended.

Hummingbirds nest in a range of climates, from high-latitude maritime, to montane areas through to lowland hot desert. During incubation recesses, unattended eggs may equilibrate with temperatures that approach the lower limit for continuing development, or if not shaded from direct solar radiation, too hot for embryo survival or fledging of the offspring. Inherent in being both homeothermic and very small is being 'tightly coupled' to the physical environment (Porter and Gates 1969). Small birds have high surface to volume ratios, thin surface insulation, and their bodies have low 'thermal inertia'. During hummingbird incubation the response to this condition is largely behavioural, through careful evaluation of the incubation site so as to take advantage of natural features that might reduce the potential for heat-loss via conduction, convection, and radiation from the nest. These features are then supplemented by building a nest from materials which also favour maintaining an energy-conserving microclimate. See Chapter 2 for more details of nest morphology in hummingbirds.

The result is that some hummingbirds can incubate with amazing success under quite adverse conditions. The Anna's hummingbird in the California chaparral is the only hummingbird that lays its first egg in December (Stiles 1973; Russell 1996). At an elevation of 2,900 m in the Elk Mountains of Colorado, early melting of winter snow-pack allows some broad-tailed hummingbirds to begin incubation as early as late May. Although late storms may still occur in early June, bringing more snow and dropping the minimum temperature down to −6°C, this does not interrupt incubation nor does it delay successful fledging (Calder unpublished). Crucial to this early activity in both species is the availability of nectar from the first flowers, an essential energy supply.

The radiative environment appears to be a primary consideration in nest site selection in cooler climates. At higher elevations the nocturnal sky temperature can drop to below −20°C which acts as a sink for heat radiated from the female's dorsal surface. Broad-tailed hummingbirds reduce this loss significantly by positioning the nest under a branch (Figure 14.1), which also shelters their nests from

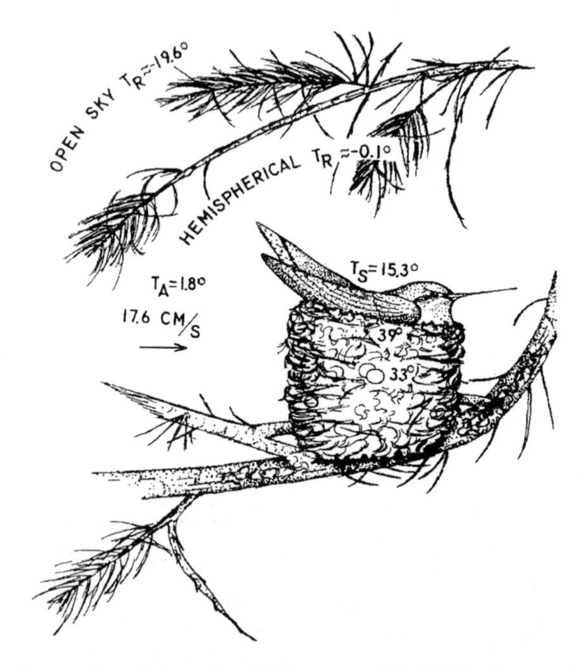

Fig. 14.1 An incubating female broad-tailed hummingbird (*Selaphorus platycercus*) in her nesting microclimate, with the potential gradient for conductive heat loss. She selected a site that is sheltered from the radiative heat sink of the cold night sky, which at the 2,910 m elevation, has one-fourth of the 'greenhouse effect' observed at sea level. (Data from Calder 1973b; reproduced with permission © Nuttall Ornithological Club.)

precipitation and potential predators (Calder 1973b). If branches are offset the nest can be built along the side of the lower branch with cantilever support of the nest with spider-web strands as guy wires. Nesting under overhangs was also noted for the Ecuadorian hillstar (*Oreotrochilus chimborazo*) by Corley Smith (1969).

In addition to selecting nest sites under branches to reduce net radiation loss, hummingbirds may also select for topographical blocking of the lower part of the sky. Whether fortuitous or by behavioural design, a female Anna's hummingbird built her nest down between the steep section of a canyon in early February in Arizona. During the night the radiation temperature of the celestial hemisphere descended to −20°C but the canyon walls, which blocked 29.6% of the sky, were considerably warmer, descending to +10–11°C at dawn. Trees on the horizon blocked radiative exchange with another 10% of the cold sky and overhead branches shielded another 3%. Thus warmer radiation temperatures were substituted for about 43% of the cold sky, reducing the cost of maintaining egg temperature overnight (Calder 1974).

Incubation behaviour

Females accomplish all of their foraging and off-nest self-maintenance in about one quarter of the daytime and so they are on their nests for an average of 73.9% (±5.88 SD; range 64–84%) of the time (Baltosser 1996). Attentiveness for six temperate species averaged 77.2% (±3.13 SD, range 74–83%) compared with a mean of 71.8% (±6.46 SD, range 64–84%) for nine species of tropical and subtropical species. On this basis Baltosser (1996) concluded that incubation attentiveness by hummingbirds was independent of latitude and species. See Figure 6.10 for length of incubation sessions and recesses.

In four independent studies of the temperate Anna's hummingbird, incubation attentiveness not only spans most of the range reported for other species but also shows a fair correlation with local physical environmental conditions and relative ease in getting food. Attentiveness of 69% was recorded at a nest in south-eastern Arizona incubated in early February (air temperature 4–22°C; feeding day 689 min). The nearest feeder was 1.6 km to the south-west, but during recesses the female ascended north-east, into the Santa Catalina Mountains, where manzanita (*Arctostaphylos pungens*) was in bloom (Calder 1974). The highest attentiveness of 82% was observed in milder surroundings during late April (10–22°C; 22% longer feeding day of 843 min), with feeders and a profusion of flowers nearby in a garden in West Los Angeles (Howell and Dawson 1954). The other studies, with 75 and 79% attentiveness, were also in April and in Southern California (Smith *et al.* 1974; Vleck 1981a).

Long incubation periods may be a function of lack of egg turning (Chapter 11). Similar to the broad-tailed hummingbird, the palm swift (*Cypsiurus parvus*) has an egg of 1.2 g but an incubation period of 19 d some 160% longer than anticipated. Being stuck to the nest these eggs are not turned. Perhaps hummingbird eggs are also not turned, which contributes to their relatively slow rate of development.

There appear to be no studies of hummingbird development and growth in the egg but often upon return to the nest after feeding, the female engages in what appears to be a treading movement by her feet within the nest cup. In the process, this seems to thrust her body back towards the wall opposite her facing direction. This could be serving either to redistribute the nest floor lining, and/or to turn the eggs.

Nest sites chosen also offer a good lateral view, from which the incubating female can watch the environs and she will usually leave the nest to harass potential predators. As they depart on recess, female broad-tailed hummingbirds often issue a few dry cheeps as if to distract a potential predator from the nest. When it is hot and the sun reaches the nest, the female may raise herself above direct contact with the eggs, to shade the nest with her body and wings.

Temperature regulation of females and eggs during incubation

Hummingbird thermobiology and energetics

Laboratory measurements have yielded a detailed understanding of the physiology of torpor in hummingbirds, including its timing, thresholds, metabolic reduction, hypothermic body temperature regulation, and respiration (Bucher and Chappell 1989, 1992; Hiebert 1990, 1991, 1992). This tactic for energy conservation can be crucial to survival during imposed fasting, but such conservation benefits are in trade-off because a torpid hummingbird with a body temperature of 12.2–16.6°C (Bucher and Chappell 1992) cannot rapidly respond to threats of predation and environmental caprice.

Laboratory studies cannot, however, describe when hummingbirds resort to hypothermic torpor in nature. One hypothesis, the 'emergency hypothesis' (Hainsworth *et al.* 1977), explains the pattern of torpor under natural conditions by suggesting that torpor occurs only during energy shortage. By contrast, the 'routine hypothesis' (Kruger *et al.* 1982) contends that torpor occurs routinely, not just when food intake is insufficient. Records of the daily cycles in body masses of free-ranging broad-tailed hummingbirds show that body mass declines between dusk and dawn feedings. When data on overnight mass decrease are converted to their stored energy equivalents [estimated energy density of accumulated hummingbird fat: 34.2 $kJ\,g^{-1}$ (87% of 39.3 $kJ\,g^{-1}$ for pure fat) which adjusts for water and connective tissue associated with fat storage], the amount of energy is equivalent to the calculated cost of maintaining high nocturnal body and nest temperatures (Calder 1994).

Furthermore, the routine hypothesis also has not been supported by the relative infrequency of torpor, being confined to only a few situations in the annual cycles of broad-tailed and rufous hummingbirds: (1) upon arrival in the nesting locale before flowers are abundant; (2) during mid-season, as a last resort during incubation and brooding, when daytime feeding was interrupted by storms which halted foraging and storage of reserves adequate for nocturnal fasting (Calder 1994); and (3) to conserve fat and expedite preparation for a long migration segment

(Carpenter and Hixon 1988). Possible routine torpor has also been suggested in cave-roosting Ecuadorian hillstars of the high Andes, but the conditions of prior foraging or resources were unknown (French and Hodges 1959).

Maintenance of nest temperature

Nest temperatures of an Anna's hummingbird, as tracked with fine thermocouples during daytime incubation periods and recesses, showed maintenance of nests at about 10°C higher than the temperature of the air (Howell and Dawson 1954). Hence a bird as small as 4 g had the heating capacity for maintaining a high temperature (normothermia) in the nest for all the night. This routine avoidance of nocturnal torpor was favoured in the relative warmth of southern California. Vleck (1981a) did record torpor in a female Anna's hummingbird nesting in southern California during a period of record low temperatures and almost constant light rain. The calliope hummingbird has a body mass of 2.8 g and its high surface to mass ratio makes temperature regulation a particular challenge. However, normothermia was also observed during incubation in two calliope hummingbird nests located at an elevation of 2,077 m in north-western Wyoming where mean daily minimum air temperature was 5.9°C at dawn (Calder 1971).

Daytime storage of energy by female broad-tailed hummingbirds is generally adequate to fuel normothermic nest-warming through most, if not all nights, on a 2,900 m valley-bottom exposed to a thermal inversion due to nightly settling of cold air from the mountain-tops. When possible the females fed extensively at dusk, especially when artificial feeders were available. With limited fat gain from daytime feeding, the final load of unprocessed nectar amounted to a total energy storage of 11 kJ, slightly exceeding the cost-estimate for maintaining a high body temperature all night (Calder 1994). If this apparent excess was real, it might have been to offset the extra heat loss via the brood patch, as yet not quantified in hummingbirds. Even within an insulated nest, nocturnal heating costs of the female were 12% more than for a male, which flies up the slope to roost above the inversion.

Temperatures were monitored in nests of the broad-tailed hummingbird located at 2,900 m in Colorado using casts of broad-tailed hummingbird eggs (Calder 1975). These eggs were made from Silastic®, which has a thermal conductivity similar to real eggs, and contained either a thermocouple (36 gauge) or a 1 mm thermistor, connected to a potentiometric strip-chart recorder. Twelve episodes of hypothermic torpor were recorded from 8 nests during two summers. Five torpor episodes occurred during incubation (Figure 14.2). Each followed a day in which rains kept the females on the nests, inhibiting them from regular feeding for an hour or more. Other torpor episodes occurred at hatching or post-hatching and were associated with either rain, human presence, or impending abandonment due to flower failure. Torpor occurred in an average of 2 out of 5 nests recorded, thus some females were able to store enough energy to maintain normal high nest temperatures.

Incubation in hummingbirds seems, therefore, to proceed as in other birds with normal body temperature being maintained throughout embryonic development.

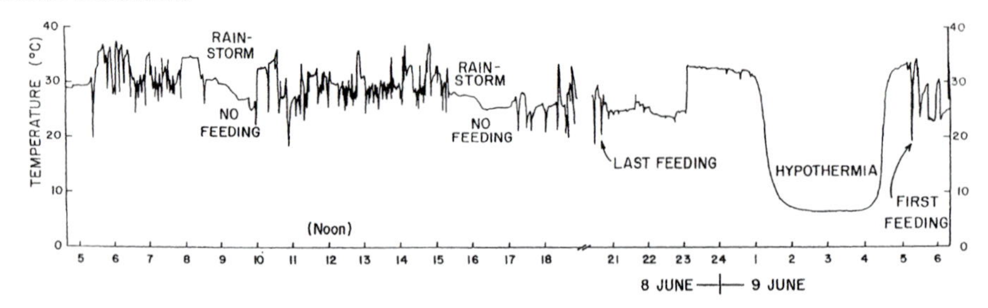

Fig. 14.2 Nest temperatures measured with a 36 gauge thermocouple in a Silastic cast of an egg from the broad-tailed hummingbird (*Selaphorus platycercus*). Recesses for feeding and self-maintenance between 0520 and 0830, 0945 to 1520, and 1710 to ~2050 were marked by 'egg' cooling. Rainstorms kept the bird on her nest for a total of three hours, one fifth of her potential feeding day. At ~2300 she apparently shifted or turned the eggs, registering a higher 'egg' temperature until ~0100, when the daytime feeding deficit apparently necessitated entry into hypothermic torpor to conserve enough energy for feeding at dawn. Such torpor bouts were not regular, nor did they correlate with especially cold nights, only with rain-induced constant attentiveness. (From Calder 1996; Reproduced with permission from Dover Press.)

However, periods of rain prevent feeding and if net energy gain during the day falls short of the nocturnal heating cost, the female responds by entering torpor. This is relatively rare and eggs, chicks, and hens can all tolerate the hypothermia.

Incubation heat input and loss

A primary function of avian incubation is to maintain egg temperature within the range suitable for normal embryonic development. When the female takes an incubation recess (Chapter 6) an egg will cool at a rate proportional to the temperature difference between egg (T_{egg}) and environmental temperatures (T_{env}). During cooling, T_{egg} exhibits exponential decay towards a horizontal asymptote. The smaller the egg, the more rapidly it cools and this will slow and suspend embryonic development. The time constant (τ in s) is the time for cooling to 63% of equilibrium (Bakken and Gates 1975) and this scales positively with egg mass (Turner 1985); for a 0.5 g egg, τ is 377 s. The rate of cooling is thus inversely related to egg mass and directly related to the temperature difference ($T_{egg} - T_{env}$).

Whereas incubation times, metabolic rates, and thermal conductance of plumage are size-dependent, properties of intensity or potential energy (e.g. resting blood pressure and temperature) are generally size-independent within a class or subclass of animals (Calder and King 1974). This is seen in a comparison of the bird-nest incubation unit in two species of widely differing sizes; the eider duck (*Somateria mollisima*) and the broad-tailed hummingbird. The former has a body mass 547 times larger the mass of the latter and egg mass differs by 220 times. Although their nests differ greatly in size and placement, normal body temperatures of alert, resting hummingbirds are like those of larger birds, so it follows that hummingbird incubation temperatures are also similar. The temperature drop from her body to the centre of an egg (Figure 14.1) is almost the same as reported

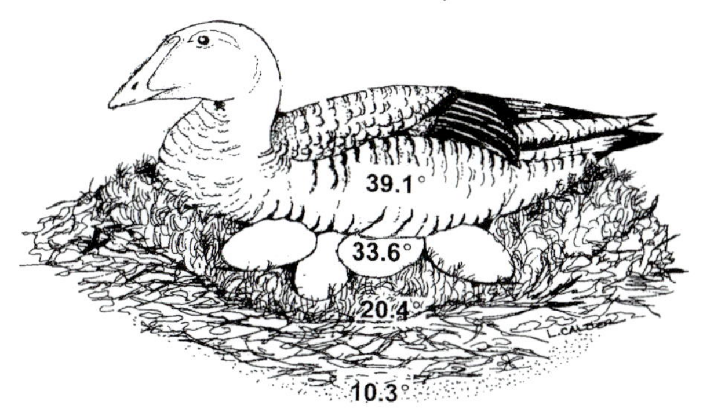

Fig. 14.3 Nest temperatures and the heat-flow gradients in the nest of the eider (*Somateria mollisima*; data from Rahn *et al.* 1983).

for the incubating eider (Figure 14.3; Rahn *et al.* 1983). However, this temperature change lies along less than a quarter of the comparable distance, so the thermal gradient is steeper by a factor of four. Then this, in turn, speeds the rate of heat flow if other things are assumed to be equal (assuming that the thermal conductivity of spider-web is similar to that of eider down).

The cup of most temperate hummingbird nests consists mostly of spider webbing and/or plant down, which is often robbed by other birds when not attended (Calder 1972). Large nest size and thick insulation of nest walls of hummingbirds has been observed repeatedly where night are cold (Corley Smith 1969; Drent 1975). However, it is obvious from visual inspection that nests of hummingbirds species breeding in warmer surroundings, either at lower elevations or later in the season, would have a higher conductance because they have thinner walls, sometimes to the extent that daylight can be seen through the wall (Calder unpublished).

During incubation, heat energy entering an egg is conserved by the insulating material and the enclosed air space within the nest. Since the nest lining of hummingbirds differ in thickness or thermal gradient, this can be further standardised as the conductivity ('specific conductivity'). Conductivity is measured as rate of heat flow (mW) per cm^2 of heat flow path area, per cm thickness of the insulating or conducting material, per °C of temperature difference. Thermal conductance, combining conductivity and thickness, is given in $mW\,{}^{\circ}C^{-1}\,cm^{-2}$. The calculated conductivity of a perfect nest (abandoned before hatching) of a broad-tailed hummingbird at an elevation of 2,910 m in Colorado was $0.55\,mW\,{}^{\circ}C^{-1}\,cm^{-2}\,cm^{-1}$ with a conductance of $8.78\,mW\,{}^{\circ}C^{-1}cm^{-2}$ (Calder 1981). Conductivity of a nest of the Anna's hummingbird near San Diego was $0.88\,mW\,{}^{\circ}C^{-1}cm^{-2}cm^{-1}$ (re-calculated from data presented by Smith *et al.* [1974]). From the 'surface area of nest exterior' of $7.68\,cm^2$ the conductance would have been $6.76\,mW\,{}^{\circ}C^{-1}\,cm^{-2}$. However, the nest was drawn with an outside diameter of 4 cm, and so the hemispherical surface would have an area of $25.1\,cm^2$, with a conductance of $22.12\,mW\,{}^{\circ}C^{-1}cm^{-2}$.

Clearly more determinations are needed to make general conclusions about thermal properties of hummingbird nests. A hypothesis to motivate measurement of nest conductance or insulation in various climates is as follows. Nest conductance for hummingbirds of similar size will be inversely proportional (increased insulation) to the mean, nocturnal temperature difference between the nest and the ambient air. Thus choice of environmental features, nesting material and energy for gathering it, and nest construction, combine to determine the nest microclimate and provide heat conservation. This investment will be paid back by reducing the energetic cost of incubation over the incubation period. Energy not spent to generate heat for thermoregulation can be used to heat the nest for longer attentive periods, before the female needs to leave and feed.

Estimated nocturnal heat flow from an incubating female broad-tailed hummingbird was 263 mW when the air temperature was 1.8°C (Calder 1981). This value was the sum of flows of conduction through the nest (55%), convection by the wind (23%), net radiation from one quarter of the female's body surface that was exposed above the insulating nest cup (18%), and respired and cutaneous evaporation (a low 4% in that cool situation). This total was 82% of the 320 mW loss rate interpolated from a rufous hummingbird perched in a 1.8°C metabolic chamber, without an insulating nest cup (Lasiewski 1963). Had this broad-tailed female nested directly under a clear sky (−19.6°C), her net radiation would have doubled, raising the total heat loss rate to 313 mW. Total heat loss rate calculated for a nest of the Anna's hummingbird by Smith *et al.* (1974) was 357 mW, 95% of the rate for a male bird perched in a metabolic chamber for a night [interpolated for a mean temperature of 12°C, between 5 and 25°C from Lasiewski's (1963) regression at the same mean temperature of 12°C]. A female broad-tailed hummingbird weighs 12% less than the female Anna's hummingbird, but the 263 mW rate during exposure to a mean T_{env} cooler by 10°, was 74% of the heat loss rate by the Anna's hummingbird (Smith *et al.* 1974). By contrast, the estimated metabolic rate for a female zebra finch (*Poephila guttata*; 11.6 g mass) was 20% higher when incubating her eggs in a domed nest, compared to being outside at the same temperature (Vleck 1981b).

Incubation to fitness: hummingbird persistence was unlikely!

The close association between hummingbirds and their physical environment makes it seem highly unlikely that evolution of this group could have occurred in areas where cold seasonal extremes prevailed. Various phylogenetic and physical factors must have come together in a gentler world of the tropics, where they were blessed by longer seasons of flowering and insect abundance. This idea is supported by the fact that hummingbirds attained their richest species diversity in the New World tropics. Once that had been accomplished, and carrying capacity was approached at the lowest latitudes, extensions in ranges could follow, with acquisition of adaptations that facilitated survival in the seasonal climates of higher latitudes. Despite the handicap to reproductive fitness posed by a clutch size of only two eggs, the hummingbirds expanded to the limits of temperate habitats in the New World. In the process, the family has radiated into 329 species, becoming the second-most species-rich family of birds (Johnsgard 1997a; Schuchmann 1999).

Southbound migrant hummingbirds, re-fueling in dense aggregations of flowers, have been popular subjects for studies of foraging behaviour. The underlying rationale was that optimal foraging strategies increase reproductive fitness. At that time, any reproductive fitness would not be expressed for ten months and hundreds of energy-store-turnovers in the future. It might be more appropriate to study this during the breeding season. However, acquisition of enough data on nesting success would take many years. Incubation in the Trochilidae is slow and careful, emphasising quality, not quantity. This is to be expected, even from knowing only that the family is characterised by polygyny, the females perform all of the nesting responsibility, and that the clutch size is fixed at two. At higher elevations or latitudes, the nesting season is too brief for completion of two clutches per year, and the nesting success rate, at least for the broad-tailed hummingbirds at 2,910 m in Colorado, was only 46% (Calder *et al.* 1983).

This limited reproductive potential of hummingbirds raises two points: (1) 'Why don't [humming]birds lay more eggs?' (Monaghan and Nager 1997). Why has natural selection been unable to produce larger clutches? Is this because the female [hummingbird] has to incubate the eggs unassisted and must divide energy and time between the mutually exclusive behaviours of supplying heat to the eggs and foraging for self-maintenance?' (2) Why has there been no selection for faster incubation and fledgling development to cope with the urgency of short flowering seasons followed by long migrations?

However, one other life history trait remains to be considered. Sunbirds and todies are both paired, rather than polygynous. The female sunbird does most of the incubating, but could have mate relief at difficult times, either in incubation or feeding of nestlings. Both male and female todies incubate. Thus, two chicks may be the limit for effective feeding by one parent. The consequences of small body size leave the incubation process with little, if any, margin for error. The male can spread his reproductive fitness over several females but females, as single parents, have no obvious options except to emphasise quality care in site selection and construction, attentiveness, and energy management. Fortunately, eggs, chicks, nest, and hen, being small and therefore having slight heat capacity, can cool and re-heat rapidly. Adult, click, and embryos all tolerate hypothermia. When energy stores are too low to support regulations of high tissue temperature, they can survive some fasting, resorting to the energy economy of hypothermic torpor. This is infrequent (see p. 216).

Future research

This extremely interesting family of small birds offers several opportunities for future inquiry. Hopefully, this will provide missing information on the following aspects of reproductive biology: (1) Nest composition and thermal conductance as a function of environmental temperature and heat exchange regimens; (2) Rate of egg mass decrease during incubation, eggshell porosity and vapour conductance as functions of egg size and incubation period; (3) Nesting success, fledgling success, and adult survivorship in, tropical humid environments; (4) Nest density, number of clutches, and incubation periods in the context of habitat, biome, and physical climate; (5) Incubation duration versus aggregate of time in hypothermia; (6) Affects on the timing, success and distribution of breeding with climate change and el Niño and la Niña cycles. These small birds may be sensitive indicators of environmental change; and (7) Direct measurement of the female's metabolism during incubation, by standard metabolic or doubly-labelled water techniques, to confirm the cost of incubation estimates and put into perspective the energy costs of reproduction.

Acknowledgements

My research on hummingbird incubation, ecophysiology, and population biology has been supported in part by grants from the National Science Foundation and the National Geographic Society, by the accommodations and facilitation by the staff of the Rocky Mountain Biological Laboratory, especially Mr. billy barr, and by the labour and enthusiasm of Lorene, Billy (now M.D.), and Suzy Calder (Shaw, now Ph.D.) and countless student assistants over three decades of the long-term hummingbird project.

15 *Intermittent incubation: predictions and tests for time and heat allocations*

F. R. Hainsworth and M. A. Voss

Parental behaviour to maintain a stable thermal environment is important for normal development of avian embryos (White and Kinney 1974; Hainsworth *et al.* 1998). Thus, single-sex intermittent incubation is a particularly interesting parental behaviour because when the parent leaves the nest there are cycles of egg cooling and heating over often short periods of time. Temperature variation should influence rate of embryo development because of lower average egg temperatures (White and Kinney 1974; Hainsworth 1995; Hainsworth *et al.* 1998). Since continuous incubation should be most beneficial for developing embryos, intermittent incubation has been thought to represent a compromise in development forced by requirements for parental survival (White and Kinney 1974; Hainsworth *et al.* 1998). It would be interesting to investigate if this compromise can be measured quantitatively so as to test the validity of this belief.

The manner in which female birds allocate time to incubation and other activities varies widely between species (Chapter 6; Conway and Martin 2000a). At one extreme of the incubation spectrum are mound-building species of the Megapodiidae that use warm nest microclimates created by fermenting vegetation and non-biotic heat sources to free them from incubation entirely (Chapter 13). At the other extreme of the continuum are high frequency intermittent incubators such as hummingbirds, which may leave the nest over 100 times a day (Vleck 1981a; Baltosser 1996). There are also species that use intermediate strategies, such as intermittently incubating the egg in well-sheltered cavity nests on the ground. One strategy employed by Procelliformes and other seabirds is interrupted incubation and neglect of the egg; both parent birds can spend days away from the nest after incubation has been initiated leading to chilling of the eggs (Whittow 1980; Warham 1990). The superb lyrebird (*Menura novaehollandiae*) also uses egg neglect to reduce the incubation time of its single egg to 27.0% of daylight hours during the early phases of embryonic development (Lill 1986). However, the reduced time for contact incubation results in an extended developmental period of 50.0 ± 2.0 days (Lill 1986). This suggests that true intermittent incubators, species that cannot rely on microclimate to maintain stable egg temperature, must be energetically restricted from spending much of the available daylight hours incubating and that there will be some cost for this energetic constraint in embryonic development. In spite of this, there are some species, such as the gray jay (*Perisoreus canadensis*), in which females are up to 97.0% attentive at the nest (Strickland and Oullet 1993). Species such as this are of particular interest for the study of factors contributing to time allocation during incubation. These include nest predation rates, female condition at the beginning of incubation, ambient temperature, and food availability.

Reviewing the massive body of literature describing the time allocation patterns of the world's intermittently incubating species would be a daunting and lengthy exercise (see Chapter 6). However, the commonality of constraints on all of these species (self-maintenance and embryonic development) and the factors affecting time allocation during incubation permit a more general time and temperature based analysis that may shed light on the consequences for both parent and developing embryo. The example analysis presented here will make use of data for yellow-eyed juncos (*Junco phaeonotus*), a species which spends 78.0% of daylight hours on the nest, suffers from moderate nest predation (26.0%), incubates at fairly moderate ambient temperatures (15.0°C), and has an incubation period of 13.0 ± 1.1 (SE) days (Weathers and Sullivan 1989a).

Modelling of time and heat allocations during intermittent incubation

A variety of ectotherms periodically heat and cool when they engage in shuttling behaviours (Dreisig 1985; Hainsworth 1995). A distinctive feature of the temperature cycles is that heating and cooling occur at different rates. This has been found for lizards (Bartholomew and Lasiewski 1965; Bartholomew 1982), insects (Dreisig 1984), turtles (Pages *et al.* 1991), fish (Brill *et al.* 1994), as well as incubated bird eggs (Turner 1994a,1994b). Several theories have been proposed for slower cooling compared with heating. These include: minimising heating time in cycles of heating and cooling (Buttemer and Dawson 1993), keeping temperature close to optimal values during cooling and heating (Bartholomew and Lasiewski 1965; Bartholomew 1982), and maximising either the proportion of time foraging or rate of net energy gain over cycles of heating and cooling (Hainsworth 1995). Investigating these ideas with intermittent incubation has the advantage of having two distinct functions in a trade-off: parental foraging for self-maintenance during egg cooling, and effects on developmental rate of the embryo within an egg which has a cycling temperature.

Developing methods to precisely predict temperatures as they change has been a substantial challenge for thermal biologists. In a sample of 53 papers concerned with variation in temperature of ectotherms published between 1962 and 1999, 25 (47.0%) used a linear model, 17 (32.0%) used a first-order negative exponential model, and only 11 (21.0%) used some form of a higher-order negative exponential model. The latter were core-shell models, which are complex and hard to use because of many heat exchange parameters that are difficult to verify for natural situations (e.g. O'Connor 1999). Although much easier to use, simpler first-order negative exponential models of heating and cooling involve errors in predicting times and temperatures (Robertson and Smith 1981; Turner 1987b; Voss and Hainsworth 2001). Linear models, of course, ignore the basic exponential nature of temperature change.

Voss and Hainsworth (2001) developed methods to avoid the complexity of core-shell models yet maintain precision with a second-order analysis. These methods were used to analyse temperature changes for house wren (*Troglodytes aedon*) eggs intermittently incubated under natural conditions. The methods are used here to develop predictions and tests for theories of time allocation during intermittent incubation. Specifically, a test is developed for maximisation of proportion of time spent foraging while eggs cool, which is thought to compete with embryo development during natural incubation cycles. This procedure was carried out using data for temperatures measured at the centre of yellow-eyed juncos eggs over many natural cycles of heating and cooling (Weathers and Sullivan 1989a).

In the following sections an optimal foraging model is developed using each time component of incubation cycles. In doing so a previously poorly understood component, the time spent after heating eggs before nest departure, is emphasised. The optimisation criterion developed for evaluation is for a female to adjust egg temperature when she exits the nest to maximise her percent time foraging over a cycle of heating and cooling. The model, with average second-order heat flux times for junco eggs is used to show how variation in five time and temperature components of the incubation cycle would influence both optimal foraging times and average egg temperature over a cooling-heating cycle. These components included egg heating time, the time the egg was held at an equilibrium temperature, egg cooling time, the return egg temperature, and the round-trip travel time of the parent. This analysis showed that an increase in brood patch temperature, which increased egg heating rate, increased both optimal foraging times and average egg temperature. Other variables influenced a trade off between foraging and average egg temperatures, but in different ways. This resulted in considerable variability in predicted optimal foraging performances relative to average egg temperatures. Data for junco eggs are then used to test for exit temperatures relative to return temperatures that would maximise percent times foraging across a series of incubation cycles that incorporate variation for each part of the trade-off. The tests are quantitative and so falsification of egg temperatures to optimise foraging are used to understand the importance of development for observed temperature patterns. Thus, the eventual goal of the model is to evaluate the extent to which the parent leaving the nest actually compromises embryo development.

Development of the model

Orians and Pearson (1979) developed a simple way to use the marginal value theorem for 'central-place' foragers, or those, such as an incubating bird, that periodically cease foraging to return to a location that has no food. A graph of diminishing cumulative energy gain versus patch time was extended on the X-axis to the left of the origin to represent round-trip travel time (τ) from the central place to a food patch (Figure 15.1; see Appendix 15.1 for all symbol definitions). The slope of the tangent from τ to the diminishing energy gain function is the marginal, or maximum rate of gross energy gain accounting for travel and food patch times. Food patch time at the tangent is the optimal time to spend for that criterion. Kacelnik and Houston (1984) modified this to account for energy costs

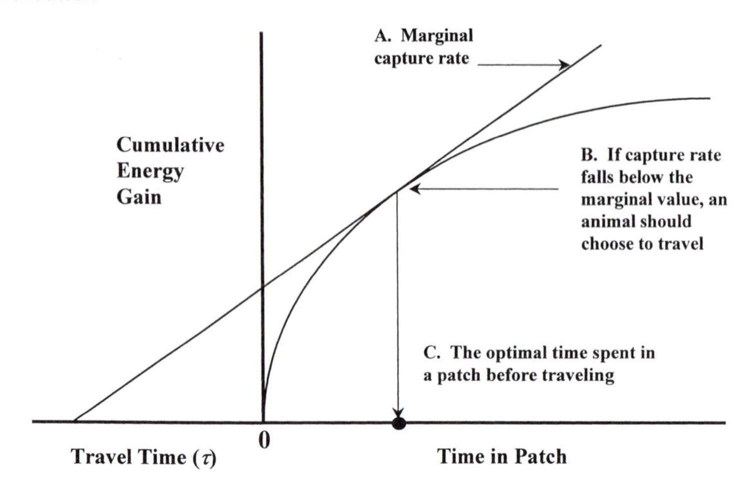

Fig. 15.1 The slope of the tangent drawn from round-trip travel time from a central place (τ) to the cumulative diminishing return function for energy gain from a patch is the marginal or maximum rate of energy gain over travel and patch times. Note that the prediction for optimal patch time (C) does not consider time or energy at a central place.

for a criterion of maximising rates of net energy gain. However, this and earlier methods did not account for time and energy use at the central place (Houston and McNamara 1985).

The importance of time and energy use at a central place became clear from optimal time allocation models of animals that dive for food and return to the surface for oxygen (Kramer 1988; Houston and Carbone 1992). Oxygen, or any resource obtained at a central place, can set a limit to travel and foraging times. Animals must cease foraging to return to the central place to replenish the resource found there. Thus, to predict and test for optimal foraging in repeating cycles of behaviour it is necessary to account for time and energy at the central place. This must result in some compromise for both foraging and the central place resource as an animal cannot be in two places at the same time.

Kramer (1988) and Houston and Carbone (1992) accounted for each cycle time component for divers by proposing that natural selection would favour those maximising the percent (P) of time foraging (t) over a cycle composed of foraging time plus round trip travel time plus surface time (s). Hence:

$$P = (100)\frac{t}{t + \tau + s}. \tag{15.1}$$

Other energy rate based models are possible (Houston and Carbone 1992) but space precludes consideration here. There is an optimal surface time to maximise P that increases with travel time (Houston and Carbone 1992). Tests showed qualitative agreement (Carbone and Houston 1994, 1996; Walton *et al.* 1998), but lack of information on rates of oxygen gain and use prevented quantitative analysis. With precise models for temperature change in incubated eggs it may be possible to

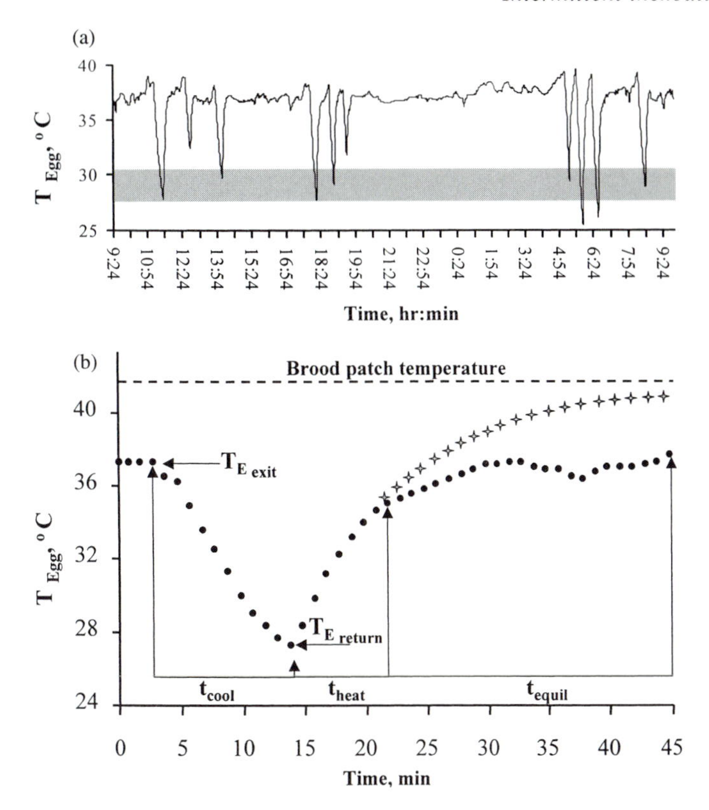

Fig. 15.2 Times and temperatures for a typical yellow-eyed junco (*Junco phaeonotus*) egg over (a) a 24-hour period and (b) for a single incubation cycle of a recess and session. (Data from Weathers and Sullivan 1989a.) In (a) note the variation in $T_{Ereturn}$ (shaded area = SD either side of mean) compared with the relative constancy of T_{Eexit}. In (b) eggs cooled for a period, t_{cool} from the egg temperature when the female left, T_{Eexit} to the temperature when she returned, $T_{Ereturn}$. Eggs heated during t_{heat} from application of the higher temperature brood patch (dashed line) to the cooler egg. The star symbols show egg temperatures if heating were to continue at the same rate. Heating rate decreased at the start of equilibrium time, t_{equil} when egg temperature was maintained close to T_{Eexit} until the female departed.

achieve quantitative tests of an analogous model for incubation. This was a major reason for adapting this model to intermittent incubation, but an additional time in incubation cycles (not present for divers) had to be accounted for.

A pattern of intermittent incubation over 24 hours is shown for junco eggs in Figure 15.2a. A drop in egg temperature indicates periods when the bird was off the nest. A typical heating and cooling cycle for a junco egg over one incubation recess and session is shown in Figure 15.2b. Eggs cool over the time interval t_{cool} from the egg temperature at departure of the female (T_{Eexit}) to the egg temperature at which she returns ($T_{Ereturn}$). She then reheats the egg toward her brood patch temperature (T_{bp}) for a period of time (t_{heat}). Heating rate decreases as egg temperature approaches T_{bp}, and egg temperature is held relatively constant below T_{bp} for a period of time (t_{equil}). Brood patch temperature is the asymptote during heating found from a least squares residual curve fitting method for negative exponential change in the absolute difference between estimated brood patch temperature and measured egg temperature ($|T_E$-$T_{bp}|$) (Bakken 1976; Voss and Hainsworth 2001). The estimated value for brood patch temperature that produced the minimum sum of squares in the regression analysis was taken to be the actual brood patch temperature. Star symbols in Figure 15.2b show the increase in egg temperature toward T_{bp} expected if heating rate during t_{heat} were to continue unchanged. Measured egg temperatures diverged from this pattern at the start of t_{equil}. This poorly understood component of the incubation cycle will have an important influence on the trade-off between foraging and average egg temperature.

Percent time allocated to foraging during an incubation cycle

In the model it was assumed that foraging was the function governing the time that eggs are allowed to cool. When this is true foraging time is $t_{cool} - \tau$. Hence:

$$P = (100)\frac{t_{cool} - \tau}{t_{cool} + t_{heat} + t_{equil}}. \qquad (15.2)$$

Data from 69 junco incubation cycles (a recess followed by a session) allowed determination of t_{cool} and t_{heat} from average values for two rate constants, k_1, k_2 (min^{-1}) derived from second-order negative exponential analyses of egg temperature change (data from Weathers and Sullivan 1989a; methods from Voss and Hainsworth 2001). Times to complete proportions of temperature differences (y) were found from:

$$y = \frac{(T_E - T_{E\infty})_t}{(T_E - T_{E\infty})_0} = -\frac{k_2}{k_1 - k_2}e^{-k_1 t} + \frac{k_1}{k_1 - k_2}e^{-k_2 t}, \qquad (15.3)$$

where: $T_{E\infty}$ is either asymptotic environmental temperature during cooling or T_{bp} during heating (Voss and Hainsworth 2001). For cooling $(T_E)_0$ is T_{Eexit}, i.e. the temperature of the egg at exit of the parent, and $(T_E)_t$ is $T_{Ereturn}$, i.e. the temperature of the egg at the return of the parent. For heating $(T_E)_0$ is $T_{Ereturn}$ and $(T_E)_t$ is T_{Eexit}. T_{Eexit} was used for $(T_E)_t$ because T_E during equilibrium was close to T_{Eexit} (Figure 15.2). For cooling, mean $k_1 = 1.43$ (SE $= 0.15$) and mean $k_2 = 0.07$ (0.004) min^{-1}. For heating, mean $k_1 = 3.92$ (0.43) and mean $k_2 = 0.17$ (0.02) min^{-1} (Voss and Hainsworth 2001). Average brood patch temperature for 31 junco egg heating periods was 40.3°C (SD $= 1.40$). Unless noted otherwise, a typical cooling asymptotic temperature of 18.0°C was used for when the birds left their nests (Weathers and Sullivan 1989a). A variety of times to heat and cool were found by calculating $(T_E)_t$ in equation (15.3) using t within 0.01 min.

Maximum percent foraging time and associated average egg temperatures during an incubation cycle

To find P, and its maximum (P^*), times to heat and cool were determined using equation 15.3 for a series of egg exit temperatures (in increments of 0.5°C) starting from a return temperature of 27.0°C and with $T_{bp} = 40.3$°C and $T_{E\infty} = 18.0$°C. Cooling and heating times for each T_{Eexit} were then used in equation 15.2 to find P for $t_{equil} = 10.0$ min and $\tau = 0.5$ min, an average travel time for observations of junco foraging cycles (Sullivan personal communication).

Percent time foraging was calculated for 0.5°C increments of T_{Eexit}. Once a value for P^* was determined, egg heating time, equilibrium time, return temperature, egg cooling time, and travel time were independently varied to find how each would influence both P^* and the associated average egg temperatures for a cycle ($\overline{T}_{Ecycle}$). Values for $\overline{T}_{Ecycle}$ were found from time-weighted averages over each cycle component (see Appendix 15.1):

$$\overline{T}_{Ecycle} = \frac{(\overline{T}_{Ecool})(t_{cool}) + (\overline{T}_{Eheat})(t_{heat}) + (\overline{T}_{Eequil})(t_{equil})}{t_{cool} + t_{heat} + t_{equil}}. \tag{15.4}$$

Incubating birds maintained egg temperatures near exit temperatures during t_{equil} (Figure 15.2) and so $\overline{T}_{Eequil}$ was assumed to be equal to the exit temperature that would maximise P. $(\overline{T}_{Ecool})(t_{cool})$ and $(\overline{T}_{Eheat})(t_{heat})$ were determined from integration over t_{cool} and t_{heat} of equation 15.3 solved for $(T_E)_t$ (Riggs 1963; see Appendix 15.1).

To vary egg heating times, brood patch temperature was decreased incrementally from 40.3 to 37.0°C keeping all other variables for equations 15.2 and 15.3 the same. An increase in equilibrium time increases cycle average egg temperature (Hainsworth *et al.* 1998), so T_{bp} was reset to 40.3°C and equilibrium time was varied from 0.0 to 25.0 min. An increase in return temperature, relative to exit temperature, also increases $\overline{T}_{Ecycle}$ (Hainsworth *et al.* 1998), so equilibrium time was reset to 10.0 min and return temperature was varied between 23.0 and 33.0°C. $T_{E\infty}$ was then varied from 8.0 to 25.0°C so as to vary t_{cool}. $T_{E\infty}$ was then reset to 18.0°C, and travel time was varied from 0.5 to 4.0 min. The ranges selected for each variable were used to quantify the linear changes in P^* versus average egg temperatures during a cycle from respective linear least squares regressions.

Physiological zero egg temperatures

Egg temperatures have been linked to a potential trade-off in development through a poorly understood thermal reference point, the 'physiological zero' egg temperature (PZT) below which it is thought embryos do not develop (White and Kinney 1974). The value and time scale for this temperature are poorly understood. Some consider it to be 26.0°C, the approximate lower value at which many intermittently incubating females return to the nest (Haftorn 1988; Conway and Martin 2000a). However, short-term decreases below 26.0°C are not detrimental as they occur with some frequency (Haftorn 1988). Others consider PZT to be a long-term, or daily average of about 35.0°C based on data for continuously incubated chicken eggs (Lundy 1969; White and Kinney 1974) and this is used here. Junco eggs averaged 35.8°C for a period of 10.0 hours overnight (Weathers and Sullivan, 1989a), so an overall incubation cycle average of 34.4°C for 14.0 h would be required for a daily PZT of 35.0° C. Uncertainty about a true daily PZT meant that a range of values from 32.0 to 35.0°C was used. If daily PZT was 32.0°C, the overall cycle average egg temperature should be 29.3°C for 14.0 h with an overnight average of 35.8°C.

Predictions from the model

Times for egg cooling, egg heating, and percent time foraging (t_{cool}, t_{heat}, and P)

Birds gain more time off the nest while eggs cool the more eggs are heated above $T_{Ereturn} = 27.0$°C, yet the more eggs are heated towards T_{bp}, the longer heating takes (Figure 15.2b). A female gained 1.4 min in cooling time if T_{Eexit} increased from 37.0 to 39.0°C, but heating from 37.0 to 39.0°C required 5.8 min. At low egg temperatures eggs heat more rapidly and cool more slowly (Figure 15.2b). The time to heat from 27.0 to 29.0°C was only 1.2 min, while cooling from 29.0 to 27.0°C took 3.6 min. Rapid heating with slow cooling should increase P (see equation 15.2).

P^* reached 36.6% at an intermediate temperature (T^*_{Eexit}) of 35.5°C (Figure 15.3a). This was an effect of parental time allocations for both travel and holding the egg at an equilibrium temperature. It became apparent when values of $t_{cool}(t_{cool} + t_{heat})^{-1}$ never reached 1.0 but were instead maximised at only 0.7 at an exit temperature of 28.0°C. Maximum for $t_{cool}(t_{cool} + t_{heat})^{-1}$ did not approach 1.0 at the return temperature of 27.0°C because of the second-order term for heating. Since travel time is low for birds that fly, the exit temperature to maximise P is shifted upwards by the longer equilibrium time. As T_{Eexit} increased above T^*_{Eexit}, cooling time increased slightly, but P declined because heating time increased disproportionately.

Travel time is short relative to other cycle times and so could be neglected allowing equation 15.2 to represent percent time cooling in a recess-session cycle. Predictions for similar T^*_{Eexit}s given below would then involve a criterion of 100 minus percent of cycle time spent on the nest (attentiveness).

Tests for effects of variables on model results

(1) *Egg heating time.* Three groups were found within the five variables influencing P^* and average egg temperatures during a cycle. The first was that faster egg heating increased both P^* and $\overline{T}_{Ecycle}$, so there was no trade-off between them. With a T_{bp} of 37.0 instead of 40.3°C, T^*_{Eexit} decreased 1.5°C to 34.0°C and P^* declined from 36.6 to 33.2%. With the lower optimum exit temperature, average egg temperature decreased by 1.2°C to 32.0°C. Faster heating raised both P^* and $\overline{T}_{Ecycle}$. Thus, a brood patch temperature close to body temperature and effective heat transfer during contact incubation means that heating time has evolved to be as short as possible. Minimising cycle heating time (Buttemer and Dawson 1993) clearly would contribute to optimising necessary self-maintenance functions during cooling, such as the proportion of time available for foraging. In a related way, the high body temperatures of birds allow proportionately more time to be spent away from the nest through decreased egg heating times while simultaneously increasing average egg temperature over an incubation session.

(2) *Equilibrium time and return egg temperature.* The remaining four variables influencing P^* and average egg temperature fell into two categories based on the extent of decrease in P^* for an increment in average egg temperature. Figure 15.3b shows the trade-off from variation in equilibrium time. $\overline{T}_{Ecycle}$ was lowest (29.0°C) and P^* highest (67.0%) when $t_{equil} = 0$, where the optimum egg exit temperature was closest to return temperature (Figure 15.3b) because it was influenced only by the short travel time. P^* declined linearly with cycle average egg temperature as t_{equil} increased (Table 15.1).

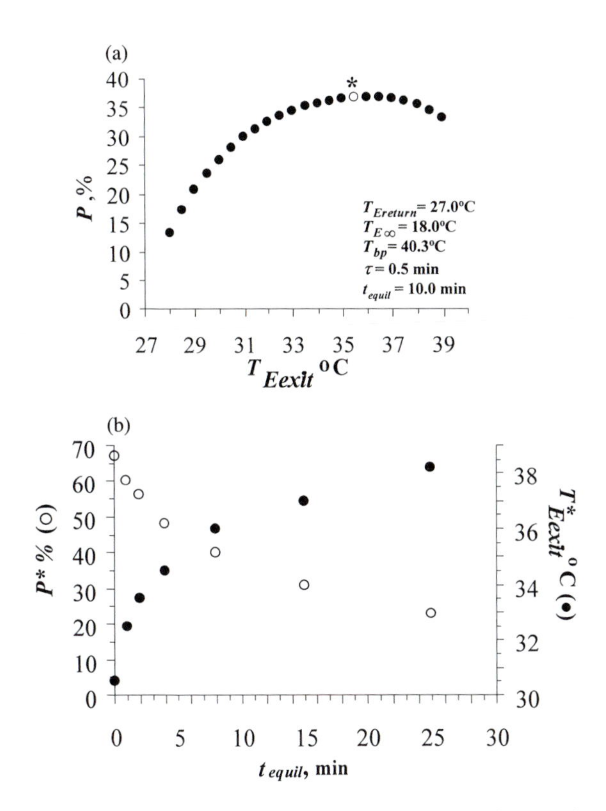

Fig. 15.3 (a) There is an optimal T_{Eexit} (*) for a maximum P for incubation cycles consisting of $[t_{cool}/(t_{cool} + t_{heat} + t_{equil})]$. P was found for a travel time of 0.5 min and an equilibrium time of 10.0 min. Cooling and heating times were found with equation 15.3 with a return egg temperature of 27.0°C. The asterisk denotes the optimum exit temperature of 35.5°C that produced the maximum $P = 36.63\%$. This temperature was relatively high because of equilibrium time since travel time was short. P declined above T^*_{Eexit} because t_{heat} increased disproportionately as T_E approached T_{bp}. (b) Maximum percent times foraging (P^*, ○) decreased, but corresponding optimum exit temperatures (T^*_{Eexit}, ●) increased as t_{equil} increased. Calculations were for $\tau = 0.5$ min for $T_{Ereturn} = 27.0$°C. Note that T^*_{Eexit} increased rapidly as t_{equil} initially increased above zero, but this was associated with a rapid decline in P^*.

Table 15.1 Changes in time and temperature components in the optimal foraging model result in trade-offs between P^* and average cycle egg temperature. The least squares linear regression equations and associated values for r^2 describe predicted relationships between the maximum percent time that an incubating bird can devote to foraging and the average cycle egg temperature that can be achieved over a incubation session and recess when the model variables are altered as described in the text.

Variable altered	Regression equation	r^2
t_{equil}, min	$P^* = 251.2 - 6.4(\overline{T}_{Ecycle})$	0.999
$T_{Ereturn}$, °C	$P^* = 178.5 - 4.2(\overline{T}_{Ecycle})$	0.997
t_{cool} (min)	$P^* = 441.8 - 12.2(\overline{T}_{Ecycle})$	0.997
τ (min)	$P^* = 356.1 - 9.6(\overline{T}_{Ecycle})$	0.998

When $t_{equil} = 0$ in Figure 15.3b, $\overline{T}_{Ecycle}$ was close to an overall cycle PZT of 29.3°C if daily PZT was as low as 32.0°C. However, $\overline{T}_{Ecycle}$ increased rapidly with equilibrium times above zero. This role for equilibrium time was not previously recognised because time on the nest was assumed to be entirely composed of heating time until a 'release' temperature was reached (e.g. White and Kinney 1974; Turner 1997). This could only be true for $t_{equil} = 0$, which is rare for incubation cycles (Figure 15.2a; Hainsworth *et al.* 1998).

A useful reference point occurs when $t_{equil} = t_{cool} + t_{heat}$. From equation 15.2, P is reduced by a half compared with when $t_{equil} = 0$. Average cycle egg temperature is mid-way between $\overline{T}_{Ecycle}$ when $t_{equil} = 0$ and $\overline{T}_{Ecycle}$ when $t_{cool} + t_{heat} = 0$ (continuous incubation) (equation 15.4, Hainsworth *et al.* 1998). Both decreasing P and increasing $\overline{T}_{Ecycle}$ are at their midpoints when $t_{equil} - (t_{cool} + t_{heat}) = 0$, and so this difference will be used as the independent variable to evaluate predictions with data (see below).

P^* also declined with increasing $\overline{T}_{Ecycle}$ as return temperature increased (Table 15.1). A lower $T_{Ereturn}$ occurs where cooling is slower and heating is faster. Both produced a lower T^*_{Eexit} that contributed to a much lower $\overline{T}_{Ecycle}$. Haftorn (1988) found that return egg temperatures seldom were less than 25.0–27.0°C for 14 species of intermittently incubating birds. For these return temperatures $\overline{T}_{Ecycle}$ would be 31.9–33.2°C, or somewhat less than 34.4°C, the over-all cycle average for a daily PZT of 35.0°C. P^* would be 44.0–35.0% for these return temperatures.

(3) *Egg cooling rate and travel time.* White and Kinney (1974) suggested a decrease in environmental temperature would lower average egg temperature. The opposite would be true for variation in egg temperatures to maximise P. When asymptotic environmental temperature was decreased to 8.0°C, T^*_{Eexit} increased to 37.0°C, and P^* declined from 36.6 to 25.4%. Faster cooling relative to heating shifted T^*_{Eexit} higher and reduced P^*. Increases in $\overline{T}_{Ecycle}$ were associated with a pronounced decline in maximal % time foraging (Table 15.1). The pattern is opposite

to that predicted by White and Kinney (1974) because their model did not include an increase in exit temperature relative to return temperature to maximise P.

An increase in travel time also lowered P^* and increased T^*_{Eexit} (Hainsworth 1995) and $\overline{T}_{Ecycle}$. When travel time was raised to 4.0 min, P^* was 25.4% with $T^*_{Eexit} = 38.0°C$, and $\overline{T}_{Ecycle}$ was higher by 1.3°C at 34.5°C. As for cooling rate, maximal % time foraging also decreased linearly with cycle average egg temperature with a relatively steep slope as τ increased (Table 15.1).

Tests of the model predictions for maximal % time foraging for juncos

Natural egg temperature records ($n = 69$) for five junco eggs (Weathers and Sullivan 1989a) were used to construct second-order negative exponential models for cooling and heating (Voss and Hainsworth 2001). This allowed us to find predicted egg temperature at departure which maximised P (T^*_{Eexit}) for a variety of cycles across the entire incubation period to compare with the observed exit temperatures.

Amongst the four cycle variables predicted to influence the trade-off between parental foraging and embryo development, increases for equilibrium time and return egg temperatures showed the least decrease in P^* for an increase in average egg temperature. For this reason, and because there was adequate variation in data for these variables, they were considered most likely to be used by birds to adjust short-term allocations within the trade-off. Egg cooling rate and travel time, which had more pronounced effects in the trade-off, are more likely to be influenced by longer-term decisions, such as time of year for breeding, nest location, and construction. As a midpoint occurred for decreasing P and increasing cycle average egg temperature when $t_{equil} - (t_{cool} + t_{heat}) = 0$, this difference was used as the independent variable to test for optimal exit temperatures to maximise P.

Figure 15.4 shows the observed 95% predictive interval for T_{Eexit} and the individual optimal exit temperatures that would have maximised P as a function of $t_{equil} - (t_{cool} + t_{heat})$ for the junco incubation cycles. In every cycle T^*_{Eexit} was found for the observed $T_{E\infty}$, $T_{Ereturn}$, and t_{equil} using $T_{bp} = 40.3°C$ and $\tau = 0.5$ with second-order values for t_{cool} and t_{heat}. By doing this across the variety of incubation cycles, the test incorporated variations in all rate constants, $T_{E\infty}$, $T_{Ereturn}$, and t_{equil}.

Cycles in which the T^*_{Eexit} predicted to maximise P statistically corresponded to the observed T_{Eexit} are the points within the 95% predictive interval of the observed T_{Eexit} values (Figure 15.4 dashed lines; category C). These points were roughly clustered around the trade-off mid-point of $t_{equil} - (t_{cool} + t_{heat}) = 0$, and they represented 19.0% of all of the bouts analysed. The observed P for these bouts averaged 28.1% and $\overline{T}_{Ecycle}$ was 38.2°C (Table 15.2).

The remaining incubation cycles were statistically different for T^*_{Eexit} from those observed but in two different ways. In 52.0% of incubation cycles the predicted optimal T_{Eexit} was lower than those observed. These cycles occurred primarily

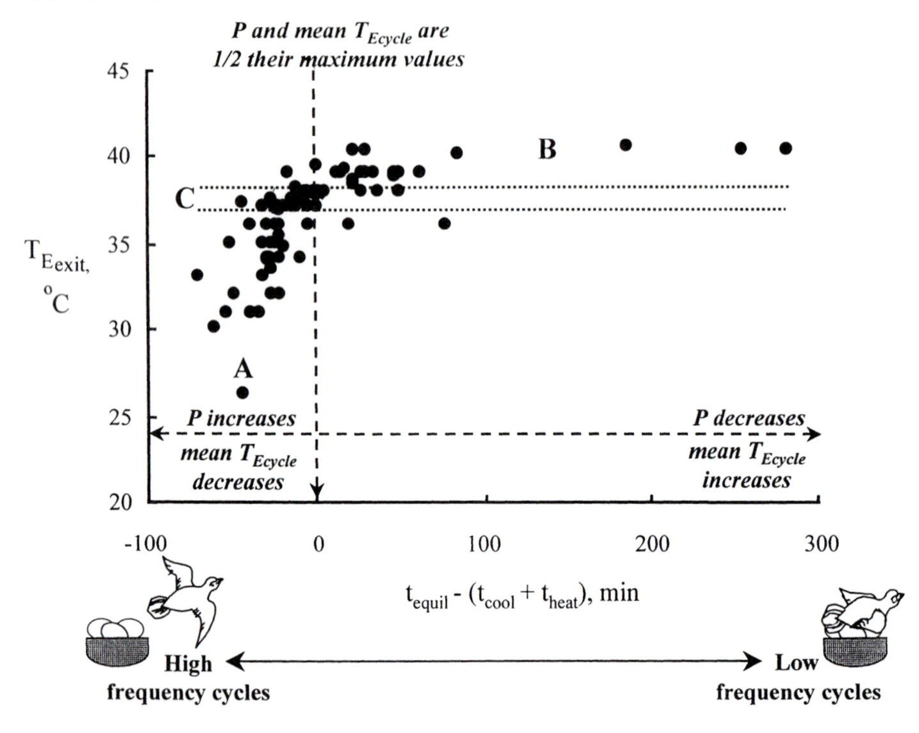

Fig. 15.4 Predicted values for T^*_{Eexit} (°C) to maximise P as a function of the difference $t_{equil} - (t_{cool} + t_{heat})$ (min) for yellow-eyed juncos (*Junco phaeonotus*) are shown as points to compare with the 95% predictive interval based on all observed values of T_{Eexit} (dotted lines). Three categories of foraging bouts are apparent: (A) those that have T^*_{Eexit} lower than observed, (B) those that have T^*_{Eexit} higher than observed, and (C) those where T^*_{Eexit} was statistically the same as observed.

when $t_{equil} - (t_{cool} + t_{heat}) < 0$. They represented cycles in which the parent devoted more time to self-maintenance (foraging) than to embryo development (Figure 15.4; category A). Had the female behaved as the model predicted she could have increased P, but at the cost of decreased $\overline{T}_{Ecycle}$. The observed $\overline{T}_{Ecycle}$ for these cycles was 37.3°C. If the parent had maximised P, the developing embryos would have experienced a $\overline{T}_{Ecycle}$ of 34.3°C. The observed average P was 37.5% . If she had lowered T_{Eexit} to the points shown, she would have attained on average a P of 45.7% (Table 15.2). Thus, in general, when the parent left the nest frequently for her self-maintenance, she did not maximise her foraging performance in deference to the development of the embryo.

In 29.0% of the cycles analysed, the predicted T^*_{Eexit} was higher than observed (Figure 15.4; category B). These cycles occurred when $t_{equil} - (t_{cool} + t_{heat}) > 0$ and represented incubation cycles in which the parent devoted more time to embryo development than to self-maintenance. The $\overline{T}_{Ecycle}$ observed for these cycles was 37.9°C and the observed average P was 17.3% . If she had raised T_{Eexit} to the points shown in category B, her average P would have been 18.3% and $\overline{T}_{Ecycle}$

Table 15.2 Observed and predicted $\overline{T}_{Ecycle}$ and $\overline{P}$ for categories in Figure 15.4. The predicted values for optimal exit temperatures (T^*_{Eexit}, °C) to maximise P for yellow-eyed juncos (*Junco phaeonotus*) are plotted in Figure 15.4 as a function of observed incubation sessions with different time structures of $t_{equil} - (t_{cool} + t_{heat})$ (min). These predicted values of T^*_{Eexit} are shown as points compared with the 95% predictive interval of all observed values of T_{Eexit}. This delineates three categories of foraging bouts: (A) those that have T^*_{Eexit} lower than observed, (B) those that have T^*_{Eexit} higher than observed, and (C) those where T^*_{Eexit} was statistically the same as observed. Values for the average cycle egg temperature and average P associated with each of these categories are reported here for comparison across categories.

Category	n	$\overline{T}_{Ecycle} \pm$ S.D. °C		$\overline{P} \pm$ S.D. %	
		Observed	Predicted	Observed	Predicted
A	36	37.3 ± 1.8	34.3 ± 2.5	37.5 ± 12.1	45.7 ± 16.6
B	20	37.9 ± 1.6	39.4 ± 0.7	17.3 ± 8.1	18.3 ± 8.2
C	13	38.2 ± 1.2	37.8 ± 0.4	28.1 ± 10.2	29.2 ± 9.8
		$\overline{P} \pm 95\%$ C.I.		p for 2 tailed t-test, $\alpha = 0.05$	
A + B predicted	56	35.90 ± 5.1		$p = 0.1629$	
C observed	13	28.06 ± 5.6		∴ means are not statistically different	

would have been 39.4°C, only a slight increment for both P and $\overline{T}_{Ecycle}$. Thus, we could falsify the predictions when t_{equil} was long, but the consequences for both parent and embryo were minor when the parent invested considerably in embryo development.

Significance of the test

It was notable that females maintained a relatively high T_{Eexit} when they left the nest frequently for their self-maintenance. This was when t_{equil} was low relative to $t_{cool} + t_{heat}$ (category A in Figure 15.4). If maximising self-maintenance through foraging determined egg temperatures when parents left the nest, they should have heated their eggs to the much lower values shown by the points in category A. By doing so an extra 8.2 average % time foraging would have been achieved, but cycle average egg temperature would have decreased by 3.0°C. Since this was not observed, the females were not compromising embryo development as much as expected for the reason for leaving the nest.

Predicted exit temperatures to maximise P were low when t_{equil} was low because cooling was slow and heating was rapid at low temperatures. Thus, $t_{cool}(t_{cool} + t_{heat})^{-1}$ was maximum close to return temperature when t_{equil} was negligible in the equation for P. Both the predicted and observed values for P decreased as t_{equil} increased (Figure 15.3b; Table 15.2), and there was wide variation in t_{equil} across cycles (Figure 15.4). Although only the few category C cycles were statistically optimal, the over-all performance of the parents across all cycles depended on how the varieties of cycles average for observed Ps. Observed Ps in category A were relatively high, while those in category B were low (Table 15.2). They averaged to be statistically the same as the predicted average maximum P for those in category C (Table 15.2). Thus, the birds did the same on average over a long-term incubation period as they would have if all cycles had been their respective 'best possible'. This occurred because P is heavily influenced and time weighted by t_{equil}. By the same result, the over-all average observed exit temperature of juncos can be said to be optimal even though values for most individual cycles clearly were not.

What is the 'best possible' for juncos need not be what is optimal for other species. The variation in t_{equil} across cycles may reflect the relative availability of food at different times. Published records of egg temperatures show considerable variation in t_{equil} (Vleck 1981a; Davis *et al.* 1984a; Zerba and Morton 1983b; Morton and Pereyra 1985), and this could reflect major differences in relative food availability during incubation. If average food availability is low, so average t_{equil} across cycles is relatively short, then average exit temperatures should be lower relative to return temperatures to achieve a somewhat higher average maximum value for P. Additional studies of other species should be interesting to test for long-term average optimal functions.

For single-gender intermittent incubation it can be concluded that it may not represent as severe a problem for embryo development as once thought. This conclusion is tentative because we only have evaluated data for juncos, and there is little information at this time on the effects of the average and variance in egg temperatures on embryo development. The over-all daily observed average egg temperature for juncos was 35.45°C (95% confidence interval ±0.13°C). This exceeds the highest daily PZT value, but it must be less than what the average would be with continuous incubation (Hainsworth *et al.* 1998). The difference will influence rate of development, but we do not now know by how much.

This is the first use of data from a natural situation to quantitatively test theories that integrate multiple functions to explain design of temperature variation. Although there have been a number of attempts to measure and account for temperature variations of ectotherms in their natural environments (Hertz *et al.* 1993; Belliure *et al.* 1996; Christian and Weavers 1996; Diaz 1997), few have attempted to account for time use for multiple functions, and those that did used first-order models for the dynamics (Dreisig 1985; Vispo and Bakken 1993; Hainsworth 1995). The extent to which we now understand how incubation temperatures are structured to influence both foraging and development as mutual constraints has depended on the quantitative features of the models and test. Perhaps this will have some heuristic value for others interested in testing hypotheses about the design of functions. At minimum, there would appear to be an important role for quantitative models and tests in future work designed to evaluate how and why variations in animal behavior, such as those seen in intermittent incubation, might be adaptive.

Acknowledgements

We would like to thank Kim Sullivan and Wes Weathers for graciously providing data used in the model and subsequent analysis presented here.

Appendix 15.1

Incubating birds maintained egg temperatures near exit temperatures during t_{equil} (Figure 15.2a) and so $\overline{T}_{Eequil}$ for equation 15.4 was assumed equal to the exit temperature that would maximise P. The average temperatures during cooling and heating were found by taking the respective integrals of equation (15.3) solved for $(T_E)_t$. These gave the areas under each function, or the average temperature times the respective times for use in equation 15.4. Integration of equation 15.3 solved for $(T_E)_t$ over a cooling or heating interval from $t_1 < t_2$ gave:

$$- [({k_2}^2 e^{-k_1 t_2} - {k_2}^2 e^{-k_1 t_1} - {k_1}^2 e^{-k_2 t_2} + {k_1}^2 e^{-k_2 t_1})(T_{E0} - T_{E\infty}) + ({k_1}^2 k_2 - k_1 {k_2}^2)(T_{E\infty})(t_2 - t_1)]/k_1 k_2(-k_1 + k_2).$$

Symbol definitions given in the order presented in the text. Abbreviations for time components are lower case and for temperature are upper case.

τ	Time required to travel round trip from a nest or a central place to a foraging patch (min).
t	Foraging time (min).
s	Surface time (min).
P	Percent time foraging over a cycle of diving or incubation.
t_{cool}	Time during which the egg cools (min).
T_E	Egg temperature (°C).
T_{Eexit}	Egg temperature at departure from nest (°C).
$T_{Ereturn}$	Egg temperature at return to nest (°C).
T_{bp}	Brood patch temperature (°C).
t_{heat}	Time during which the eggs heat (min).
t_{equil}	Time at which the eggs are kept at an equilibrium temperature before departure (min).
P^*	Maximum percent time foraging.
k	The rate constant associated with heating or cooling (min^{-1}).
$T_{E\infty}$	Asymptotic environmental temperature (°C).
$\overline{T}_{Ecycle}$	Average egg temperature over a cycle of incubation and foraging (°C).
T^*_{Eexit}	Egg temperature at departure from nest that maximises P (°C).
$\overline{T}_{cool}$	Average egg temperature during egg cooling (°C).
$\overline{T}_{heat}$	Average egg temperature during egg heating (°C).
$\overline{T}_{Eequil}$	Average egg temperature during t_{equil} (°C).

16 *Incubation in extreme environments*

C. Carey

Some birds can nest successfully in some of the most inhospitable environments on earth, such as cold regions, high altitudes, and deserts. These accomplishments are particularly striking because the embryos are incubated outside the body of the adult where they are more vulnerable to fluctuations of the ambient environment than are embryos protected within the body of viviparous vertebrates. Embryonic requirements for successful development (application of external heat, periodic turning, defence from predators, and exchange of appropriate levels of gases with the environment) must be met in all environments in which birds breed (Drent 1975). Adults breeding in hostile environments must meet these embryonic demands despite major challenges to meet their own needs for nutrition, thermoregulation, water and gas exchange (Schmidt-Nielson 1964; Dawson and Bartholomew 1968; Serventy 1971; Immelman 1971; Irving 1972; Remmert 1980; Marsh and Dawson 1989; Bech and Reinertsen 1989; Faraci 1991). These physiological challenges, added to the demands of reproduction, can create situations in which the needs of the adults conflict with those of their offspring.

Many behavioural, physiological and morphological attributes of species breeding in cold, high altitude, desert and wet environments are shared with species breeding in more moderate areas (Dawson and Bartholomew 1968; Morton 1976; Webb 1987; Ward 1990). This chapter reviews the specialisations that allow successful breeding in hostile regions. These specialisations principally involve adult behaviours that protect eggs from overheating or excessive chilling, and structural features of the eggshell that regulate embryonic gas exchange. Although this review will emphasise general trends rather than exceptions to the rule, it must be noted that birds have utilised numerous solutions to common challenges.

Heat flows from a warmer object to a colder one at a rate proportional to the temperature difference between the objects. This principal is reflected in the challenges to adults for thermoregulation in cold or hot environments. At colder air temperatures, metabolic production of heat must increase to offset increased rates of heat loss. In warmer conditions, particularly when air temperatures exceed adult body temperatures, water must be evaporated to balance heat loss with heat gain and heat production (Dawson and O'Connor 1996). Principles of heat transfer also relate to the maintenance of egg temperatures. Optimal incubation temperatures are similar among all avian species (Webb 1987). Embryos produce such little heat that it must be supplied externally by the adult or some other external source of heat. In cold climates, including high altitudes, absences from the nest cause eggs to cool at rates that are determined by factors like the numbers of eggs in the nest, sizes of the eggs, nest insulation, the proportion of eggs exposed to ambient air, and air temperature. If air temperature is higher than optimal egg temperature, then the eggs gain heat from the environment and must be cooled by the adults.

Embryonic gas exchange occurs through the eggshell by the process of diffusion (Paganelli *et al.* 1975). Oxygen diffuses into, and carbon dioxide (CO_2) moves out, of the egg down concentration gradients established by embryonic metabolism. Water vapour diffuses from the saturated interior into the less saturated nest environment (Wangensteen and Rahn 1970/71; see review of basic principles by Carey 1983). The gas conductance of the egg (G) describes the diffusive capacity of the shell, which is a function of the functional pore area (total number and sizes of pores), the shell thickness, a gas constant, and the diffusion coefficient (D) for each gas (Ar *et al.* 1974; Chapter 3). Shell morphology of each species has evolved under opposing selection pressures to allow sufficient oxygen to diffuse into the egg while preventing excessive losses of CO_2 and water vapour (Rahn *et al.* 1974; Rahn and Ar 1980). It is generally accepted that selective pressures have favoured conservation of water vapour over those for oxygen delivery in most instances (Rahn *et al.* 1974; Rahn and Ar 1980). Water vapour losses, amounting to 10–20% of the initial egg mass during incubation, are essential for maintaining the relative water proportion of the egg contents and for creating the air cell in the blunt end of the egg (Rahn and Ar 1980). Excessive losses of water vapour will cause embryonic dehydration and death.

Reduction of the avian adaptations that foster breeding in hostile environments to these basic physical principles aids understanding about how birds breed in such difficult situations. However, this approach tends to obscure the heroics involved in successful breeding in these environments. Perhaps these achievements can be appreciated fully only when one experiences first-hand the conditions under which the birds nest.

Cold environments

For this review, 'cold' environments include Antarctica, the geographically ill-defined Arctic, and other lowland, high latitude terrestrial areas in both hemispheres. These localities are characterised by severely cold winters, generally with snow cover, and summers with cool air temperatures, strong winds, and late-season snowstorms that can disrupt breeding (Spellerberg 1969; Irving 1972; Remmert 1980; Astheimer *et al.* 1992; Wingfield *et al.* 1995; Bradley *et al.* 1997). However, overheating can be a problem even in cold environments. Full exposure to sunlight and lack of wind may occasionally force incubating adults to increase heat loss by panting (Spellerberg 1969).

Of the roughly 8,600 avian species in the world, relatively few breed in cold climates. Approximately 24 bird species are resident year-round and at least 90 others migrate into trophically rich niches that open up in the summer in the Alaskan tundra (Irving 1972). Density of breeding migrants can be quite high. For instance, single colonies of snow geese (*Anser caerulescens*) in the Canadian Arctic may reach over 200,000 individuals (Remmert 1980). The breeding avifauna of

Antarctica is limited to birds that feed on marine food or that prey on eggs and young of other birds. Only 8 species of seabirds and one penguin species breed from Cap Bienvenue (66°43′ S, 140°31′ E) to Moyes Island (67° S, 143°56′ E) on the Terre Adélie coast of Antarctica (Barbraud *et al.* 1999).

Compression of breeding activities

In general, avian breeding seasons are co-ordinated with seasonal cycles with the result that hatching is synchronised with the peak in food availability (Lack 1968; Immelman 1971). In cold regions, reproductive activity (establishment of territories, pair formation, nest building, egg laying, incubation, care of nestlings, and fledglings, post-nuptial moult, and fattening in preparation for migration or winter survival) must be compressed into a short summer of four months or less (Irving 1972). Although preparations for breeding must be synchronised with the environment, birds must be able to adjust to unpredictable local variation in snow cover, food supplies, etc. (Silverin 1995). Although some migrants may arrive on the Alaskan interior tundra as early as April, breeding does not commence until late May or June (Irving 1972). Compared with the time allocated for various activities by conspecifics breeding at low latitudes, white-crowned sparrows (*Zonotrichia leucophrys*) breeding in Alaska reduce the amount of time spent in the pre-nesting interval (territorial establishment and pair formation), nest building, the period between the completion of the nest and the laying of the first egg, the duration of the nestling period, and the time between fledging and testicular regression (Morton 1976). Incubation period and the time involved in egg formation do not vary over a latitudinal gradient and the egg laying period is longer at high latitude because the clutch size is larger than at low latitude (Morton 1976). The nestling periods of several other species of arctic passerines are also shorter than comparative groups in more moderate climates (Irving and Krog 1956; Williamson and Emmison 1971).

Prevention of egg chilling and/or freezing

The thermal requirements of avian embryos in cold environments mirror those in other environments (Webb 1987). Eggs can be the most vulnerable to freezing during the laying period when adult attentiveness is low. Attentiveness of females, like the greater snow goose (*Chen caerulescens atlantica*), during the laying period is generally adequate to prevent freezing but not necessarily sufficient to raise egg temperature to the minimal level for development (Poussart *et al.* 2000). Both adults sharing incubation duties prevents eggs from freezing and minimises cooling of eggs during absences of adults from the nest. Eggs of the McCormick skua (*Catharacta maccormicki*) at Cape Royds, Antarctica (77°33′ S) cool by 2.5–8°C during the switchover of adults (Spellerberg 1969). When adult were prevented from attending the nest for 14 minutes, egg temperature dropped to 0°C. Re-warming the eggs requires considerable metabolic heat expenditure by the adult (Spellerberg 1969).

Nests of many species breeding in cold climates are simply shallow scrapes in the substrate that provide no insulation against heat loss for either adults or unattended eggs (Spellerberg 1969; Remmert 1980). Brünnich's guillemot (*Uria lomvia*) lays eggs on the frozen ground in spring and as the thaw progresses, the eggs become surrounded with ice water. Although the adult may brood the eggs on their feet in this situation, the temperature of the lower side of the egg is at 0–1°C while the top of the egg in contact with the brood patch is around 39°C. Frequent turning of the egg is considered to ensure thermal equilibrium of the egg contents (Belopolski 1957 in Remmert 1980; but see Chapter 11). Chinstrap penguins (*Pygoscelis antarctica*) line their nest scrapes with pebbles which keeps eggs above the level of melt water (Moreno *et al.* 1999a). Other birds, particularly waterfowl, insulate their nests with down and may cover the eggs when the adult leaves for a recess (Remmert 1980; McCracken *et al.* 1997). The insulative layer of nests increases with latitude in some species, such as magpies (*Pica pica*; Remmert 1980).

Energy utilisation during incubation

Body temperatures of adult birds breeding in cold regions do not differ from those in warmer areas (Irving 1972). Metabolic production of heat is necessary to offset heat losses at cold temperatures, particularly when gradients between body and air temperatures are large. The largest differential recorded is about 90°C for emperor penguins (*Aptenodytes forsteri*) incubating eggs in the Antarctic winter (Le Maho 1977). Incubation poses conflicting difficulties for adults, because attendance at the nest precludes foraging at a rate that might be necessary to meet the challenges of high rates of heat production. The conflict is particularly acute for small birds that have higher mass specific metabolic rates and lower somatic energy reserves than larger birds (Williams 1996). Inclement weather may force small birds to curtail foraging to the extent that the incubating adult may become unable to maintain homeothermy during the night, even in moderate climates (Calder and Booser 1973; Williams 1993c; see also Chapter 14). Many breeding birds have hormonally-mediated stress responses that trigger behavioural changes, including abandonment of nests, during severe storms. These changes promote survival of the adults until the threat is over. A few arctic birds, however, suppress their stress response so that breeding can continue during storms despite the possibility that adult mortality might result (Wingfield *et al.* 1995).

When food is near the nest, plentiful, and easy to consume without a great deal of handling, a minimum of foraging is necessary to maintain body reserves (Drent *et al.* 1985). Some small birds may be able to forage sufficiently that they can maintain body mass throughout incubation, while others do not (see Williams 1996). Waterfowl make fewer foraging trips per day than necessary to maintain body reserves and lose up to 30% of body mass during incubation (Peters 1983; McCluskie and Sedinger 2000; Chapter 6).

The ability to utilise body reserves during incubation is developed to an extreme level in some species of penguins. Emperor penguins establish rookeries on sea ice far from the open sea. After feeding on crustaceans and fish during the austral summer, they walk 50–120 km to rookeries. The female lays one egg and then departs for the sea, leaving the male to incubate. He incubates through the austral winter at temperatures averaging about −28°C and dipping as low as −48°C. A low critical temperature (at least −10°C) and group behaviours, such as huddling to protect from wind, serve to minimise energy expenditure but approximately 40% of his initial mass is lost during the approximately 155-day fast before his mate returns (Le Maho 1977). During this fast, lipid accounts for about 62% of the mass loss and 96% of the calories that are burned (Groscolas 1988; Robin *et al.* 1988). Minimisation of protein loss is also a feature of fasting in king penguins (*Aptenodytes patagonica*), which fast for 5–6 weeks from the beginning of breeding (Cherel *et al.* 1993, 1994). Adélie penguins (*Pygoscelis adeliae*) breed during the austral summer in Antarctica (Williams 1995) although much closer to the sea than emperor penguins do. Patterns of nest attendance vary with population, but generally the bout lengths vary from 3–20 days (Bucher and Vleck 1997). When the incubating adults of these species are not relieved before energy supplies reach critical levels, the eggs or young are abandoned (Le Maho 1977; Vleck *et al.* 1999). Fasts up to 40 days do not cause stress in this species, as judged by an increase in plasma corticosterone (Vleck *et al.* 2000) although corticosterone does rise after 50 days of fasting and may serve as part of the internal regulatory mechanism that leads to nest abandonment.

Eggshell conductance

Cold environments are amongst the driest on earth. The average absolute ambient humidity, measuring only a few torr, creates one of the largest overall differences in water vapour between the ambient air and the interior of an avian egg (Morgan *et al.* 1978). This gradient could result in excessively high rates of water loss that might compromise the ability of the embryo to develop and hatch. However, eggs of Adélie penguins laid on Ross Island, Antarctica, and arctic terns (*Sterna paradisaea*) found near Homer, Alaska, lose an optimal amount of water during incubation (13 and 15% of initial mass respectively). This is due to significant reductions in the numbers of pores of the eggshell (Rahn *et al.* 1976; Rahn and Hammel 1982). In common with other penguins (Tyler 1965), eggshells of Adélie and king penguins are covered at laying with a waxy covering that limits gas conductance and reduces water loss in the early stages of incubation. As the covering is gradually abraded off by contact with the substrate and the incubating adults, gas conductance rises in concert with the metabolic needs of the embryo (Handrich 1989a; Thompson and Goldie 1990).

High altitude environments

The definition of 'high' altitude is ambiguous. Generally, altitudes above 3,000 m are considered sufficient to require some degree of physiological acclimation, if not genetic adaptation, by birds and mammals (Bouverot 1985). Birds breed up to 6,500 m in the Himalaya and 4,900 m in the Andes (Rahn 1977). In mountain ranges such as the Rocky Mountains and Sierra Nevada of the United States, a number of species breed from sea level to at least 3,000 m. In the Andes and the

Himalaya, most species breeding at low altitudes are replaced by other species above 3,500 m. Only a few species, like the rufous-collard sparrow (*Zonotrichia capensis*), breed over the entire elevational range from sea level to over 4,500 m (Parker *et al.* 1982). The upper altitudinal limit of nesting for each species is determined by various factors, such as the height of the mountain range, presence or lack of a permanent snow line, availability of trees or other suitable vegetation for nest placements, presence of a specific type or quantity of food, and the ability of embryos to cope with abnormal gas fluxes between the interior and exterior of the egg.

Effects of low barometric pressure

The reduction in barometric pressure with increasing altitude presents major challenges for birds breeding at altitude. Furthermore, the intensity of solar radiation, relatively unattenuated by the air column at high altitudes, can lead to overheating of embryos during absences of adults from the nest (Morton and Carey 1971; Zerba and Morton 1983a). Low density of mountain air fosters large re-radiational losses of heat and results in perpetually cold air temperatures and eggs may require incubation during the laying stage to prevent freezing (Zerba and Morton 1983a). The growing season in which the ground is snow-free is limited in many mountain ranges to only a few months. Except in tropical mountains where little or no snow accumulates, snow depths and the timing of the snowmelt determine when birds can begin breeding activities, the length of the breeding season, and even the choice of nest sites (Morton 1978). Snowstorms can occur in any month (Ehrlich *et al.* 1972; Morton *et al.* 1972; Hendricks and Norment 1992) and high winds are common in mountains, making selection of protected microclimates for nests especially important (Zerba and Morton 1983b).

Many similarities exist between breeding characteristics of species that breed at high altitudes and high latitudes. In common with high latitude birds, some components of the breeding period, such as the duration of the nest building period, can be compressed. For instance, the amount of time spent by mountain white-crowned sparrows in various stages of breeding at 3,000 m resembles the pattern of lowland sparrows breeding much farther north, around 52° N (Morton 1976). Some characteristics, like the incubation periods and the average %attentiveness by the incubating adult, are not compressed in white-crowned sparrows and red-winged blackbirds (*Agelaius phoeniceus*) breeding between 290 and 2,900 m (Morton 1976; Carey *et al.* 1983). Both high altitude and high latitude groups must deal with inclement weather, cold air temperatures and the occasional necessity to abandon nests and breeding territories during snowstorms (Morton *et al.* 1972; Astheimer *et al.* 1992; Morton *et al.* 1993). Clutch sizes of several species increase from sea level to about 3,000 m (Carey *et al.* 1983). However, these patterns are not uniform for all species. White-tailed ptarmigan (*Lagopus leucurus*) nesting between 3,500 and 4,200 m in the Colorado Rockies have smaller clutch sizes, slower laying rates, smaller overall reproductive investments and longer incubation periods than willow ptarmigan (*Lagopus lagopus*) nesting at sea level in sub-arctic Canada (Martin *et al.* 1993; Wiebe and Martin 1995).

Reduction in barometric pressure is the primary environmental factor that affects avian breeding differently at high altitude than in low altitude, cold climates. After hatching, birds breathe by convection of air over respiratory surfaces. The reduction in the partial pressure of oxygen (P_{O_2}) at high altitudes reduces the gradient for diffusion between alveolar gases and the blood. Birds have a superior ability to tolerate hypoxia compared to mammals and a few have been observed flying at altitudes up to 11,300 m (Laybourne 1974; Faraci 1991). Birds breeding above 3,000 m have circulatory and haematological specialisations that foster oxygen transfer from lungs to tissues (Carey and Morton 1971; Clemens 1990). Increases in haemoglobin oxygen affinity in species that breed at high altitudes or that fly at very high altitudes, like Andean geese (*Chloephaga melanoptera*) or black vultures (*Aegypius monachus*) respectively, are due in part to genetic modifications of haemoglobin (Hiebl *et al.* 1987; Jessen *et al.* 1991). Birds are faced with the conflicting problems that cold temperatures require heightened levels of aerobic heat production in conditions in which oxygen availability is limited. However, specialisations for oxygen delivery during flight, which requires metabolic expenditures 3- to 4-times those measured during maximal shivering, are so effective that it is doubtful that lack of oxygen compromises thermoregulatory ability (Faraci 1991).

Eggshell conductance to gases and embryonic gas exchange

Reduced barometric pressure presents a quite different challenge to avian embryos than to adults. The diffusion coefficient of each gas is inversely proportional to barometric pressure. Therefore, gases diffuse more rapidly at high altitude than at sea level at a rate directly proportional to the reduction in barometric pressure (Rahn and Ar 1974; Paganelli *et al.* 1975). As a result, effective conductance of the shell to all gases increases in direct proportion of the reduction in barometric pressure (Figure 16.1). This effect ameliorates, in part, the reduction in ambient P_{O_2}. Although there are fewer O_2 molecules in air at altitude, the increase in effective conductance will cause them to diffuse more rapidly through the shell. On the other hand, the increase in effective conductance could cause excessive losses of CO_2 and water vapour that could disrupt acid–base balance and cause embryonic dehydration, respectively (Rahn and Ar 1974).

A number of studies have examined how birds nesting at high altitude have dealt with this problem (Packard *et al.* 1977; Rahn *et al.* 1977; Sotherland *et al.* 1980; Taigen *et al.* 1980; Carey *et al.* 1983, 1987, 1989a, 1989b, 1991, 1993, 1994). Conductance of eggshells (corrected to 760 torr) of eggs laid up to 3,000 m gradually decreases in approximate proportion to the reduction in barometric pressure. The reduction in conductance is caused by a decrease in the numbers of pores, rather than shell thickness (Packard *et al.* 1977; Rahn *et al.* 1977; Sotherland *et al.* 1980; Carey *et al.* 1983, 1987, 1989a, 1989b). This modification counteracts the increase in effective conductance at altitude, resulting in a rate of water loss that is independent of altitude in eggs laid to about 3,000 m (Figure 16.2).

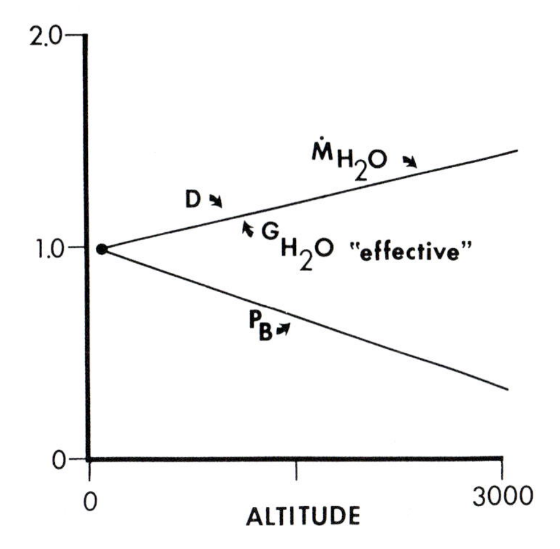

Fig. 16.1 Illustration of the way in which barometric pressure (P_B), the diffusion coefficient for any gas (D), the 'effective' G_{H_2O} (the actual conductance of the egg to water vapour at a given altitude), and daily water loss from the egg (M_{H_2O}) in the field change with an increase in altitude (in m). P_B, D, 'effective' G_{H_2O}, and M_{H_2O} are designated as '1.0' at sea level. Lines represent the value relative to 1.0 for each parameter as altitude increases. For instance, imagine that an egg laid at sea level is transported to 2400 m. P_B at that altitude is roughly 0.75 of the value at sea level. Since D for any gas varies inversely with P_B, D would be 1.25 times that at sea level. As a result of the effect of D on G_{H_2O}, the effective G_{H_2O} and M_{H_2O} at 2400 m would also be 1.25 times their respective values. Reprinted from Carey (1994) with permission of Journal of Biosciences.

However, the P_{O_2} inside the shell, which forms the upper end of the gradient for diffusion of O_2 into the embryonic blood, drops as altitude increases (Carey *et al.* 1993, 1994). Although the rate of O_2 diffusion is enhanced at lower barometric pressure, the drop in P_{O_2} is due to the reduction in the numbers of pores in the shell and the decrease in absolute numbers of O_2 molecules, the ambient P_{O_2}. Embryonic puna teal (*Anas versicolor puna*) and American coots (*Fulica americana peruviana*) develop at 4,150 m in the Andes at blood oxygen tensions (10–12 torr) inside the eggshell that would disrupt growth in lowland mammalian and avian embryos (Carey *et al.* 1993, 1994).

The lack of adequate oxygen supplies has apparently exerted sufficient selective pressure on shell conductance that gas conductances of eggs laid above 3,000–3,500 m are not reduced to the same extent as the reduction in barometric pressure. In fact, conductance in some species laying eggs between 4,000 and 4,500 m approaches the value of eggs laid by sea level congeneric species (Carey *et al.* 1987, 1989, 1991; Figure 16.2). This modification improves the conductance to O_2 diffusion but also increases rates of water and CO_2 losses above sea level values. At these altitudes, modifications in the eggshell cannot create gaseous conditions inside the egg that resemble those at sea level. As a result, physiological and biochemical specialisations are required by embryos so they can develop under these conditions. Compared to embryos of low altitude relatives, montane embryos possess

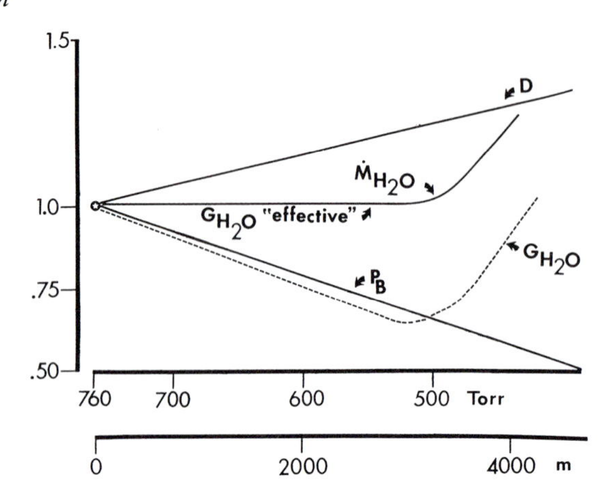

Fig. 16.2 This figure, based on data from eight species of birds breeding between sea level and 4,478 m (Carey *et al.* 1983, 1987, 1989a, 1989b, 1991), describes how changes in eggshell conductance affects gas exchange in eggs collected at various altitudes. Changes in barometric pressure (P_B), the diffusion coefficient for any gas (D), the 'effective' G_{H_2O} (the actual conductance of the egg to water vapour at a given altitude), and daily water loss from the egg (M_{H_2O}) in the field change with an increase in altitude (in meters). P_B, D, 'effective' G_{H_2O}, G_{H_2O} (standardised to 760 torr) and M_{H_2O} of eggs at the altitude at which they are laid as a function of altitude (in m). All parameters are standardised to 1.0 at sea level. The G_{H_2O} (standardised to 760 torr) of eggs laid at altitude decreases in approximate proportion to the decrease in barometric pressure and the increase in diffusion coefficients for gases to about 3,500 m. This modification (usually due to a reduction in the numbers of pores) offsets the increase in diffusion coefficients, with the result that the effective conductance and daily water loss of the eggs at the altitude they are laid are the same as at sea level. Above 3,500 m, eggs are laid with a G_{H_2O} (standardised to 760 torr) that is not reduced to the same degree as the reduction in barometric pressure, probably due to selection for improved oxygen diffusion. As a result, the effective conductance of the eggs increases, as does daily water loss. Reprinted by permission from Carey (1994) Journal of Biosciences.

an enhanced oxygen capacity of the blood and a rapid developmental transition of red blood cell organic phosphates, which affect haemoglobin affinity for oxygen (Carey *et al.* 1993, 1994; Dragon *et al.* 1999). These adjustments foster the ability of high altitude embryos to maintain a 'normal' level of metabolism and growth, with the result that incubation periods of some species do not differ from those of sea level congeners (Carey *et al.* 1982, 1994). Physiological properties of montane white-tailed ptarmigan that enhance blood oxygen carrying properties disappear if the embryos are incubated at lower altitudes, indicating that these adjustments are acclimatory responses to low O_2 rather than genetically fixed characteristics (Carey unpublished data). Unfortunately, no data are available on physiological properties of embryos laid above 4,400 m, but it might be hypothesised that genetic specialisations become essential for development at very high altitudes.

Whether females have the physiological ability to modify the gas conductance of eggshells in response to a change in altitude, or whether the conductance is genetically fixed, has been investigated with conflicting results. Rahn *et al.* (1982) found a significant increase in shell conductance following transfer of chickens from 3,800 to 1,600 m. On the other hand, Leon-Velarde *et al.* (1984) found no difference between shell conductance of eggs produced by chickens transported from sea level to 2,800 m and that of siblings held at sea level for an equivalent amount of time. No significant variation in conductance existed in eggs of Japanese quail (*Cotunrix japonica*) and Bengalese finches (*Lonchura striata*) after transfer to 2,900 m compared to their own eggs laid at sea level (Carey *et al.* 1984).

Desert environments

Deserts are commonly characterised as hot and dry (generally less than 250 mm rainfall per year, with low levels of primary productivity (Serventy 1971; Polis 1991). Most areas classified as 'desert' occur in temperate regions (15–35° latitudes) and so deserts can be cold as well as hot. Even in the summer, night-time temperatures can require thermogenic activity. For instance, temperatures in the Colorado Desert of California rise above 40°C during summer days but night-time temperatures can drop below the lower critical temperature (28°C) of very small birds, like verdins (*Auriparus flaviceps*; Webster and Weathers 2000).

Numbers of species that breed in deserts are relatively low compared to more moderate environments. About 40 species breed in the Sonoran desert of the southwestern United States region (Walsberg and Voss-Roberts, 1983) and about 25 and 17 species nest in the Sahara and in the deserts of Australia, respectively (Serventy 1971). A few birds take advantage of burrows to escape the heat (Williams and Tieleman 1999).

Birds breeding during the hottest months face the conflicting challenges of high ambient temperatures and low water availability. When body temperatures exceed air temperature, metabolically produced heat flows from the bird to the environment at a rate proportional to the difference between body and air temperature. As this differential decreases to only a few degrees, the rate of heat flow is inadequate to dissipate all metabolically produced heat. As a result, the rate of heat loss must be increased through evaporation of water by panting, gular fluttering, etc. When body temperatures equal air temperature, no net heat flow results. As air temperatures exceed body temperatures, the bird gains heat at a rate directly proportional to the differential between air and body temperature. Although some birds reduce their need for evaporative cooling by allowing body temperatures to rise during the day (Dawson and Bartholomew 1968) and by having lower than average basal levels of heat production (Williams and Tieleman 2000), many birds require additional water for thermoregulation in an environment in which water is generally scarce or non-existent. The metabolic costs of regulating body temperature in hot climates have a minimal impact on energy requirements, but it can have an important impact on water balance (Dawson and O'Connor 1996).

For instance, the water turnover of verdins increased 16% in summer compared with that observed during the winter Webster and Weathers 2000). Although birds can fly to water (Thomas and Robin 1977), a few types of desert birds do not drink water, even if it is available. The urine concentrating ability of the kidney, use of uric acid as the major nitrogenous waste, reduction of activity during the hottest part of the day, and utilisation of food, such as insects or cactus fruit that contain pre-formed water, allow adult birds generally to avoid dehydration (Carey and Morton 1971; Thomas and Robin 1977; Vleck 1993; Webster and Weathers 2000). However, increases in plasma concentration of corticosterone, an indicator of stress, during record-breaking temperatures (over 50°C) suggest that birds can occasionally have difficult maintaining water balance (Wingfield *et al.* 1992).

Tolerance of embryos to heat and prevention of overheating of eggs

Lethal limits of embryos of desert breeding birds do not vary from those of species breeding in cooler conditions (Bennett and Dawson 1979; Grant 1982; Webb, 1987). Despite the fact that avian eggshells are highly reflective of near infrared radiation (Bakken *et al.* 1978), temperatures of eggs exposed to mid-day sunlight in desert environments can rise over lethal limits within several minutes during voluntary departures of the adults from the nest (Howell *et al.* 1974; Bennett and Dawson 1979; Grant 1982). As a result, attendance at the nest by an incubating adult during the hottest parts of the day is critical to embryonic survival, but adults may become severely dehydrated during this period.

Since the upper thermal tolerances of embryos of desert species do not differ from those of species nesting in more moderate climates (Webb 1987), the challenge for breeding adults is to prevent overheating of the young when they leave the nest to forage. The problem may be particularly acute for species that exhibit single-gender incubation (Chapter 6), and for ground-nesting birds that lay eggs in scrapes on the soil with no vegetative cover. Some species prevent excessive exposure to the sun by covering eggs with soil or sand when the incubating adult is absent (Maclean 1988).

Like most species in other climates, desert birds incubate the eggs when air temperatures are below the optimal incubation temperature. When air temperatures equal optimal egg temperatures, shading behaviour occurs in many ground-nesting birds, even those that breed in environments that would not be classified as deserts (Ward 1990). Gray gulls (*Larus modestus*) stand over the nest and shade the eggs, which equilibrate with the air temperature. Shading is necessary because the eggs will overheat if exposed to direct sunlight. When air temperatures fall in the evening, the adult resumes incubation (Howell *et al.* 1974). Shading behaviour of crowned plovers (*Vanellus coronatus*) not only protects the eggs from overheating but also benefits the adult. The adult standing over the nest minimises contact of its body with the hot soil (50°C) and achieves an 11% improvement in convective heat transfer by elevating the body into the wind above the surface boundary layer (Downs and Ward 1997). Costa's hummingbird (*Calypte costae*) maintains temperatures of eggs between 35 and 36°C in nests in full sunlight with a combination of nest attentiveness and shading behaviour (Vleck 1981b).

When air temperatures exceed optimal incubation temperatures, the adult must actively cool the eggs because shading the eggs will not prevent overheating.

If water is available nearby, some birds, like the blacksmith plover (*Vanellus armatus*), squat in the water, soak the ventral feathers, and then transport the water to the eggs (Grant 1982; Ward 1988). Although Egyptian plovers (*Pluvianus aegyptius*) which breed on the sandy banks of rivers in lowlands of tropical Africa, are not desert birds, their incubation behaviour deserves mention. A lethal rise in egg temperature is prevented by a combination of egg burying and water transport. Eggs are contact incubated at night but are covered with several mm of sand during the day. As solar radiation heats the sand surface to 46°C, the adults wet their feathers in the nearby water and allow it to drip onto the sand covering the eggs. Evaporation of the water maintains egg temperature near 37–39°C even though air temperatures may reach 50°C (Howell 1979).

Egg shading and egg wetting are metabolically inexpensive for the adult. A different, more costly process of egg cooling is used by Hermann's gulls (*Larus heermani*), which sit tightly on the eggs at soil temperatures up to 55°C and use a combination of convective heat exchange and evaporation to cool both adult and eggs (Bartholomew and Dawson 1979). Mourning doves (*Zenaida macroura*) build nests in trees above the soil boundary layer. They also sit tightly on the eggs during the hottest part (40–45°C) of the day and maintain egg temperatures near 40°C by establishing a thermal gradient in favour of heat flow from the eggs to the adult (Walsberg and Voss-Roberts 1983). Panting, gular flutter, and the use of an oesophageal heat exchanger reduce body temperature to mildly hypothermic levels, around 38°C. These processes produce several problems for the adult. Muscle contractions involved in panting and gular flutter raise the heat production of the adult and its own requirement for evaporative water loss, the increased rate of respiratory activity accompanying panting can cause metabolic alkalosis through excessive losses of CO_2 (Dawson and Hudson 1970). The adult may also become quite dehydrated before it is relieved by its mate (Walsberg and Voss-Roberts 1983). Under extreme conditions, adults may be forced to desert the eggs if dehydration reaches dangerous levels.

Nest placement and orientation

Nest-building behaviour can promote the survival of young during the summer heat through microclimate selection for the nest. Cactus wrens (*Campylorhynchus brunneicapillus*) build enclosed nests in cholla cacti (*Opuntia*) or other spiny shrubs in the Sonoran Desert of Arizona. The adults vary the directional orientation of the nest entrance during the prolonged breeding season (Ricklefs and Hainsworth, 1969b). Cactus wrens experience superior hatching success when the nest entrance faces into the prevailing winds (Austin 1974). Presumably, successful fledging was influenced by an increase in airflow through the nest that facilitates thermoregulation by the young.

Gila woodpeckers (*Melanerpes uropygialis*) and northern flickers (*Colaptes auratus*) are primary hole nesters that excavate cavities in saguaro cacti (*Carnegiea gigantea*; Kerpez and Smith, 1990). The entrances of nests in the northern Sonoran Desert are randomly oriented, possibly because nest orientation might not measurably improve the air flow into a closed cavity. Secondary hole-nesters, like elf owls (*Micrathene whitneyi*), select existing holes with no preference for compass orientation of the entrance (Goad and Mannan 1987). However, a study at a more southerly location than that used by Kerpez and Smith (1990) found that nest entrances of Gila woodpecker were oriented significantly to the north (Inouye *et al.* 1981). Detailed information about the microclimate of the nest and fledging success of primarily hole nesters in these two regions is necessary before conclusions can be reached about the significance of these disparate findings.

Eggshell conductance

The low absolute humidity of the desert air creates a large gradient for water vapour that could cause excessive water loss (Walsberg 1983a). The conductance to water vapour of eggs of a number of species nesting in desert environments has been shown to be significantly below values predicted on the basis of allometric relationships or below values for related species breeding in more humid climates (Grant 1982; Ar and Rahn 1985; Arad *et al.* 1988). These adjustments serve to regulate the water loss, judged as a proportion of the initial egg mass, within limits observed in other species nesting in more mesic environments (Ar and Rahn 1980, 1985). Despite suggestions that nest humidity should be regulated by the adult (Rahn *et al.* 1976), a study of house finches (*Carpodacus mexicanus*) and phainopeplas (*Phainopepla nitens*) nesting in the Sonoran desert convincingly showed that the adults did not vary incubation behaviour when nest humidity was experimentally varied (Walsberg 1983a).

Timing of breeding

Time is not as critical for birds breeding in deserts as it is for those in cold climates or high altitudes. Breeding seasons may be as long as six months in the Sonoran desert (Davis and Russell 1984). Zebra finches (*Taenopygia guttata*) have been observed breeding in every month of the year in the deserts of Australia (Zaan 1996). Birds breeding in deserts which have relatively predictable precipitation breed at roughly the same time each year. Although reproduction of these birds may be timed by photoperiodic cues (Vleck 1993), reproduction may not occur in years of low or absent rainfall. Anecdotal information indicates that chemicals synthesised by particular plants during dry years may inhibit reproduction of the birds that eat those plants (Leopold *et al.* 1976).

By contrast, in some deserts like the interior of Australia and South Africa, sporadic rainfall is separated by droughts of unpredictable lengths with no precipitation occurring for some years. Birds in these regions breed opportunistically in the irregular, but favourable conditions following good rains (Serventy 1971). Zebra finches are the most well-studied opportunistic breeders (Zaan 1996). By conserving water and utilising metabolic water production and the pre-formed water in seeds, zebra finches can subsist without drinking water under moderate laboratory conditions (Cade *et al.* 1965). However, under severe conditions of high heat and low humidity in the field, they require access to liquid water. Initial reports indicated that this species exhibited courtship and copulation within hours of rainfall; eggs appeared within two weeks (Immelman 1971). However, a long-term study indicated that zebra finches do not breed following every rainfall; in fact, rain had a negative effect on breeding in the month in which it fell (Zaan *et al.* 1994). Breeding in zebra finches appears to be most highly correlated with rainfall four months prior to egg laying and hatching. This interval is required for germination, growth and maturation of grass seed, the food most commonly fed to the young (Zaan *et al.* 1994). Therefore, timing of breeding by zebra finches is currently thought to be associated with the maturation of grass seed, not the onset of rainfall.

A laboratory study found that zebra finches maintain ovaries in a medium-developed resting state, from which they can be rapidly prepared to lay eggs within 1–2 weeks (Sossinka 1980). However, severe dehydration in the laboratory impairs complete functioning of the testes (Vleck and Priedkalns 1985). If gonads are maintained year-round in a quasi-functional state in wild zebra finches, this pattern contrasts with that of birds breeding on a regular seasonal schedule in temperate and high latitude areas, which allow their gonads to collapse completely between breeding seasons (Morton *et al.* 1972).

Wet-nesters

Members of the order Podicipediformes nest in more moderate climates than other species featured in this chapter. However, they are included in this chapter because their nests pose some interesting challenges for eggshell design. Grebes build nests on islands of floating vegetation in aquatic situations (Palmer 1962). The nests are composed of saturated, rotting vegetation collected from the surrounding water. The placement of the nest away from shore provides some degree of protection from predators, particularly when then the adult covers the eggs with wet vegetation during absences from the nest (Broekhuysen and Frost 1968). On sunny days, adults leave the nest unattended for extended periods of time, during which solar radiation, plus heat generated by the rotting vegetation, heat the eggs to optimal incubation temperature during absences from the nest (Bochenski 1961; Davis *et al.* 1984b). Overheating is prevented by evaporation from the wet nest material. Little of the nest rises above the water surface, and so the eggs in the nest cup are in contact with, or are even partially submerged in, water (Sotherland *et al.* 1984).

Avian eggs lose water vapour during incubation, amounting to an average of about 15% of their initial mass. This water loss results in maintenance of the initial hydration of egg contents throughout incubation. The water vapour that is lost is replaced by the formation of an air cell in the blunt end of the egg (Ar and Rahn 1980). This air cell provides the volume of air necessary for the embryo to inflate its lungs prior to the pipping of the shell. Grebe eggs are not exempt from the general requirement that an appropriate level of water loss be achieved during incubation. However, the warm, extremely humid environment of grebe nests decreases the gradient for water vapour diffusion to the point that, all other factors held equal, water loss could be severely restricted (Davis *et al.* 1984b). Furthermore, the potential hazard exists in grebe nests that eggs in contact with water might actually absorb water from the nest.

Water loss from naturally incubated pied-billed grebe *(Podilymbus podiceps)* eggs averaged 16.4% of their initial mass due to an eggshell conductance that averaged 2.7-fold the value predicted for other avian eggs of similar mass and incubation period (Davis *et al.* 1984b). An exceptionally high rate of water loss of eggs incubated in controlled conditions confirmed this trend in great crested grebes *(Podilymbus cristatus*; Lomholt 1974). Absorption of liquid water by partially submerged eggs of eared grebes (*Podiceps nigricollis*) does not occur, although fowl and American coot eggs treated similarly do absorb water (Sotherland *et al.* 1984). Grebe eggs have a shell with large pores that is covered with a thick cuticle perforated with small fissures. Water vapour can diffuse through the small fissures, but the interaction of the fissures with the surface tension of liquid water prevents water entry into the pores (Board 1982; Sotherland *et al.* 1984).

The unusual nests of grebes result in a few problems for the embryos. Larger thermal gradients can exist between the top and the bottom of grebe eggs compared to those of most other birds and eggs immersed in water can cool more rapidly than eggs incubated in air. Furthermore, the pipped embryo can face suffocation or drowning if the pip hole becomes submerged in water. Vocalisations of eared grebe embryos appear to stimulate an increase in attentiveness to the nest in the last few days of incubation in comparison with earlier stages (Brua 1996). Adults respond to vocalisations by turning eggs, a behaviour that could minimise the risk of suffocation and help to equilibrate temperatures in the egg (Brua *et al.* 1996; Chapter 7).

Prospects for birds breeding in extreme environments

Despite the fact that few humans live in deserts, cold regions, and high altitudes, anthropogenic alteration of these environments occurs through numerous factors, such as deposition of windblown or waterborne pollutants, climate change, tourism, and introduction of predators. Presence of humans in a colony of emperor penguin chicks causes increases in energy expenditure and apparent 'nervous apprehension' (Regel and Pütz 1997; Giese and Riddle 1999). Population sizes of certain penguin colonies at sites frequently disturbed by humans have decreased (Wilson *et al.* 1991; but see Cobley and Shears 1999). Additionally, climate patterns in the arctic and Antarctica are not protected from global climate change (Serreze *et al.* 2000). Such change can alter sea temperatures that affect marine food supplies and the distribution and timing of melt of sea ice (Boersma and Stokes, 1995). Weather patterns in deserts and at high altitudes could also become more extreme and more unpredictable in the future. If birds breeding in these environments are at the edges of their tolerances for stress, these additional stresses may prove debilitating.

This chapter summarises only a fraction of the considerable progress that has been made in the understanding of adaptations of birds to deserts, high altitudes, and cold climates since the pioneering observations of George Bartholomew, William Dawson, Lawrence Irving, and Knut Schmidt-Neilson and F. Hall (references in Schmidt-Nielsen 1964; Irving 1972; Bartholomew and Dawson 1968;

Bouverot 1985; Dawson and O'Connor 1996). Although much remains to be learned in every area covered by this review, there are two areas in which additional research would be useful. First, no data exist on eggshell structure or embryonic physiology of species laying eggs at altitudes over 4,700 m. The mechanisms that foster successful development in conditions of extreme hypoxia and high rates of water and CO_2 losses could identify novel adaptations. Secondly, it is presently unclear to what extent population differences in adult behaviour, embryonic and adult physiology, and eggshell morphology in geographically widespread species result from genetic differences or from physiological acclimatisation and behavioural adjustments. Use of new technologies to identify novel gene products or differences in DNA composition might prove interesting in this regard.

17 *Tactics of obligate brood parasites to secure suitable incubators*

S. G. Sealy, D. G. McMaster and B. D. Peer

Obligate brood parasites lay all of their eggs in the nests of other species (Payne 1977a; Rothstein 1990), but some females of many other species that typically incubate their own eggs and care for their young also lay parasitically, in nests of their own or of other species (Yom-Tov 1980a; Rohwer and Freeman 1989; Sorenson 1991; Saylor 1992; Slagsvold 1998). A remarkable array of adaptations enables these brood parasites to lay their eggs in nests of unrelated individuals, delegating incubation, feeding and early defence of all or some of their young to foster parents, or hosts.

Obligate brood parasitism is an uncommon (about 1% of bird species) but successful life history strategy that has evolved independently in several groups of birds (Table 17.1). Obligate brood parasites exhibit several major departures from the normal reproductive physiology and behaviour typical of birds. Females have completely lost the ability to construct nests and to incubate eggs, and they neither feed nor tend their young (but see Hahn and Fleischer 1995; Lorenzana and Sealy 1998). Even with treatments of oestrogen and prolactin, female brown-headed cowbirds (*Molothrus ater*) did not develop brood patches or broodiness behaviour (Höhn 1959; Selander and Kuich 1963). Höhn (1972) suggested that a lack of prolactin led to the parasitic habit and resulted in failure to produce brood patches in the parasitic species. The pituitaries of brown-headed cowbirds, however, contained as much prolactin as those of related non-parasitic icterids (Höhn 1972; Chapter 8). This suggests secondary loss of sensitivity to prolactin after brood parasitism evolved. Prolactin was absent, however, in the breeding female common koel (*Eudynamis scolopacea*), a parasitic cuckoo (Höhn 1972). These differences in the production of prolactin likely reflect the independent evolution of brood parasitism in these groups.

Many hosts of brood parasites produce fewer young of their own, whereas others rear only the parasite (Friedmann 1929; Payne 1977a; Rothstein 1990; Rothstein and Robinson 1998; Lorenzana and Sealy 1999). In response, some hosts have evolved adaptations that reduce the cost of parasitism or the likelihood that parasitism will occur. Cycles of adaptations and counter-adaptations have resulted in some brood parasites countering host adaptations such as egg rejection with the development of mimetic eggs (Davies and Brooke 1998).

In this chapter, incubation of the eggs of obligate brood parasites is described with the focus on the tactics employed by brood parasites to obtain a suitable incubator for their eggs, a choice that must facilitate effective dominance between

Table 17.1 Occurrence of obligate brood parasitism in birds (after Johnsgard 1997b; Davies 2000).

Order	Family	Subfamily/species
Anseriformes	Anatidae	Black-headed Duck (*Heteronetta atricapilla*)
Cuculiformes	Cuculidae	Old world cuckoos (Cuculinae)
		American ground cuckoos (Neomorphinae)
Passeriformes	Indicatoridae	Honeyguides (Indicatorinae)
	Icteridae	Cowbirds (Icterinae)
	Estrildidae	Whydahs (Viduinae), Cuckoo finch (*Anomalospiza imberbis*)

the nestlings of the parasite and the host. After briefly reviewing some of the attributes of suitable hosts, tactics are described that brood parasites have adopted to: (1) ensure hosts accept the parasite's egg; (2) lay in nests before incubation has advanced too far; (3) ensure adequate incubation of the parasite's egg; and (4) ensure the parasite's egg hatches early enough for the nestling to gain an advantage in competition with host nestlings.

Suitable hosts

A suitable host species is one that: (1) normally cannot defend its nest from a would-be parasite (Sealy *et al.* 1998); (2) builds a penetrable nest (Payne 1973a; Carter 1986); (3) lays eggs similar in size to that of the parasite's to facilitate efficient incubation (Mills 1987; McMaster and Sealy 1997; Peer and Bollinger 2000); (4) accepts the parasite's egg (Rothstein 1975b, 1982b; Brooke and Davies 1988); (5) brings enough food (Slagsvold 1997) that is suitable for the parasite (Seel and Davis 1981; Middleton 1991; Kozlovic *et al.* 1996); (6) can feed fledgling parasites (Payne 1973b; Eastzer *et al.* 1980); and (7) is abundant (Payne 1977a).

Certain species are better hosts than others and not all nests of favoured host species will be discovered or are accessible. Finding host nests therefore is critical and making the right choices should be important (Brown and Brown 1991; Grant 1998). For parasites that preferentially parasitise only one or a few species, such as the common cuckoo (*Cuculus canorus*), selectivity of host nests is necessary (Wyllie 1981; Davies 2000). High fecundity has been believed to enable females of generalist parasites, such as the shiny cowbird (*Molothrus bonariensis*), to pursue opportunistic strategies (Kattan 1993, 1995), laying in nests of poor hosts in the absence of higher-quality nests (Rothstein 1975b; Weatherhead 1989). Support for this strategy has been weakened, however, as recent use of molecular genetic techniques has revealed that the generalist brown-headed cowbird apparently lays fewer eggs than previously believed (see below). Selectivity may also be a part of the generalist parasite's strategy of host use.

Do hosts vary in their quality? Are suitable hosts selected on the basis of host size, egg size, diet, or a combination of factors (Wiley 1988)? Some species are rarely parasitised, but apparently make good hosts (Eastzer *et al.* 1980; Ortega and Cruz 1991; Peer and Bollinger 1997), whereas others that are moderately or frequently parasitised fare poorly at rearing cowbirds or their own young in parasitised nests (Finch 1983). High frequencies of parasitism on such species may be due to a shortage of high-quality nests, among other factors (Ortega 1998). Ortega (1998) reviewed fledging success of brown-headed cowbirds and found an average of 22% (SD = 22%, range = 0–100%). These data reveal an absence of a consistent fledging rate or clearly superior hosts. On the other hand, cowbirds and cuckoos survive better, grow faster, and are larger at fledging when reared by larger hosts (Fraga 1985; Carter 1986; Mason 1986; Weatherhead 1989; Kleven *et al.* 1999; but see Scott and Lemon 1996). Thus, selection should favour some degree of selectivity in brood parasites in their use of host nests.

Tactics to ensure hosts accept the parasite's egg

Many species of cuckoo that preferentially parasitise one species or a group of species lay eggs of a single type that match the size and appearance of their host's eggs (Baker 1942; Brooke and Davies 1988; Higuchi 1989; Moksnes *et al.* 1995; Gibbs *et al.* 2000; Chapter 19). Hosts of many of these parasites eject non-mimetic eggs. By contrast, no species of cowbirds lay mimetic eggs (Rothstein 1990; Fleischer and Smith 1992; but see Jaramillo 1993; Peer *et al.* 2000). Individual females of the generalist brown-headed cowbird are not host specific (Fleischer 1985; Alderson *et al.* 1999), but the relatively few species that reject foreign eggs tend not to be parasitised (Sealy and Bazin 1995; Peer and Bollinger 1997).

A case of sensitive host discrimination has affected the size of shiny cowbird eggs in Uruguay (Mason and Rothstein 1986). Frequently parasitised in Uruguay, rufous horneros (*Furnarius rufus*) accept cowbird eggs depending on size. Eggs with widths less than 88% of their own egg breadth were ejected. In Uruguay, cowbird eggs have unusually large breadths, larger than those in Argentina, and this has allowed more eggs to be accepted by horneros in Uruguay. Mason and Rothstein (1986) interpreted this change in shape as a response to selection exerted by the rufous hornero to circumvent rejection. No such response is seen in Argentina because interactions there between the cowbird and rufous hornero are probably of recent origin.

Many birds recognise female brood parasites as threats to their reproductive success and respond intensively when parasites approach their nests (Sealy *et al.* 1998). After encountering parasites at the nests, some hosts are more likely to eject foreign, even mimetic eggs (Davies and Brooke 1988; Moksnes *et al.* 1993; but see Hill and Sealy 1994; Sealy 1995). Parasites therefore tend to visit nests when they are least likely to be seen by the hosts. Cowbirds lay around sunrise (Scott 1991; Neudorf and Sealy 1994; Peer and Sealy 1999a), and cuckoos from mid-morning to the afternoon, ostensibly when hosts are least attentive (Brooker *et al.* 1988; Davies and Brooke 1988). Laying usually takes only a few seconds, which may permit parasites to avoid being seen by the hosts (Sealy *et al.* 1995).

Many brood parasites remove a host egg in association with parasitism (Davies and Brooke 1988; Lombardo *et al.* 1989; Sealy 1992). Early workers (Savage 1897) believed that host egg removal duped hosts into accepting parasitic eggs by making it difficult for hosts to detect parasitism through changes in clutch size (Hamilton and Orians 1965; Moksnes and Røskaft 1987). Numerous experiments have shown, however, that a host's response to a parasite's egg is consistent whether or not a host egg is removed (Rothstein 1975b; Davies and Brooke 1988; Moksnes and Røskaft 1987; Sealy 1992).

With the exception of parasitic waterfowl (Mallory and Weatherhead 1990), eggs of brood parasites have stronger shells than expected for their body size because they are spherical and have thicker shells (Hoy and Ottow 1964; Blankespoor *et al.* 1982; Spaw and Rohwer 1987; Rahn *et al.* 1988; Picman 1989; Brooker and Brooker 1991; Johnsgard 1997b; Picman and Pribil 1997). Based on data on strength of cowbird eggs in Picman (1989) and on a regression of egg volume and egg strength for a group of control species (Picman *et al.* 1996), brown-headed cowbird eggs were 2.4 times stronger than expected for their size (Picman unpublished data). From the inside, the strength of brown-headed cowbird eggs is 1.7 times greater than that expected from body mass (Picman 1997). The spherical shape increases outside strength of the egg (Picman 1989), although it may decrease its inside strength because this increases as eggs become more elongated (Picman 1997). These observations suggest that hatching is more difficult for cowbirds than it is for other birds with eggs the same size but with shells of normal strength (Picman 1997).

Stronger eggshells may resist puncture-ejection by hosts (Hoy and Ottow 1964; Spaw and Rohwer 1987; Picman 1989). Presumed to be unable to grasp a cowbird egg, small hosts may attempt to puncture-eject the egg but risk damaging their own eggs if their bills deflect off the strong-shelled egg. As such, such hosts are expected to accept parasite eggs (Spaw and Rohwer 1987; Lotem and Nakamura 1998). Recent evidence, however, does not support this hypothesis. Weighing only 15 g, the warbling vireo (*Vireo gilvus*) puncture- or grasp-ejects cowbird eggs, rarely damaging its own eggs (Sealy 1996; Underwood and Sealy unpublished data).

Tactics to lay in nests before incubation has advanced too far

Egg production and size

The production by brood parasites of large numbers of eggs each breeding season should make the need for precise timing of laying less critical. Early studies revealed high numbers of eggs were produced annually, with host-specific parasites generally laying fewer eggs than the generalist parasites (Table 17.2). For example, brown-headed cowbirds were estimated to produce on average 0.73 eggs d^{-1} for a large sample of shot birds (Scott and Ankney 1980; Fleischer *et al.* 1987; Jackson and Roby 1993), and one captive female laid 77 eggs (Holford and Roby 1992). However, molecular genetic techniques, used to track eggs laid in different host nests by individual females, suggest the number of eggs produced is much

Table 17.2 Estimated annual egg production (number of eggs per female per breeding season) in obligate brood parasites.

Parasite species	Number of eggs laid[1]	Method of determination[2]	Location	Reference[#]
Molothrus bonariensis	Up to 120	POFs	Colombia	1
Molothrus ater	11.3	POFs	Michigan	2
	24.1	POFs	California	2
	25.0	POFs	Oklahoma	2
	Up to 40	POFs	Ontario	3
	16.4	Captivity	Illinois	4
	26.3	Captivity	Illinois	5
	1.7	Genetics	New York	6
	2.8[3]	Genetics	Manitoba	7
Cuculus canorus	Up to 15	Similar eggs	UK	8
	Up to 15	Similar eggs	UK	9
Cuculus spp.	22	POFs	Africa	10
Chrysococcyx spp.	16–21	POFs	Africa	10
Clamator spp.	19–25	POFs	Africa	10
Vidua spp.	22–26	POFs	Africa	10

[1]Single numbers are means. [2]POFs = post-ovulatory follicles. Ovaries of collected females were examined under a dissecting microscope. The number of eggs within the preceding 10 days was determined by counting the post-ovulatory follicles visible in the gross dissected ovary (see Payne 1976); Captivity = eggs collected from females held in outdoor pens; Similar eggs = similar egg-types assumed to have been laid by the same female observed in different host nests. [3]Maximum number of eggs laid by a single female was 13, all in red-winged blackbird (*Agelaius phoeniceus*) nests. [#]References: 1 – Kattan (1993); 2 – Payne (1976); 3 – Scott and Ankney (1980); 4 – Jackson and Roby (1992); 5 – Holford and Roby (1993); 6 – Hahn *et al.* (1999); 7 – Alderson *et al.* (1999); 8 – Chance (1940); 9 – Wyllie (1981); 10 – Payne (1973b).

lower (Table 17.2). These estimates, however, do not include the frequency of females that do not produce eggs in a given season and the number of young cowbirds recruited into the breeding population. The discrepancy between the various estimates of annual egg production has not been reconciled. Workers may not find all host nests on a particular site, and so this problem possibly will be resolved by tracking eggs using molecular genetic techniques backed by radio-tracking female parasites.

Eggs of parasitic cuckoos are smaller than those of non-parasitic cuckoos of similar body mass (Payne 1974). The egg of the common cuckoo is 3.4 g on average, whereas a non-parasitic cuckoo of the same body mass (100 g) would be expected to lay a 10-g egg (Davies 2000). Disproportionately small eggs apparently are laid more rapidly (Ortega 1998). By laying quickly, parasites may go unnoticed at host nests, increasing the likelihood hosts will accept their eggs (Sealy *et al.* 1995; Moksnes *et al.* 2000). In addition, the small size of cuckoo eggs is important for host egg mimicry because some hosts use size as a cue for egg recognition (Moksnes and Røskaft 1995; Marchetti 2000). Eggs of cowbirds, honeyguides and viduine finches are not smaller relative to body size, and thus are not specialised for brood parasitism (Payne 1977b, 1989; Briskie and Sealy 1990).

Monitoring host behaviour and egg laying

Careful monitoring and selection of host nests followed by appropriately timed laying relative to the host's laying cycle ensure that the parasite eggs hatch early and gain an advantage in competition with host nestlings. Brood parasites discover hosts and their nests by watching birds build and engage in related activities (Øien *et al.* 1996; Grieef and Sealy 2000). Coincidentally they determine the stage of the host's nesting cycle as clutches are initiated. Laying generally occurs during the host's laying stage (Figure 17.1) and the parasite's egg will likely hatch before or simultaneously with host eggs. Searching for and monitoring nests, however,

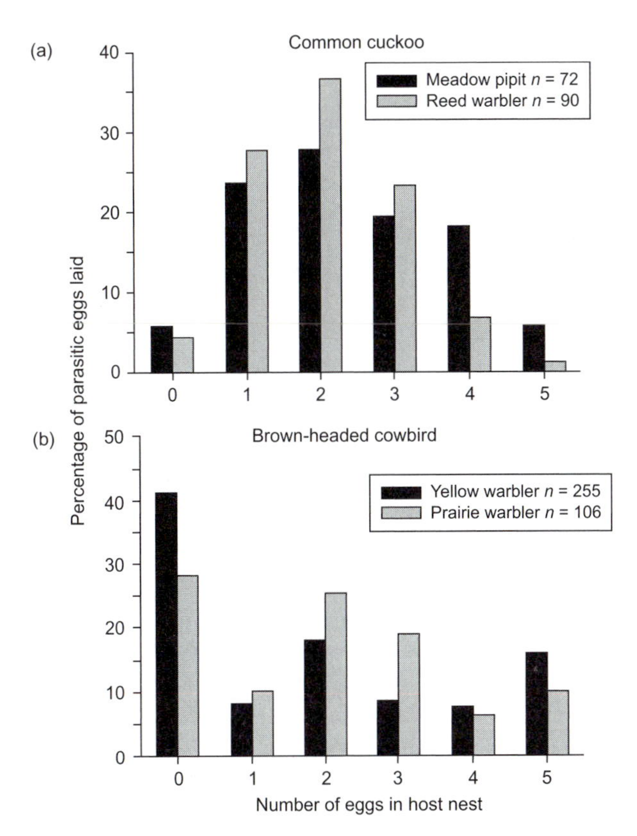

Fig. 17.1 Number of host eggs in nests when parasitised by (a) common cuckoos (*Cuculus canorus*) and (b) brown-headed cowbirds (*Molothrus ater*). Data for common cuckoo eggs laid in nests of meadow pipits (*Anthus pratensis*) and reed warblers (*Acrocephalus scirpaceus*) are modified from Chance (1940) and Wyllie (1981), respectively; data for brown-headed cowbird eggs laid in yellow warbler (*Dendroica petechia*) and prairie warbler (*Dendroica discolor*) nests are modified from Sealy (1995) and Nolan (1978), respectively. Most cowbird eggs laid through the second yellow warbler eggs were buried under a new nest (Sealy 1995; Mico 1998). In both host species, some cowbird eggs were laid after clutch completion, up to day 20 in yellow warbler nests and through to hatching in prairie warblers.

takes time from such activities as foraging (Curson *et al.* 2000). In the end not all nests discovered are accessible or suitable.

Parasites monitor nests (Mayfield 1961; Wyllie 1981) and usually lay before the host's incubation has advanced too far (Figure 17.1). The parasites may pay a price if they fail to lay at the appropriate time, whether too late or too early. Eggs laid before host clutches are initiated or early in the laying stage may be deserted (Briskie and Sealy 1987; Hill and Sealy 1994; Fraga 1998), buried under a new nest (Wyllie 1975; Brooker and Brooker 1989; Sealy 1995; Mico 1998) or ejected (Stouffer *et al.* 1987) by hosts that normally accept an egg. Nests found during incubation or with broods, on the other hand, are likely too advanced for parasitism.

Depredation by parasites of nests unsuitable for parasitism would create new opportunities for parasitism by forcing potential hosts to nest again (Livesey 1938; Gärtner 1981; Davies and Brooke 1988; Alvarez 1994; Arcese *et al.* 1996; Peer and Sealy 1999b; Hauber 2000; Nakamura and Cruz 2000). Support for the predation hypothesis comes from natural and experimental removal of female brown-headed cowbirds during the laying season of an important host, the song sparrow (*Melospiza melodia*), which resulted in lower predation on all nests (Arcese *et al.* 1996; Arcese and Smith 1999). Statistical comparisons also revealed parasitised nests failed less often than unparasitised nests when cowbirds were present (Arcese *et al.* 1996; Arcese and Smith 1999; Hauber 2000). By contrast, McLaren and Sealy (2000) recorded more failures among parasitised nests than unparasitised nests in a population of another frequent host, the yellow warbler (*Dendroica petechia*). Across seven populations of yellow warblers, predation and parasitism were significantly negatively correlated, in contrast to what has typically been documented (Johnson and Temple 1990; Arcese and Smith 1999). Clearly, more tests are required on other locally abundant hosts of cowbirds and other brood parasites. Ideally, the strongest support for this hypothesis will be the identification of females that destroy unparasitised clutches or broods and then linking them to eggs laid in replacement clutches.

Pecking host eggs to test stage of development

Livesey (1936) suggested that the removal and ingestion of a host egg in association with parasitism allowed common cuckoos to determine which clutches are suitable for parasitism. Clutches at advanced stages of incubation should be avoided. However, cuckoos generally remove the whole egg during the laying visit (Wyllie 1975; Brooker *et al.* 1988) and brown-headed cowbirds often wait until after they have parasitised a nest before returning to it to remove a host egg (Sealy 1992). Many parasites remove host eggs after parasitism and so determining the stage of development of eggs by 'tasting' (Livesey 1936) generally would not be practical.

Nevertheless, Massoni and Reboreda (1999) reported that shiny cowbirds punctured one or two host eggs before they parasitised yellow-winged blackbird (*Agelaius thilius*) nests, either later the same day or the day after. Nests with incubated eggs were seldom parasitised. The authors suggested that by pecking the eggs cowbirds were able to assess the stage of development of the clutch, although how the freshness of eggs was ascertained, by tactile, visual or by some other means, was not determined. The key test of this hypothesis, Massoni and Reboreda (1999) pointed out, will be to switch clutches between nests with freshly laid eggs and old nests with eggs close to hatching. Parasitism should follow punctures at old nests containing fresh eggs, but not at new nests containing old eggs.

Tactics to ensure adequate incubation of the parasite's egg

Importance of egg and clutch size

Eggs of some brood parasites have smaller length-to-width ratios than eggs of non-parasitic birds, which gives them a rounded appearance (Morel 1973; Picman 1989; Brooker and Brooker 1991). Spherical eggs may facilitate greater contact with the brood patch (Peer and Bollinger 1997) and retention of heat when hosts are off the nest (Johnsgard 1997b). Eggs of *Clamator* cuckoos are more rounded compared to the host eggs but eggs of some other groups of *Cuculus, Cacomantis* and *Chrysococcyx* cuckoos (Brooker and Brooker 1991), as well as those of the parasitic waterfowl (Mallory and Weatherhead 1990), are not as round. Thus, rounded eggs may not be a characteristic feature of eggs of all brood parasites.

Among the several features of a good host listed by Davies (2000) is that the host be a suitable size for efficient incubation. The smallest regular host of the common cuckoo are warblers of the genus *Phylloscopus*, such as the chiffchaff (*Pylloscopus collybita*), whose eggs are 35% the volume of a cuckoo's egg (Davies 2000). Egg volumes of the smallest hosts of Australian cuckoos are 38% of that of the cuckoo (Brooker and Brooker 1989). This suggests that there is a lower limit to host size and that a cuckoo's egg that is too large may not be adequately incubated by a small host. This would exclude the goldcrest (*Regulus regulus*) as a common cuckoo host (Davies 2000). The smallest hosts of the brown-headed cowbird are the irregularly parasitised golden-crowned kinglet (*Regulus satrapa*) and frequently parasitised blue-gray gnatcatcher (*Poliotila caerulea*), whose eggs are 22 and 29% the volume of a cowbird egg, respectively (Johnsgard 1997b). These hosts are the smallest of the suitable hosts available to this brood parasite. At the other extreme, a host may be too large, and the parasite egg cannot be properly heated lying amid a clutch of large eggs. Nevertheless, the egg of the great spotted cuckoo is only 46% the volume of a carrion crow (*Corvus corone*) egg, yet hatchability is similar to that observed in magpie (*Pica pica*) nests, where its egg is similar in size to the host egg (Soler 1990). The brown-headed cowbird's egg is only 47.5% the volume of the egg of its largest, frequently parasitised host, the western meadowlark (*Sturnella neglecta*). These figures may reflect an upper limit to host size for each parasite although in larger hosts, hatching early is not always enough to overcome the hosts' larger size (Lichtenstein 1998).

Unhatched eggs in parasitised nests of smaller hosts often belong to the host. In the common yellowthroat (*Geothlypis trichas*), whose eggs are 48% the volume of cowbird eggs (Johnsgard 1997b), Hofslund (1957) estimated that the heat received by one cowbird egg prevented two host eggs from hatching per nest. If nests received more than one cowbird egg, no host eggs hatched although the cowbird eggs generally did. Cain and McCuistion (1977) suspected two host and one cowbird embryo died in a Carolina wren (*Thyrothorus ludovicianus*) nest because of insufficient heat during incubation. Finally, cowbird eggs had 1.8 times the volume of ovenbird (*Seiurus aurocapillus*) eggs, which when laid on top of host eggs, may have resulted in reduced heat to host eggs and contributed to reduced hatchability (Hann 1947). Carefully controlled experiments are needed to quantify the dynamics of incubation in clutches containing parasite and host eggs.

Increases in clutch size by the addition of parasitic eggs may interfere with heat exchange during incubation and increase energy expenditure to maintain egg temperature in large clutches (Biebach 1981; Moreno *et al.* 1991). However, the amount of heat that can be produced and transferred to the clutch may be limited physiologically (Tøien 1989) and, hence, egg temperatures may fall as clutch size increases (Batt and Cornwell 1972; Mertens 1977). Extended cooling associated with incubation of a supernumerary clutch may slow the rate of embryonic development and increase the incubation time of both host and parasite eggs (Zimmerman 1983) and reduce the hatching success of these eggs (Hofslund 1957; Fraga 1983; Wiklund 1985; Kendra *et al.* 1988; Davies and Brooke 1988). Removal of a host egg may reduce the clutch volume to more closely match the host's incubation capability and reduce competition for heat during incubation. The parasite's egg, therefore, may be more likely to hatch and to hatch sooner than the host's eggs (Peer and Bollinger 2000). Contrary to speculation by Berger (1951), simulation of host egg removal by female brown-headed cowbirds did not promote hatching of cowbird eggs before host nestlings (McMaster and Sealy 1997).

Two studies have supported the role of host egg removal in improving the incubation environment of parasitic eggs (Davies and Brooke 1988; Peer and Bollinger 2000), whereas other studies indicated variation in effect among years (McMaster and Sealy 1997) or the results were negative (Wood and Bollinger 1997). Egg removal may promote hatching of parasitic eggs in clutches with large eggs, in most or all years (Peer and Bollinger 1997, 2000), but egg removal may be important only in some years for clutches with smaller eggs (McMaster and Sealy 1997). In clutches of smaller hosts, cowbird eggs may contact the brood patch essentially continuously, rendering egg removal less crucial for successful incubation (Peer and Bollinger 2000).

Avoidance of multiple parasitism

Brood parasites may parasitise clutches once, more than once, or lay in nests already parasitised by other females. Multiple parasitism may occur in response to competition for limited numbers of host nests (Davies and Brooke 1988; Arcese *et al.* 1996; Payne 1977b; Øien *et al.* 1996; Mermoz and Reboreda 1999). Hahn *et al.* (1999) and McLaren (2000) suggested that many females that lay in nests already parasitised by other females are yearlings without established egg-laying ranges, but this needs to be confirmed.

Multiple parasitism is common on hosts of some brood parasites, but infrequent on others. Nearly half of the nests of firefinchs (*Lagonistica senegala*) parasitised by village indigobirds (*Vidua chalybeata*) were multiple parasitised (Payne 1977b). About 14% of all host clutches in Britain were multiple parasitised by common cuckoos (Davies and Brooke 1988). Of 3,950 nests parasitised by eight species of cuckoos (*Cuculus*, *Chrysococcyx* and *Eudynamis*) in Australia (Brooker and Brooker 1989), 97% received one cuckoo egg.

In the great spotted cuckoo (*Clamator glandarius*), magpie nests were re-parasitised by the same female independently of their availability (Martinez *et al.* 1998). The nests may have been preferred in some way, possibly related to host quality (Soler *et al.* 1995). In other years, availability influenced host choice and typically some nests were parasitised by more than one female and females apparently responded to environmental conditions (Martinez *et al.* 1998).

On average about 35% of all nests parasitised by brown-headed cowbirds received more than one cowbird egg (Friedmann 1963). In Kansas, analysis of yolk proteins showed that most multiple parasitised clutches involved different females (Fleischer 1985). Use of DNA microsatellite markers revealed multiple parasitism on 31 song sparrow nests by the same female at 10 nests, by different females parasitising the same nest ($n = 16$), and at five nests by a combination of the same and different females (McLaren 2000). Almost half the nests were parasitised more than once by the same female, whereas nests in which more than one female laid were suspected to involve yearlings laying indiscriminately (McLaren 2000). In New York State, only one of 15 multiple parasitised nests involved laying by a single female, although one-third of all host nests received more than one cowbird egg (Hahn *et al.* 1999).

Hatching success of brood parasite eggs generally declines as the number of parasitic eggs in nests increases. In wood thrush (*Hylocichla mustelina*) nests parasitised by brown-headed cowbirds, Trine (2000) calculated a decline in hatching success of cowbird eggs from 81% (clutches with 1 cowbird egg) to 41% (6 cowbird eggs), or 8–10% for each additional cowbird egg. Overall fledging success of wood thrushes was reduced by about 18% for each brown-headed cowbird egg received. Hatching success of great spotted cuckoo eggs was 95.8% in clutches with one cuckoo egg and 75% in clutches with three cuckoo eggs (Soler 1990). When two nestling common cuckoos occupy the same nest, the first to hatch usually evicts the other (Wyllie 1981). More data are needed on the success of parasitic eggs in multiple parasitised nests, in addition to data on hatching and fledging success of host young.

Use of hosts with compatible incubation environments

Do attributes of hosts and host nests, such as nest insulation and incubation environment, influence host quality? With the exception of cavity-nesting birds, CO_2 and O_2 levels in most species' nests differ little from atmospheric levels (Walsberg 1980). Nest humidity, however, varied significantly among the few passerine species studied (Walsberg 1980). Black-headed ducks (*Heteronetta atricapilla*) parasitise gulls (*Larus* spp.) and coots (*Fulica* spp.) that nest near or on water

(Weller 1968) where humidity levels are presumably similar to ancestral pre-parasitic conditions. Parasitism on gulls is interesting given the gull's constant clutch size and 3-lobed brood patch because the eggs must contact enough of the gull's brood patch for effective heat transfer (*c.f.* Drent 1975). Hatching success of the gull eggs is not known (see Weller 1968).

Other brood parasites tend to parasitise relatively closely related hosts, possibly with similar incubation environments. For example, related to woodpeckers, honeyguides parasitise cavity nesters, viduine finches parasitise the domed nests of closely related finches, and cowbirds often use the cup nests of other closely related passerines. Cuckoos parasitise more distantly related hosts, but many use species with cup nests (Wyllie 1981; Rowan 1983) that are similar in type and incubation environment to those of ancestral non-parasitic cuckoos. As in their cavity-nesting hosts, the short incubation periods of honeyguides may reflect low O_2 and high CO_2 around the eggs, which results in poor gas exchange in and out of the egg and may have selected for an early transition to pulmonary respiration, as in woodpeckers (Yom-Tov and Ar 1993).

Strong eggs to resist accidental damage

Clamator cuckoos lay eggs that are stronger than expected for their size and this, too, is due to thicker shells and spherical shape (Brooker and Brooker 1991). By contrast, the eggshells of *Cuculus, Cacomantis* and *Chrysococcyx* cuckoos are not thicker than expected, but their eggshells, and those of *Chalcites* cuckoos are stronger apparently because they are more dense (Picman and Pribil 1997). Shells of *Clamator* and the other parasitic cuckoos (*Penthoceryx*, *Eudynamis*, *Scythrops*, *Tapera*, and *Dromococcyx*) are less dense than the aforementioned cuckoo species.

Picman and Pribil (1997) suggested eggshell density may be increased in two different ways. First, the inorganic portion of the shell, which determines strength, may comprise a greater proportion of the shell, or there may be more heavy elements within the inorganic portion. Second, *Cuculus* cuckoos lay eggs every two-days and the slower passage of the egg through the oviduct may allow the oviducal muscles to compress the shell more compactly, possibly increasing shell density (Wyllie 1981).

Strong eggshells could protect parasite eggs from accidental damage during incubation (Blankespoor *et al.* 1982). This hypothesis relies on the observation that *Clamator* and *Molothrus* species often lay more than one egg per nest and often remove or damage eggs in the process (Sealy 1992; Soler *et al.* 1997). Although not tested (Spaw and Roher 1987), parasitic eggs should not be more vulnerable to damage than host eggs. By contrast, eggs of *Cuculus* cuckoos have strong shells but clutches are seldom multiple parasitised (Wyllie 1981; Brooker and Brooker 1989), thus weakening support for this hypothesis (Picman and Pribil 1997).

Stronger eggshells may reduce the chances of breakage when parasites lay from above at nests into which they do not fit (Lack 1968). Struck by the falling egg, host eggs may be damaged but parasite eggs should be unharmed (Soler *et al.* 1997; Peer and Sealy 1999b). Cracked eggs may hatch but dented or more seriously damaged eggs normally fail or are removed by the hosts (Gaston 1976; Kemal and Rothstein 1988).

Tactics to ensure the parasite's egg hatches first

Hatching asynchrony and short incubation periods

Feeding success of brown-headed cowbird nestlings hatched one or two days before yellow warbler nestlings was significantly greater than that of cowbirds that hatched on the same day as the warbler young (Lichtenstein and Sealy 1998). The earlier a common cuckoo chick hatches, the easier it can evict the host's nestlings (Wyllie 1981; Payne and Payne 1998; Gill 1998). The likelihood of hatching before the host nestlings hatch depends in part upon the timing of laying by the female parasite, usually during laying or the first few days of incubation of the host as incubation periods of many hosts are also short (Briskie and Sealy 1990). When laid much later, parasite eggs may not hatch, especially if the egg temperature falls when host nestlings are brooded (*c.f.* Drent 1975). A cuckoo that hatches after the host's nestlings may need a few days to grow large enough before it can evict the more slowly growing host's nestlings (Payne and Payne 1998).

In many passerine species, incubation begins and increases gradually as the eggs are laid, becoming full time as laying ceases (Haftorn 1978, 1981; Mead and Morton 1985; Magrath 1992; Anderson 1997). The rate at which incubation increases is governed largely by the size of clutches females will lay (Haftorn 1981; Meijer 1990). The amount of incubation during laying largely determines the hatching spread of the clutch (Haftorn 1981; Bortolotti and Wiebe 1993; Wiebe *et al.* 1998). Experimental addition of conspecific eggs to clutches early in the laying period resulted in earlier onset of female incubation behaviour (Hébert and Sealy 1992), likely via increased hormonal secretions due to tactile stimulation of the brood patch (Chapter 8).

Therefore, if a brood parasite egg is added to a host clutch early in the host's laying period, without host egg removal, the host may be stimulated to incubate sooner. An earlier onset of incubation should speed the development of the parasite's embryo and increase hatching asynchrony of the host's clutch. Early hatching of brown-headed cowbird chicks may delay the hatch date of the last-laid host egg (McMaster and Sealy 1999), which may create a further size disadvantage for last-hatched nestlings. The addition of a brown-headed cowbird egg on the first day of laying, however, did not promote an earlier onset of incubation during laying in yellow warblers relative to control clutches (McMaster 1997). Given the already short incubation period of 11.9 days (range 10–14 days; Briskie and Sealy 1990; McMaster and Sealy 1998a), the benefit to the cowbird from earlier onset of incubation by hosts is questionable.

Brown-headed cowbird eggs may gain their head start by prolonging the incubation needed by eggs of smaller hosts. McMaster and Sealy (1998a) found a cowbird egg added about one and one-half days to the normal 11-day incubation period of yellow warblers. Tests in incubators supported this conclusion. Although host clutches with a cowbird nestling experienced reduced hatching success, and more eggs disappeared just prior to or while young were hatching, these effects were more likely due to the presence of artificial cowbird eggs in the clutches throughout incubation (McMaster and Sealy 1999).

Brood parasites are believed to have evolved incubation periods that on average permit them to hatch sooner than host nestlings. Indeed, the incubation periods of shiny, bronzed (*Molothrus aeneus*), and brown-headed cowbirds are short compared to those of non-parasitic icterids, due to shortened development times, not to the production of smaller eggs (Briskie and Sealy 1990; Kattan 1995).

Although few accurate determinations of incubation periods of any brood parasite have been made (Briskie and Sealy 1990), parasites typically hatch one or more days before the host nestlings (Hamilton and Orians 1965; Lack 1968; Payne 1977a). Smith (1968) determined that giant cowbird nestlings (*Molothrus oryzivorus*) hatched in 5–7 days less time than nestlings of the hosts, oropendolas (*Psarocolius* spp.) and caciques (*Cacicus* spp.). The incubation period of the screaming cowbird (*Molothrus rufoaxillaris*) is at least 12 days, close to that expected for the size of its egg, yet it hatches about one day before its usual host, the non-parasitic bay-winged cowbird (*Molothrus badius*; Briskie and Sealy 1990).

Patterns of embryonic development

Common cuckoos generally lay on alternate days (Chance 1940; Wyllie 1981). This is also observed in some other species of cuckoo (Liversidge 1961; Vernon 1970; Payne 1973b; Steyn and Howells 1975; Wyllie 1981) and possibly in honeyguides (Payne 1988). The shelled egg may be retained in the oviduct for up to 24 hours (Payne 1973b), about twice as long as it is in the domestic fowl or in any other species that lays every day (Romanoff and Romanoff 1949). The embryos in these eggs could begin developing inside the female before being laid, thus contributing to the short incubation period of the parasitic egg (Perrins 1967). The egg must be fertilised shortly after the previous egg is laid to give the newly fertilised egg sufficient time to develop before being laid. Consistent with this, Liversidge (1961) observed copulation attempts by the jacobin cuckoo (*Chrysoccocyx jacobinus*) in the morning, shortly after laying.

Evidence for pre-incubation development comes from examinations of embryonic development in freshly laid cuckoo eggs. In the jacobin cuckoo, Liversidge (1961) showed microscopically that the embryo of an unincubated egg, laid no more than three hours before fixation, was at the same stage as the embryo of a domestic fowl after 17–20 hours of incubation. In the same species, Vernon (1970) noted two eggs collected several days after laying were further advanced in their embryonic development than those of the hosts. The primitive streak was visible, the headfold was indicated, and in cross-section the primitive groove showed quite clearly but there were no somites.

Non-parasitic cuckoos lay at irregular intervals, often 48 hours or more (Hughes 1999), but it has not been determined whether pre-incubation development occurs in these species. It is therefore premature to conclude that pre-incubation development evolved specifically for brood parasitism. Parasitic cuckoos lay smaller eggs than nesting cuckoos of the same body size (Payne 1974), but the incubation periods of parasitic cuckoos are not shorter than non-parasitic cuckoos with eggs of the same size (Payne 1977a; Davies and Brooke 1988). With short incubation periods, parasitic cowbirds lay every day but apparently there is no additional oviducal development of embryos (Briskie and Sealy 1990).

The energy content of 10 shiny cowbird eggs on average was 14.5 kJ (±0.19), which was lower than expected (19.1 kJ, 95% prediction interval = 16.3–22.2) from the mean egg mass (Kattan 1995). Smaller yolk reserves may trigger earlier hatching compared with species that lay eggs of comparable size, because the cowbird embryos are forced to hatch before they run out of energy. McMaster and Sealy (1998a), however, noted that as energy investment per unit egg mass is expected to be lower in species with shorter incubation periods, the low energy content of shiny cowbird eggs may be due to the short incubation period, rather than a cause of it. Unlike shiny cowbirds, energy investment in brown-headed cowbird eggs does not differ significantly from that predicted by egg mass (Ankney and Johnson 1985). Eggs of other species such as the song sparrow have caloric contents ($kJ\,g^{-1}$) lower than brown-headed cowbirds (Carey *et al.* 1980), which suggests reduced energy content in eggs is not a specific adaptation of parasitism.

Recent experimental evidence suggests that parasites may synchronise their hatching by picking up auditory cues from host eggs. Calculations have shown that the development of brown-headed cowbirds outpaced that of the yellow warbler, possibly because they were stimulated to hatch earlier when in contact with host eggs. McMaster and Sealy (1998a) suggested that cowbirds might even be able to accelerate their hatching after picking up clues that host nest mates were on the verge of hatching. Cowbirds possibly cue into the clicking sounds embryos make when they are nearly ready to hatch and have started to breathe (Vince 1966b; Chapter 7).

Parasitic waterfowl can lay their eggs over more days than it takes hosts to complete their clutches (Saylor 1992). Waterfowl start incubating during laying, hence, although disparities are created in the development of embryos (Caldwell and Cornwell 1975; Afton 1980), in the end eggs hatch together because intra-clutch communication accelerates development of delayed embryos and retards the development of others (Vince 1968; Chapter 7). Snow goose (*Chen caerulescens*) eggs hatched even if laid in other nests up to four days after the start of incubation (Davies and Cooke 1983). However, Saylor (1992) noted that Davies and Cooke's (1983) methods underestimated the percentage of eggs that could synchronise with the host's and suggested that even if nests were parasitised at random, then about one-third of all parasite eggs would hatch, depending upon nesting chronology and how many hosts are laying or incubating at the time.

A similar mechanism may account for hatching of eggs laid parasitically by cliff swallows (*Petrochelidon pyrrhonota*). Some incubating females transfer eggs in their beaks from their own nests to neighbouring nests any time after laying up to one or two days before the eggs are due to hatch (Brown and Brown 1988). About 6% of nests receive parasitic eggs, and most hatch. Additional careful studies of clutches with numbered eggs may reveal this tactic is employed in other species (see Trost and Webb 1986).

Despite the advantage of hatching before host nest mates, hatching too early may be costly. Kemal and Rothstein (1988) found that red-winged blackbirds (*Agelaius phoeniceus*) ejected 83% of simulated broken eggs from their nests up to day 9 of incubation. They considered this behaviour adaptive because it ensured that cracked or broken eggs would be removed and thus not soil the nest or attract nest parasites. After 10 days of incubation and the day before blackbird clutches started to hatch, only 22% of broken eggs were ejected. It was suggested that if cracked eggs are ejected, a parent might mistakenly remove its own pipped eggs at this time.

Conceivably, eggs of brood parasites with incubation periods shorter than 10 days (the shortest recorded for the brown-headed cowbird) could be mistaken for damaged eggs and removed (Briskie and Sealy 1990). However, McMaster and Sealy (1998b) recorded acceptance by red-winged blackbirds of eight cowbird eggs about 25 hours from hatching when added to blackbird nests and which hatched midway through incubation (see also Holcomb 1967). Another potential cost of short incubation periods in the brood parasites may involve the immune system of embryos. Birds with shorter incubation periods are more frequently infected with blood parasites, which suggests that immune function or development may be compromised when the incubation period is reduced (Ricklefs 1992).

Nestling cuckoos and honeyguides that eject or kill host nestlings after hatching (Friedmann 1955; Payne and Payne 1998) appear to have little to gain by altering host incubation after hatching. Likewise, alteration of host incubation behaviour would appear to have little benefit for the precocial young of black-headed ducks (Weller 1968). For parasitic cowbirds, however, whose nestlings compete with host nestlings, early hatching by the parasite could divert the host female's attention away from incubation of the remaining eggs, thereby reducing the hatching success or slowing the development of the eggs (Nolan 1978; Dolan and Wright 1984; Fraga 1985; Petit 1991). The experimental addition of a cowbird nestling to yellow warbler clutches the day before hatching, however, did not reduce incubation attentiveness or prolong incubation periods of any but the third host egg (McMaster and Sealy 1999).

Priorities for future research

Despite the upsurge in research on avian brood parasitism over the last three decades, few studies on the incubation of eggs of parasites and hosts have been conducted and much of what has been done is biased toward brown-headed cowbirds. Hatchability of parasitic eggs varies widely among hosts of different sizes, especially in clutches parasitised more than once, but the optimal temperatures for normal embryonic development, extremes of embryonic temperature tolerance, gaseous requirements of embryos, and maximum hatchability generally have not been recorded for parasitic eggs. Consideration should be given to the possible role of the incubation environment in host selection as many parasites with different-sized eggs lay over wide ranges of temperature, humidity and elevation. Data are needed on the timing of laying relative to host laying and hatching for more parasitic species. The influence of parasitic eggs on synchronisation with hatching of host eggs and its implications for timing of parasitism are poorly known. Host egg removal in association with parasitism remains enigmatic but results of experiments suggest egg removal enhances hatching success of both parasite and host eggs, but the extent may differ among parasites and hosts. Controlled experiments involving hosts of different sizes and an array of brood parasites are needed. Experiments also should focus on the function of host egg destruction by parasites, an apparently risky way to reduce nestling competition or possibly force re-nesting. The adaptive significance of parasitic eggs and eggshells should be explored more thoroughly; trade-offs between increased shell strength and hatching ability of chicks need more attention.

The variable estimates of the annual production of eggs by most brood parasites based on field and laboratory techniques has not been reconciled. It is not known with certainty how many eggs female parasites produce, the distribution of eggs among host nests, and the proportion of eggs laid that produce young that are recruited into breeding populations. Do all females produce eggs, every year? Is egg production tied to availability of host nests? Demographic implications of the recently appreciated depredation by parasites of unparasitised clutches and broods to create new opportunities for parasitism are far-reaching. Tested on only two hosts of the brown-headed cowbird, with contradictory results, the predation hypothesis should be tested on other hosts of this and other species of brood parasites. Future research will undoubtedly refine our knowledge of the tactics used by brood parasites to obtain incubators for their eggs and also will elucidate adaptations that facilitate effective competition between parasite and host eggs.

Acknowledgements

We thank D. C. Deeming for his editorial assistance and P. N. Hébert, C. M. McLaren, J. Picman, T. J. Underwood and K. L. Wiebe for constructive comments on various drafts or sections of the manuscript, and for useful discussion. Financial support during the preparation of this review was provided by a grant from the Natural Sciences and Engineering Research Council of Canada.

18 *Ecological factors affecting initiation of incubation behaviour*

P. N. Hébert

Contact incubation behaviour ensures that bird eggs are maintained at a temperature sufficient to promote embryonic development (Nice 1954). Although the dynamics of heat exchange vary during the incubation period (Turner 1991; Chapter 9), the thermal requirements of the clutch generally remain constant from the start of incubation through to hatching. Incubation temperatures, once attained, must be maintained within narrow limits to assure optimal embryonic development (Webb 1987). As such, the thermal requirements of the clutch may conflict (*sensu* Trivers 1974) with the nutritional and metabolic requirements of the parent (Nilsson 1993a).

If the ecological constraints incurred by adults vary then the resolution of the trade-off between the metabolic requirements of the parent and the thermal requirements of the developing embryo should vary in a similar manner. Indeed, considerable variation has been observed in the length of the incubation period within and between populations (Slagsvold 1986), seasonally (Mead and Morton 1985), and yearly (Hébert and Sealy 1992). Within a species, variation in the length of the incubation period can be attributed to two main components of incubation: first, variation in the onset of incubation during laying (Clark and Wilson 1981), and second, variation in %attentiveness once full incubation has been initiated (Sealy 1984; Astheimer 1991; Chapter 6). In this chapter the main ecological constraints responsible for variation in the onset of incubation during laying are reviewed.

Physio-ethology of the onset of incubation

The onset of incubation involves the development and expression of certain behaviours through the singular and interactive influence of several internal and external stimuli. Initially, interactions between photoperiod and the endocrine system (Silverin *et al.* 1989; Meijer *et al.* 1990), stimulate nest building. Incubation behaviour develops through the expression of appetitive behaviours associated with nest building (Baerends 1970), and is characterised by increases in blood titres of hormones such as prolactin and oestrogen. In addition to mediating the onset of incubation behaviour (Chapter 5), these hormones also stimulate processes associated with the development of the brood patch, such as defeathering and vascularisation (Lloyd 1965a, 1965b; Selander and Kuich 1963; Chapter 8).

Production of prolactin is further stimulated by visual and tactile stimuli associated with the presence of eggs in the nest (Hall and Goldsmith 1983).

Transition to full incubation involves a decrease in the frequency of interruptive behaviours (Baerends 1970). Full expression of incubation behaviour, typically requires the presence of a complete clutch, which apparently serves to inhibit appetitive behaviours associated with nest-building, and thus promotes full incubation (Beer 1963; Baerends 1970; Meijer 1995). In many species the onset of full incubation is apparently triggered by the rapid increase in prolactin associated with the ovulation of the last ovum (Mead and Morton 1985; Chapter 5). However, there are species of owl (Marti 1991) and parrot (Enkerlin-Hoefich and Hogan 1997; Beissinger *et al.* 1998) that initiate full incubation well before ovulation of the last ovum. Additional data on hormone levels during nest-building and laying are required to assess the causal and functional importance of hormone levels in mediating incubation attentiveness during laying.

Patterns of onset of incubation

Among birds, patterns of initiation of incubation fall into two broad categories. First, the onset of incubation is delayed until completion of the clutch. If ambient temperatures remain below physiological zero (Webb 1987) during laying, embryonic development is initiated at approximately the same time in all eggs, and the eggs hatch synchronously, i.e. the first and last-hatched nestlings are separated by 24 hours or less (Ricklefs 1993). While it is often assumed that embryonic development proceeds at a similar pace in all eggs of the clutch, in Anseriformes there is considerable variability in rates of embryonic development within a clutch and hatching synchrony may be a function of co-ordination by embryo-embryo communication (Afton and Paulus 1992; see Chapter 7). A delay in the onset of incubation until clutch completion is apparently uncommon, and is documented in only a few, yet diverse orders, particularly those with nidifugous young, although some groups with nidicolous young also exhibit a delay in the onset of incubation (Table 18.1). For species with nidicolous young, a delay in the onset of incubation and the subsequent hatching synchrony, is presumably a secondarily derived characteristic given that their common ancestors exhibited hatching asynchrony (Stoleson and Beissinger 1995).

In the second pattern, incubation is initiated with the laying of the penultimate egg or earlier (Beissinger and Waltman 1991). Earlier laid eggs receive a developmental head start (Enemar and Arheimer 1989) and eggs in a clutch hatch over a period greater than 24 hours and exhibit hatching asynchrony (e.g. Bryant 1978; Zerba and Morton 1983a; Skagen 1987; Smith 1988; Wiebe and Bortolotti 1994). Hatching asynchrony is believed to be a derived phenotype (Stoleson and Beissinger 1995), suggesting that incubation prior to clutch completion is also a derived characteristic. Early onset to incubation is the more common pattern of incubation during laying, especially among taxa with nidicolous young (Table 18.2). Incubation may also be initiated during laying in some precocial species (Afton and Paulus 1992; Meijer and Siemers 1993). In this instance the

Table 18.1 Examples of some avian orders containing species where incubation has been documented to be delayed until clutch completion.

Order	Family	Species	Reference#
Anseriformes*	Anatidae	*Chen canagica*	1
		Chen caerulescens	2
		Melanitta nigra	3
Galliformes	Phasianidae	*Perdix perdix*	4
Gruiformes*	Rallidae	*Rallus elegans*	5
Charadriiformes*	Scolopacidae	*Calidris melanotos*	6
	Charadriidae	*Charadrius semipalmatus*	7
Columbiformes	Columbidae	*Columba livia*	8
Apodiformes*	Apodidae	*Chaetura vauxi*	9
Coraciiformes	Alcedinidae	*Ceryle alcyon*	10
Piciformes*	Picidae	*Melanerpes carolinus*	11
Passeriformes*	Tyrannidae	*Sayornis phoebe*	12
	Turdidae	*Sialia currucoides*	13
	Sturnidae	*Acridotheres cristatellus*	14
	Mimidae	*Mimus polyglottos*	15
	Alaudidae	*Eremophila alpestris*	16
	Parulidae	*Dendroica virens*	17
	Cardinalidae	*Cardinalis cardinalis*	18

*Orders that also contain species that initiate incubation prior to clutch completion (see Table 18.2).
#References: 1 – Peterson *et al.* (1994); 2 – Poussart *et al.* (2000); 3 – Bordage and Savard (1995); 4 – Carroll (1993); 5 – Meanley (1992); 6 – Holmes and Pitelka (1998); 7 – Nol and Blanken (1999); 8 – Johnston (1992); 9 – Baldwin and Zaczkowski (1963); 10 – Hamas (1994); 11 – Jackson (1976); 12 – Weeks (1994); 13 – Hébert (1999); 14 – Johnson and Campbell (1995); 15 – Derrickson and Breitwisch (1992); 16 – Beason and Franks (1974); 17 – Pitelka (1940); 18 – Laskey (1944).

hatching synchrony, typical of precocial species, is facilitated by synchronising mechanisms (Vince 1964, 1968; Forsythe 1971; Flint *et al.* 1994; Chapter 7).

Unlike species that delay incubation until clutch completion, the tendency to initiate incubation during laying varies greatly among individuals, populations, and species. Variability among individuals is likely due to differences in motivational factors, learning, and adaptation (Beer 1963; Baerends 1970). In turn, ecological factors can moderate individual predispositions to incubate during laying. Given the extent of variation in incubation patterns during laying among individuals, and among species, it would be important to determine how much of this variation is of genetic origin and what proportion is environmentally induced. There are many hypotheses that propose an adaptive function to initiating incubation prior to clutch completion, and yet it is still not known what proportion of the observed variation in the onset of incubation is genetically determined (but see Ricklefs and Smeraski 1983, and references therein). In addition, the effect of partial incubation on the rate of embryonic development is not documented. In some cases, incubation attentiveness of less than 50% on the day the ante-penultimate egg is laid, is nonetheless sufficient to produce hatching spreads of 48 hours or more (Hébert and Sealy 1992).

Table 18.2 Examples of some avian orders containing species where incubation has been observed to be initiated prior to the laying of the last egg.

Order	Family	Species	Reference#
Gaviiformes	Gaviidae	*Gavia stellata*	1
Pelecaniformes	Phalacrocoracidae	*Phalacrocorax brasilianus*	2
Anseriformes*	Anatidae	*Aix sponsa*	3
Ciconiiformes*	Threskiornithidae	*Ajaia ajaja*	4
Falconiformes	Falconidae	*Falco sparverius*	5
	Accipitridae	*Ictinia mississippiensis*	6
Gruiformes	Gruidae	*Grus canadensis*	7
Charadriiformes*	Laridae	*Larus argentatus*	8
		Sterna caspia	9
Psittaciformes	Psittacidae	*Rhynchopsitta pachrhyncna*	10
		Amazona viridigenalis	11
Cuculiformes	Cuculidae	*Coccyzus americanus*	12
Strigiformes	Strigidae	*Asio otus*	13
	Tytonidae	*Tyto alba*	14
Apodiformes	Trochilidae	*Selasphorus sasin*	15
Piciformes*	Picidae	*Dryocopus pileatus*	16
		Melanerpes formicivorus	17
Passeriformes*	Tyrannidae	*Empidonax minimus*	18
		Empidonax oberholseri	19
	Vireonidae	*Vireo atricapillus*	20
	Corvidae	*Pica hudsonia*	21
	Bombycillidae	*Bombycilla cedrorum*	22
	Sturnidae	*Sturnus vulgaris*	23
	Paridae	*Parus major*	24
	Hirundinidae	*Riparia riparia*	25
	Passeridae	*Passer domesticus*	26
	Motacillidae	*Ficedula hypoleuca*	27
	Muscicapidae	*Regulus regulus*	28
	Emberizidae	*Plectrophenax nivalis*	29
	Parulidae	*Dendroica petechia*	30
	Ptilogonatidae	*Phainopepla nitens*	31
	Icteridae	*Dolichonyx oryzivorus*	32
		Xanthocephalus xanthocephalus	33
		Agelaius phoeniceus	34
		Qiuscalus quiscula	35

*Orders that also contain species that delay incubation until clutch completion (see Table 18.1). #References: 1 – Bundy (1976); 2 – Telfair and Morrison (1995); 3 – Kennamer *et al.* (1990); 4 – White *et al.* (1982); 5 – Bortolotti and Wiebe (1993); 6 – Parker (1999); 7 – Tacha *et al.* (1992); 8 – Drent (1970); 9 – Fetterolf and Blokpoel (1983); 10 – Snyder *et al.* (1999); 11 – Enkerlin-Hoeflick and Hogan (1997); 12 – Potter (1980); 13 – Marks *et al.* (1994); 14 – Marti (1991); 15 – Mitchell (2000); 16 – Bull and Jackson (1995); 17 – Koenig *et al.* (1995); 18 – Briskie and Sealy (1989); 19 – Morton and Peryera (1985); 20 – Grzybowski (1995); 21 – Buitron (1988); 22 – Puttnam (1949); 23 – Meijer (1990); 24 – Haftorn (1981); 25 – Peterson (1955); 26 – Lowther and Cink (1992); 27 – Potti (1998); 28 – Haftorn (1978); 29 – Lyon and Montgomerie (1995); 30 – Hébert and Sealy (1992); 31 – Rand and Rand (1943); 32 – Martin (1974); 33 – Fautin (1941); 34 – Yasukawa and Searcy (1995); 35 – Howe (1976).

Ecological factors affecting the onset of incubation

Interest in incubation attentiveness during laying has been limited relative to that garnered by hatching patterns and their impact on reproductive success (Stenning 1996). Currently, there are at least 19 hypotheses (e.g. Lack 1954; Hussell 1972; Clark and Wilson 1981; Slagsvold and Lifjeld 1989; Pijanowski 1992) regarding the significance of hatching patterns in birds (Viñuela 2000), with the majority assuming that hatching patterns reflect patterns of incubation during laying. However, incubation attentiveness during laying can be influenced by several ecological factors (Magrath 1990), the effects of which are relatively unexplored.

For example, hatch spreads tend to vary with clutch size (Haftorn and Reinertson 1985), and date of clutch initiation (Beason and Franks 1974; Hébert 1999), but it is not known whether this is related to different incubation patterns during laying (Stoleson and Beissinger 1995). Within a species with altricial young, hatch spreads are typically correlated positively with clutch-size (e.g. Howe 1976; Zack 1982; Hébert and Sealy 1992), such that hatching is more synchronous in smaller compared with larger clutches. In addition, food availability at the time of laying may have an effect on the ability of the female to produce eggs (Krementz and Ankney 1986), which in turn could affect incubation patterns and subsequent hatch spreads (Mertens 1980; Slagsvold 1986; Magrath 1992). Between species, hatch spreads tend to be relatively synchronous in species producing relatively small clutches compared to species that produce larger clutches (reviewed by Clark and Wilson 1981). At present there are few data to indicate if such variation is the result of differences in incubation behaviour during laying.

Hatch spreads have also been observed to vary with date of clutch initiation (reviewed by Stoleson and Beissinger 1995). Spread of hatch is relatively more synchronous early in the breeding season than later in the breeding season. It is unknown whether seasonal variation in hatch spreads is associated with a concomitant change in incubation behaviour during laying, or is the result of associated changes in ecological factors such as food availability, temperature, predation risk. Confounding seasonal variations in hatch spreads are correlated changes in clutch size (Perrins and McCleery 1985).

It is also unclear whether incubation patterns during laying are influenced by different mating systems. For example, in polygynous species such as the pied flycatcher (*Ficedula hypoleuca*; Slagsvold and Lijfeld 1989), house wren (*Troglodytes aedon*; Harper *et al.* 1994), or the bobolink (*Dolichonyx oryzivorous*; Martin 1974), does the incubation pattern during laying exhibited by a primary female differ from that of a secondary female, or a monogamous female? These questions are particularly relevant to hypotheses suggesting that hatching asynchrony is a female strategy to maximise paternal investment (Slagsvold and Lifjeld 1989, but see Hébert and Sealy 1993a). Furthermore, there is also some indication that the onset of incubation can be influenced by life-history strategies. For example, Nilsson (1993b) reported that hatch spreads tended to be relatively longer in swallow species with shared incubation compared to species where only the female performed all incubation duties. Presumably in species where both sexes incubate, the effects of environmental variation during laying are less constraining than in species with single-sex incubation. In some cavity nesters, the availability of nest sites may be limited, and may thus influence the onset of incubation (Grenier and Beissinger 1999). In the green-rumped parrotlet (*Forpus passerinus*), a Neotropical parrot, there is evidence that an early onset to incubation is at least partly stimulated by the availability of nest sites. However, there are other secondary cavity nesters that delay incubation until clutch completion (Johnson and Campbell 1995; Hébert 1999).

There is also evidence that clutches tended by older western gulls (*Larus occidentalis*) tend to hatch relatively synchronously compared to clutches tended by younger pairs. Sydeman and Emslie (1992) argue that older pairs are more experienced and thus less likely to require recourse to the competitive asymmetries among brood mates to facilitate brood reduction (Lack 1954). Although the observation is consistent with the brood reduction hypothesis (Lack 1954) and other hypotheses concerning the evolution of hatching asynchrony (Hussell 1972), there are no data on incubation attentiveness and age or experience of the parent birds. Shorter spread of hatch in early season clutches may be influenced by relationships between motivational state, learning, and adaptation (Baerends 1970), or due to variation related to ecological constraints occurring during laying (e.g. temperature), such that any correlated advantages may be incidental. Clearly, our understanding of hatching patterns would benefit from additional studies on incubation patterns during laying that have a natural history perspective.

Predation

Predation imposes important ecological constraints on reproductive success (Ricklefs 1969). During the breeding period, the risk of predation in birds varies with nest type (Ricklefs 1969), and nest height (Nilsson 1984). Variation in the risk of predation on incubating adults (Magrath 1989), or nest contents (Clark and Wilson 1981) can influence the onset of incubation (e.g. Hébert and McNeil 1999a, 1999b). Clark and Wilson (1981, 1985) argued that some species may be more prone to predation during the egg-laying/incubation period, whereas others are more susceptible during the nestling/fledgling period. Consequently, if the risk of total nest failure due to predation is concentrated during the egg stage, selection will favour individuals that initiate incubation relatively early to minimise the time their nests contain only eggs. As a result, eggs within a clutch would hatch asynchronously. Conversely, if total nest failure is greatest during the nestling/fledgling stage, individuals should delay the onset of incubation to maximise the time the nest contains eggs, in order to minimise the amount of time the nest contains nestlings (Clark and Wilson 1981). In this instance hatching would be relatively synchronous.

The total nest failure hypothesis receives little support in the literature (Hussell 1985; Clotfelter and Yasukawa 1999). In most cases, productivity ratios based on daily survival probabilities predict greater hatching spreads than those observed (Bancroft 1985; Briskie and Sealy 1989; Hébert and Sealy 1993b; Hébert 1999; Persson and Goransson 1999). One factor that may select against an earlier onset to incubation is the inherent increase in the probability of adult mortality due to predation (Magrath 1989). Adding predation risk on adults, such that total nest failure results from either predation on the adult or nest contents, alters optimal incubation patterns such that the onset of incubation should be delayed.

The total nest failure hypothesis (Clark and Wilson 1981) and the adult mortality hypothesis (Magrath 1989) although difficult, may be tested experimentally. For instance, experimental eggs additions during laying can advance experimentally the onset of incubation (Slater 1967; Beukeboom *et al.* 1988; Hébert and Sealy 1992), whereas modifications of the hatching pattern can alter the length of the nestling/fledging period (Hébert and Sealy 1992, Hébert 1999). Until further experimentation is done, it will be difficult to judge the importance of predation in influencing incubation patterns during laying.

Temperature

Temperature can influence the onset of incubation in at least two ways (Beer 1963, Hébert and Sealy 1992). First there is the potential effect of ambient temperature on the thermal homeostasis of the adult. Under thermoneutral conditions incubation may represent a small but non-trivial metabolic expense (White and Kinney 1974), but incubation does carry significant metabolic costs when ambient temperatures fall outside the thermoneutral zone (El Wailly 1966; Davis *et al.* 1984; Williams 1991; Tatner and Bryant 1993; Chapter 20). However, there is little information regarding the influence, if any, of ambient temperature on incubation patterns during laying.

With few exceptions (Wilson *et al.* 1986; Beissinger and Waltman 1991) adults initially spend little time incubating during laying (Haftorn 1981; Mead and Morton 1985). Therefore, temperature variations should have little if any effect on the metabolic balance of incubating birds, and thus only a modest effect on attentiveness patterns during laying. Several studies have observed that attentiveness during laying is unaffected by ambient temperature (Haftorn 1981; Morton and Pereyra 1985; Clotfelter and Yasukawa 1999; but see Baerends *et al.* 1970). This helps explain, at least in part, the positive correlation between ambient temperature during laying and hatching spreads (Slagsvold 1986; Hébert and Sealy 1992; Viega and Viñuela 1993; see also Hébert and McNeil 1999a). In temperate regions, clutches can be initiated in early spring when ambient temperatures are well below physiological zero (Webb 1987). Thus when the female leaves after having laid an egg, or leaves the nest to forage, the eggs will cool relatively faster and will attain relatively cooler temperatures compared to later in the breeding season when ambient temperatures are closer to physiological zero. Eggs exposed to cooler temperatures during laying would incur retarded development, and this would promote hatching synchrony early in the breeding season. This would be especially important in species with female-only intermittent incubation (Williams 1991; Nilsson 1993b). Again this area would benefit from further experimental study. Additional investigation of the effects (if any) of ambient temperature on incubation patterns during laying are important given that several hypotheses regarding the adaptive significance of hatching asynchrony also predict a seasonal increase in hatch spreads (Lack 1954; Hussell 1972).

Variations in ambient temperature may also affect incubation patterns during laying as a result of effects on the viability of the clutch. Embryonic development can occur between the physiological zero of ~27°C up to ~39°C, with the optimum being between 34–39°C (Webb 1987). Once development has been initiated, embryos become more sensitive to temperatures outside the range of optimal incubation temperatures (White and Kinney 1974; Webb 1987). Arnold *et al.* (1987) and Viega (1992) proposed the 'egg viability hypothesis' that suggests patterns of initiation of incubation during laying represent a parental strategy to optimise egg viability during fluctuating environmental temperatures. The hypothesis refers especially to species that experience ambient temperatures well below 10°C (Arnold *et al.* 1987) or temperatures in the critical zone between minimal optimal incubation temperatures and physiological zero (Stoleson and Beissinger 1999). Observations have shown that viability decreased if eggs were exposed to and neglected at temperatures between 34°C and 27°C, (Webb 1987; Stoleson and Beissinger 1999; Viñuela 2000), or temperatures below physiological zero (Arnold *et al.* 1987; Viega 1992). These experiments involved removing freshly laid eggs and keeping them in a cardboard box (Viñuela 2000), unused nest-boxes (Veiga 1992; Stoleson and Beissinger 1999) or nests (Arnold *et al.* 1987). In all cases the removed eggs did not receive contact from a female laying subsequent eggs, and thus only partially mimicked the 'neglect' experienced by early laid eggs. Contact, however minimal, between the laying female and the clutch, may be important in maintaining the viability of early laid eggs, and can be accomplished without inducing hatching asynchrony (Veiga 1992; Hébert and Kahn unpublished data). Furthermore, exposure times of experimentally neglected eggs tended to exceed that which would typically be encountered under natural conditions but despite this, hatchability was only significantly reduced at periods of neglect exceeding 3–5 days (Veiga 1992; Stoleson and Beissinger 1999; Viñuela 2000). By contrast, Arnold *et al.* (1987) removed freshly laid eggs of several duck species and placed them in deserted nests to mimic neglect. Hatching success was only marginally reduced when eggs were left neglected for 5 days; a period of neglect of at least 10 days was required to significantly reduce egg viability.

Also of interest is a similar study on house sparrows (*Passer domesticus*) indicating that experimentally neglected eggs suffered reduced hatchability (Viega 1992). The house sparrow is sympatric with the mountain bluebird (*Sialia currucoides*). In Manitoba, Canada, both species can be found nesting in nest-boxes on adjacent fence posts. The house sparrow exhibits hatching asynchrony, initiating incubation on the ante-penultimate or penultimate egg (Lowther and Cink 1992), whereas bluebirds tending first clutches delay the onset of incubation until completion of the clutch (Hébert 1999). Thus although both species experience similar temperature regimens, their incubation strategies differ during laying. The relevance of the egg viability hypothesis in explaining hatching asynchrony in house sparrows must be reconsidered.

It is possible that the reduced hatching success in the neglected first-laid eggs, observed in the above mentioned studies, was in part due to the complete neglect, a situation not typical during laying. Again, the minimal contact the eggs receive prior to the onset of incubation may serve to maintain their viability without requiring the onset of full incubation. Also, future studies should take care to report egg size to establish if hatching failure was associated with egg size. This is important because egg-size often increases with laying order, e.g. common grackle (*Quiscalus quiscula*; Howe 1976) and house sparrow (Lowther and Cink 1992), and egg size can influence hatching success (Williams 1980; Rofstad and Sandvik 1985).

Finally, although temperature is undoubtedly an important element in the viability of eggs, the sensitivity of unincubated eggs to variations in environmental and incubation temperatures may be overstated. Sensitivity of freshly laid eggs to temperature fluctuations likely varies between species, and likely reflects their natural history. For example, several seabird species exhibit egg neglect of variable duration, during the incubation period, with little effect on the viability of the eggs (Boersma 1982; Sealy 1984; Chaurand and Weimerskirch 1994). Thus, it can be argued that the sensitivity of eggs to temperature fluctuations varies between species, and that incubation patterns may reflect this sensitivity. Additional experimental studies are required to determine if incubation patterns truly reflect selection pressures on egg viability.

Food availability

Costs of egg production and incubation must be met through foraging, either by the incubating birds (Nilsson 1993a, 1993b; Rauter and Reyer 1997), or by the male bird that supplements the diet of the female (Smith *et al.* 1989). If foraging success is less than optimum, then more time must be spent foraging to subsidise egg production and/or incubation. Additional time spent foraging may reduce the amount of time that an individual can spend incubating during laying, and lead to a delay in the onset of incubation (Slagsvold 1986; Moreno 1989a). Indeed, there is substantial evidence indicating that variations in hatch spreads correlate positively with food availability or female condition during laying (Slagsvold 1986; Moreno 1989a; Hébert 1993; Wiebe and Bortolotti 1994; Persson and Goransson 1999).

Nilsson (1993a) proposed the 'nutritional constraints hypothesis', arguing that variations in hatching spreads represent a trade-off between the nutritional constraints of the parent and the thermal requirements of the clutch, and ultimately reproductive success. This hypothesis was tested by providing a sample of marsh tits (*Parus palustris*) with supplemental food during laying. Pairs provided with supplemental food initiated incubation earlier in the laying sequence, and thus their eggs hatched more asynchronously compared to control pairs (Nilsson 1993a). The importance of food to incubation patterns during laying can also be inferred from differences in hatch spreads between swallow species with shared incubation and species with single-sex incubation (Nilsson 1993b). Species that share incubation exhibited greater hatch spreads compared to species with single-sex incubation, presumably reflecting their greater ability to initiate incubation earlier in the laying sequence.

However, the influence of food during laying apparently varies between species. Wiebe and Bortolotti (1994) observed that clutches tended by female American kestrels (*Falco sparverius*) receiving food supplements hatched more synchronously compared to clutches tended by females that did not receive food supplements. This was attributed to a delay in the onset of incubation by the females receiving food supplements, although incubation behaviour during laying was not monitored. Hatching in kestrels tends to be asynchronous, especially in clutches tended by females producing clutches with a volume that exceeds their incubation capacity such that not all eggs are incubated efficiently, and hence may hatch out of sequence. It is possible then that the reduced asynchrony observed in supplemented kestrel clutches was not due to an adaptive response in food availability, but rather an increase in the mass of the female, such that incubation was more efficient (Wiebe and Bortolotti 1994; see also Wiebe *et al.* 1998). In the mountain bluebird food supplements prior to and during laying did not alter the pattern of incubation during laying (Hébert and Asselin unpublished data). This is particularly telling given that mountain bluebird females typically delay incubation until clutch completion (Hébert and Khan unpublished data). It appears that, at least in mountain bluebirds, a delay in the onset of incubation may be a parental strategy, rather than the result of environmental constraints. The influence of food on incubation attentiveness during laying is another area that requires further experimental work.

Direction of future studies

The onset of incubation follows two general patterns, such that incubation can be delayed until clutch completion, or incubation can be initiated prior to clutch completion. Evidence suggests that incubation attentiveness of less than 50% during laying can nonetheless result in hatching asynchrony. There is evidence that incubation patterns during laying are influenced by proximate factors such as food availability and ambient temperature. However, the evidence is for the most part equivocal. The study of hatching asynchrony would certainly benefit from further studies of incubation patterns during laying. Indeed, if incubation patterns during laying, and the subsequent hatching spreads represent an adaptive strategy, it should also be possible to establish that incubation patterns are heritable. At present it is not known what are the relative contributions of environment and genetic factors to variations in the onset of incubation.

Acknowledgements

I thank R. Buisson, T. Dickinson, N. Flood and J. Hare for their timely logistic support. D. C. Deeming, J. Hare and S. G. Sealy also made helpful comments on an earlier draft of this paper. Some unpublished data reported here was supported by a research grant to the author from the Natural Sciences and Engineering Research Council of Canada.

19 *Adaptive significance of egg coloration*

T. J. Underwood and S. G. Sealy

The extreme variability in the colours and patterns on birds' eggs captured the attention of early naturalists and oologists. Poulton (1890) suggested, in relation to eggs, '...any description of colour and marking will be considered incomplete unless supplemented by an account of their meaning and importance in the life of the species.' Several hypotheses have been proposed to explain the adaptive significance of egg coloration. These include crypsis, egg recognition and mimicry, filtering solar radiation, eggshell strength, aposematism, and others. These hypotheses are reviewed for the adaptive significance of egg coloration as they relate to the variability in the ground colours and maculation of eggs. Except for immaculate blue and white eggs, and intraclutch variability, each hypothesis is discussed separately. Throughout the chapter, the following terminology is used. The ground colour of an egg shell is its uniform base colour, whereas maculation is the superficial markings or patterns (e.g. spots, scrawls, etc.) on top of the ground colour. Spots or spotting patterns describe a type of maculation composed of several round marks of various sizes. Pigmentation and coloration are general terms that refer collectively to the ground colour and maculation.

Crypsis

Crypsis is the resemblance in coloration of an animal to its background (Edmunds 1990). Cryptic animals should face less risk of predation than non-cryptic animals. Egg coloration is widely believed to function in concealing eggs from predators because predation strongly influences nest success (Ricklefs 1969) and life histories (Martin 1995). Based on subjective assessments, the colours and maculation of the eggs of many species, especially ground-nesters, have been assumed to be cryptic because eggs appear to blend in with their surroundings (e.g. Poulton 1890; Chapman 1924; Cott 1940). Evidence for this was observational with many examples provided of species whose eggs match their surroundings and other species that laid conspicuous white eggs in concealed nests (Cott 1940).

Tinbergen *et al.* (1962) were the first to demonstrate the cryptic value of egg coloration in black-headed gulls (*Larus ridibundus*). Brownish eggs with variously coloured spots were depredated significantly less than eggs painted white. However, two other experiments revealed no significant difference between egg colour treatments (Tinbergen *et al.* 1962). Since then, many experiments have yielded variable results (Table 19.1). Of 19 studies, only six have found a significant effect of egg coloration on predation frequency.

A closer look at these studies reveals two interesting trends. In studies where eggs were placed in (usually) artificial nest structures only 10% (n = 10) found a significant effect of coloration, whereas 56% (n = 9) of studies where eggs were placed on the ground or in trees (one study) with no nest found a significant effect of coloration. A reduction in predation levels occurred mostly in the absence of a nest structure. Furthermore, studies that used painted eggs for their cryptic treatments found a significant effect of coloration in 10% (n = 10) of experiments, but 56% (n = 9) of studies using naturally coloured eggs, or eggs with only painted spots for cryptic treatments found a significant effect. These trends suggest that coloration conceals eggs of ground-nesters, where no nest structure exists, and that predators are attracted to nest structures. Therefore, the appearance of eggs in a nest may not conceal them. However, these results also suggest that differences in experimental design may explain the variable responses to cryptic coloration.

Skutch (1976) and Götmark (1992) suggested that predators find nests rather than eggs, which removes any selective advantage for coloured eggs in conspicuous nests. Hence, birds that build conspicuous nests should lay non-cryptic eggs, whereas birds that lay in scrapes on the ground should lay cryptic eggs (Götmark 1992). Comparison of the nest and egg characteristics of 27 families of non-passerine birds showed that individuals in significantly more families with conspicuous nests laid conspicuous eggs than families that used no nest or it was inconspicuous (Götmark 1993). Cryptically coloured eggs seem to be adaptive only for species that do not build a nest. However, only one of eight experimental studies that used nests (Table 19.1) found a significant effect of coloration, using real eggs, real nests and a nest site chosen by the birds (Westmoreland and Best 1986).

Two main concerns have been raised about using artificial nests to study nest predation (Martin 1987; Haskell 1995; Major and Kendal 1996). Artificial nests are usually more conspicuous than real nests and the eggs used are often an inappropriate size or colour. Indeed, Poulton (1890) and Lack (1958) stressed that to determine the function of egg coloration, eggs must be assessed in their natural surroundings, not out of context. In addition, Ortega *et al.* (1998) found that artificial nests were depredated significantly more often than American robin (*Turdus migratorius*) nests tested in their original locations, even when baited with Japanese quail (*Coturnix coturnix*) eggs. Thus, the results of many studies may not reflect the influence of coloration on predation as it might prevail under natural conditions.

Other researchers have compared how well eggs match the background of the nest and whether this affected predation. Thomas *et al.* (1989) found that most egg characteristics did not match the nest site features of dotterels (*Charadrius morinellus*), but they did not relate this to predation. In stone curlews (*Burhinus oedicnemus*), ground colour did not differ significantly from the nest substrate colour and depredation was significantly higher on clutches that did not perfectly match the substrate (Solís and Lope 1995). By contrast, the relationship between pigmentation and substrate type did not effect predation frequency of Namaqua sandgrouse (*Pterocles namaqua*) clutches (Lloyd *et al.* 2000). Westmoreland and Kiltie (1996) found a significant correlation of pattern disparity (an index to visual

Table 19.1 Summary of experimental studies that have compared frequencies or instances of predation on different coloured egg treatments (modified and expanded from Major and Kendal 1996).*

Egg treatment[1]				Nest type	Results[2]	Reference[3]
1	2	3	4			
Speckled (herring gull)[P]	Unspeckled (herring gull)[P]			Artificial	N.S.	1
White (fowl)[P]	Brown-spotted (fowl)[P]			None	N.S.	2
White (black-headed gull)[P]	Brown-spotted (black-headed gull)			None	T2 < T1	2
White (fowl)	Khaki (fowl)[P]	Khaki-spotted (fowl)[P]	Brown-spotted (laughing gull)	None	N.S.	3
White (fowl)	Khaki (fowl)[P]	Khaki-spotted (fowl)[P]		Artificial	N.S.	3
White (fowl)	Khaki-spotted (fowl)[P]	Black (fowl)[P]		None	N.S.	3
White (quail)[P]	Beige-splotched & spotted (quail)			None	T2 < T1	3
White (fowl)	Blue (fowl)[P]	Brown (fowl)	Beige-splotched & spotted (quail)[4]	None	N.S.	4
White (fowl)	Green (fowl)[P]			Artificial	N.S.	5
White (ostrich)	Brown (ostrich)[P]			None	T2 < T1	6
White (mourning dove)	White-spotted (mourning dove)[S]	White (mourning dove)[5]	White-spotted (mourning dove)[S,5]	Natural and Active	T4 < T1, T2, T3	7

White (fowl)	Tea-stained (fowl)[P]			Artificial	N.S.	8
White (quail)[P]	Blue (quail)[P]	Beige-spotted (quail)[S]		Artificial	N.S.	9
White (quail)[P]	Blue (quail)[P]	Beige-spotted (quail)[S]		None	T3 < T1 and T2	9
Blue (ceramic)[P]	Speckled-beige (ceramic)[P]			Natural	N.S.	10
Blue-green (fowl)[P]	Cream-speckled (fowl)[P]			Natural	N.S.	11
White (clay)	Speckled (clay)[S]			Artificial	N.S.	12
White (fowl)	Brown (fowl)	Creamy-white (northern bobwhite)		None[6]	T2 < T1 and T3	13
White (fowl)	Brown (fowl)			Artificial	N.S.	14

*Scientific names of birds listed: Ostrich (*Struthio camelus*), fowl (*Gallus gallus*), herring gull (*Larus argentatus*), black-headed gull (*Larus ridibundus*), laughing gull (*Larus atricilla*), quail (*Coturnix coturnix*), mourning dove (*Zenaida macroura*), and northern bobwhite (*Colinus virginianus*). [1]All egg treatments are natural unpainted eggs unless otherwise noted; P = painted or artificially coloured to a large degree; S = spots of paint added only. Coloration descriptors are those of the authors. [2]Results indicate which treatment suffered significantly less predation than the others; N.S. = no significant differences between treatments; T = treament. [3]References: 1 – Kruijt (1958); 2 – Tinbergen *et al.* (1962); 3 – Montevecchi (1976); 4 – Janzen (1978); 5 – Slagsvold (1980); 6 – Bertram and Burger (1981); 7 – Westmoreland and Best (1986); 8 – Salonen and Penttinen (1988); 9 – Götmark (1992); 10 – Yahner and DeLong (1992); 11 – Bildstein (1993); 12 – Major *et al.* (1996); 13 – Yahner and Mahan (1996); 14 – Jobin and Picman (1997). [4]Janzen (1978) used eggs of an unspecified quail species. [5]In treatments 3 and 4, incubation was interrupted every 3 days when the adult was flushed, while treatments 1 and 2 were checked from a distance. [6]An 'artificial nest' was used, but is not considered a nest because it was a simple depression in leaves.

patterns) between clutches and their background in three species of cup-nesting icterids, but predation frequency was not significantly related to different levels of crypsis. In these studies, the degree to which eggs matched their background and escaped predation varied among species and studies. This may reflect the difficulty of accurately quantifying egg colours and maculation.

The results of experiments and observations permit a tentative conclusion that the egg coloration of many ground-nesting birds conceals them from predators. However, biases present in many experiments mean that conclusive tests on the cryptic value of coloration are lacking for the cup nests of passerines.

Egg/nest recognition

Egg recognition in colonial birds

Relocation of a nest after disturbance may be difficult in colonial birds nesting at high densities, increasing the probability of misdirected parental care. Consequently, individual clutch recognition is expected (Buckley and Buckley 1972; Birkhead 1978). In many colonial birds, there is much interclutch variability in egg coloration and maculation (Gaston and Nettleship 1981; Baerends and Hogan-Warburg 1982), which may allow birds to recognise their own clutch (Birkhead 1978).

Experiments with colonial sea birds that nest in similar sites but different densities provide support for this hypothesis. In egg-switching experiments, common guillemots (*Uria aalge*) rejected most foreign eggs that differed substantially from their own egg (Tschanz 1959). Similarly, thick-billed murres (*Uria lomvia*) identified their own egg when offered a choice between it and a foreign egg (Gaston *et al.* 1993). However, razorbills (*Alca torda*) did not differentiate between foreign eggs and their own and Birkhead (1978) concluded that razorbills did not need to recognise their eggs because they nest at lower densities than common guillemots. By contrast, other densely nesting sea birds, such as Adélie penguins (*Pygoscelis adeliae*), Laysan albatrosses (*Phoebastria immutabilis*) and black-footed albatrosses (*Phoebastria nigripes*), did not recognise their eggs and accepted a variety of foreign eggs placed in their nests (Bartholomew and Howell 1964; Frederickson and Weller 1972; Davis and McCaffrey 1989). Selection for egg recognition in penguins and albatrosses is not expected because their distinct nest bowls may prevent eggs from being dislodged and aid in nest relocation.

In addition to dense colonies, the indistinct nest scrapes of terns may select for clutch recognition (Schaffner 1990). Royal terns (*Sterna maxima*) found their own, single eggs and incubated them when they were switched with a neighbour's eggs (Buckley and Buckley 1972). Elegant terns (*Sterna elegans*) scraped out a new nest for their own eggs when they were experimentally moved, and rejected most foreign eggs added late in incubation (Schaffner 1990). By contrast, individual laughing gulls (*Larus atricilla*) and herring gulls (*Larus argentatus*) did not distinguish their own clutches from other clutches (Noble and Lehrman 1940; Tinbergen 1960). Both gulls nest less densely than the terns and build large, bulky nests (Harrison 1975) that the herring gulls, at least, recognised more strongly than their eggs (Tinbergen 1960).

These results support the hypothesis that individual clutch recognition has evolved in response to dense nesting. However, Shugart (1987) questioned the results of some studies during research on Caspian terns (*Sterna caspia*) that did not recognise their own clutches when switched with other clutches. The only evidence for clutch recognition came from Caspian terns that chose clutches that matched the ground colour of their own clutches over clutches with different ground colours. Shugart (1987) also noted that his experiments used randomly chosen clutches, whereas Buckley and Buckley (1972) chose clutches that differed conspicuously in appearance. Indeed, all studies that strongly supported individual clutch recognition used eggs or clutches that were chosen preferentially because they appeared different. When tested with eggs whose appearance was similar to their own, most individuals did not recognise their own clutches (Tschanz 1959; Schaffner 1990).

Focusing on the biases identified by Shugart (1987), re-evaluation of the results of the above studies found little conclusive evidence for individual clutch recognition in densely nesting colonial birds. Caspian terns, common guillemots and thick-billed murres still found their nest sites when their clutches and/or nests were removed (Tschanz 1959; Shugart 1987; Gaston *et al.* 1993). Apparently, individual clutch recognition is not always necessary for nest relocation but it could be important for relocating eggs that have rolled away after a disturbance. Several species exhibited a coarse level of egg recognition that may be used along with features of the landscape and nest site, or perhaps neighbour recognition (*c.f.* Tinbergen 1960), to locate their nests and eggs.

Egg recognition in response to brood parasitism

Some species of birds whose nests receive a brood parasite's egg can identify and reject the egg. Rejector species may distinguish foreign eggs by true recognition of their own eggs, or by discordancy, i.e. recognition of the odd egg (Rothstein 1982a). Rensch (1924, 1925) suggested that passerines recognise foreign eggs by discordancy, but reinterpreting Rensch's data, Rothstein (1970) showed that the experiments did not support discordancy. Support for true egg recognition derives from many experiments on host species of various brood parasites that rejected foreign eggs regardless of the proportion of foreign eggs to their own (Victoria 1972; Rothstein 1975a, 1977, 1982a; Moksnes 1992; Sealy and Bazin 1995).

Egg recognition in response to conspecific brood parasitism has been documented in only a few species (Victoria 1972; Bertram 1979; Arnold 1987; Braa *et al.* 1992; Lyon 1993; McRae 1995) out of about 162 species known to be conspecific brood parasites (Yom-Tov 1980a; Rohwer and Freeman 1989; Ortega 1998). Swynnerton (1916) and Victoria (1972) suggested that interclutch variability in egg coloration facilitates egg recognition and, therefore, is a defence against brood parasitism. Møller and Petrie (1991) formalised this hypothesis stating that brood parasitism should select for consistency in appearance among eggs within clutches and greater interclutch variability. Therefore, females laying clutches of similar eggs whose appearance differs from the population mean would more likely recognise a parasitic egg (conspecific or interspecific) in their nests. Currently, there is little support for interclutch variability as a facilitator for conspecific egg recognition.

Most support for the interclutch variability hypothesis comes from village weaverbirds (*Ploceus cucullatus*), whose eggs vary enormously in coloration and maculation among the clutches of different females (Victoria 1972). Experimentally parasitised weaverbirds accepted only those conspecific eggs that closely matched the ground colour and maculation of their own eggs (Victoria 1972). Victoria (1972) suggested that interclutch variability provided protection against parasitism by the diederik cuckoo (*Chrysococcyx caprius*). In support of this, village weavers introduced into Hispaniola lost most of their egg recognition abilities in the 200-year absence of interspecific parasitism (Cruz and Wiley 1989; but see Payne 1997). After 16 years of sympatry with the newly established shiny cowbird (*Molothrus bonariensis*), the rejection frequency of foreign eggs in this population increased dramatically from 14% to 68–89% (Cruz and Wiley 1989; Robert and Sorci 1999). By contrast, Freeman (1988) and Jackson (1992a, 1992b, 1998) argued that egg variability facilitating egg recognition in *Ploceus* weaverbirds was likely due to conspecific parasitism, which was more frequent than parasitism by diederik cuckoos. However, in two species of weaverbirds, Lindholm (1997) found that conspecific parasitism was not more common than cuckoo parasitism and argued that Jackson (1992a) overestimated conspecific parasitism by assigning odd eggs parasitic status. These somewhat contradictory results highlight the need for further fieldwork combined with genetic identification of parasitic eggs.

The interclutch variability hypothesis for conspecific egg recognition was not supported by several other studies. Møller and Petrie (1991) examined intra- and interclutch variability in egg characteristics between 'socially breeding' species, which were expected to have more conspecific brood parasitism, and closely related solitary breeding species. Social breeders had significantly lower intraclutch variability, but not significantly higher interclutch variability in their eggs, the key prediction of the hypothesis. In addition, ostriches (*Struthio camelus*) rejected conspecific eggs (Bertram 1979) although the eggs vary little in appearance (Bertram 1992). In several experiments, most species did not recognise conspecific eggs (Lanier 1982; Emlen and Wrege 1986; Briskie and Sealy 1987; Møller 1987a; Stouffer *et al.* 1987; Sealy *et al.* 1989; Briskie *et al.* 1992; Moksnes and Røskaft 1992; Lyon and Hamilton 1992; Kempenaers *et al.* 1995; Peer and Sealy 2000a), regardless of whether their clutches showed a significant amount of interclutch variability (Bischoff and Murphy 1993; Gifford 1993).

Not surprisingly, interspecific egg recognition is more common than conspecific egg recognition, with more than 1,000 species identified as hosts of interspecific brood parasites (Johnsgard 1997b). Recognition in hosts of Old World cuckoos is complicated by the evolution of host egg mimicry. Several host species reject non-mimetic eggs experimentally added to their nests, and for those species tested, mimetic eggs were rejected at lower frequencies than non-mimetic eggs (e.g. Swynnerton 1916, 1918; Ali 1931; Brooke and Davies 1988; Davies and Brooke 1988,

1989a; Higuchi 1989; Moksnes *et al.* 1990, 1991; Moksnes 1992; Moksnes and Røskaft 1992).

Few studies have examined responses of hosts to cuckoo egg types other than mimetic and non-mimetic. Higuchi (1989) recorded the reaction of bush warblers (*Cettia diphone*) to model eggs painted a variety of ground colours. Bush warblers rejected significantly more eggs that were farthest apart in ground colour from their own eggs. Similarly, chaffinches (*Fringilla coelebs*) and bramblings (*Fringilla montifringilla*) increasingly rejected model eggs that contrasted highly with their own, compared to medium or low contrasting models that more closely matched their eggs (Moksnes 1992). In addition, willow warblers (*Phylloscopus trochilus*) rejected significantly more great tit (*Parus major*) eggs that are similar in appearance but smaller than conspecific eggs (Moksnes and Røskaft 1992). Willow warblers apparently recognise foreign eggs based on differences in size (Moksnes and Røskaft 1992). Size also is the most important cue for egg recognition in a former host species of cuckoos in India (Marchetti 2000). Cuckoo host species apparently base their recognition of foreign eggs on differences in ground colour and size. However, cuckoo dummies placed beside nests along with parasitism significantly increased the rejection frequency of model cuckoo eggs (Moksnes and Røskaft 1989; Moksnes *et al.* 1993). Thus, egg characteristics likely play a major, but not complete role in the recognition of cuckoo eggs.

In addition to individual characteristics used for egg recognition, interclutch variability may also facilitate common cuckoo (*Cuculus canorus*) egg recognition. As a counteradaptation to egg mimicry, Davies and Brooke (1989b) suggested that host species should evolve distinct eggs that vary less within and more between clutches. These differences should make it easier for hosts to distinguish cuckoo eggs from their own eggs and more difficult for cuckoos to match hosts' eggs. A comparison of the egg variability of seven species of hosts with long histories of cuckoo parasitism with five species with shorter histories of parasitism revealed no difference between intra- and interclutch variability (Davies and Brooke 1989b). They also found no difference in the egg variability of host populations isolated from parasitism (Iceland) and those in parasitised populations (Britain).

By contrast, Øien *et al.* (1995) tested the interclutch variability hypothesis for 34 host species of the common cuckoo and found a significant amount of interclutch variability when suitable and unsuitable host species were compared. Interclutch variability was also positively related to rejection frequencies. A re-analysis of these data that incorporated phylogenetic history confirmed these trends and showed that intraclutch variation was significantly negatively related to egg rejection frequency (Soler and Møller 1996). Similarly, Stokke *et al.* (1999) found a negative relationship between intraclutch variation and rejection frequency in the reed warbler (*Acrocephalus scirpaceus*). Thus, there is strong support for differences in intraclutch and interclutch variability that facilitates cuckoo egg recognition.

Despite the well-developed rejection behaviour of about 10% of brown-headed cowbird (*Molothrus ater*) hosts (Rothstein 1975b, 1992; Sealy 1996; Peer and Sealy 2000b; Peer *et al.* 2000), most of the 220 hosts (Friedmann and Kiff 1985) behave non-optimally and accept the reproductive costs of brood parasitism (Rothstein 1982a). Why do so few species reject parasitism? Egg characteristics of host species in relation to those of cowbirds may play a role in recognition and, hence, may influence the evolution of rejection behaviour.

Fretwell (1973) suggested that the blue eggs of American robins and gray catbirds (*Dumetella carolinensis*) evolved to facilitate recognition of cowbird eggs. Rothstein (1975b) tested the hypothesis that rejector species should lay eggs easily distinguishable from cowbird eggs, whereas the eggs of acceptor species should be similar to cowbird eggs, with eight rejector and 23 acceptor species. Eggs of one rejector and six acceptor species subjectively appeared similar to cowbird eggs but, to the human observer, cowbird eggs were easily distinguished from eggs even in clutches of species whose eggs appeared similar to the cowbird's. Rothstein (1975b) suggested that these hosts should also be able to identify a cowbird egg in their nests and concluded that egg characteristics cannot be used to differentiate between rejector and acceptor species. Fretwell's (1973) hypothesis also seems unlikely because several acceptor species lay blue eggs. Current evidence suggests that general host egg characteristics may not promote brown-headed cowbird egg recognition. An evolutionary lag in the response of hosts to cowbird eggs is a more likely explanation for the large number of acceptor species (see Rothstein 1990).

Eggs of species that reject brown-headed cowbird eggs vary widely in ground colour, maculation, and size compared with cowbird eggs (Figure 19.1). Thus, the characteristics that ejectors use to identify foreign eggs should vary. Experiments with model eggs that varied in ground colour, maculation, and size revealed that American robins recognised cowbird eggs by their difference in at least two of the three characteristics from robin eggs, whereas catbirds recognised cowbird eggs based only on their white ground colour (Rothstein 1982b). Based on these results, Rothstein (1982b) predicted that ejectors with eggs more similar to a cowbird's will tolerate foreign eggs less to maximise ejection of cowbird eggs and minimise mistakenly ejecting their own eggs. Thus, ejectors with eggs similar to cowbird's should eject foreign eggs, relying on fewer differences between the foreign eggs and their own.

Egg recognition related to nest usurpation

Egg recognition also may have evolved as an adaptation for nest usurpation (Peer and Bollinger 1998). Nest usurpers take over the active nests of other species for their own breeding attempts (Lindell 1996). Peer and Bollinger (1998) argued that nest usurpation is the selective pressure responsible for egg recognition in mourning doves (*Zenaida macroura*) because they nest solitarily, were rarely parasitised by brown-headed cowbirds or conspecifics, are unsuitable cowbird hosts, did not reject conspecific eggs, and are known to usurp nests. This is the only evidence for nest usurpation as a selective pressure for egg recognition.

Egg mimicry

Common cuckoo eggs have long been believed to match the colours and maculation of their hosts' eggs (Davies 2000; Figure 19.2). Subjective assessments of the appearance of cuckoo and host eggs have supported this idea (e.g. Baker 1942) and recent quantitative studies have confirmed these trends. After measuring

Plate 1 Brown-headed cowbird (*Molothrus ater*) eggs and host eggs. The cowbird egg is shown on the left of the passerine clutches. From top downward the clutches are: wood thrush (*Hylocichla mustelina*), American robin (*Turdus migratorius*), gray catbird (*Dumetella carolinensis*), yellow warbler (*Dendroica petechia*), grasshopper sparrow (*Ammodramus savannarum*), song sparrow (*Melospiza melodia*), and red-winged blackbird (*Agelaius phoeniceus*). American robins and gray catbirds eject cowbird eggs, whereas all other hosts listed here accept cowbird eggs.

Plate 2 Common cuckoo (*Cuculus canorus*) eggs and host eggs. The cuckoo egg is shown on the left of the passerine clutches. From top downward the clutches are: redstart (*Phoenicurus phoenicurus*), followed by the five principal hosts in Britain: meadow pipit (*Anthus pratensis*), reed warbler (*Acrocephalus scirpaceus*), dunnock (*Prunella modularis*), pied wagtail (*Motacilla alba*), and robin (*Erithacus rubecula*). Cuckoos parasitising redstarts in Finland lay pale blue eggs that match host eggs (Moksnes *et al.* 1995). In the dunnock, there is no mimicry between the cuckoo and host eggs because dunnocks cannot recognise their eggs (Brooke and Davies 1988).

Fig. 19.1 Brown-headed cowbird (*Molothrus ater*) eggs and host eggs. The cowbird egg is shown on the left of the passerine clutches. From top downward the clutches are: wood thrush (*Hylocichla mustelina*), American robin (*Turdus migratorius*), gray catbird (*Dumetella carolinensis*), yellow warbler (*Dendroica petechia*), grasshopper sparrow (*Ammodramus savannarum*), song sparrow (*Melospiza melodia*), and red-winged blackbird (*Agelaius phoeniceus*). American robins and gray catbirds eject cowbird eggs, whereas all other hosts listed here accept cowbird eggs.

Fig. 19.2 Common cuckoo (*Cuculus canorus*) eggs and host eggs. The cuckoo egg is shown on the left of the passerine clutches. From top downward the clutches are: redstart (*Phoenicurus phoenicurus*), followed by the five principal hosts in Britain: meadow pipit (*Anthus pratensis*), reed warbler (*Acrocephalus scirpaceus*), dunnock (*Prunella modularis*), pied wagtail (*Motacilla alba*), and robin (*Erithacus rubecula*). Cuckoos parasitising redstarts in Finland lay pale blue eggs that match host eggs (Moksnes *et al.* 1995). In the dunnock, there is no mimicry between the cuckoo and host eggs because dunnocks cannot recognise their eggs (Brooke and Davies 1988).

almost 12,000 clutches, Moksnes and Røskaft (1995) found a significant correlation between common cuckoo eggs and their hosts' eggs in ground colour, spot size and density, and volume. In addition, blue cuckoo eggs (Figure 19.2) were found more often in nests of host species laying blue eggs than in nests of species laying eggs of other colours (Moksnes *et al.* 1995). Host egg matching is not coincidental because females lay eggs of various ground colours and maculation, but each female lays a single egg type and preferentially parasitises one species (Chance 1922, 1940) or group of similar species whose eggs usually match (Wyllie 1981). These host-specific races of cuckoos are believed to be maintained through sex linkage to the W chromosome specific to females (Punnett 1933; Jensen 1966; Gibbs *et al.* 2000) and through host or habitat imprinting on nestling cuckoos (Brooke and Davies 1991; Teuschl *et al.* 1998).

Wallace (1889) proposed that cuckoo eggs mimic host eggs to prevent losses to predators because the eggs of most birds are cryptically coloured and an odd egg would attract the attention of predators. By contrast, Baker (1913) suggested that egg mimicry was selected for by the discriminating ability of hosts, which reject cuckoo eggs different from their own. More recently, Davies and Brooke (1988) and Brooker and Brooker (1989, 1990) hypothesised that egg mimicry or crypsis was a response to competition among cuckoos seeking to parasitise the same nest. The first cuckoo that parasitises a nest risks losing its egg to another cuckoo that subsequently parasitises the nest and preferentially removes the cuckoo's egg instead of a host's egg. Cuckoo eggs that mimic the host's eggs or blend in to the nest should be less likely to be removed.

Most experiments support Baker's (1913) host discrimination hypothesis. There is no support for Wallace's (1889) predation hypothesis (Mason and Rothstein 1987; Davies and Brooke 1988; but see Verbeek 1988, 1990) and limited evidence for the competition hypothesis. Brooker and Brooker (1989) found that Horsfield's bronze-cuckoo (*Chrysococcyx basalis*) laid mimetic eggs and the shining bronze-cuckoo (*Chrysococcyx lucidus*) laid cryptic eggs but their hosts did not reject non-mimetic eggs. In addition, some common cuckoos have been recorded removing and replacing a previous cuckoo's egg with their own egg (Brooker and Brooker 1990). Despite the lack of host discrimination and limited evidence for egg switching, the competition hypothesis remains critically untested.

For the common cuckoo, several studies have shown that mimetic eggs added to host nests are rejected significantly less frequently than non-mimetic eggs (Brooke and Davies 1988; Davies and Brooke 1988; Moksnes 1992; Moskát and Fuisz 1999). The eggs of many other cuckoo species also match the appearance of their hosts' eggs (Johnsgard 1997b). The magpie (*Pica pica*), the main host of great-spotted cuckoos (*Clamator glandarius*), and bush warblers, hosts of Himalayan cuckoos (*Cuculus saturatus*) and little cuckoos (*Cuculus poliocephalus*), rejected non-mimetic eggs significantly more frequently than mimetic eggs (Higuchi 1989; Soler *et al.* 1994, 1999). Similarly, ploceid hosts of diederik cuckoos rejected significantly more non-mimetic than mimetic eggs (Noble 1995; Lawes and Kirkman 1996; Lindholm 1997). Thus, host discrimination appears to have been the main selective pressure for the evolution of egg mimicry in Old World cuckoos.

Egg mimicry is suspected in New World cuckoos as well. Hughes (1997) proposed that two *Coccyzus* cuckoos, once obligate brood parasities, evolved blue eggs to mimic the eggs of their hosts. Their blue eggs matched a significant proportion of the egg types of their reported hosts (Hughes 1997). However, experimental support for host discrimination is not available.

Limited evidence exists for egg mimicry in cowbirds (*Molothrus* spp.). The eggs of brown-headed cowbirds do not appear to mimic the wide variety of colours and maculation of eggs of their numerous hosts (Figure 19.1). Eggs of some hosts may be similar to cowbird eggs, but this similarity is more likely due to their close phylogenetic relationships (Johnsgard 1997b), or to convergent evolution of a cryptic egg type. However, egg mimicry may be present in some grassland hosts that have been sympatric with brown-headed cowbirds for a long time (Peer *et al.* 2000). In parts of their range, shiny cowbirds lay eggs of two obvious colour morphs, immaculate white or heavily spotted, that may mimic the eggs of at least two hosts who reject immaculate eggs but accept spotted eggs (Fraga 1985; Mason 1986; Mermoz and Reboreda 1994; see Mason and Rothstein 1986 for size mimicry). Host discrimination is more likely the selective force for spotted eggs because these were depredated as frequently as immaculate eggs (Mason and Rothstein 1987). In addition, the eggs of giant cowbirds (*Molothrus oryzivora*) appear similar to eggs of their hosts (Smith 1968), but a morphometric analysis did not support egg matching (Fleischer and Smith 1992). Although the eggs of the screaming cowbird (*Molothrus rufoaxillaris*) and their main host, the bay-winged cowbird *(Molothrus badius*), are similar, the 'background' coloration and maculation of eggs of these two species were significantly different (Fraga 1983), and the host did not discriminate against non-mimetic or mimetic eggs (Mason 1980; Fraga 1986; but see Jaramillo 1993 for possible size or shape mimicry).

Filtering solar radiation

Eggs of ground-nesting birds are often exposed to direct sunlight for brief periods while adults are away. McAldowie (1886) proposed that pigmentation shields developing embryos from excessive solar radiation. A comparison of yolk temperatures of white and pigmented eggs placed in direct sunlight for several hours showed that yolk temperatures were significantly higher in pigmented eggs than in white eggs (Montevecchi 1976). Similarly, ostrich eggs covered in dark crayon warmed significantly more than natural white eggs (Bertram and Burger 1981). Montevecchi (1976) suggested that the increased heat load caused by egg pigmentation should be detrimental because overheating can threaten the eggs of ground-nesting birds. Assuming that the eggshell pigmentation of these birds has been selected for crypsis, Montevecchi concluded that there must be a compromise of selection pressures on eggs of ground-nesting birds in response to predation and the potential for overheating.

Bakken *et al.* (1978) demonstrated, however, that eggshell pigmentation reduces the heat load of the egg in comparison to plumage pigmentation. Eggshell pigmentation in 25 species reflected a high percentage of near-infrared wavelengths of light, which contain almost half of the solar energy in incident light. In addition, the white eggshells of the ostrich reflect 98% of radiated and near infra-red light as well as 99.9% of violet and ultraviolet light (Deeming and Ar 1999). These results appear to be due to the unique properties of the eggshell pigments, protoporphyron and bilverdin, in comparison to the melanin pigments of feathers (Bakken *et al.* 1978).

These results are somewhat contradictory but both studies that compared the temperature of dark- and light-coloured eggs used artificially pigmented (paint or crayon) eggs (Montevecchi 1976; Bertram and Berger 1981). Artificially coloured eggs likely do not accurately simulate natural pigmentation. Nevertheless, to some degree, heavily pigmented eggs probably gather more heat than lightly pigmented eggs. Indeed, the degree of pigmentation may be a trade-off between predation risk and the influence of solar radiation.

Heat reflection and absorbance properties of eggshell pigments are also likely to influence incubation. The length of incubation bouts may mitigate the trade-off between overheating and predation risk. For those species less vulnerable to predation, pale or unpigmented eggs may allow for longer inattentive periods. In addition to overheating, Wink *et al.* (1985) suggested that the darker last-laid eggs of some falcons may gather more heat to increase embryonic development and reduce hatching asynchrony. However, the effect of degree of pigmentation on the rate of embryonic development or incubation attentiveness has yet to be examined.

Eggshell strength

Poultry researchers have been interested in eggshell pigmentation because it may increase eggshell strength (Solomon 1987). In domestic fowl (*Gallus gallus*), eggshell colour was significantly correlated with specific gravity and/or shell thickness (Grover *et al.* 1980; Campo and Escudero 1984). In addition, white Japanese quail eggs broke more easily than normal, heavily pigmented eggs (Briggs and Williams 1975). By contrast, other workers have not found a relationship between eggshell colour and thickness (Anderson *et al.* 1970), breaking strength (Briggs and Teulings 1974), or measures of eggshell strength (Potts and Washburn 1974). Carter (1975) showed that colour had a significant effect on eggshell defects in fowl, although colour had a significantly negative relationship with eggshell thickness. In ring-necked pheasant (*Phasianus colchinus*) eggs, blue shells were thinner and defective compared with typical olive-brown shells (Richards and Deeming 2001). Despite the contrasting results, any relationship between colour and eggshell strength is likely due to other characteristics correlated with or genetically linked to colour and is not the product of pigmentation itself (Briggs and Teulings 1974; Carter 1975; Shoffner *et al.* 1982).

Aposematism

Swynnerton (1916) proposed that the coloration of some eggs functioned as an aposematic signal to warn predators of their unpalatability and, hence, minimise predation. In several experiments, eggshell coloration was correlated with palatability and most species with eggs rated as unpalatable laid conspicuously coloured eggs (Cott 1948, 1951, 1952, 1953, 1954). The palatability of birds' eggs varied widely and preferences or aversions were found for particular eggs (Cott 1948). A closer look at Cott's data raises serious doubts concerning his conclusions. Cott (1948) considered most species in the unpalatable group to have relatively conspicuous eggs. However, many species have egg colours and maculation that match the description of cryptic eggs. Thus, there does not appear to be a strong correlation between unpalatability and conspicuous egg coloration.

Even if unpalatable eggs do not appear to be conspicuously coloured, they still could function to deter predators because some predators have learned to avoid cryptically coloured eggs injected with toxins (Nicolaus *et al.* 1983; Nicolaus 1987). It is doubtful, however, that wild bird eggs are toxic or completely unpalatable and there is no evidence to show definite avoidance indicative of truly unpalatable eggs. Toxic eggs are rare in animals because several barriers appear to limit the evolution of toxicity (Orians and Janzen 1974). Thus, egg coloration likely does not function as an aposematic signal.

Intraclutch variability

Throughout this chapter, hypotheses that mostly concern interspecific variation in egg coloration are discussed. However, much intraclutch variability in the ground colour and/or maculation of eggs has been observed in several groups of birds: falcons (Horváth 1955; Wink *et al.* 1985), pheasants (Labisky and Jackson 1966), moorhens (McRae 1997), gulls and terns (Preston 1957; Gochfeld 1977; Baerends and Hogan-Warburg 1982; Verbeek 1988), corvids (Holyoak 1970; Verbeek 1990), swallows (Brown and Sherman 1989), house wrens (*Troglodytes aedon*; Kendeigh *et al.* 1956), icterids (Neufeld 1998; Peer 1998), and Old World sparrows (Seel 1968; Yom-Tov 1980b; Lowther 1988). Intraclutch variability is sometimes attributed to one 'odd' egg that is usually distinctly paler than the other eggs in the clutch (Lowther 1988). Often this is the last egg in the laying sequence of the clutch (Nice 1937; Preston 1957; Seel 1968; Gochfeld 1977; Lowther 1988; Verbeek 1988; 1990; Kilpi and Byholm 1995).

The simplest explanation for intraclutch variability and odd eggs is that females run out of pigment before completing their clutch (Seebohm 1896; Nice 1937; Holyoak 1970). Most evidence for this is circumstantial. Some authors have noted a correlation between pale colour and last-laid eggs (Nice 1937; Seel 1968; Holyoak 1970), whereas others have quantitatively determined that most last-laid eggs are pale (Gochfeld 1977; Baerends and Hogan-Warburg 1982; Lowther 1988; Verbeek 1988, 1990). Lowther (1988) found that as clutch size in house sparrows (*Passer domesticus*) increased so did the proportion of clutches with a pale last egg. In house wrens, eggs tended to become lighter as the laying sequence progressed (Kendeigh *et al.* 1956). Egg removal experiments that induced indeterminate laying in a few females showed that eggs continued to lighten beyond the normal clutch size (Kendeigh *et al.* 1956). By contrast, eggs darken as laying progresses in some falcons with the last egg the most heavily pigmented (Horváth

1955; Wink *et al.* 1985). Other explanations for the pigment depletion hypothesis have been proposed. Kilpi and Byholm (1995) suggested that in herring gulls last-laid eggs are retained in the uterus for less time than earlier eggs and receive less pigment. However, this trend does not hold in gulls where the laying intervals were measured (Drent 1970; Parsons 1976; Verbeek 1986). Furthermore, physiological changes may inhibit pigment production because incubation often begins before clutch completion (Lowther 1988; Verbeek 1988, 1990).

The appearance of odd-coloured eggs in clutches may be related to brood parasitism. Odd eggs may have been laid by interspecific parasites that mimic their hosts' eggs (Freeman 1988) or by conspecific parasites (e.g. Gibbons 1986; Møller 1987a; Thomas *et al.* 1989; Hötker 2000). However, Brown and Sherman (1989) and McRae (1997) have demonstrated that odd eggs often reflect natural intraclutch variation and that assigning eggs as parasitic based solely on appearance is unreliable in some species. Parasitic eggs should be identified on the basis of appearance only when intraclutch versus interclutch variability has been quantified.

Odd eggs may not only be the product of brood parasitism, they may prevent conspecific parasitism as well. Yom-Tov (1980b) hypothesised that the paler, last-laid eggs of the dead sea sparrow (*Passer moabiticus*) and other ploceids signal would-be conspecific parasites that their clutches are complete and, thus, eggs laid thereafter are not likely to hatch. Håland (1986) argued that a differently coloured, last-laid egg should promote rather than restrain conspecific brood parasitism. The absence of the signal egg would reveal to the parasite that there exists a favourable opportunity for parasitism. Being parasitised early in laying would be more detrimental to the host than having an egg added after incubation has started but this hypothesis requires rigorous testing.

Verbeek (1988, 1990) proposed that a pale last egg draw a predator's attention to the egg with the lowest probability of survival in a clutch. In studies of glaucous-winged gulls (*Larus glaucescens*) and northwestern crows (*Corvus caurinus*), most last-laid eggs were pale and smaller than the other eggs in the clutch and they had the poorest survival (Verbeek 1988, 1990). When clutches were partially depredated, the pale egg was taken significantly more often than other eggs in both species. Two other studies tested the predator attraction hypothesis, in herring gulls and Brewer's blackbirds (*Euphagus cyanocephalus*). Pale eggs were uncommon in the herring gull colony examined by Kilpi and Byholm (1995) but most were laid last. The last-laid eggs were not smaller, had an equal survival probability, and pale eggs were not taken preferentially by predators. Most Brewer's blackbird clutches contained odd eggs, but there was no difference in egg volume over the laying sequence and predation on odd eggs was infrequent (Neufeld 1998). These two tests may not be adequate because predation was extremely rare in the herring gull colony (Kilpi and Byholm 1995) and most predators on Brewer's blackbirds took entire clutches (Neufeld 1998). The predator attraction hypothesis may apply only in species subject to a high level of partial predation.

Opposite to predator attraction, Hockey (1982) suggested that intraclutch variability may enhance clutch crypsis because more than one similarly coloured and patterned object in a nest may stand out more than divergent ones. Namaqua sandgrouse clutches with intraclutch diversity in ground colour and pigment pattern suffered significantly less predation than uniform clutches (Lloyd *et al.* 2000), which provides initial support for this hypothesis.

White eggs

Many authors have assumed that immaculate white is the ancestral coloration of bird eggs because reptiles lay white eggs (e.g. Wallace 1889; Chapman 1924; Romanoff and Romanoff 1949; Pettingill 1985; Welty and Baptista 1988). White eggs are also assumed to lack pigment. Contrary to these assumptions, no study has examined the ancestral origin of egg coloration in birds, and the white eggs of many species (but not all) contain the pigment protoporphyron (Kennedy and Vevers 1976).

Hewitson (1846) suggested that most species that lay white eggs nest in cavities. This idea was expanded by observations that open-nesting species with white eggs covered their eggs when off the nest or the incubating bird itself was cryptic and rarely left the nest (Wallace 1889; Cott 1940; Westmoreland and Best 1986). Comparative studies have confirmed that white eggs are most often found in species that nest in cavities or enclosures (von Haartman 1957; Lack 1958; Oniki 1985). The lack of pigmentation in these eggs has been interpreted as a response to the absence of selection for cryptic coloration. However, this implies that pigmentation is costly to produce, and this has not been investigated.

A single hypothesis has been proposed to explain the significance of white eggs in cavity nests. After observing high losses of jackdaw (*Corvus monedula*) eggs that became soiled with mud after a heavy rain, Holyoak (1969) suggested that white eggs allow the parents to see them in a dark nest. A comparison of the survival of eggs blackened with ink and white eggs found that the former disappeared significantly more than white eggs (Holyoak 1969), which is the only test of this hypothesis. In general, immaculate white eggs remain an enigma.

Why so blue?

The blue colour of eggs laid by some thrushes and other birds is perhaps the most striking coloration. Wallace (1889) suggested that blue eggs blend in with the green leaves around nest sites. Circumstantial evidence for this idea has come from comparisons of egg coloration to nest site characteristics of several species. Oniki (1979, 1985) found that blue eggs of Amazonian birds were significantly associated with green foliage in bushes that received direct sunlight. It was concluded that blue eggs are cryptic because they blend in with the sunspots on the green leaves surrounding the nest. Nests of Holarctic and Indian turdids that lay blue eggs tended to be placed in dense bushes or other sites with low light, which may provide cryptic conditions for blue eggs (Lack 1958). Nevertheless, Lack (1958) noted that blue egg colour is puzzling.

Götmark (1992) tested the cryptic blue egg hypothesis by comparing predation frequencies on eggs painted white, beige with dark spots, or blue at artificial nests that simulated song thrush (*Turdus philomelos*) nests. There was no significant difference in predation on nests with differently coloured eggs. Coloration also had no effect on predation frequencies on nests differently concealed. When eggs were placed in trees without nests, blue and white eggs were depredated significantly more than beige-spotted eggs (Götmark 1992). Three other studies indirectly tested this hypothesis under mostly artificial nesting conditions (Janzen 1978; Yahner and DeLong 1992; Bildstein 1993). All of these studies found that blue ground colour had no effect on predation frequencies compared to other colours. Similarly, an observational study of crow tits (*Paradoxornis webbiana*), which lay clutches of either blue or white eggs in Korea, found no difference in nest success between the two egg morphs (Kim *et al.* 1995). Collectively, these results suggest that blue eggs are not cryptic and may even be conspicuous to predators.

Support is lacking for the cryptic blue egg hypothesis, although experimental studies share the bias of artificial nesting conditions. Egg mimicry of blue host eggs by common cuckoos is the only supported function of blue egg colour (Moksnes *et al.* 1995). For most species, the adaptive significance of blue eggs remains a mystery.

Priorities for future research

Despite the large body of literature concerning egg coloration, relatively few hypotheses in explanation of the significance of egg coloration have been adequately tested. Future research is needed to address most hypotheses. Due to the biases present in many experiments, the cryptic value of egg coloration in cup nests of passerines requires further attention. Similarly, the cryptic blue egg hypothesis has not been adequately tested. Egg recognition has received the most attention, especially in relation to brood parasitism. However, individual clutch recognition in colonial birds is not completely understood. Further experiments with mimetic and non-mimetic eggs are warranted in cowbirds and *Coccyzus* cuckoos to address egg mimicry. The ability of eggshell pigments to filter solar radiation has only been preliminarily investigated. Studies are needed to confirm these trends, especially in other wavelengths of light, and to determine whether they relate to embryonic development and incubation inattentiveness. Several hypotheses and mechanisms explaining intraclutch variation have been proposed, but few experiments have been conducted. The energetic cost of producing eggshell pigments has not been addressed. The role of white eggs in enclosed nests has received little attention. Finally, the ancestral origins of egg pigmentation are unknown and should be investigated. Future researchers also should consider that egg coloration in some species may be selectively neutral or negative traits (Götmark 1992). The importance of egg coloration has been revealed for some species, but there is still much to be learned.

Acknowledgments

We thank A. A. Andruschak, D. C. Deeming, F. Götmark, P. N. Hébert, P. E. Lowther, B. D. Peer and R. M. Underwood for reviewing the manuscript. A. J.

Gaston kindly shared a translation of Tschanz's (1959) paper on murres. We thank C. M. McLaren and V. A. McLaren for translating Rensch's (1924, 1925) papers. We are grateful to R. Y. McGowan, National Museums of Scotland, Edinburgh, and P. E. Lowther, Field Museum of Natural History, Chicago, for photographs of clutches parasitised by common cuckoos and brown-headed cowbirds, respectively. Financial support during the preparation of this review was provided by a grant from the Natural Sciences and Engineering Research Council of Canada.

20 *Energetics of incubation*

J. M. Tinbergen and J. B. Williams

All birds lay eggs which must be maintained at suitable temperatures for proper embryological development to occur (Lundy 1969; Anderson *et al.* 1987). Contact incubation, the act of applying heat to eggs, has captured the imagination of investigators for decades, but the constraints imposed by the incubation period on life-history evolution have only begun to be appreciated (Huggins 1941; Kendeigh 1952; Drent 1975; Williams 1996; Bryan and Bryant 1998).

A pioneer in thinking about energy expenditure during incubation, Kendeigh (1963) developed a biophysical model that predicted that female house wrens (*Troglodytes aedon*) expended energy at a rate 1.2–2.1 times their basal metabolic rate while sitting on eggs. Kendeigh (1963) concluded that birds augmented their energy expenditure during the incubation period, especially during episodes of low ambient temperature (Williams 1996). An antagonist of Kendeighs' additional cost hypothesis, King (1973) argued for the substitution hypothesis that most, if not all, of the heat required to incubate eggs could potentially be supplied from basal heat production. In support of this view, Walsberg and King (1978b) constructed a biophysical model for heat flux in incubating white-crowned sparrows (*Zonotrichia leucophrys*) that predicted a reduced metabolic rate for the incubating female owing to the increased resistance to heat flow provided by the nest. Further calculations of the cost of re-warming eggs convinced these authors that during incubation the total energy expenditure of a female sparrow was likely substantially reduced, a perspective that gained wide acceptance (Freed 1981; Mugaas and King 1981; Walsberg 1983b; Ricklefs and Hussell 1984; Murphy and Haukioja 1986; Gill 1990).

In a comprehensive review on the energetics of avian incubation, Williams (1996) concluded that: (1) the energy expenditure of birds while sitting on eggs was increased over that of non-incubating individuals when they experienced temperatures below their thermal neutral zone; (2) the cost of re-warming eggs could be significant; (3) larger clutch sizes mandated increased energy expenditure; and (4) for systems in which only the female intermittently incubated, birds could have elevated field metabolic rates (FMR) and potentially face serious energy shortage. Williams could find no evidence that for intermittently incubating species the FMR of incubating females differed from that when they were feeding young, and suggested that investigators pay more attention to the influences of energy expenditure during incubation on parameters of life-history.

A cornerstone of life-history theory is that a trade-off exists between current reproduction and future survival; a relatively high FMR during breeding is thought to translate into reduced survival or future reproduction, the cost of reproduction

(Williams 1966; Bryant 1979; Reznick 1992; Daan *et al.* 1996; Leroi 2001). Central to this thinking for the last several decades was that the nestling period was a time of peak energy demand (Lack 1947; Walsberg 1983a, 1983b; Nur 1988; Daan *et al.* 1990; Dijkstra *et al.* 1990) whereas the incubation period was viewed as period when energy expenditures were low, and its impact on life-history was largely ignored (Williams 1991). However, if during the incubation period energy demands are as high as when parents are feeding young, energy constraints may be significant at this time because parents spend a considerable portion of each day on eggs, a behaviour that precludes foraging.

A number of studies have been published that provide additional data allowing us to revisit some of the questions posed by Williams (1996). Here new issues are approached in light of recent studies that provide additional insights into the connection between energy expenditure during the incubation period and avian life-history evolution. Energy expenditure during incubation, both within and between species, is examined in order to look for patterns and associations with ecological factors. The following notions are explored: (1) that clutch size may be influenced by energy expenditure during incubation among terrestrial birds; (2) that incubating birds elevate their energy expenditure above resting birds experiencing the same ambient temperature; (3) that the FMR of incubating birds is less than when they are feeding young; (4) that birds increase their energy intake efficiency during incubation; and (5) that seabirds have a relatively low cost of incubating eggs.

Terminology

In over half of the families of birds (Figure 6.1) (Van Tyne and Berger 1976), both parents share incubation of eggs (biparental continuous incubation; Williams 1996). Female-only incubation (gyneparental incubation; Williams 1996) is relatively common (Table 6.2) although there is considerable variation on this theme (Chapter 6). The spectrum of behaviours ranges from the male providing food to his mate but does not incubate (assisted gyneparental incubation; Williams 1996) to the situation where the male does not provide food and the female must balance maintenance of egg temperature with her own energy needs by frequently leaving the eggs to forage during the day (gyneparental intermittent incubation; Williams 1996). The pattern of incubation for each species described below is shown in Table 20.1.

To reduce ambiguity, the terms used in this chapter to describe energy expenditure during incubation are described here. FMR_{inc} is the field metabolic rate ($kJ\,d^{-1}$) during the incubation phase. Estimates of FMR_{inc} can be divided into energy expenditure when birds are sitting on eggs or absent. MR_{inc} is the metabolic rate ($kJ\,d^{-1}$) while incubating the eggs (equivalent to incubation metabolic rate [IMR; Williams 1996]). $MR_{off-nest}$ is the metabolic rate ($kJ\,d^{-1}$) during period of nest recess during the incubation period. FMR_{nestl} is the field metabolic rate ($kJ\,d^{-1}$) during the nestling phase. Measurements of FMR obtained by the double labelled water method (Speakman 1997), represent energy expenditure integrated

Table 20.1 Double labelled water measurements of terrestrial species and studies used. For the analysis values were averaged over the relevant stages, gender or system of incubation.

Species	Stage[$]	Gender*	System[#]	Mass (g)	BMR ($kJ\,d^{-1}$)	FMR ($kJ\,d^{-1}$)	Sample size	Reference
Arctic nesting								
Calidris alpina	I	3	BCI	53.0	51.4	192.7	7	1
Calidris minuta	I	3	BCI	29.0	33.1	184.4	3	1
Calidris maritima	I	3	BCI	64.0	58.9	236.0	10	2
Calidris maritima	I	3	BCI	79.0	68.7	307.7	1	1
Calidris canutus	I	3	BCI	142.0	105.4	373.5	6	1
Charadrius hiaticula	I	3	BCI	57.0	54.2	218.8	3	1
Arenaria interpres	I	3	BCI	108.0	86.3	348.1	16	1
Calidris alba	I	3	?	59.0	55.5	229.2	4	1
Calidris fuscicollis	I	3	GII	39.0	41.1	189.6	1	1
Stercorarius longicaudus	I	3	BCI	336.0	156.9	635.0	2	1
Non-Arctic nesting								
Recurvirostra avosetta	I	1	BCI	309.7	186.0	428.5	3	3
Recurvirostra avosetta	I	2	BCI	337.1	197.9	422.5	3	3
Actitis hypoleucos	I	3	BCI	54.7	52.5	135.5	3	4
Actitis hypoleucos	N	3	BCI	48.5	48.1	156.9	1	4
Charadrius alexandrinus	I	1	BCI	41.2	42.7	106.3	9	5
Charadrius alexandrinus	I	2	BCI	40.0	41.8	100.4	6	5
Haematopus ostralegus	I	2	BCI	491.0	260.3	400.0	1	6
Haematopus ostralegus	N	2	BCI	492.3	260.8	708.0	3	6
Charadrius hiaticula	I	3	AGI	74.8	66.0	301.5	4	7
Falco tinnunculus	I	1	AGI	275.0	135.4	324.0	–	8
Falco tinnunculus	N	1	AGI	235.0	120.7	346.0	–	8
Merops viridas	I	3	BCI	35.4	43.4	90.4	9	4
Merops viridas	N	3	BCI	33.2	41.6	80.3	18	4
Mirafra erythroclamys	I	1	GII	26.4	35.5	88.1	6	9
Mirafra erythroclamys	N	1	GII	25.5	34.7	88.5	6	9
Cinclus cinclus	I	1	AGI	63.0	64.6	204.5	8	7
Cinclus cinclus	N	1	AGI	57.1	60.4	250.1	7	7
Cinclus cinclus	N	2	AGI	68.1	68.2	296.2	9	7
Passerculus sandwichensis	I	1	GII	18.7	28.0	73.8	16	10, 11
Passerculus sandwichensis	I	1	GII	20.2	29.5	85.2	10	12
Passerculus sandwichensis	N	1	GII	17.0	26.2	67.7	10	10, 11
Passerculus sandwichensis	N	1	GII	18.3	27.6	80.6	13	13
Junco phaeonotus	I	1	GII	20.5	29.8	66.7	6	14, 15
Junco phaeonotus	N	1	GII	18.9	28.2	74.4	7	14, 15
Hirundo rustica	I	1	GII	20.5	29.8	120.6	1	16
Hirundo rustica	N	1	GII	19.2	28.5	108.5	8	16
Hirundo rustica	N	2	GII	18.9	28.2	111.0	6	16
Delichon urbica	I	1	BCI	18.7	28.0	80.7	5	16
Delichon urbica	I	2	BCI	19.3	28.6	79.6	5	16
Delichon urbica	N	1	BCI	18.4	27.7	95.1	28	16
Delichon urbica	N	2	BCI	18.6	27.9	96.6	28	16
Hirundo tahitica	I	1	GII	14.7	23.7	53.2	8	4
Hirundo tahitica	N	3	GII	14.1	23.1	76.6	6	17
Riparia riparia	I	1	BCI	14.3	23.3	81.7	3	16
Riparia riparia	N	2	BCI	12.8	21.6	93.7	6	16

Table 20.1 *Continued.*

Species	Stage[$]	Gender*	System[#]	Mass (g)	BMR ($kJ\,d^{-1}$)	FMR ($kJ\,d^{-1}$)	Sample size	Reference
Riparia riparia	N	1	BCI	13.0	21.8	90.0	4	16
Tachycineta bicolor	I	1	AGI	22.6	31.9	118.9	9	18
Tachycineta bicolor	N	1	AGI	19.4	28.7	136.4	11	18
Ficedula albicollis	I	1	GII	15.5	24.6	72.1	3	19, 20
Ficedula albicollis	N	1	GII	12.9	21.7	71.7	20	19, 20
Ficedula hypoleuca	I	1	AGI	14.4	23.4	64.4	3	21
Ficedula hypoleuca	I	1	AGI	14.7	23.7	56.6	2	22
Ficedula hypoleuca	I	1	AGI	15.0	24.0	68.0	6	22
Ficedula hypoleuca	N	1	AGI	12.5	21.2	60.9	1	4
Erithacus rubecula	I	1	AGI	20.6	29.9	75.1	3	4
Erithacus rubecula	N	2	AGI	17.8	27.1	81.8	3	4
Erithacus rubecula	N	1	AGI	17.9	27.2	58.8	4	4
Nectarinia violacea	I	1	GII	9.5	17.6	66.2	10	23
Parus caeruleus	I	1	AGI	12.0	20.7	61.1	4	24
Parus caeruleus	N	3	AGI	11.0	19.4	66.9	2	24
Parus major	I	1	AGI	21.9	31.2	111.2	8	25
Parus major	N	1	AGI	17.2	26.4	97.5	15	26
Parus major	N	1	AGI	17.7	27.0	96.0	3	27
Parus major	N	1	AGI	17.7	27.0	98.7	22	28
Parus major	N	2	AGI	18.1	27.4	89.2	19	26
Mimus polyglottos	I	1	GII	50.9	55.8	112.4	1	29
Mimus polyglottos	N	1	GII	46.7	52.6	122.5	4	29
Sturnus vulgaris	I	1	BCI	81.4	77.1	183.2	6	30
Sturnus vulgaris	I	1	BCI	85.0	79.4	212.3	4	31
Sturnus vulgaris	N	1	BCI	74.0	72.2	301.4	7	31
Sturnus vulgaris	N	3	BCI	84.3	78.9	247.8	26	30

[$]Reproductive stage: I = incubation; N = Nestling; *Gender measured: 1 = female, 2 = male, 3 = unknown (or averaged); [#]Incubation system: BCI – Shared incubation (biparental continuous incubation; Williams 1996); AGI – Male-fed, female-only incubation (Assisted gyneparental incubation; Williams 1996); GII – Female-only intermittent incubation (Gyneparental intermittent incubation; Williams 1996). References: 1 – Piersma *et al.* (2001); 2 – Pierce in Drent and Piersma (1990); 3 – Hotker *et al.* (1996); 4 – Tatner and Bryant (1993); 5 – Amat *et al.* (2000); 6 – Kersten (1996); 7 – Bryant and Tatner (1988); 8 – Masman *et al.* (1988); 9 – Williams (2001); 10 – Williams and Dwinnel (1990); 11 – Williams and Nagy (1985); 12 – Williams (1987); 13 – Willaims (1991); 14 – Weathers and Sullivan (1989b); 15 – Weathers (personal communication); 16 – Westerterp and Bryant (1984); 17 – Bryant *et al.* (1984); 18 – Williams (1988); 19 – Moreno *et al.* (1991); 20 – Moreno (personal communication); 21 – Moreno and Sanz (1994); 22 – Moreno and Carlson (1989); 23 – Williams (1993); 24 – R. Prys-Jones (in Tatner and Bryant 1993); 25 – Bryan and Bryant (1998); 26 – Verhulst and Tinbergen (1997); 27 – Sanz and Tinbergen (1999); 28 – Tinbergen and Dietz (1994); 29 – Utter (1971); 30 – Westerterp and Drent (1985); 31 – Ricklefs and Williams (1984).

over periods on and off the nest mostly over a 24-h cycle. BMR is the rate of metabolism for post-absorptive individuals in their thermal neutral zone at rest during the rest phase of their diurnal cycle ($kJ\,d^{-1}$). RMR ($kJ\,d^{-1}$) is similar except that measurements are made during the active phase of the diurnal cycle.

Energy constraints during incubation and clutch size

Documented in a variety of organisms, costs of reproduction are thought to arise from decreased longevity, reduced future reproduction or from the increased risk of predation (Cichon and Kozlowski 2000; Leroi 2001). Reproductive costs are viewed as trade-offs among life-history traits, such as between fecundity and longevity (Reznick 1985; Bell and Koufopanou 1986; Roff 1992; Chapter 21), and it is thought that maintenance and reproduction compete for some common limiting resource such as energy.

Although most studies on the cost of reproduction in birds have focussed on the nestling period (Drent and Daan 1980; Lessells 1991; Stearns 1992), some evidence exists that implicates the incubation period as a time when parents suffer energy stress, potentially translating into increased mortality (Siikamäki 1995). In great tits (*Parus major*), a species in which the male feeds the incubating female, Yom-Tov and Hilborn (1981) constructed an energy budget for the various stages of reproduction using parameters from studies on several species of passerines. The model identified the incubation and nestling periods as times of energy limitation, and predicted an inverse relationship between clutch size and energy limitation. The latter result suggests that birds incubating small clutches have higher energy expenditure than individuals with large clutches. Their model assumed that MR_{inc} was dependent on ambient temperature (T_a), and that birds incurred no additional cost to keep eggs warm or to re-warm them. In addition, they assumed that birds with smaller clutches started incubation earlier in the season and hence experienced lower T_a causing higher MR_{inc}. If these authors had assumed a cost of incubation, as more recent information might suggest (Weathers 1985; Biebach 1986; Williams 1991), and had they assumed an added cost to parents to re-warm eggs, they might have concluded tits have even greater energy limitation during incubation, and that as clutch size increased, so did energy constraints.

Experiments in which clutch size was manipulated (Siikamäki 1995; Thomson *et al.* 1998b), in which supplemental food was provided to laying (Nilsson 1994; Sanz and Moreno 1995) or incubating parents (Moreno 1989a; Sanz 1996), or in which nest boxes were heated to reduce night-time energy expenditure (Bryan and Bryant 1998), collectively provide insights into the potential role that energy expenditure during the incubation period has played in the life-history evolution of birds. If the incubation period is energetically stressful, it is predicted that an increase in clutch size will be associated with reduced hatching success, longer incubation periods, increased FMR_{inc}, and declines in body condition of incubating parents. Periods of low temperature or bad weather negatively affected the condition of females (Jones 1987; Mertens 1987) and the length of incubation was associated with male feeding frequency (Nilsson and Smith 1988). Smith (1989) showed that female blue tits (*Parus caeruleus*) with enlarged clutches spent less time incubating presumably owing to their increased energy demand, and as a result, the length of the incubation period increased (Figure 20.1). Moreno (1989a) provided supplemental food to incubating female wheatears (*Oenanthe oenanthe*) for three consecutive years, but only during the breeding season with the harshest weather conditions did provisioning lead to a decrease in the length of incubation.

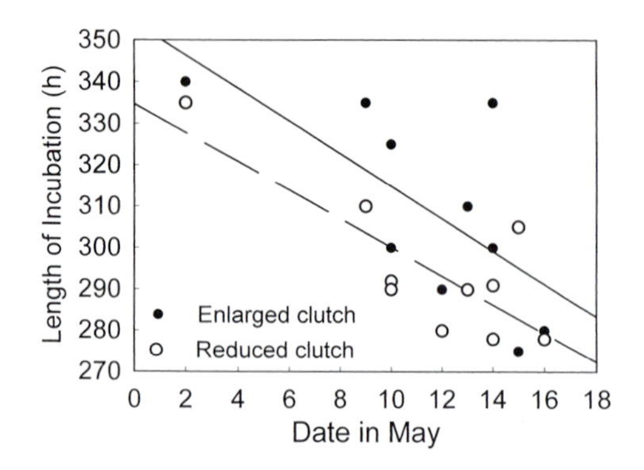

Fig. 20.1 Length of the incubation period for blue tits (*Parus caeruleus*) with enlarged and reduced clutches as a function of date (Redrawn from Smith 1989). Lines represent regressions for enlarged and reduced clutches.

Bryan and Bryant (1998) heated nest boxes of great tits at night thereby reducing female energy expenditure by 6%. Experimental females increased time on the eggs by 55 min d^{-1} indicating that tits face a trade-off between parental energy needs and the need to provide heat to developing embryos.

In a review of studies that manipulated clutch size in birds, Thomson *et al.* (1998b) concluded that experimentally enlarged clutches resulted in lower hatching success, in greater loss of body condition and in higher energy expenditure. It is concluded, therefore, that considerable evidence exists that energy expenditure during incubation could be a significant determinant of clutch size.

Energetic costs of incubation in terrestrial birds

Energy expenditure while sitting on eggs

The energy expenditure of parents while sitting on eggs (MR_{inc}) is typically elevated over that of resting individuals when ambient temperatures are below their thermal neutral zone (Williams 1996; Thomson *et al.* 1998b; Table 20.2). Measuring instantaneous oxygen consumption for female starlings (*Sturnus vulgaris*) at night, Biebach (1986) concluded that incubating birds had a higher rate of metabolism than did resting individuals experiencing the same ambient temperature (T_a), and that in this species MR_{inc} varied inversely with T_a. For blue tits Haftorn and Reinertsen (1985) showed that MR_{inc} was elevated by 40–50% over resting individuals in the same nest box at ambient temperatures below the thermal neutral zone. MR_{inc} also varies with clutch size, at least in two passerines (Biebach 1981; Haftorn and Reinertsen 1985). For each additional egg, night time MR_{inc} increased by 3–5% in starlings (Figure 20.2) and by 6–7% in blue tits.

During the daytime intermittent incubators leave the nest and have to re-warm cooled eggs when they return. Vleck (1981a) and Biebach (1979) both showed that cooling eggs during continuous incubation increased the energy expenditure of the female. Biebach (1986) reported that the costs of re-warming the eggs

Table 20.2 Overview of the measurements of RMR, MR_{inc} at night and during the day for a number of passerine birds. Estimates of MR_{inc} at night and during the day are also given in multiples of BMR.

Species	Body mass (g)	Clutch size	RMR ($kJ\,d^{-1}$)	BMR ($kJ\,d^{-1}$)	MR_{inc} ($kJ\,d^{-1}$)		MR_{inc}/BMR		References
					Night	Day	Night	Day	
Sturnus vulgaris	80	6	86.4	82.1	129.6	190.1	1.6	2.4	1,2
Parus caeruleus	11.5	13	32.8	19.8	51.8		2.6		3
Taeniopygia guttata	11.6	3–5	38.0	19.8		45.8		2.3	4
Serinus canaria	20.8	4–5	50.9	29.4	59.6		2.0		5
Parus major	18.0	9	47.5	27.6	82.1		3.0		6
Average							2.3	2.3	

References: 1 – Biebach (1979); 2 – Biebach (1984); 3 – Haftorn and Reinertsen (1985); 4 – Vleck (1981a); 5 – Weathers (1985); 6 – Mertens (1980).

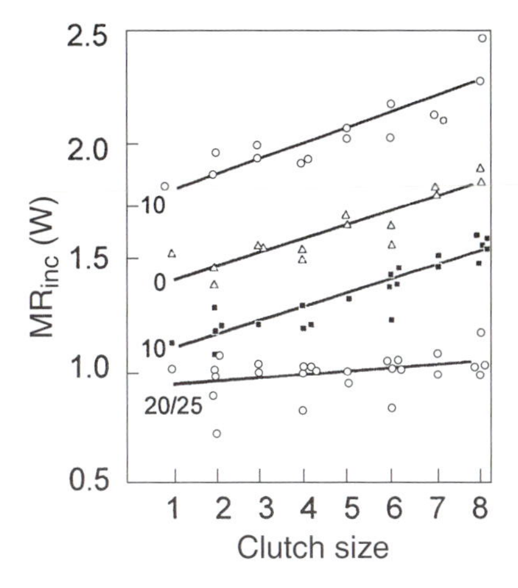

Fig. 20.2 Relationship between MR_{inc} and clutch size for female starlings at four different temperatures measured at night. MR_{inc} is dependent on clutch size below the lower critical temperature (<20°C). (Redrawn from Biebach 1984.)

for the starling depended on T_a and on the inattentive period. He speculated that incubating starlings used excess heat from activity to cover these costs. For starlings, MR_{inc} at moderate T_a (10 and 25°C) was 50–60% higher than during the night, probably due to the energy needed for digestion (Biebach 1984).

To ascertain the importance of ambient temperature on MR_{inc}, results of five studies were collated (Table 20.2). In four studies oxygen consumption was estimated whereas in one study heat production was measured (Mertens 1980). MR_{inc} was standardised for these studies at $T_a = 8°C$, a reasonable estimate of T_a during incubation for temperate zone birds. Estimates of resting metabolic rate were taken from the studies, while BMR was calculated from Kendeigh *et al.* (1977). Calculations indicated that MR_{inc} exceeded resting metabolism with the ratio of MR_{inc} to BMR ranging between 2 and 3 (Table 20.2).

It is concluded that MR_{inc} is elevated above that of resting birds when ambient temperatures are below the thermal neutral zone, which is often the case for temperate zone birds nesting in the spring. Furthermore, MR_{inc} ranges from 2 to 3 times BMR, but estimates are strongly dependent on T_a. Lastly, the fact that clutch size is an important determinant of MR_{inc} suggests that large clutch size could pose energetic stress on the birds.

Field metabolic rate during incubation

Thomson *et al.* (1998b) argued that MR_{inc} is the parameter of choice when looking for correlates of energy expenditure and life-history variables. However, it is stressed here that MR_{inc} is of little importance unless it translates into an increase in FMR and a concomitant decline in energy reserves of the parent. Data on FMR of incubating birds are shown in Table 20.1. Where more than one estimate was available for a species, average values have been used, except for the ringed plover (*Charadrius hiaticula*) where measurements on arctic and temperate birds were kept separate. The equation that describes FMR_{inc} ($kJ\,d^{-1}$) as a function of body mass (BM, g) for all birds is:

$$FMR_{inc} = 14.79\ BM^{0.61}, \tag{20.1}$$

($F_{1,31} = 174.0$, $r^2 = 0.84$, $p < 0.001$, $N = 33$). Birds breeding on the arctic tundra have a higher FMR_{inc} (Figure 20.3) compared to species breeding in more southern latitudes (Ancova, $F_{1,30} = 29.8$, $p < 0.001$, full model $r^2 = 0.91$) but the slope does not differ between these regressions (interaction of arctic*logmass, $p > 0.9$):

$$FMR_{inc} = 16.56\ BM^{0.545} \quad \text{Non-arctic breeding birds,} \tag{20.2}$$

$$FMR_{inc} = 25.64\ BM^{0.545} \quad \text{Arctic breeding birds.} \tag{20.3}$$

Arctic waders appear to have a higher FMR_{inc} compared to other birds from lower latitudes (Piersma *et al.* 2001). The difference in metabolic rate may be attributable to several factors like the fact that arctic species experience colder ambient temperatures when attending eggs, or it may be associated with their experiencing 24 hours of daylight, which increases their feeding time (Lindström and Kvist 1995). That the FMR_{inc} for the ringed plover (Tatner and Bryant 1993) taken in the temperate zone is also high remains puzzling because other waders from the temperate region do not show elevated levels of FMR_{inc}. Thus the higher BMR of wading birds (Charadriiformes) alone cannot explain the elevation of FMR_{inc} in the arctic (Kersten and Piersma 1987).

Some temperate breeding birds that experienced unusually low ambient temperatures, such as orange-breasted sunbirds (*Nectarinia violacea*; Williams 1993c), tree swallows (*Iridoprocne bicolor*; Williams 1988), and great tits (Bryan and Bryant 1998) also have unusually high FMR_{inc}.

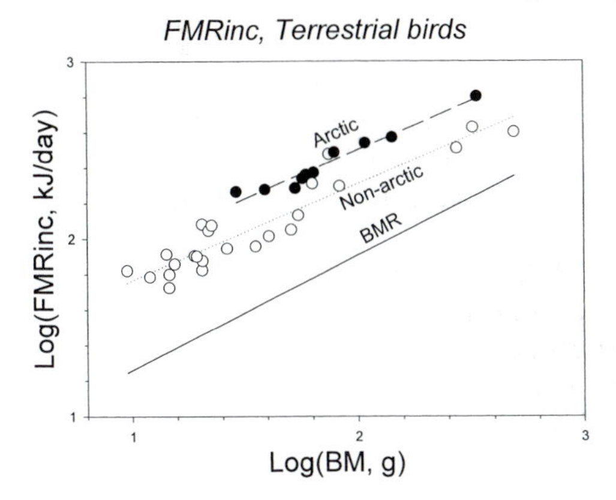

Fig. 20.3 Log FMR_{inc} as a function of log body mass for arctic and non-arctic breeding terrestrial birds. The relation between log BMR and log body mass is given as a solid line for reference. Data from Table 20.1.

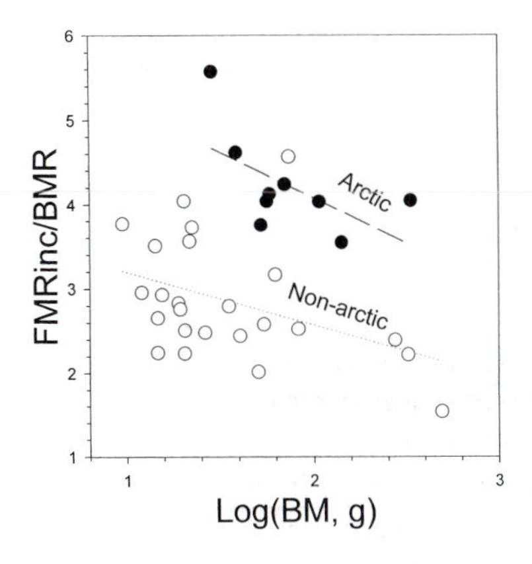

Fig. 20.4 Log FMR_{inc}/log BMR in relation to body mass for arctic and non-arctic terrestrial birds. Data from Table 20.1.

Examination of the curve for FMR_{inc} versus body mass suggests smaller birds have relatively higher FMR_{inc}. The ratio FMR_{inc}/BMR declined with body mass (Figure 20.4; Ancova in a model including arctic breeding as a factor: $F_{1,28} = 7.588$, $p = 0.01$). FMR_{inc}/BMR for birds breeding outside the arctic was on average 2.9 BMR ($N = 24$) whereas for birds breeding in the arctic it was 4.2 BMR ($N = 9$).

Comparison between the incubation and the nestling phase

The idea that a peak in energy expenditure occurs when parents are delivering food to young, the 'peak demand hypothesis', has become entrenched in discussions about avian life-history evolution (Drent and Daan 1980; Walsberg 1983b; Bennett and Harvey 1987). An alternative view, the 'reallocation hypothesis'

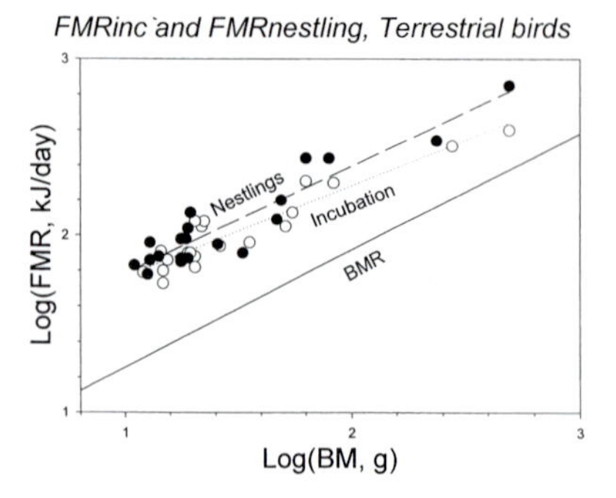

Fig. 20.5 Log FMR as a function of log body mass for incubating birds and for birds tending nestlings (both genders included). The relation between log BMR and log body mass is given as a solid line for reference. Data from Table 20.1.

(West 1968), which suggests that FMR is relative constant over the annual cycle, has also received support (Bryant and Tatner 1988; Weathers and Sullivan 1993).

The idea that FMR_{inc} is significantly lower than FMR_{nestl}, as is suggested by the 'peak demand hypothesis' (Wijnandts 1984; Masman *et al.* 1988; Gales and Green 1990), is tested here using data in Table 20.1. Firstly measurements of both genders were averaged so as to test for the relationship between FMR and stage of breeding in a model containing log mass, arctic breeding, and incubation or nestling stage, as covariates. In this analysis, arctic breeding was again highly significant but the effect of stage of breeding only approached significance ($p = 0.06$). Species in which both FMR_{inc} and FMR_{nestl} was measured were also compared by combining all species irrespective of incubation behaviour pattern (Figure 20.5). A repeated measures Anova on the log FMR with species as subject showed a significant difference between the stages with FMR_{nestl} having higher values (within species contrast; $F_{1,19} = 5.673$, $p = 0.028$). However, the difference between the stages depended on body mass (within species interaction stage*log body mass; $F_{1,18} = 9.197$, $p < 0.001$). The difference between FMR_{inc} and FMR_{nestl} is less pronounced in small birds.

Male and female FMR_{inc} might be expected to differ in species where only one sex incubates (male assisted or intermittent female-only incubation) and so this analysis was repeated for the species exhibiting shared-incubation ($N = 21$). In a model controlling for arctic breeding and body mass, there was a significant effect of breeding stage on the FMR_{inc} ($F_{1,17} = 5.718$, $p < 0.05$). FMR_{inc} for species with shared-incubation was lower than FMR_{nestl}.

Differences between FMR_{inc} and FMR_{nestl} were tested using data measured in females of all species (Figure 20.6). The difference between FMR_{inc} and FMR_{nestl} was affected by the incubation system (repeated measures Anova, effect of body mass, $F_{1,11} = 5.366$, $p < 0.05$, effect of incubation system, $F_{2,11} = 4.665$, $p < 0.05$). Females which shared incubation had a higher FMR during nestling rearing than during incubation, but females that incubate alone have equivalent FMRs during these two periods.

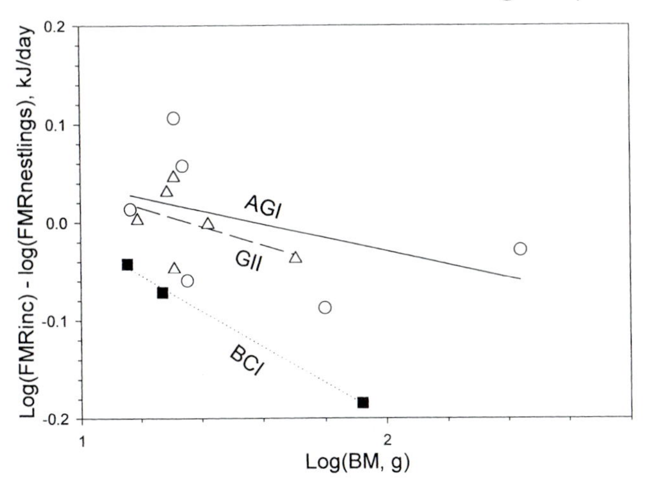

Fig. 20.6 The change in FMR of females between nesting and incubation as a function of body mass for shared incubation (BCI) and female-only incubation (GII, AGI). During incubation the larger females sharing incubation spend less energy than females incubating alone relatively to their expenditure during the nestling period. Data from Table 20.1.

Recent data collected on dune larks (*Mirafra erythroclamys*) living in the Namib desert indicated a uniform energy expenditure for females incubating eggs alone and feeding nestlings (Williams 2001). These data are especially interesting because primary production in the Namib is remarkably low and ambient temperatures during incubation and the nestling phase are moderate (30–35°C).

In terrestrial birds, the difference between FMR_{inc} and FMR_{nestl} as estimated by double labelled water measurements on both sexes depends on body mass. Small birds show little difference in energy expenditure between the two stages while large birds tend to have lower FMR_{inc}. Within the group exhibiting shared incubation FMR_{inc} was lower than FMR_{nestl}. For females, FMR_{inc} is lower than FMR_{nestl} for birds with shared incubation, and this effect is also more pronounced for the heavier birds. Comparison of the FMRs during incubation and the nestling period provides limited support to the peak demand hypothesis, and suggests that models of life-history theory need to incorporate events throughout the breeding cycle, if not the entire year.

Effects of ambient temperature

In temperate regions ambient temperatures are typically lower during incubation than during the nestling phase. That this may be part of the reason that FMR is not so different between these phases is illustrated by data of FMR_{inc} (Bryan and Bryant 1998) and FMR_{nestl} (Tinbergen and Dietz 1994) plotted against the minimum temperatures experienced by the birds during the measurement (Figure 20.7). The level of FMR_{inc} relative to FMR_{nestl} seems to depend on T_a. The data suggest that, for great tits, FMR_{inc} is equal to FMR_{nestl} when measured at high T_a but may be higher at ambient temperatures approaching freezing (interaction minimum

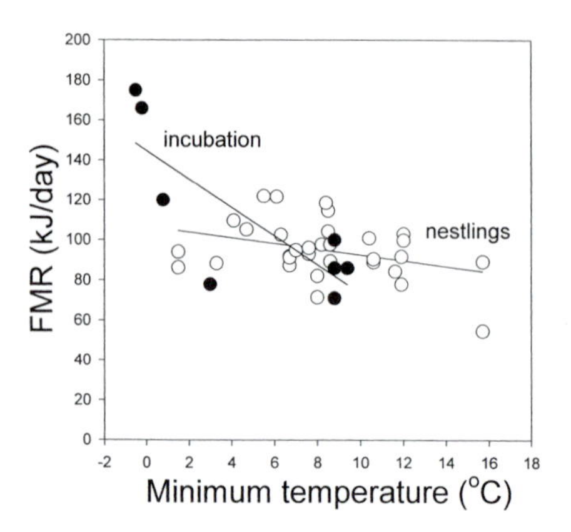

Fig. 20.7 FMR as a function of minimum temperature for great tits (*Parus major*) incubating (closed points) and great tits feeding nestlings (open points). Regressions are drawn for each group separately. Data from Bryan and Bryant (1998) and Tinbergen and Dietz (1994).

temperature*phase; $F_{1,36} = 10.985$, $p < 0.005$). These data illustrate the importance of detailed knowledge of microclimate and other ecological variables when comparisons are made of FMR between stages. Temperature effects are probably very important factors determining FMR_{inc}, and are therefore potentially important agents in selection on the timing and intensity of reproduction.

FMR/BMR and reproductive stage

Body masses of individuals differed between incubation and the nestling period, and so the ratio of FMR/BMR was compared by repeated measures Anova on log body mass; within species comparison $F_{1,19} = 48.7$, $p < 0.001$). BMR was calculated for the stages separately on the basis of three different relationships: waders (Kersten and Piersma 1987), non-passerines, and passerines (Kendeigh *et al.* 1977). Within the subset of terrestrial species where measurements of FMR were available in both stages (averaged for both genders), FMR/BMR during incubation was lower than during the nestling phase (2.77 BMR, SD = 0.61, N = 20 versus 3.25 BMR, SD = 0.66, N = 20) and repeated measures Anova on log ratio: within species comparison ($F_{1,19} = 20.7$, $p < 0.001$), showed that no additional effect of body mass existed. However, it has to be emphasised that BMR may differ between stages not only because of mass differences but also because ovarian and other reproductive tissues are being absorbed during incubation. Using allometric estimates of BMR is therefore a crude approach. It would be best if those who measure FMR also measure BMR in the same birds.

Energy expenditure, time budgets and intake rates

If FMR_{inc} is relatively high, and foraging time is reduced then energy intake per unit time must be increased to meet parental need during incubation. From knowledge of MR_{inc}, FMR_{inc} and nest attentiveness, it is possible to calculate $MR_{off\text{-}nest}$

Table 20.3 Predicted $MR_{off\text{-}nest}$ and intake rate given in multiples of BMR, on the basis of FMR and MR_{inc}. Data are given seperately for the incubation (female-only and shared) and the nestling period. $MR_{on\text{-}nest}$: metabolic rate while on the nest in the nestling phase.

Phase	FMR	MR_{inc}	Duration night (hr)	% of day on nest	$MR_{off\text{-}nest}$ (predicted)	Intake rate (predicted)
Incubation	2.85	2.3	10	75 (Female-only)	5.5	19.5
	2.85	2.3	10	50 (Shared)	3.9	9.8
		$MR_{on\text{-}nest}$ (kJ d^{-1})				
Nestling	3.25	1.9	8.5	10	4.2	11.2

(Table 20.3) under the assumption that the body mass is constant. Calculations indicate that a female exhibiting intermittent incubation, assuming 75% of the daytime on the nest (Chapter 6), expends energy at a rate 4.9 times the BMR when off the nest, whereas for parents sharing incubation the estimate is 3.6 BMR. These levels of energy expenditure are comparable to those during the nestling period when energy expenditure is around 4.2 BMR (Tinbergen and Dietz 1994; Tinbergen unpublished data). If it is assumed that a female with nestlings delivers the same amount of food that she needs herself, as is true for great tits (Tinbergen unpublished data), it is also possible to compare intake rates between the stages.

For birds that share incubation, intake rates between incubation and nestling phases are similar, but for females that receive no assistance from their mates, energy intake during incubation is estimated to be 100% higher than during the nestling phase. This is remarkable because, at least in the temperate zone, food availability is thought to be lower during the incubation phase, especially where females are not assisted by the males in terms of incubation or additional food. Weathers and Sullivan (1989b) report that female yellow-eyed juncos (*Junco phaeonotus*) that incubate eggs indeed have a higher intake rate than during all other stages in the nesting cycle. How incubating birds achieve this remains unclear. Apparently birds can increase their foraging efficiency dependent on the needs. This may be achieved by a different prey choice and/or foraging technique or by assistance of the male through feeding the female or by directing her to the better foraging sites.

Energetic cost of incubation in seabirds

Williams (1996) and Thomson *et al.* (1998b) both suggested that MR_{inc} was elevated above resting levels in seabirds, but disagreed on the magnitude of the increase. For the same species (Procellariidae, Spheniscidae, Laridae, Sulidae) Williams found an increase of 1.6 BMR whereas Thomson *et al.* (1998b) calculated an increase of 1.2 BMR. This difference is attributable to Thomson *et al.* (1998b) using mean estimates of MR_{inc} as given by Croxall (1982) based on mass loss during incubation, whereas Williams (1996) used minimum estimates of MR_{inc}. This latter selection criterion was based on the finding of Croxall (1982) that for penguins (Spheniscidae) minimum estimates of MR_{inc} based on mass loss closely corresponded to energy expenditure based on oxygen consumption.

In his analysis Williams (1996) concluded that measurements of MR_{inc} in seabirds based on oxygen consumption were systematically lower than estimates of MR_{inc} as given by the other two methods (mass loss and double labelled water). When measurements of MR_{inc} are made using flow through masks or plastic chambers fitted over the incubating bird in the field, it is highly likely that these experimental procedures markedly influence the metabolic rate of the individual, thereby biasing results. More confidence can be placed on estimates obtained by the double labelled water (DLW) method.

For seabirds there are a number of DLW estimates for the period when incubating parents are away from the nest foraging (data from Williams 1966). A regression model revealed significant differences between MR_{inc} and $MR_{off\text{-}nest}$ ($F_{1,8} = 9.823$, $p = 0.014$, $N = 11$; Figure 20.8). These estimates differ somewhat from Williams (1996) because only direct estimates were used (no estimate for MR_{inc} was available for the gannet *Morus bassana*). The resulting regressions from this model were:

$$FMR_{inc} = 3.48\ BM^{0.749}, \qquad (20.4)$$

and

$$FMR_{off\text{-}nest} = 8.65\ BM^{0.749}. \qquad (20.5)$$

In these seabirds the period off the nest was energetically more expensive than incubation on the nest, and on average $MR_{off\text{-}nest}$ was 4.8 BMR ($SD = 2.1$, $N = 6$) while MR_{inc} for this group was 1.7 BMR ($SD = 0.7$, $N = 5$).

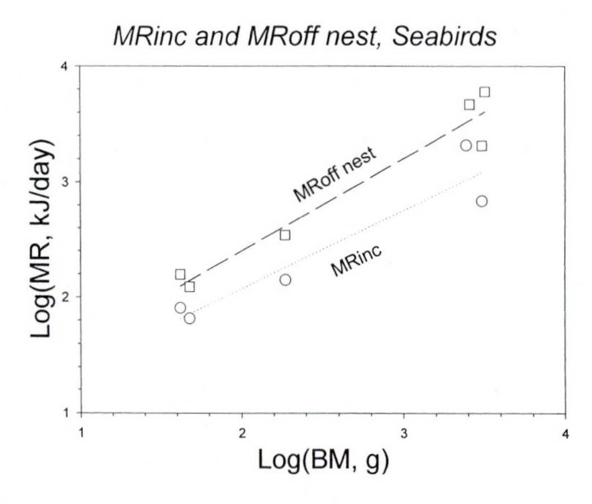

Fig. 20.8 Log $MR_{off\text{-}nest}$ and log MR_{inc} as function of log body mass for the species where a double labelled water estimate of the off-nest period was available. The relation between log BMR and log body mass is given as a solid line for reference. Data from Williams (1996).

Calculation of FMR_{inc} based on the assumption that 50% of the time is spent incubating amounts to 3.25 BMR for seabirds. The direct DLW estimate of FMR_{inc} of the kittiwake (*Rissa tridactyla*), integrating $MR_{off\text{-}nest}$ and MR_{inc} (Thomson *et al.* 1998a) was 4.6 BMR, and is relatively high. Thomson *et al.* (1998a) also compared FMR_{inc} and FMR_{nestl} in this species and found no differences. This is interesting because species which share incubation do not fit the idea developed here for terrestrial birds. More direct estimates of FMR_{inc} and FMR_{nestl} in seabirds would be welcome.

Future perspectives for research into energetics of incubation

It is concluded that the energy expenditure of terrestrial birds whilst sitting on eggs is increased over that of non-incubating individuals when they experienced temperatures below their thermal neutral zone, that larger clutch sizes mandate increased energy expenditure, and that, especially for systems in which only the female incubates, birds could have elevated field metabolic rates and potentially face serious energy shortage.

Both correlational and new experimental evidence suggest that terrestrial birds face a trade-off between incubation and foraging and consequently changes in food availability and energy expenditure affect the outcome of incubation. Furthermore energy expenditure is strongly affected by ambient temperature. In concert with seasonal patterns in food availability, this must affect the optimal timing and intensity of reproduction.

Researchers that manipulated energy expenditures and looked at the effects on incubation behaviour have shown that there is trade-off between incubation and foraging (Bryan and Bryant 1998). This finding makes it very likely that the energy balance during incubation is important to understand selection on the timing and intensity of reproduction.

Evidence for the peak demand hypothesis in terrestrial birds is limited. This implies that incubating terrestrial birds, especially those exhibiting single-gender intermittent incubation must have high intake rates because they have limited feeding times. How these birds achieve this remains an intriguing question.

The approach in this chapter has been to analyse between species variation in energy expenditure during incubation to come to a general prediction of the level of energy expenditure during incubation. For seabirds MR_{inc} is somewhat lower than for terrestrial birds (1.7 BMR versus 2.3 BMR respectively). FMR_{inc} for seabirds (3.3 BMR) was somewhat higher than FMR_{inc} in non-arctic terrestrial birds (2.7 BMR) but lower than in arctic terrestrial birds (4.2 BMR). However, these estimates suffer from the lack of measurements of FMR and BMR in the same birds. Continued detailed experimental studies on incubation in different species of birds manipulating expenditure, energy intake, or energy reserves to reach a better understanding of the factors that have shaped current patterns in energetics of avian incubation should prove fruitful.

21 *Incubation and the costs of reproduction*

J. M. Reid, P. Monaghan and R. G. Nager

Life-history theory is concerned with explaining the diversity of life cycles observed in nature. It provides a framework within which intra- and inter-specific variation in how organisms allocate resources amongst growth, self-maintenance and reproduction can be interpreted.

A central tenet of life-history theory is that organisms have finite resources at their disposal, and thus that the allocation made to one aspect of life will reduce the resources available for investment in others (Roff 1992; Stearns 1992). The consequent competition for resources amongst life-history traits will constrain life-history evolution, with the allocation made to any particular trait reflecting the compromise that maximises the individual's lifetime fitness. In terms of resource allocation during reproduction, such constraints and compromises translate into costs of reproduction, defined as the extent to which investment in one reproductive event, or component or phase of that event, reduces capacity of a parent to invest elsewhere. Thus, in order for overall resource allocation patterns and consequent life-history strategies to be interpreted, an understanding of the resource limitations and fitness costs associated with each component part of reproduction is required.

A high proportion of studies on resource allocation trade-offs and reproductive costs has been carried out on birds. Avian breeding behaviour can be studied with relative ease, birds generally have the capacity to reproduce more than once, and each breeding event is divided into the relatively discrete phases of courtship and mating, egg production, incubation and chick rearing. These factors mean that birds are suitable organisms for investigating resource trade-offs both between different reproductive attempts and between different phases of the same attempt (Lessells 1991; Heaney and Monaghan 1995, 1996; Monaghan and Nager 1997). However, until recently, such studies have focussed almost exclusively on trade-offs involving the demands of rearing chicks (Lindén and Møller 1989; Dijkstra *et al.* 1990), the assumption being that the resource requirements of earlier reproductive stages, such as incubation, are relatively trivial in comparison (Monaghan and Nager 1997). In this chapter, it is shown that the evidence that the time and energy demands of the incubation phase of avian reproduction can be substantial, and can represent an important component of reproductive costs in birds.

Why might incubation be costly?

The way in which avian embryos develop, and hence the success of a breeding event, is greatly influenced by the physical environment that embryos experience whilst inside the egg. Specific temperatures, humidities and degrees of physical disturbance are required to ensure that embryos hatch successfully and in good condition (Lundy 1969; Carey 1980; Webb 1987; Chapters 9–11). However, given fluctuating external conditions and the limited ability of embryos to modify their own environment, parents must generally regulate the developmental conditions experienced by their offspring. In the majority of species, this is done by contact incubation of the clutch. Exceptions include brood parasites which exploit the incubation environment created by other parents (Davies 2000; Chapter 17), and megapodes that have discarded contact incubation entirely (Chapter 13). White-rumped swiflets (*Aerodramus spodiopygius*) exhibit an unusual system whereby eggs are warmed by the body heat of older chicks still present in the nest (Tarburton and Minot 1987).

Early models of incubation dynamics suggested that, due to deployment of basal metabolic heat to warm the clutch and the relative lack of physical activity involved in sitting on a nest, incubation may be a relatively undemanding phase of avian reproduction in terms of energy expenditure (Walsberg and King 1978b). However, recent studies have shown that maintaining a gradient between the physical conditions within a nest and the surrounding environment can impose considerable energetic demands on parents (Williams 1996; Chapter 20). The demand of re-warming cold eggs may be particularly high (Vleck 1981a; Biebach 1986), and overall demands may approach those experienced whilst rearing chicks (Moreno *et al.* 1991; Williams 1996; Ward 1996). Furthermore, regulating the embryos' developmental environment requires parents to spend considerable time sitting on the nest (Chapter 6), restricting the time available for other activities, such as foraging or attracting further mates.

Parents can reduce the resource requirement of incubation to some extent. Nests can be constructed to minimise the rate of heat loss from the clutch (Schaefer 1980; Møller 1984, 1987b; Nager and von Noordwijk 1992), and nests and nest sites selected for their thermal quality (Inouye *et al.* 1981; Walsberg 1981; Hoi *et al.* 1994). Brood patches are likely to have evolved to facilitate efficient transfer of heat to eggs (Bailey 1952; Chapter 8), and eggs could be shaped to facilitate efficient packing under the brood patch (Barta and Székely 1997). Division of incubation duties between partners or other helpers at the nest (Chapter 6) can reduce the time and energy demands imposed upon any one individual. Variation in the effectiveness of these measures, and in the discrepancy between external conditions and the optimal developmental conditions for the embryos, means that the resource requirements of incubation will vary within and between species. However in general, provision of a high quality environment for development requires deployment of parental resources, in terms of time and energy, to incubation.

In the following sections, correlative and experimental evidence is examined to test the idea that the time and energy invested in incubation can limit the resources available for investment elsewhere, and thus that resource allocation trade-offs and fitness costs of incubation occur. The circumstances under which costs are observed, and investigate the ways in which they are manifested through effects on parents and offspring are also discussed.

Costs of incubation: time spent away from the nest

Allocating time to activities away from the nest and leaving clutches of eggs unattended may allow the conditions experienced by embryos to diverge rapidly from developmental optima. Such divergence may adversely affect offspring and hence parental fitness in several ways. There will be a direct and immediate fitness cost to spending time away from the nest if the conditions to which eggs are consequently exposed are severe enough to kill the embryos outright, an occurrence most likely in extreme climates (Chapter 16). However, as embryos are relatively resistant to short periods of exposure to sub-optimal conditions (Drent 1975; Webb 1987; Sockman and Schwabl 1998), sub-lethal effects on development may be more common than embryonic mortality. In contrast to reptiles, in which incubation temperatures have been shown to profoundly affect offspring phenotype (Burger 1989; Angilletta *et al.* 2000), relatively little is known about how developmental conditions influence post-hatching phenotype and fecundity in birds. However, poor incubation conditions may impair future performance of the offspring and hence have intergenerational consequences (Lindstrom 1999; Metcalfe and Monaghan 2001).

The commonest consequence of leaving a clutch unattended is the slowing of embryonic development and hence the elongation of the incubation period (Webb 1987). Intraspecific variation in the duration of incubation has frequently been related to variation in parental attentiveness (Lifjeld and Slagsvold 1986; Lifjeld *et al.* 1987; Nilsson and Smith 1988; Aldrich and Raveling 1983), and the conditions experienced within the nest can have a greater effect on development rates than intrinsic egg quality (Ricklefs and Smeraski 1983; Reid *et al.* 2000a). A prolonged incubation period can adversely affect offspring in several ways. Embryos may expend more energy prior to hatching (Vleck and Hoyt 1980) resulting in greater depletion of resources and poorer hatchling condition. The time for which a clutch is vulnerable to predation can also be increased (Perrins 1977; Bosque and Bosque 1995; Tombre and Erikstad 1996). Delayed hatching can itself have a fitness cost if late-fledging offspring are disadvantaged when competing for food or territories, or suffer a reduced chance of accumulating sufficient resources for over-wintering, moult or migration (Pettifor *et al.* 1988; Verboven and Visser 1998; Visser and Verboven 1999). Hence, via detrimental effects on their offspring, incubating parents may accrue considerable fitness costs by allocating too much time to activities away from the nest.

Costs of incubation: energetic limitations

Spending time away from a nest may affect parental fitness via effects on current offspring. However, spending too much time incubating and thereby failing to allocate enough time to other activities may affect parental fitness directly, by affecting the ability to complete the current breeding event, or to breed again either simultaneously or subsequently.

Incubating parents must meet the energetic demand of regulating the nest environment either by foraging or by depleting stored resources (Chapter 6). Since the body condition of a parent bird can influence both its ability to provision dependent offspring and its chance of surviving to breed again (Bolton 1991; Jones 1992; Golet *et al.* 1998; Wendeln and Becker 1999), allowing body condition to deteriorate during incubation may be costly, reducing the fitness benefits accrued from both current and future broods. However, foraging and incubation are mutually exclusive activities and, unless provisioned on the nest by partners or helpers, incubating parents must leave the clutch unattended if they need to feed during the incubation period. A fitness cost due to energetic limitation would arise if the time and energy demands imposed during incubation were substantial enough to prevent parents from maintaining either the nest environment or their own body condition at optimal levels, reducing the survival or fecundity of either themselves or their offspring.

Analysis of time-energy models suggests that incubation may indeed be a marked period of energetic limitation during reproduction (Yom-Tov and Hilborn 1981; Moreno and Hillstrom 1992). This may be because incubation restricts foraging time rather than because the absolute energetic requirement of regulating clutch temperature is prohibitively high (Moreno and Hillstrom 1992; Reid *et al.* 2002). Although these models have been analysed using species-specific parameter values, their conclusions may apply more generally, at least to those passerines in which only one parent incubates (Yom-Tov and Hilborn 1981). Thus, especially given that incubation periods often fall before seasonal peaks in food availability, incubation may commonly be a period of time and energy limitation in birds.

Given time and energy limitations, a trade-off between the time that parents spend incubating and the time spent foraging is predicted. Indeed, there is considerable correlative evidence that the way in which incubating birds allocate time between incubation and other activities depends on the balance between endogenous and exogenous resources and hence on their current state (*sensu* McNamara and Houston 1997). Time spent incubating can increase with increasing food availability (Drent *et al.* 1985; Rauter and Reyer 1997), and when mates provision incubating birds on the nest (Lifjeld *et al.* 1987; Nilsson and Smith 1988; Halupka 1994; Hatchwell *et al.* 1999). Indeed, Martin and Ghalambor (1999) found a positive relationship between the level of incubation provisioning and nest attentiveness across species, although Conway and Martin (2000b) suggest that the risk of predation may have a greater overall influence on incubation behaviour. Furthermore, individuals that commence breeding in better body condition spend more time on the nest (Lifjeld and Slagsvold 1986; Afton and Paulus 1992; Aldrich and Raveling 1983; Hegyi and Sasvari 1998), suggesting that parents' incubation strategies may often be energy-limited. As changes in attentiveness can affect the duration of the incubation period, such energetic limitation during incubation may have fitness consequences for offspring and parents. However, conclusive evidence for the existence of a trade-off between incubation and foraging and a consequent fitness cost of incubation must generally be provided by experiment.

Appendix 21.1 summarises the rationale for such experiments, and outlines the manipulations and measures of fitness that could be used to demonstrate costs of incubation. The results of experiments that have been undertaken are discussed below.

Parental incubation effort has frequently been manipulated by adding or removing eggs from a clutch, the underlying assumption being that incubation demands vary with clutch size. This assumption has been experimentally confirmed in terms of both the energy required for steady-state incubation (Biebach 1981, 1984; Haftorn and Reinertsen 1985) and the total daily energy expenditure of an incubating bird (Coleman and Whittall 1988; Moreno *et al.* 1991). However, incubation demands do not change with all clutch size changes, and the extra energy required to incubate a single additional egg may be relatively small (Moreno *et al.* 1991; Moreno and Sanz 1994). Furthermore, the number of eggs in a clutch influences the intrinsic thermal properties of the clutch, with large clutches cooling down more slowly than small clutches when left unattended (Frost and Siegfried 1977; Reid *et al.* 2000a). This reduction in cooling rate may be sufficient to enable parents to forage for the extra time required to meet a higher energetic demand of incubating a large clutch without mean clutch temperature falling too low (Reid *et al.* 2002). Since experimentally altering clutch size can alter both the energetic demand of incubation and the time that parents have available to meet that demand, clutch size manipulations may not be the most rigorous way to investigate resource trade-offs and costs of incubation. Whilst positive evidence would still be convincing, given the possibility of a null manipulation, an apparent absence of costs can not necessarily be accepted as evidence that such costs do not exist.

Nevertheless, several studies have examined the consequences of enlarging a clutch for incubation performance (reviewed in Thomson *et al.* 1998b). Clutch enlargement has frequently been found to prolong the incubation period (Coleman and Whittall 1988; Moreno and Carlson 1989; Smith 1989; Székely *et al.* 1994; Siikamäki 1995; Sandercock 1997; Reid *et al.* 2000a; Wiebe and Martin 2000), increase hatching asynchrony (Moreno and Carlson 1989; Reid *et al.* 2000a) and reduce hatching success (Andersson 1976; Moreno *et al.* 1991; Siikamäki 1995; Reid *et al.* 2000a). Such changes have been interpreted as evidence of fitness costs arising from increased incubation effort (Moreno and Carlson 1989; Thomson *et al.* 1998b). However, parents may be physically unable to cover all eggs within artificially enlarged clutches equally, and thus clutch enlargement may affect incubation performance by constraining incubation ability of a parent physically rather than energetically. Peripheral eggs within enlarged clutches may be incubated inefficiently, leading to within-clutch temperature differences that could prolong incubation, increase hatching asynchrony and reduce hatching success in the absence of increased incubation demands (Sandercock 1997; Reid *et al.* 2000a). Indeed, many clutch enlargements have failed to demonstrate changes in either adult time budget, or adult mass, that would be expected had adult energy balance been affected (Jones 1987; Moreno *et al.* 1991; Székely *et al.* 1994; Siikamäki 1995; Sandercock 1997; Cichon 2000). Furthermore, observed increases in adult mass loss may be due to prolonged incubation periods arising from physical constraints on efficient incubation rather than to increases in the energetic demand imposed per unit time (e.g. Moreno and Carlson 1989). Hence, in the majority of clutch enlargement experiments, it is difficult to exclude the possibility that physical rather than energetic constraints were responsible for observed changes in incubation performance.

Few studies have assessed the consequences of clutch enlargement during incubation for measures of post-hatching breeding performance. Experimental clutch enlargements have frequently not been reversed at hatching, and fitness costs of incubating enlarged clutches have thus been confounded with costs of rearing resulting enlarged broods (Monaghan and Nager 1997). However, three studies that have examined consequences of increased clutch size during incubation only have suggested that there may be energetic consequences of clutch enlargement for parents rather than solely incubation effects on offspring. Common tern (*Sterna hirundo*) chicks belonging to parents that had incubated experimentally enlarged clutches but reared their original brood size showed significantly reduced growth rates compared to chicks belonging to control parents (Heaney and Monaghan 1996). This effect was attributed to the energetic consequences of incubating an enlarged clutch for subsequent provisioning ability of the parents. Similarly in starlings (*Sturnus vulgaris*) and collared flycatchers (*Ficedula albicollis*), chicks reared by parents that had incubated enlarged clutches were in poorer condition at fledging than chicks reared by parents that had incubated their natural clutch size (Cichon 2000; Reid *et al.* 2000a). Poor nestling growth may negatively affect post-fledging survival (Nur 1984; Tinbergen and Boerlijst 1990; Hochachka and Smith 1991; Magrath 1991). However, these experiments do not completely exclude the possibility that poor fledgling condition resulted from long-term physiological consequences of experiencing poor developmental conditions during incubation, rather than from reduced parental provisioning during the rearing period. Experimental fostering of control broods to parents that had experienced increased incubation demands would be required to eliminate this possibility.

Clearer demonstration of energetic costs of incubation may be achieved by directly manipulating either the energy obtained or the energy expended by incubating parents. Several studies have manipulated rates of parental energy expenditure or energy intake during incubation, and observed consequent changes in the division of time between foraging and incubation. Reducing the demand of incubation by providing heat to nests increased the duration of incubation sessions and overall nest attentiveness in great tits (*Parus major*) and starlings (Bryan and Bryant 1998; Reid *et al.* 1999), but reduced attentiveness in pied flycatchers (*Ficedula hypoleuca*; von Haartman 1956). Incubation sessions were also prolonged by experimentally reducing the rate of heat loss from storm petrel (*Hydrobates pelagicus*) nest cavities (Bolton personal communication). Moreno (1989a) showed that attentiveness was increased by providing supplementary food to incubating female wheatears (*Oenanthe oenanthe*). Furthermore, nest attentiveness of female snow buntings (*Plectrophenax nivalis*) was decreased by removing males and thus depriving females of being provisioned on the nest (Lyon and Montgomerie 1985). Hence there is experimental evidence that an energetic trade-off between time spent incubating and time spent foraging exists across a range of species.

Fewer studies have provided clear evidence that manipulating adult energy balance actually has fitness consequences. Provision of heat to starling nests during incubation increased fledging success in the same brood and hatching success in subsequent unmanipulated second clutches (Reid *et al.* 2000b). Increasing overall incubation demands by experimentally prolonging incubation periods affected hatching success and female condition in barnacle geese (*Branta leucopsis*; Tombre and Erikstad 1996), current hatching success and future laying date in storm petrels (Minguez 1998) and the time until the next breeding attempt in black swans (*Cygnus atratus*; Brugger and Taborsky 1994). Removal of provisioning males reduced hatching success in snow buntings (Lyon and Montgomerie 1985) and Nilsson and Smith (1988) showed that provision of supplementary food increased hatching success in blue tits (*Parus caeruleus*). However, neither of these two studies found a change in fledging success, and supplementary feeding did not affect wheatear breeding performance (Moreno 1989a). Thus fitness costs due to energetic limitation can arise during incubation, but may not always do so (see below).

Costs of incubation: limitations on mating opportunities

If the time allocated to incubating one clutch of offspring significantly reduces a parent's chance of obtaining further mating opportunities, incubation will impose non-energetic fitness costs on parents. As reproductive rate of a male may often be limited by the number of mates acquired (Clutton-Brock 1991; Andersson 1994), such costs may apply particularly to males.

Incubation may preclude activities such as song and display that function in mate attraction (Whitfield and Brade 1991; Albrecht and Oring 1995; Catchpole and Leisler 1996), and thus reduce the ability of the male to attract additional mates. Whilst hormone implants have been used to experimentally increase male investment in mate attraction and conclusively demonstrate a trade-off with paternal care during chick-rearing (Hunt *et al.* 1999; Moreno *et al.* 1999b), similar experiments have yet to be carried out during incubation. However, experimentally increasing the chances of male starlings attracting secondary females by providing some males with additional nest boxes caused those males to incubate less and display more (Smith 1995), suggesting a trade-off between mate attraction and incubation.

Incubating one clutch may also reduce the ability of polygamous parents to care for other simultaneous broods. Correlative evidence from facultatively polygynous species suggests that parents with multiple clutches contribute less time to incubating each one, decreasing the total time for which each clutch is attended (Pinxten *et al.* 1993; Smith *et al.* 1995; Fitzpatrick 1996; Reid *et al.*, unpublished data). Reduced care for one brood due to the allocation of resources to incubating another will be costly if it significantly reduces the brood's survival or reproductive value. Indeed, lack of male incubation can prolong incubation periods and reduce hatching and fledging success (Smith *et al.* 1995; Reid *et al.*, unpublished data). Furthermore, in their review of male removal experiments, Bart and Tornes (1989) concluded that the loss of male help may have particularly severe fitness consequences if it occurs during the incubation period. However, the absence of a male contribution to incubation did not affect breeding success in starlings (Pinxten *et al.* 1993) or reed warblers (*Acrocephalus scirpaceus*; Duckworth 1992), although these studies did not examine the consequences of reduced male help for female condition or future fitness.

Finally, time spent incubating may reduce a male's opportunity to obtain extra-pair copulations. Extra-pair paternity rates are generally low in species where males contribute extensively to incubation (Schwagmeyer *et al.* 1999). Furthermore, male fairy martins (*Hirundo ariel*) were found to adjust the time allocated to incubation in response to the availability of fertile females, suggesting the existence of a trade-off between incubation and attempting extra-pair copulations. The consequent low male contribution to incubation prolonged incubation periods, although no reduction in hatching success was found (Magrath and Elgar 1997). Equally, incubating the first-laid eggs in a clutch may reduce mate-guarding ability of a male, involving a potential cost in terms of lost paternity. Whilst this hypothesis has not been rigorously tested, it may explain why, in species where males assist females with incubation, males often do not incubate until after the clutch has been completed, whilst females commence incubation on the penultimate egg (Power *et al.* 1981).

Occurrence and distribution of incubation costs

The studies reviewed in the previous sections demonstrate that experimentally manipulating the resource requirements of incubation can alter the ways in which parents allocate resources amongst other activities, both within and between reproductive events. This evidence, and that from correlative studies, suggests the existence of resource trade-offs and hence resource limitation during incubation. Furthermore, experimentally induced changes in allocation patterns have affected measures of both current and future reproductive success. Thus there is evidence that the resources required for incubation can be sufficient to limit parental fitness, and hence that incubation can be a costly stage of avian reproduction.

Who pays?

The way in which costs are distributed between parents and current offspring depends on how incubating parents resolve trade-offs between investment in current and future reproduction, and hence on their allocation of time between incubation and alternative activities. Largely due to links with mating and social systems, the determinants of time allocations of breeding birds to mating activities have attracted considerable attention. Patterns of investment are likely to depend on multiple ecological factors that vary amongst individuals and species (reviewed in Clutton-Brock 1991). Patterns of time allocation between foraging and incubation also vary greatly between species. In relatively large and long-lived species with precocial young, parents often remain on the nest for long periods during incubation, meeting energetic requirements predominantly from stored endogenous resources (Aldrich and Raveling 1983; Afton and Paulus 1990; Hepp *et al.* 1990; Erikstad and Tveraa 1995; Chapter 6). Continuous incubation may be the most efficient incubation strategy, as it is likely to minimise the incubation period and the frequency of energetically demanding clutch re-warming. In such species, the resolution of the trade-off between foraging and incubation is towards the incubation extreme, and costs of incubation may therefore operate via parents to a large degree. Indeed, whilst experimentally manipulating incubation demands in such species can affect measures of offspring survival, e.g. hatching success (Tombre and Erikstad 1996; Minguez 1998), increased costs are often manifested through changes in subsequent adult performance (Heaney and Monaghan 1996; Tombre and Erikstad 1996; Minguez 1998).

By contrast, continuous incubation is unlikely to be physiologically possible for small-bodied passerines (but see Chapter 6), as the resources required to sustain long incubation sessions equate to large proportions of normal body mass (Moreno 1989b). In small-bodied species, parents often incubate intermittently, resolving the trade-off between foraging and incubation so as to spend considerable time away from the nest. When only one parent incubates, this resolution frequently results in clutches being left unattended, and costs of incubation may often affect current offspring directly. Indeed, the average temperatures recorded by passerine eggs are typically well below the optimum for embryonic development (Haftorn 1983, 1988; Webb 1987; Williams 1996). In such species, experimental manipulations have provided little clear evidence that increased energetic costs of incubation are borne by parents. This is partly because few studies have rigorously investigated the consequences of manipulated incubation demands for estimators of adult fitness other than body mass. In particular, there is a lack of studies that have measured adult survival or future reproductive performance, or more rigorous indicators of physiological condition (although see Cichon 2000; Reid *et al.* 2000b). However, in passerines, there is experimental evidence that parents may often preserve their own condition at the expense of their offspring. Patterns of parental mass loss are difficult to interpret directly as indicators of resource limitation during incubation, because mass loss may reflect the post-laying atrophy of reproductive organs (Ricklefs and Hussell 1984) or be an adaptation to reduce flight costs during chick-rearing (Moreno 1989b; Merilä and Wiggins 1997) rather than reflecting physiological stress. However, in experiments designed to test these alternatives, provision of supplementary food to incubating parents did not alter mass loss rate (Hillström 1995; Merkle and Barclay 1996; Slagsvold and Johansen 1998). This suggests that parents allocate resources during incubation so as to optimise their own condition, passing supplementary resources on to offspring when they become available. Indeed, the only study that has assessed the consequences of supplementary feeding for both adult mass and incubation strategy found increases in nest attentiveness but not in adult mass (Slagsvold and Johansen 1998).

Hence energetic trade-offs during incubation can be resolved so that costs are manifested in both parents and current offspring to some degree. Costs may be manifested extensively in current offspring in small intermittently incubating species, and by parents to a greater degree in larger species that incubate continuously. However, more experiments are clearly needed across a wider range of species in order to tease apart the ecological, physiological and phylogenetic relationships underlying such patterns. The suggestion that intermittently incubating passerines invest preferentially in their own condition during incubation whereas larger species invest more in current offspring is perhaps contrary to expectation (Stearns 1992), and to the conclusions of studies carried out during the chick-rearing period (Sæther *et al.* 1993; Mauck and Grubb 1995; Weimerskirch *et al.* 1995). However, such allocation patterns make more sense in the context of within-brood and within-season trade-offs involving incubation effort, as breeding passerines may have to invest in their own condition during incubation in order to successfully rear the current brood and breed again within the same season (Moreno 1989b).

Condition-dependence of costs

The majority of the studies that have demonstrated clear energetic fitness costs to incubation have been carried out in relatively cold environments (e.g. Lyon and Montgomerie 1985; Nilsson and Smith 1988; Heaney and Monaghan 1996; Tombre and Erikstad 1996; Reid *et al.* 2000a; Bolton personal communication). However, as egg temperature regulation may be less demanding in warm climates, more studies are required to assess whether similar costs apply in other environments. Furthermore, even in cold climates, costs of incubation may only become apparent during particularly harsh conditions. For example, supplementary feeding of female wheatears reduced incubation periods only during the coldest of three springs (Moreno 1989a), incubating goldeneyes (*Bucephala clangula*) lost more mass in cold weather (Mallory and Weatherhead 1993) and male assistance with incubation increased hatching success in moustached warblers (*Acrocephalus melanopogon*) during April but not during May (Kleindorfer *et al.* 1995).

Even when costs of incubation are evident, their impact may vary with parent and offspring condition, not affecting all parents, or all offspring within a brood. Costs of incubation due to energetic limitation may only affect parents in poor condition (Moreno and Sanz 1994; Heaney and Monaghan 1996; Minguez 1998). Investing in incubation rather than in attracting mates or seeking extra-pair copulations may entail little cost for poor quality individuals that would anyway have achieved little additional mating success. There may also be intra-specific variation in the thermal tolerance of embryos. Embryos within relatively well-provisioned eggs may be better able to withstand chilling or prolonged incubation periods, without adverse consequences for their subsequent development or survival. Egg composition could therefore influence incubation costs, and the possibility that the resources that parents are able to invest in egg-production may influence those required for incubation as yet remains unexplored. Furthermore, little is known about how embryo gender may affect incubation demands, and sensitivity to varying conditions. Thus there is now clear evidence that incubation can be a costly activity. What is needed is further experimental research into state-dependent effects and interactions with the demands of other reproductive stages.

Acknowledgements

Mark Bolton kindly allowed us to quote his unpublished data, and Graeme Ruxton provided useful discussion of the ideas presented in this chapter.

Appendix 21.1

A fitness cost of any reproductive event, or phase within that event, arises when the resources required are sufficient to significantly constrain the resources available for investment in other phases or events (Roff 1992; Stearns 1992). Individuals, however, differ in their capacity to acquire resources. Hence, it may often be the case that individuals investing most in one activity also invest most in another, resulting in positive rather than the expected negative relationships between resource allocations. To eliminate such correlations, conclusive evidence for the existence of resource trade-offs and consequent fitness costs can only be provided by experiment. Quantitative genetic experiments can be used to demonstrate genetic correlations between life-history components, and phenotypic consequences of experimentally manipulated resource allocation patterns can demonstrate changes in individual fitness (Reznick *et al.* 1986; Partridge 1992). Although both genetic methods and phenotypic manipulations have limitations (Reznick 1992), only phenotypic methods have as yet been used to estimate fitness costs of avian incubation.

Phenotypic consequences of manipulating resource allocation patterns may be manifested in parents or offspring, during current or subsequent reproductive phases. Whilst the inclusive fitness cost of a resource allocation is the quantity that selection will act to minimise, this cost is determined by the sum of all individual phenotypic consequences, and empirical measurement is difficult. Thus in practice, fitness costs of reproduction are usually estimated by examining a small number of phenotypic parameters.

In the context of demonstrating incubation costs, the possible means of manipulating resource trade-offs and measures of subsequent reproductive performance are shown in Figure 21.1. Details and outcomes of all studies are discussed in the main text. To conclusively demonstrate costs of incubation, manipulations must be confined to incubation periods. Whilst the consequences of manipulating incubation demands for incubation parameters are reasonably well documented (reviewed in Thomson *et al.* 1998b), their impact on the subsequent reproductive phases that may more accurately reflect overall fitness consequences remain relatively unexplored.

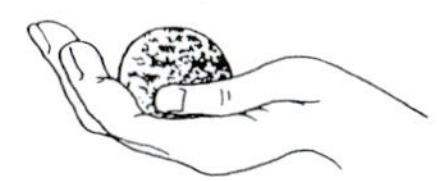

Fig. 21.1 Examples of the means by which incubation demands can be manipulated and of the current and future reproductive parameters that could be inspected for evidence of costs. Manipulations that have been performed and consequences that have been observed are in bold type.

22 *Perspectives in avian incubation*

D. C. Deeming

Throughout the 20th Century avian eggs and incubation has been an area of considerable research effort. Studies concentrated on domestic fowl (*Gallus gallus*) driven, particularly in the first 50–60 years of the century, by a need to improve hatchability of poultry eggs under artificial incubation. These studies were largely successful and modern commercial incubation is able to achieve high hatchability within some very large incubators. Interest in natural incubation has been particularly strong during the last 30–40 years, and although the domestic fowl remained the species of choice for many laboratory studies, incubation in nests was studied for a wide range of species. In particular, the physiology of the avian egg during incubation was extensively studied (see Rahn and Paganelli 1990). The contents of this book are a reflection of our broad knowledge of incubation in nests. However, the book has also revealed our lack of knowledge and understanding about many aspects of contact incubation. Each chapter has highlighted at least one area of research where further work is needed to extend our comprehension of natural incubation. Some of these (albeit a personal choice) are briefly discussed below.

Areas for further research into incubation

The evolution of reproduction in birds, and in particular contact incubation, is certainly of great interest. It is unfortunate that eggs and eggshell fragments appear to be the only fossil evidence of avian reproduction (Chapter 1). However, if information about water vapour conductance could be gleaned from fossil eggshells, as has been the case in dinosaurs, then this would provide a useful insight into the nesting environment when such eggs were laid. There is certainly scope for a systematic study of eggs throughout the evolutionary history of birds.

Nests perform many functions in avian reproduction ranging from containers for eggs and offspring through to night time roosts (Chapter 2). How nest morphology is integrated into the ambient temperature, egg mass and parental behaviour is poorly understood. Further research could focus on how the thermal characteristics of nests are linked with parental behaviour and energetics, and how egg characteristics mould the bird-nest incubation unit. Avian eggs are extremely well studied, especially for the fowl (Chapter 3) but questions still remain. For instance, it is far from clear how egg size and shape are influenced by the amounts of yolk and albumen. Such knowledge could assist in understanding how the avian embryo is able to assess the size of the egg in which it develops (Chapter 4). How far do genetics or resources determine rates of embryonic growth?

Incubation behaviour could be seen as being rather uninteresting to study given that birds spend much of their time simply sitting on eggs. However, the mechanisms controlling the initiation and maintenance of incubation are varied and can be species specific (Chapter 5). Studies of the reproductive endocrinology in a wider range of species of birds will surely provide a greater insight into the control systems of incubation behaviour. How thermal characteristics of different sized eggs affect incubation attentiveness (Chapter 6) is certainly a new area for further investigation. It would be extremely interesting to test the idea that the cooling rate of the upper surface of the eggshell is correlated with mass in contact incubated eggs. The idea that there is gaseous communication between embryos and incubating adults is intriguing (Chapter 7). The idea that the concentration of nitrous oxide can influence parental behaviour is certainly worthy of considerable further study.

Although our understanding of the morphology and development of the brood patch is good (Chapter 8), there remain many aspects of its function which are unclear. In particular, there is little data on how the application of the brood patch certainly produces heterogeneity in the thermal characteristics of the egg contents and the control mechanisms behind brood patch function need to be investigated further. As our understanding of the thermal relations between the brood patch and the egg improves (Chapter 9), it is becoming clear that studies of avian energetics during incubation may need to take into account the thermal characteristics of eggs under contact incubation rather than under heating by convection. The interaction between the brood patch and the egg within a nest environment (Chapter 10) has not been fully considered. This merits further investigation in a wide variety of types of nests with particular emphasis in determining much of the eggshell is covered by a brood patch, and how this varies between species, as well as how this coverage of the egg affects gaseous exchange across the shell. Compilation of reports of egg turning in nests (Chapter 11) highlighted the considerable variation between species and the correlation with egg contents. Further study of the patterns of egg turning throughout incubation in a wide variety of species is certainly merited. The review of the microbiology of natural incubation (Chapter 12) certainly revealed a wide range of microbial flora surviving within the nesting environment. It would be very interesting to see whether organisms such as *Bacillus licheniformes* are widespread in avian orders or whether they are limited to waterfowl. It would also be interesting to see whether microbial degradation of the shell accessory material is widespread in birds and whether this is important in changing gas conductance as incubation progresses.

Of those species exhibiting unusual nesting behaviour, the megapodes are most divergent from the basic avian pattern (Chapter 13). As such, incubation in some species has been extensively studied but there are many other species that could yield further invaluable information. In particular, burrow nesting is relatively poorly investigated and it would be interesting to compare the relationship between shell morphology and porosity and the nesting environment, with that found in mound-nesting species. Hummingbirds represent another extreme in avian biology (Chapter 14). Incubation in these birds has been influenced by their small size and although resorting to torpor during incubation may be rare, their embryos seem adapted to survive this significant drop in incubation temperature. It would be interesting to compare the incubation strategies of other small birds that inhabit different climatic and geographical areas. Modelling of the time and heat allocations of intermittent incubation (Chapter 15) has provided a useful insight into the incubation pattern of one passerine species but such techniques need to be applied to other species. It would be particularly interesting to examine the effects of body mass or egg mass on the model. Birds nesting in extreme climates have developed numerous behavioural and physiological adaptations to cope with adverse temperature or humidity (Chapter 16). It remains unclear whether differences between populations of the same species nesting under different conditions represent genetic or phenotypic changes. Despite the many studies of incubation strategies of brood parasites (Chapter 17) almost nothing is known about the rate of embryonic development in these species. For instance, cuckoo eggs can often be larger than the host eggs and yet will hatch first. Are brood parasite embryos more advanced at oviposition or do they have a high genetic rate of growth?

Of the ecological and evolutionary aspects of incubation (Chapter 18) the relationships between the timing of initiation of incubation and hatching synchrony have not been fully investigated. It would be very interesting to assess rates of embryonic development in clutches under natural incubation which exhibit hatching synchrony to see whether this trait is really established because all eggs initiate embryonic development at the onset of incubation, or is promoted more by embryo–embryo communication close to hatching. Egg coloration has long fascinated oologists and many hypotheses have been developed to explain the variation in egg colour (Chapter 19). Many of these ideas have yet to be fully tested. The energetics of incubation has only been examined in a few species (Chapter 20) and this has to be extended in as many species as possible so that a better understanding of the energetic costs of incubation in different environments can be fully assessed. Fitness costs during incubation range from time spent away from the nest, energetic costs and limitations on mating opportunities (Chapter 21). Further study of this relatively new approach could concentrate on the effects on overall reproductive fitness of the involvement of male birds in contact incubation or feeding of mates.

The future for incubation research?

Often research emphasis is placed on the more unusual aspects of avian reproduction, e.g. brood parasitism or nesting in adverse climatic conditions, or studies are concentrated on species which are relatively easy to keep in captivity or to study in their natural habitat. As such, there are considerable data on incubation in Passeriformes, Charadriiformes, Galliformes and Anseriformes. There is sometimes a danger that researchers dealing with each type of bird place too much emphasis on their own species and fail to appreciate the wider picture. For instance, a new look at the relationships across a wide groups of birds strongly suggests that minimum %attentiveness in single-gender incubation is a function of egg mass irrespective of bird type (Chapter 6). Indeed, the conservative nature of avian egg characteristics strongly suggests that on the whole, the nest environment is very similar in most species investigated (Chapter 10; Chapter 16). It can be hoped that future research into avian incubation will aim to investigate the numerous 'ordinary' species in order to determine whether the unusual species are representative.

It is noteworthy that this volume has little data on the molecular genetics. This is largely because, at present, avian incubation does not readily lend itself to this type of study. It does not indicate that research into incubation is not of high scientific value or interest. There are far too many gaps in our knowledge of the basic biology of incubation for a refined tool, like molecular biology, to yield much information of significance. There are some interesting exceptions to this trend, e.g. the phylogenetic relationships between nest type in swallows (Winkler and Sheldon 1993), and the use of DNA marker to determine parentage of brown-headed cowbird (*Molothrus ater*) eggs (Alderson *et al.* 1999), where molecular techniques enhance the more holistic approach required in incubation science. It is likely, however, that as research progresses, molecular techniques will perhaps become more useful but first our understanding of the basic biology of this critical part of avian reproduction has to vastly improve. I do hope that the relative unsuitability of avian incubation for molecular techniques does not preclude it from receiving funding for it to continue in the future.

I hope that this book will serve as a useful review text for researchers in the next few years as they develop new areas of research hopefully following many of the suggestions made by the contributors here. It is my sincerest hope that this research will continue in many places around the world for, as is shown above and throughout the book, we still have a lot to learn. A better understanding of natural incubation may prove invaluable when it comes to conservation of birds because it could help in ensuring captive breeding programmes are successful, and in conserving habitats which promote breeding in the wild.

References

Acharya, A. V. R. L. N. and Menon, G. K. (1984). Observations on 17b-hydroxysteroid dehydrogenase in the brood patch of house sparrow, *Passer domesticus* L. *Current Science*, **53**, 160–2.

Ackerman, R. A. (1994). Temperature, time, and reptile egg water exchange. *Israel Journal of Zoology*, **40**, 293–306.

Ackerman, R. A., Dmi'el, R. and Ar, A. (1985). Energy and water vapor exchange by parchment-shelled reptile eggs. *Physiological Zoology*, **58**, 129–37.

Ackerman, R. A. and Seagrave, R. C. (1984). Parent–egg interactions. Egg temperature and water loss. In: *Seabird Energetics* (eds G. C. Whittow and H. Rahn), pp. 73–88. Plenum Press, New York.

Ackerman, R. A., Whittow, G. C., Paganelli, C. V. and Pettit, T. N. (1980). Oxygen consumption, gas exchange and growth of embryonic wedge-tailed shearwaters (*Puffinus pacificus chlororhynchus*). *Physiological Zoology*, **53**, 210–21.

Addicott, A. B. (1938). Behavior of the bush-tit in the breeding season. *Condor*, **40**, 49–63.

Afton, A. D. (1977). *Aspects of reproductive behavior in the northern shoveler*. MSc thesis, University of Minnesota, Minneapolis.

Afton, A. D. (1979). Incubation temperatures of the northern shoveler. *Canadian Journal of Zoology*, **57**, 1052–6.

Afton, A. D. (1980). Factors affecting incubation rhythms of northern shovelers. *Condor*, **82**, 132–7.

Afton, A. D. and Paulus, S. L. (1992). Incubation and brood care. In: *Ecology and Management of Breeding Waterfowl* (eds B. D. J. Batt, A. D. Afton, M. G. Anderson, C. D. Ankney, D. H. Johnson, J. A. Kadlec and G. L. Krapu), pp. 62–108. University of Minnesota Press, Minneapolis.

Ahmadjian, V. (1993). *The Lichen Symbiosis*. New York. John Wiley and Sons.

Aichorn, A. (1989). Nestbautechnik des Sneefinken (*Montifringilla nivalis*). *Egretta*, **32**, 49–63.

Ainley, D. G., LeResche, R. E. and Sladen, W. J. L. (1983). *Breeding Biology of the Adélie Penguin*. University of California Press, Berkeley.

Albrecht, D. J. and Oring, L. W. (1995). Song in Chipping Sparrows *Spizella passerina* – structure and function. *Animal Behaviour*, **50**, 1233–41.

Alderson, G. W., Gibbs, H. L. and Sealy, S. G. (1999). Determining the reproductive behaviour of individual brown-headed cowbirds using microsatellite DNA markers. *Animal Behaviour*, **58**, 895–905.

Aldrich, T. W. and Raveling, D. G. (1983). Effects of experience and body weight on incubation behavior of Canada geese. *Auk*, **100**, 670–9.

Ali, S. A. (1931). The origin of mimicry in cuckoo eggs. *Journal of the Bombay Natural History Society*, **34**, 1067–70.

Alvarez, F. (1994). Cuckoo predation on nests of nearest neighbours of parasitised nests. *Ardea*, **82**, 269–70.

Amat, J. A., Visser, G. H., Pérez-Hurtado, A. and Arroyo, G. M. (2000). Brood desertion by female shorebirds: a test of the differential parental capacity hypothesis on Kentish plovers. *Proceedings Of The Royal Society Of London Series B*, **267**, 2171–6.

Andersen, Ø. and Steen, J. B. (1986). Water economy in bird nests. *Journal of Comparative Physiology B*, **156**, 823–8.

Anderson, D. J., Stoyan, N. C. and Ricklefs, R. E. (1987). Why are there no viviparous birds? A comment. *American Naturalist*, **130**, 941–7.

Anderson, G. B., Carter, T. C. and Morley Jones, R. (1970). Some factors affecting the incidence of cracks in hens' egg shells. *British Poultry Science*, **11**, 103–16.

Anderson, S. J. and Deeming, D. C. (2001). Dimensions and composition of eggs from captive-bred houbara (*Chlamydotis undulata*), rufous-crested (*Eupodotis ruficrista*), and Kori (*Ardeotis kori*) bustards (Gruiformes: Otididae). *Zoo Biology*, in press.

Anderson, T. R. (1997). Intermittent incubation during egg laying in house sparrows. *Wilson Bulletin*, **109**, 324–8.

Andersson, M. (1976). Clutch size in the Long-tailed Skua *Stercorarius longicaudus*: some field experiments. *Ibis*, **118**, 586–8.

Andersson, M. (1994). *Sexual Selection*. Princeton University Press, Princeton.

Andrén, H. (1991). Predation: an overrated factor for over-dispersion of birds' nests? *Animal Behaviour*, **41**, 1063–9.

Angilletta, M. J., Winters, R. S. and Dunham, A. E. (2000). Thermal effects on the energetics of lizard embryos: implications for hatchling phenotypes. *Ecology*, **81**, 2957–68.

Ankney, C. D. and Johnson, S. L. (1985). Variation in weight and composition of brown-headed cowbird eggs. *Condor*, **87**, 296–9.

Ar, A. (1991a). Roles of water in avian eggs. In: *Egg Incubation: Its Effects on Embryonic Development in Birds and Reptiles* (eds D. C. Deeming and M. W. J. Ferguson), pp. 229–43. Cambridge University Press, Cambridge.

Ar, A. (1991b). Egg water movements during incubation. In: *Avian Incubation* (ed. S. G. Tullett), pp. 157–73. Butterworth-Heinemann, London.

Ar, A. (1993). Gas exchange of the avian embryo at altitude – the half empty glass. *Funktionsanalyse biologischer Systeme*, **23**, 339–50.

Ar, A., Arieli, B., Belinsky, A. and Yom-Tov, Y. (1987). Energy in avian eggs and hatchlings: utilization and transfer. *Journal of Experimental Zoology*, **Supplement 1**, 151–64.

Ar, A. and Gefen, E. (1998). Further improving hatchability in artificial incubation of ostrich eggs. In: *Ratites in a Competitive World. Proceedings of the 2nd International Ratite Congress* (ed. F. W. Huchzermeyer), pp. 160–6. Oudtshoorn, South Africa.

Ar, A. and Girard, H. (1989). Anisotropic gas diffusion in the shell membrane of the hen's egg. *Journal of Experimental Zoology*, **251**, 20–6.

Ar, A., Girard, H. and Rodeau, J. L. (1991). Oxygen uptake and chorioallantoic blood flow changes during acute hypoxia and hyperoxia in the 16 d chicken embryo. *Respiration Physiology*, **83**, 295–312.

Ar, A., Ifergan, O., Reizis, A., Zelik, L. and Feldman, A. (2000). Does nitric oxide (NO) play a role in embryo-bird communication during incubation? *Avian and Poultry Biology Reviews*, **11**, 284.

Ar, A. and Mover, H. (1994). Oxygen tensions in developing embryos: system inefficiency or system requirement? *Israel Journal of Zoology*, **40**, 307–26.

Ar, A., Paganelli, C. V., Reeves, R. B., Greene, D. G. and Rahn, H. (1974). The avian egg: water vapour conductance, shell thickness, and functional pore area. *Condor*, **76**, 153–8.

Ar, A. and Piontkewitz, Y. (1992). Nest ventilation explains gas composition in the nest-chamber of the European bee-eater. *Respiration Physiology*, **87**, 407–18.

Ar, A. and Rahn, H. (1978). Interdependence of gas conductance, incubation length and weight of the avian egg. In: *Respiratory Function in Birds, Adult and Embryonic* (ed. J. Piiper), pp. 227–38. Springer, Berlin.

Ar, A. and Rahn, H. (1980). Water in the avian egg: overall budget of incubation. *American Zoologist*, **20**, 373–84.

Ar, A. and Rahn, H. (1985). Pores in avian eggshells: gas conductance, gas exchange and embryonic growth rate. *Respiration Physiology*, **61**, 1–20.

Ar, A., Rahn, H. and Paganelli, C. V. (1979). The avian egg: mass and strength *Condor*, **81**, 331–7.

Ar, A. and Sidis, I. (1991). The role of the nest in egg and bird thermoregulation. *International Hatchery Practice*, **5**(**8**), 22.

Ar, A. and Yom-Tov, Y. (1978). The evolution of parental care in birds. *Evolution*, **32**, 655–69.

Ar, A. and Yom-Tov, Y. (1985). Thermal conductance of bird nests – generalizations based on body mass and ambient temperature. *The Physiologist*, **28**, 273.

Arad, Z., Gavrieli-Levin, I. and Marder, J. (1988). Adaptation of the pigeon egg to incubation in dry hot environments. *Physiological Zoology*, **61**, 293–300.

Arcese, P. and Smith, J. N. M. (1999). Impacts of nest predation and brood parasitism on the productivity of North American passerines. In: *Proceedings of the 22nd International Ornithological Congress* (eds N. J. Adams and R. H. Slotow), pp. 2953–66. Durban, 1998. BirdLife South Africa, Johannesburg.

Arcese, P., Smith, J. N. M. and Hatch, M. I. (1996). Nest predation by cowbirds and its consequences for passerine demography. *Proceedings of the National Academy of Sciences USA*, **93**, 4608–11.

Arieli, B. (1983). *Water and calories in eggs and hatchlings*, MSc thesis, Tel Aviv University, Israel.

Arnold, T. W. (1987). Conspecific egg discrimination in American coots. *Condor*, **89**, 675–6.

Arnold, T. W., Rohwer, F. C. and Armstrong, T. (1987). Egg viability, nest predation, and the adaptive significance of clutch size in prairie ducks. *American Naturalist*, **130**, 643–53.

Aschoff, J. and Pohl, H. (1970). Der Ruheumsatz von Vögeln als Funktion der Tageszeit und der Körpergrösse. *Journal für Ornithologie*, **111**, 38–47.

Astheimer, L. B. (1991). Embryo metabolism and egg neglect in Cassin's auklets. *Condor*, **93**, 486–95.

Astheimer, L. B., Buttemer, W. B. and Wingfield, J. C. (1992). Interactions of corticosterone with feeding, activity, and metabolism in passerine birds. *Ornis Scandinavica*, **23**, 355–65.

Astheimer, L. B., Buttemer, W. A. and Wingfield, J. C. (1995). Seasonal and acute changes in adrenocortical responsiveness in an Arctic-breeding bird. *Hormones and Behavior*, **26**, 442–57.

Astheimer, L. B. and Grau, C. R. (1990). A comparison of yolk growth rates in seabird eggs. *Ibis*, **132**, 380–94.

Astheimer, L. B., Prince, P. A. and Grau, C. R. (1985). Egg formation and the prelaying period of black-browed and grey-headed albatrosses *Diomedea melanophris* and *D. chrystostoma* at Bird Island, South Georgia. *Ibis*, **127**, 523–9.

Aulie, A. and Moen, P. (1975). Metabolic thermoregulatory responses in eggs and chicks of willow ptarmigan (*Lagopus lagopus*). *Comparative Biochemistry and Physiology*, **51A**, 605–9.

Aulie, A. and Tøien, O. (1988). Threshold for shivering in aerobic and anaerobic muscles in bantam cocks and incubating hens. *Journal of Comparative Physiology B*, **158**, 431–5.

Austin, G. T. (1974). Nesting success of the cactus wren in relation to nest orientation. *Condor*, **76**, 216–7.

Austin, O. L., Jr and Singer, A. (1961). *Birds of the World*. Golden Press, New York.

Babiker, E. M. and Baggott, G. K. (1991). The acid-base status of the sub-embryonic fluid of the Japanese quail, *Coturnix coturnix japonica*. In: *Avian Incubation* (ed. S. G. Tullett), p. 326. Butterworths-Heinemann, London.

Babiker, E. M. and Baggott, G. K. (1992). Effect of turning upon the sub-embryonic fluid and albumen of the egg of the Japanese quail. *British Poultry Science*, **33**, 973–91.

Babiker, E. M. and Baggott, G. K. (1995). The role of ion transport in the formation of sub-embryonic fluid by the embryo of the Japanese quail. *British Poultry Science*, **36**, 371–83.

Badyaev, A. V., Gibson, D. D. and Kessel, B. (1996). White Wagtail (*Motacilla alba*) and Black-backed Wagtail (*Motacilla lugens*). In: *The Birds of North America, No. 236–7* (eds A. Poole and F. Gill). The Academy of Natural Sciences, Philadelphia and, The American Ornithological Union, Washington DC.

Baerends, G. P. (1970). A model of the functional organisation of incubation behaviour. *Behaviour*, **Supplement 17**, 265–312.

Baerends, G. P. and Drent, R. H. (1970). The herring gull and its egg. *Behaviour*, Supplement **17**, 1–312.

Baerends, G. P., Drent, R. H., Glas, P. and Groenewold, H. (1970). A ethological analysis of incubation behaviour in the herring gull. *Behaviour*, **Supplement 17**, 137–235.

Baerends, G. P. and A. J. Hogan-Warburg. (1982). The external morphology of the egg and its variability. In: *The Herring Gull and Its Egg. Part II The Responsiveness to Egg-features* (ed. G. P. Baerends and R. H. Drent). pp. 1–32. *Behaviour*, **82**.

Baggott, G. K. and Graeme-Cook, K. (1997). Variable shell conductance during natural incubation. *Poultry and Avian Biology Reviews*, **8**, 158.

Baicich, P. J. and Harrison, C. J. O. (1997). *A Guide to the Nests, Eggs and Nestlings of North American Birds, 2nd Ed.* Academic Press, New York.

Bailey, E. D. and Ralph, K. M. (1975). The effects of embryonic exposure to pheasant vocalizations in later call identification by chicks. *Canadian Journal of Zoology*, **53**, 1028–34.

Bailey, R. E. (1952). The incubation patch of passerine birds. *Condor*, **54**, 121–36.

Bailey, R. E. (1955). The incubation patch in tinamous. *Condor*, **57**, 301–3.

Baker, E. C. S. (1913). The evolution of adaptation in parasitic cuckoos' eggs. *Ibis*, **1**, 384–98.

Baker, E. C. S. (1942). *Cuckoo Problems*. Witherby, London.

Baker, G. C. (1999). Temporal and spatial patterns of laying in the Moluccan megapode *Eulipoa wallacei* (G.R. Gray). In: *Proceedings of the Third International Megapode Symposium, Nhill, Australia* (eds R.W.R.J. Dekker, D.N. Jones and J. Benshemesh), *Zoologische Verhandelingen* **327**, 53–9.

Baker, G. C. and Dekker, R. W. R. J. (2000). Lunar synchrony in the reproduction of the Moluccan megapode *Megapodius wallacei. Ibis* 142, 382–8.

Bakken, G. S. (1976). An improved method for determining thermal conductance and equilibrium body temperature with cooling curve experiments. *Journal of Thermal Biology*, **1**, 169–75.

Bakken, G. S. and Gates, D. M. (1975). Heat-transfer analysis of animals: Some implications for field ecology, physiology, and evolution. In: *Perspectives of Biophysical Ecology* (eds D. M. Gates and R. B. Schmerl), pp. 255–90. Springer-Verlag, New York.

Bakken, G. S., Vanderbilt, V. C., Buttemer, W. A. and Dawson, W. R. (1978). Avian eggs: Thermoregulatory value of very high near-infrared reflectance. *Science*, **200**, 321–3.

Baldwin, P. H. and Zaczkowski, N. K. (1963). Breeding biology of the Vaux swift. *Condor*, **65**, 400–6.

Ball, G. F., Dufty, A. M., Goldsmith, A. R. and Buntin, J. D. (1988). Autoradiographic localization of brain prolactin receptors in a parental and non-parental songbird species. *Society for Neuroscience, Abstracts*, **14**, 88.

Baltin, S. (1969). Zur biologie und ethologie des Talegalla-Huhnes (*Alectura lathami* Gray) unter besonderer Berucksichtigung des verhaltens wahrend der brutperiode. *Zeitschrift fur Tierpsychologie* **26**, 524–72.

Baltosser, W. H. (1996). Nest attentiveness in hummingbirds. *Wilson Bulletin*, **108**, 228–45.

Baltosser, W. H. (1996). Costa's Hummingbird. In: *Birds of North America, No. 251* (eds A. Poole and P. Stettenheim). American Ornithologists' Union and Academy of Natural Sciences, Philadelphia.

Baltosser, W. H. and Russell, S. M. (2000). Black-chinned Hummingbird. In: *Birds of North America, No. 495* (eds A. Poole and P. Stettenheim). American Ornithologists' Union and Academy of Natural Sciences, Philadelphia.

Bancroft, G. T. (1985). The influence of total nest failures and partial losses on the evolution of asynchronous hatching. *American Naturalist*, **126**, 495–504.

Barbraud, C., Delord, K. C., Micol, T. and Jouventin, P. (1999). First census of breeding seabirds between Cap Bievenue (Terre Adélie) and Moyes Islands (King George V Land), Antarctica: New records for Antarctic seabird populations. *Polar Biology*, **21**, 146–51.

Barkow, H. C. L. (1830). Anatomisch-physiologische Untersuchungen über das Schlagadersystem der Vögel. Leipzig.

Bart, J. and Tornes, A. (1989). Importance of monogamous male birds in determining reproductive success: evidence for house wrens and a review of male removal studies. *Behavioural Ecology and Sociobiology*, **24**, 109–16.

Barta, Z. and Székely, T. (1997). The optimal shape of avian eggs. *Functional Ecology*, **11**, 656–62.

Bartholomew, G. A. (1982). Physiological control of body temperature. In: *Biology of the Reptilia, Volume 12* (eds C. Gans and F. H. Pugh), pp. 167–212. Academic Press, London.

Bartholomew, G. A. and Dawson, W. R. (1979). Thermoregulatory behavior during incubation in Heermann's gulls. *Physiological Zoology*, **52**, 422–37.

Bartholomew, G. A. and Howell, T. R. (1964). Experiments on nesting behaviour of Laysan and black-footed albatrosses. *Animal Behaviour*, **12**, 549–59.

Bartholomew, G. A. and Lasiewski, R. C. (1965). Heating and cooling rates, heart rate and simulated diving in the Galapagos marine iguana. *Comparative Biochemistry and Physiology*, **16**, 573–82.

Bassett, S. M., Potter, M. A. and Fordham, R. A. (1996). Internal temperature and egg rotation in naturally incubated emu eggs. In: *Improving Our Understanding Of Ratites In A Farming Environment* (ed. D. C. Deeming), pp. 148–9. Ratite Conference Books, Oxfordshire.

Batt, B. D. J. and Cornwell, G. W. (1972). The effects of cold on mallard embryos. *Journal of Wildlife Management*, **36**, 745–51.

Beason, R. C. and Franks, E. C. (1974). Breeding behavior of the horned lark. *Auk*, **91**, 65–74.

Bech, C. and Reinertsen, R. E. (1989). *Physiology of Cold Adaptation in Birds*. NATO ASI Series A: Life Sciences, Volume 173. Plenum Press, New York.

Bédécarrats, G., Guémené, D. and Richard-Yris, M. A. (1997). Effects of environmental and social factors on incubation behavior, endocrinological parameters, and production traits in turkey hens *(Meleagris gallopavo)*. *Poultry Science*, **76**, 1307–14.

Beer, C. G. (1961). Incubation and nest building behaviour of black-headed gulls. I: Incubation behaviour in the incubation period. *Behaviour*, **18**, 62–106.

Beer, C. G. (1963). Incubation and nest-building behaviour of black-headed gulls. IV: Nest-building in the laying and incubation period. *Behaviour*, **21**, 155–76.

Beer, C. G. (1965). Clutch size and incubation behaviour in black-billed gulls (*Larus bulleri*). *Auk*, **82**, 1–18.

Beer, C. G. (1966). Incubation and nest-building behaviour of Black-headed Gulls. V: The post-hatching period. *Behaviour*, **26**, 189–214.

Beer, C. G. (1980). The communication behavior of gulls and other seabirds. In: *Behavior of marine animals, Volume 4* (eds J. Burger, B. L. Olla, and H. E. Winn), pp. 169–205. Plenum Press, New York.

Beissinger, S. R., Tygielski, S. and Elderd, B. (1998). Social constraints on the onset of incubation in a neotropical parrot: a nestbox addition experiment. *Animal Behaviour*, **55**, 21–32.

Beissinger, S. R. and Waltman, J. R. (1991). Extraordinary clutch size and hatching asynchrony of a neotropical parrot. *Auk*, **108**, 863–71.

Bell, G. and Koufopanou, V. (1986). The cost of reproduction. *Oxford Surveys in Evolutionary Biology*, **3**, 83–131.

Bellairs, R. and Osmond, M. (1998). *The Atlas of Chick Development*. Academic Press, London.

Belliure, J., Carrascal, L. M. and Diaz, J. A. (1996). Covariation of thermal biology and foraging mode in two Mediterranean Lacertid lizards. *Ecology*, **77**, 1163–73.

Bennett, A. F. and Dawson, W. R. (1979). Physiological responses of embryonic Heermann's gulls to temperature. *Physiological Zoology*, **52**, 413–21.

Bennett, P. M. and Harvey, P. H. (1987). Active and resting metabolism in birds: allometry, phylogeny and ecology. *Journal of Zoology, London*, **213**, 327–63.

Benshemesh, J. (1992). *The Conservation Ecology Of Malleefowl, With Particular Regards To Fire*. PhD thesis, Monash University, Australia.

Berger, A. J. (1951). The cowbird and certain host species in Michigan. *Wilson Bulletin*, **63**, 26–34.

Bergstrom, P. W. (1989). Incubation temperatures of Wilson's plovers and killdeers. *Condor*, **91**, 634–41.

Berlin, K. E. and Clark, A. B. (1998). Embryonic calls as care-soliciting signals in budgerigars, *Melopsittacus undulatus*. *Ethology*, **104**, 531–44.

Bertram, B. C. R. (1979). Ostriches recognise their own egg and discard others. *Nature*, **279**, 233–4.

Bertram, B. C. R. (1992). *The Ostrich Communal Nesting System*. Princeton University Press, Princeton.

Bertram, B. C. R. and Burger, A. E. (1981). Are ostrich *Struthio camelus* eggs the wrong colour? *Ibis*, **123**, 207–10.

Beukeboom, L., Dijkstra, C., Daan, S. and Meijer, T. (1988). Seasonality of clutch size determination in the kestrel *Falco tinnunculus*: an experimental approach. *Ornis Scandinavica*, **19**, 41–8.

Biebach, H. (1979). Energetik des Bruetens beim Star (*Sturnus vulgaris*). *Journal für Ornithologie*, **120**, 121–38.

Biebach, H. (1981). Energetic costs of incubation on different clutch sizes in starlings (*Sturnus vulgaris*). *Ardea*, **69**, 141–2.

Biebach, H. (1984). Effect of clutch size and time of day on the energy expenditure of incubating starlings (*Sturnus vulgaris*). *Physiological Zoology*, **57**, 26–31.

Biebach, H. (1986). Energetics of re-warming a clutch in starlings (*Sturnus vulgaris*). *Physiological Zoology*, **59**, 69–75.

Bildstein, K. L. (1993). *White Ibis. Wetland Wanderer*. Smithsonian Institution Press, Washington.

Birchard, G. F. and Kilgore, D. L. (1980). Conductance of water vapor in eggs of burrowing and nonburrowing birds: implications for embryonic gas exchange. *Physiological Zoology*, **53**, 284–92.

Birchard, G. F., Kilgore, D. L., Jr. and Boggs, D. F. (1984). Respiratory gas concentrations and temperatures within the burrows of three species of burrow-nesting birds. *Wilson Bulletin*, **96**, 451–6.

Birkhead, T. (1978). Behavioural adaptations to high density nesting in the common guillemot *Uria aalge*. *Animal Behaviour*, **26**, 321–31.

Birkhead, T. R. and Del Nevo, A. J. (1987). Egg formation and the pre-laying period of the common guillemot *Uria aalge*. *Journal of Zoology, London*, **211**, 83–8.

Birks, S. M. (1996). *Reproductive behavior and paternity in the Australian Brush-turkey*, Alectura lathami. PhD thesis, Cornell University, USA.

Bischoff, C. M. and Murphy, M. T. (1993). The detection of and responses to experimental intraspecific brood parasitism in eastern kingbirds. *Animal Behaviour*, **45**, 631–8.

Bishop, K. D. (1980). Birds of volcanoes – the scrubfowl of West New Britain. *World Pheasant Association*, **5**, 80–90.

Blagosklononv, K. N. (1977). Experimental analysis of the rhythm of incubation in the pied flycatcher (*Ficedula hypoleuca*). *Soviet Journal of Ecology*, **8**, 340–4.

Blankespoor, G. W., Oolman, J. and Uthe, C. (1982). Eggshell strength and cowbird parasitism of red-winged blackbirds. *Auk*, **99**, 363–5.

Board, R. G. (1980). The avian eggshell: a resistance network. *Journal of Applied Bacteriology*, **48**, 303–13.

Board, R. G. (1982). Properties of avian egg shells and their adaptive value. *Biological Review*, **57**, 1–28.

Board, R. G., Ayres, J. C., Kraft, A. A. and Forsythe, R. G. (1964). The microbiological contamination of egg shells and egg packing materials. *Poultry Science*, **43**, 584–95.

Board, R. G., Clay, C., Lock, J. and Dolman, J. (1994). The egg: compartmentalized, aseptically, packaged food. In: *Microbiology of the Avian Egg* (eds R. G. Board and R. Fuller), pp. 43–61. Chapman and Hall, London.

Board, R. G. and Fuller, R. (1974). Non-specific antimicrobial defences of the avian egg, embryo and neonate. *Biological Reviews*, **49**, 15–49.

Board, R. G., Loseby, S. and Miles, V. R. (1979). A note on microbial growth on hen egg-shells. *British Poultry Science*, **20**, 413–20.

Board, R. G., Perrott, H. R., Love, G. and Seymour, R. S. (1982). A novel pore system in the eggshell of Mallee Fowl, *Leipoa ocellata*. *Journal of Experimental Zoology*, **220**, 131–4.

Board, R. G. and Scott, V. D. (1980). Porosity of the avian eggshell. *American Zoologist*, **20**, 339–49.

Board, R. G. and Sparks, N. H. C. (1991). Shell structure and formation in avian eggs. In: *Egg Incubation: Its Effects on Embryonic Development in Birds and Reptiles* (eds D. C. Deeming and M. W. J. Ferguson), pp. 71–86. Cambridge University Press, Cambridge.

Board, R. G., Tullett, S. G. and Perrott, H. R. (1977). An arbitrary classification of the pore systems in avian eggshells. *Journal of Zoology, London*, **182**, 251–65.

Bochenski, Z. (1961). Nesting biology of the Black-necked Grebe. *Bird Study*, **8**, 6–15.

Boersma, P. D. (1982). Why some birds take so long to hatch. *American Naturalist*, **120**, 733–50.

Boersma, P. D. and Stokes, D. L. (1995). Conservation: threats to penguin populations. In: *The Penguins Spheniscidae* (ed. T. D. Williams), pp. 127–39. Oxford University Press, Oxford.

Boersma, P. D., Wheelwright, N. T., Nerini, M. K. and Wheelwright, E. S. (1980). The breeding biology of the fork-tailed storm-petrel (*Oceanodroma furcata*). *Auk*, **97**, 268–82.

Boles, W. E. and Ivison, T. J. (1999). A new genus of dwarf megapode (Galliformes: Megapodiidae) from the Late Oligocene of Central Australia. In: *Avian Paleontology at the Close of the 20th Century* (ed. S. L. Storrs), *Smithsonian Contributions to Paleontology*, **89**, 199–206.

Bolhuis, J. J. (1991). Mechanisms of avian imprinting: A review. *Biological Reviews*, **66**, 303–45.

Bolhuis, J. J. (1999). Early learning and the development of filial preferences in the chick. *Behavioural Brain Research*, **98**, 245–52.

Bolhuis, J. J. and Van Kampen, H. S. (1992). An evaluation of auditory learning in filial imprinting. *Behaviour*, **122**, 195–230.

Bolton, M. (1991). Determinants of chick survival in the lesser black-backed gull: relative contributions of egg size and parental quality. *Journal of Animal Ecology*, **60**, 949–60.

Bond, G. M., Board, R. G. and Scott, V. D. (1988*a*). A comparative study of changes in the fine structure of avian eggshells during incubation. *Zoological Journal of the Linnean Society*, **92**, 105–13.

Bond, G. M., Board, R. G. and Scott, V. D. (1988*b*). An account of the hatching strategies of birds. *Biological Reviews*, **63**, 395–415.

Bond, G. M., Scott, V. D. and Board, R. G. (1986). Correlation of mechanical properties of avian eggshells with hatching strategies. *Journal of Zoology, London*, **209**, 225–37.

Book, C. M. and Millam, J. R. (1991). Influence of nest-box substrate (pine showings vs. artificial turf) on nesting behaviour and prolactin levels in turkey hens (*Meleagris gallopavo*). *Applied Animal Behavior*, **33**, 83–91.

Book, C. M., Millam, J. R., Guinan, M. J. and Kitchell, R. L. (1991). Brood patch innervation and its role in the onset of incubation in the turkey hen. *Physiology and Behavior*, **50**, 281–5.

Boone, R. B. and Mesecar, R. S. (1989). Telemetric egg for use in egg-turning studies. *Journal of Field Ornithology*, **60**, 315–22.

Booth, D. T. (1984). Thermoregulation in neonate mallee fowl *Leipoa ocellata*. *Physiological Zoology*, **57**, 251–60.

Booth, D. T. (1985). Thermoregulation in neonate Brush Turkeys (*Alectura lathami*). *Physiological Zoology*, **58**, 374–79.

Booth, D. T. (1987a). Effect of temperature on development of Mallee Fowl *Leipoa ocellata* eggs. *Physiological Zoology*, **60**, 437–45.

Booth, D. T. (1987b). Metabolic response of malleefowl *Leipoa ocellata* embryos to cooling and heating. *Physiological Zoology*, **60**, 446–53.

Booth, D. T. (1987c). Home range and hatching success of malleefowl *Leipoa ocellata* Gould (Megapodiidae), in Murray mallee near Remark, S.A. *Australian Wildlife Research*. **14**, 95–104.

Booth, D. T. (1988). Shell thickness in megapode eggs. *Megapode Newsletter*, **2**, 13.

Booth, D. T. (1989). Regional changes in shell thickness, shell conductance, and pore structure during incubation in eggs of the mute swan. *Physiological Zoology*, **62**, 607–20.

Booth, D. T. (2000). Incubation of eggs of the Australian broad-shelled turtle, *Chelodina expansa* (Testudinata: Chelidae), at different temperatures: effects on pattern of oxygen consumption and hatchling morphology. *Australian Journal of Zoology*, **48**, 369–78.

Booth, D. T. and Rahn, H. (1990). Factors modifying rate of water loss from birds' eggs during incubation. *Physiological Zoology*, **63**, 697–709.

Booth, D. T. and Seymour, R. S. (1987). Effect of eggshell thinning on water vapor conductance of malleefowl eggs. *Condor*, **89**, 453–9.

Booth, D. T. and Thompson, M. B. (1991). A comparison of reptilian eggs with those of megapode birds. In: *Egg Incubation: Its Effects on Embryonic Development in Birds and Reptiles* (eds D. C. Deeming and M. W. J. Ferguson), pp. 325–44. Cambridge University Press, Cambridge.

Bordage, D. and Savard, J.-P. L. (1995). Black scoter (*Melinitta nigra*). In: *The Birds of North America, No. 177* (eds A. Poole and F. Gill). The Academy of Natural Sciences, Philadelphia: The American Ornithologists' Union, Washington, DC.

Bortolotti, G. R. and Wiebe, K. L. (1993). Incubation behaviour and hatching patterns in the American kestrel *Falco sparverius*. *Ornis Scandinavica*, **24**, 41–7.

Bosque, C. and Bosque, M. T. (1995). Nest predation as a selective factor in the evolution of developmental rate in altricial birds. *American Naturalist*, **145**, 234–60.

Bouverot, P. (1985). Adaptation to Altitude-Hypoxia in Vertebrates. Springer-Verlag, Berlin.

Bowman, D. M. J. S., Panton, W. J. and Head, J. (1999). Abandoned orange-footed scrubfowl (*Megapodius reinwardt*) nests and coast rainforest boundary dynamics during the late Holocene in monsoonal Australia. *Quaternary International*, **59**, 27–38.

Braa, A. T., Moksnes, A. and Røskaft, E. (1992). Adaptations of bramblings and chaffinches towards parasitism by the common cuckoo. *Animal Behaviour*, **43**, 67–78.

Bradfield, J. R. G. (1951). Radiographic studies on the formation of the hen's egg shell. *Journal of Experimental Biology*, **28**, 125–40.

Bradley, M., Johnstone, R., Court, G. and Duncan, T. (1997). Influence of weather on breeding success of peregrine falcons in the Arctic. *Auk*, **114**, 786–91.

Brickhill, J. (1987). Breeding success of mallee-fowl *Leipoa ocellata* in central New South Wales. *Emu*, **87**, 42–5.

Briggs, D. M. and Teulings, E. (1974). Correlation and repeatability between chicken egg shell color and breaking strength. *Poultry Science*, **53**, 1904.

Briggs, D. M. and Williams, C. M. (1975). Shell strength, hatchability, egg production and egg shell pigmentation in the Japanese quail. *Poultry Science*, **54**, 1738.

Brill, R. W., Dewar, H. and Graham, J. B. (1994). Basic concepts relevant to heat transfer in fishes, and their use in measuring the physiological thermoregulatory abilities of tunas. *Environmental Biology of Fishes*, **40**, 109–24.

Briskie, J. V. and Sealy, S. G. (1987). Responses of least flycatchers to experimental inter- and intra-specific brood parasitism. *Condor*, **89**, 899–901.

Briskie, J. V. and Sealy, S. G. (1989). Nest-failure and the evolution of hatching asynchrony in the least flycatcher. *Journal of Animal Ecology*, **58**, 653–65.

Briskie, J. V. and Sealy, S. G. (1990). Evolution of short incubation periods in the parasitic cowbirds, *Molothrus* spp. *Auk*, **107**, 789–94.

Briskie, J. V., Sealy, S. G. and Hobson, K. A. (1992). Behavioral defenses against avian brood parasitism in sympatric and allopatric host populations. *Evolution*, **46**, 334–40.

Broekhuysen, G. J. and Frost, P. G. H. (1968). Nesting behaviour of the black-necked grebe *Podiceps nigricollis* (Brehm) in southern Africa. I. The reaction of disturbed birds. *Bonner Zoologische Beitrage*, **19**, 350–61.

Brooke, M. de L. and Davies, N. B. (1988). Egg mimicry by cuckoos *Cuculus canorus* in relation to discrimination by hosts. *Nature*, **335**, 630–2.

Brooke, M. de L. and Davies, N. B. (1991). A failure to demonstrate host imprinting in the cuckoo (*Cuculus canorus*) and alternative hypotheses for the maintenance of egg mimicry. *Ethology*, **89**, 154–66.

Brooker, L. C. and Brooker, M. G. (1990). Why are cuckoos host specific? *Oikos*, **57**, 301–9.

Brooker, M. G. and Brooker, L. C. (1989). The comparative breeding behaviour of two sympatric cuckoos, Horsfield's bronze-cuckoo *Chrysococcyx basalis* and the shining bronze-cuckoo *C. lucidus*, in Western Australia: A new model for the evolution of egg morphology and host specificity in avian brood parasites. *Ibis*, **131**, 528–47.

Brooker, M. G., Brooker, L. C. and Rowley, I. (1988). Egg deposition by the bronze-cuckoos *Chrysococcyx basalis* and *Ch. lucidus*. *Emu*, **88**, 107–9.

Brooker, M. G. and Brooker, L. C. (1989). Cuckoo host nests in Australia. *Australian Zoological Reviews*, **2**, 1–67.

Brooker, M. G. and Brooker, L. C. (1991). Eggshell strength in cuckoos and cowbirds. *Ibis*, **133**, 406–13.

Brown, C. R. (1994). Nest microclimate, egg temperature, egg water loss, and eggshell conductance in Cape Weavers *Ploceus capensis*. *Ostrich*, **65**, 26–31.

Brown, C. R. and Brown, M. B. (1988). A new form of reproductive parasitism in cliff swallows. *Nature*, **331**, 66–8.

Brown, C. R. and Brown, M. B. (1991). Selection of high-quality host nests by parasitic cliff swallows. *Animal Behaviour*, **41**, 457–65.

Brown, C. R. and Sherman, L. C. (1989). Variation in the appearance of swallow eggs and the detection of intraspecific brood parasitism. *Condor*, **91**, 620–7.

Brown, J. H. and West, G. B. (2000). *Scaling in Biology*. Oxford University Press, New York.

Brown, J. L. and Vleck, C. M. (1998). Prolactin and helping in birds: has natural selection strengthened helping behavior? *Behavioral Ecology*, **9**, 541–5.

Brown, P. W. and Fredrickson, L. H. (1987). Time budget and incubation behavior of breeding white-winged scoters. *Wilson Bulletin*, **99**, 50–5.

Brua, R. B. (1996). Impact of embryonic vocalisations on the incubation behaviour of eared grebes. *Behaviour*, **133**, 145–60.

Brua, R. B., Nuechterlein, G. L. and Buitron, D. (1996). Vocal response of eared grebe embryos to egg cooling and egg turning. *Auk*, **113**, 525–33.

Bruce, J. and Drysdale, E. M. (1991). Egg hygiene: route of infection. In: *Avian Incubation* (ed. S. G. Tullett), pp. 257–76. Butterworth-Heinemann, London.

Bruce, J. and Drysdale, E. M. (1994). Trans-shell transmission. In: *Microbiology of the Avian Egg* (eds R. G. Board and R. Fuller), pp. 63–91. Chapman and Hall, London.

Brugger, C. and Taborsky, M. (1994). Male incubation and its effect on reproductive success in the Black Swan, *Cygnus atratus*. *Ethology*, 96, 138–46.

Brummermann, M. and Reinertsen, R. E. (1991). Adaptation of homeostatic thermoregulation: comparison of incubating and non-incubating Bantam hens. *Journal of Comparative Physiology B*, **161**, 133–40.

Brummermann, M. and Reinertsen, R. E. (1992). Cardiovascular responses to thoracic skin cooling: comparison of incubating and non-incubating Bantam hens. *Journal of Comparative Physiology B*, **162**, 16–22.

Bruning, D. F. (1973). Breeding and rearing rheas in captivity. *International Zoo Yearbook*, **12**, 163–72.

Bryan, S. M. and Bryant, D. M. (1998). Heating nest-boxes reveals an energetic constraint on incubation behaviour in great tits, *Parus major*. *Proceedings of the Royal Society of London B*, **266**, 157–62.

Bryant, D. M. (1978). Establishment of weight hierarchies in the broods of house martins *Delichon urbica*. *Ibis*, **120**, 16–26.

Bryant, D. M. (1979). Reproductive costs in the House Martin (*Delichon urbica*). *Journal of Animal Ecology*, **48**, 655–75.

Bryant, D. M., Hails, C. J. and Tatner, P. (1984). Reproductive energetics of two tropical bird species. *Auk*, **101**, 25–37.

Bryant, D. M. and Tatner, P. (1988). Energetics of the annual cycle of Dippers *Cinclus cinclus*. *Ibis*, **130**, 17–38.

Bucher, T. L. and Chappell, M. A. (1989). Energy metabolism and patterns of ventilation in euthermic and torpid hummingbirds, In: *Physiology of cold adaptation in birds* (eds C. Bech and R. E. Reinertsen), pp. 187–95. Plenum, New York.

Bucher, T. L. and Chappell, M. A. (1992). Ventilatory and metabolic dynamics during entry into and arousal from torpor in *Selasphorus* hummingbirds. *Physiological Zoology*, **65**, 978–93.

Bucher, T. L. and Vleck, C. M. (1997). Patterns of nest attendance and relief in Adélie penguins, *Pygoscelis adeliae*. *Antarctic Journal*, **32**, 103–4.

Buckley, P. A. and Buckley, F. G. (1972). Individual egg and chick recognition by adult royal terns (*Sterna maxima maxima*). *Animal Behaviour*, **20**, 457–62.

Bugden, S. C. and Evans, R. M. (1991). Vocal responsiveness to chilling in embryonic and neonatal American Coots. *Wilson Bulletin*, **103**, 712–7.

Buitron, D. (1988). Female and male specialization in parental care and its consequences in black-billed magpies. *Condor*, **90**, 29–39.

Bull, E. L. and Jackson, J. E. (1995). Pileated Woodpecker (*Dryocopus pileatus*). In: *The Birds of North America, No. 148* (eds A. Poole, P. Stettenheim and F. Gill). The Academy of Natural Sciences, Philadelphia: The American Ornithologists' Union, Washington, DC.

Bundy, G. (1976). Breeding biology of the red-throated divers. *Bird Study*, **23**, 249–56.

Buntin, J. D. (1996). Neural and hormonal control of parental behavior in birds. *Advances in the Study of Behavior*, **25**, 161–213.

Buntin, J. D., El Halawani, M. E., Ottinger, M. A., Fan, Y. and Fivizzani, A. J. (1998). An analysis of sex and breeding stage differences in prolactin binding activity in brain

and hypothalamic GnRH concentration in Wilson's phalarope, a sex-role-reversed species. *General and Comparative Endocrinology*, **109**, 119–32.
Buntin, J. D., Hnasko, R. M., Zuzick, P. H., Valentine, D. L. and Scammell, J. G. (1996). Changes in bioactive prolactin-like activity in plasma and its relationship to incubation behavior in breeding ring doves. *General and Comparative Endocrinology*, **102**, 221–32.
Burger, A. and Williams, A. (1979). Egg temperature of the rockhopper penguin and some other penguins. *Auk*, **96**, 100–5.
Burger, J. (1989). Incubation temperature has long-term effects on behaviour of young Pine Snakes (*Pituophis melanoleucus*). *Behavioural Ecology and Sociobiology*, **24**, 201–7.
Burke, W. and Dennison, P. T. (1980). Prolactin and luteinizing hormone levels in female turkeys *(Meleagris gallopavo)* during a photoinduced reproductive cycle and broodiness. *General and Comparative Endocrinology*, **41**, 92–100.
Burley, R. W. and Vadhera, D. V. (1989). *The Avian Egg, Chemistry and Biology*. John Wiley and Sons, New York.
Burton, F. G. and Tullett, S. G. (1983). A comparison of the effects of eggshell porosity on the respiration and growth of domestic fowl, duck and turkey embryos. *Comparative Biochemistry and Physiology*, **75A**, 167–74.
Burton, F. G. and Tullett, S. G. (1985). The effects of egg weight and shell porosity on the growth and water balance of the chicken embryo. *Comparative Biochemistry and Physiology*, **81A**, 377–85.
Burton, F. G., Stevenson, J. M. and Tullett, S. G. (1989). The relationship between eggshell porosity and air space gas tensions measured before and during the parafoetal period and their effects on the hatching process in the domestic fowl. *Respiration Physiology*, 77, 89–100.
Burtt, E. H. and Ichida, J. M. (1999). Occurrence of feather-degrading bacilli in the plumage of birds. *Auk*, **116**, 364–72.
Butler, R. G. and Janes-Butler, S. (1983). Sexual differences in the behaviour of adult Great black-backed gulls (*Larus marinus*) during the pre- and post-hatch periods. *Auk*, **100**, 63–75.
Buttemer, W. A. and Dawson, T. J. (1989). Body temperature, water flux and estimated energy expenditure of incubating emus (*Dromaius novaehollandiae*). *Comparative Biochemistry and Physiology*, **94A**, 21–4.
Buttemer, W. A. and Dawson, W. R. (1993). Temporal patterns of foraging and microhabitat use by Galapagos marine iguanas, *Amblyrhynchus cristatus*. *Oecologia*, **96**, 56–64.
Cade, T. J., Tobin, C. A. and Gold, A. (1965). Water economy and metabolism of two estrildine finches. *Physiological Zoology*, **38**, 400–13.
Cain, B. W. and McCuistion, R. D. (1977). Incubation temperature of a parasitized Carolina wren nest. *Bulletin of the Texas Ornithological Society*, **10**, 8–10.
Calder, W. A. (1971). Temperature relationships and nesting of the calliope hummingbird. *Condor*, **73**, 314–21.
Calder, W. A. (1972). Piracy of nesting material from and by the Broadtailed Hummingbird. *Condor*, **74**, 485.

Calder, W. A. (1973a). An estimate of the heat balance of a nesting hummingbird in a chilling climate. *Comparative Biochemistry and Physiology*, **46A**, 291–300.
Calder, W. A. (1973b). Microhabitat selection during nesting of hummingbirds in the Rocky Mountains. *Ecology*, **54**, 127–34.
Calder, W. A. (1974). The thermal and radiant environment of a winter hummingbird nest. *Condor*, **76**, 268–73.
Calder, W. A. III. (1975). Factors in the energy budget of mountain hummingbirds. In: *Perspectives of Biophysical Ecology*; *Ecological Studies*: *Analysis and Synthesis* (ed. D. M. Gates), pp. 431–42. Springer-Verlag, New York, Heidelberg.
Calder, W. A. (1981). Heat exchange of nesting hummingbirds in the Rocky Mountains. *National Geographic Society Research Reports*, **13**, 145–69.
Calder, W. A. (1993). Rufous Hummingbird. In: *Birds of North America, No. 53* (eds A. Poole and P. Stettenheim). American Ornithologists' Union, Washington, D.C. and Academy of Natural Sciences, Philadelphia.
Calder, W. A. (1994). When do hummingbirds use torpor in nature? *Physiological Zoology*, **67**, 1051–76.
Calder, W. A. (1996). *Size, Function, and Life History*. Dover, New York.
Calder, W. A. (2000). Diversity and convergence: Scaling for conservation. In: *Scaling in Biology* (eds J. H. Brown and G. B. West), pp. 297–324. Oxford University Press, New York.
Calder, W. A. and Booser, J. (1973). Hypothermia of broad-tailed hummingbirds during incubation in nature with ecological considerations. *Science*, **180**, 751–3.
Calder, W. A. and Calder, L. L. (1992). The Broad-tailed Hummingbird. In: *Birds of North America, No. 16* (eds A. Poole and P. Stettenheim). American Ornithologists' Union and Academy of Natural Sciences, Philadelphia.
Calder, W. A. and Calder, L. L. (1994). The Calliope Hummingbird. In: *Birds of North America, No. 135* (eds A. Poole and P. Stettenheim). American Ornithologists' Union, Washington, D.C. and Academy of Natural Sciences, Philadelphia.
Calder, W. A. III, Hiebert, S. M., Waser, N. M., Inouye, D. W. and Miller, S. (1983). Site-fidelity, longevity, and population dynamics of broad-tailed hummingbirds: a ten-year study. *Oecologia*, **56**, 359–64.
Calder, W. A. and King, J. R. (1974) Thermal and caloric relations of birds. In: *Avian Biology, Volume 4* (eds D. S. Farner, J. R. King, and K. C. Parkes), pp. 260–415. Academic Press, New York and London.
Calder III, W. A., Parr, C. R. and Karl, D. P. (1978). Energy content of the eggs of the brown kiwi *Apteryx australis*: an extreme in avian evolution. *Comparative Biochemistry and Physiology*, **60A**, 177–9.
Caldwell, P. J. and Cornwell, G. W. (1975). Incubation behaviour and temperatures of the mallard duck. *Auk*, **92**, 706–31.
Callebaut, M. (1991). Influence of daily cyclical variations of embryonic incubation temperature on ovarian morphology in the adult Japanese quail. *European Archives of Biology (Bruxelles)*, **102**, 171–3.
Camin, A. M., Chabasse, D. and Guiguen, C. (1998). Keratinophilic fungi associated with starlings (*Sturnus vulgaris*) in Brittany, France. *Mycopathologia*, **143**, 9–12.
Campbell, A. J. (1901). *Nests and eggs of Australian birds*. Private, Sheffield.
Campbell, B. (1974). *The Dictionary of Birds in Colour*. Michael Joseph Ltd., London

Campo, J. L. and Escudero, J. (1984). Relationship between egg-shell colour and two measurements of shell strength in the Vasca breed. *British Poultry Science*, **25**, 467–76.

Carbone, C. and Houston, A. I. (1994). Patterns in the diving behavior of the pochard *Aythya ferina*: a test of an optimality model. *Animal Behavior*, **48**, 457–65.

Carbone, C. and Houston, A. I. (1996). The optimal allocation of time over the dive cycle: an approach based on aerobic and anaerobic respiration. *Animal Behaviour*, **51**, 1247–55.

Carey, C. (1979). Increase in conductance to water vapor during incubation in eggs of two avian species. *Journal of Experimental Zoology*, **209**, 181–6.

Carey, C. (1980). The ecology of avian incubation. *Bioscience,* **30**, 819–24.

Carey, C. (1983). Structure and function of avian eggs. *Current Ornithology,* **1**, 69–103.

Carey, C. (1994). Structural and physiological differences between montane and lowland avian eggs and embryos. *Journal of Biosciences*, **19**, 429–40.

Carey, C., Dunin-Borkowski, O., Leon-Velarde, F., Espinoza, D. and Monge, C. (1993). Blood gases, pH, and hematology of montane and lowland coot embryos. *Respiration Physiology,* **93**, 151–63.

Carey, C., Dunin-Borkowski, O., Leon-Velarde, F., Espinoza, D. and Monge, C. (1994). Gas exchange and blood gases of puna teal *(Anas versicolor puna)* embryos in the Peruvian Andes. *Journal of Comparative Physiology B*, **163**, 649–56.

Carey, C., Dunin-Borkowski, O., Leon-Velarde, F. and Monge, C. (1989a). Shell conductance, daily water loss, and water content of puna teal eggs. *Physiological Zoology,* **62**, 83–95.

Carey, C., Garber, S. D., Thompson, E. L. and James, F. C. (1983). Avian reproduction over an altitudinal gradient. II. Physical characteristics and water loss of eggs. *Physiological Zoology*, **56**, 340–52.

Carey, C., Hoyt, D. F., Bucher, T. L. and Larson, D. F. (1984). Eggshell conductances of avian eggs at different altitudes. In: *Respiration and Metabolism of Embryonic Vertebrates* (ed. R. S. Seymour), pp. 259–70. Dr. W. Junk Publishers, Dordrecht.

Carey, C., Leon-Velarde, F., Castro, G. and Monge, C. (1987). Shell conductance, daily water loss, and water content of Andean gull and puna ibis eggs. *Journal of Experimental Zoology,* **Supplement 1**, 247–52.

Carey, C., Leon-Velarde, F., Dunin-Borkowski, O., Bucher, T. L., de la Torre, G., Espinoza, D. and Monge C. (1989b). Variation in eggshell characteristics and gas exchange of montane and lowland coot eggs. *Journal of Comparative Physiology B,* **159**, 389–400.

Carey, C., Leon-Velarde. F. and Monge, C. (1991). Eggshell conductance and other physical characteristics of avian eggs laid in the Peruvian Andes. *Condor*, **92**, 790–3.

Carey, C. and Morton, M. L. (1971). A comparison of salt and water regulation in California quail (*Lophortyx californicus*) and Gambel's quail (*Lophortyx gambelii*). *Comparative Biochemistry and Physiology,* **39A**, 75–101.

Carey, C. and Morton, M. L. (1976). Aspects of circulatory physiology of montane and lowland birds. *Comparative Biochemistry and Physiology,* **54A**, 61–74.

Carey, C., Rahn, H. and Parisi, P. (1980). Calories, water, lipid and yolk in avian eggs. *Condor*, **82**, 335–43.

Carey, C., Thompson, E. L., Vleck, C. M. and James, F. C. (1982). Avian reproduction over an altitudinal gradient: incubation period, hatchling mass, and embryonic oxygen consumption. *Auk*, **99**, 710–8.

Carpenter, F. L. and Hixon, M. A. (1988). A new function for torpor: fat conservation in a wild migrant hummingbird. *Condor*, **90**, 373–8.

Carpenter, K. (1999). *Eggs, Nests and Baby Dinosaurs. A Look at Dinosaur Reproduction*. Indiana University Press, Bloomington.

Carpenter, K., Hirsch, K. F. and Horner, J. R. (1994). *Dinosaur Eggs and Babies*. Cambridge University Press, Cambridge.

Carroll, J. P. (1993). Gray Partridge (*Perdix perdix*). In: *The Birds of North America, No. 58* (eds A. Poole and F. Gill). The Academy of Natural Sciences, Philadelphia: The American Ornithologists' Union, Washington, DC.

Carter, M. D. (1986). The parasitic behavior of the bronzed cowbird in south Texas. *Condor*, **88**, 11–25.

Carter, T. C. (1975). The hen's egg: Relationships of seven characteristics of the strain of hen to the incidence of cracks and other shell defects. *British Poultry Science*, **16**, 289–96.

Catchpole, C. K. and Leisler, B. (1996). Female Aquatic Warblers (*Acrocephalus paludicola*) are attracted by playback of longer and more complicated songs. *Behaviour*, **133**, 1153–64.

Chadwick, A. (1969). Effects of prolactin in homoeothermic vertebrates. *General and Comparative Endocrinology*, **Supplement 2**, 63–8.

Chamberlain, M. L. (1977). Observation on the red-necked grebe nesting in Michigan. *Wilson Bulletin*, **89**, 33–46.

Chance, E. (1922). *The Cuckoo's Secret*. Sidgwick and Jackson, London.

Chance, E. P. (1940). *The Truth About the Cuckoo*. Country Life, London.

Chandler, C. R., Ketterson, E. D. and Nolan, V., Jr. (1997). Effects of testosterone on use of space by male dark-eyed juncos when their mates are fertile. *Animal Behaviour*, **54**, 543–9.

Chapman, F. M. (1924). *Handbook of Birds of Eastern North America*, Revised edition. D. Appleton and Company, New York.

Chatterjee, S. (1997). *The Rise of Birds: 225 Million Years of Evolution*. Johns Hopkins University Press, Baltimore.

Chattock, A. P. (1925). On the physics of incubation. *Philosophical Transactions of the Royal Society, London, B*, **213**, 397–450.

Chaurand, T. and Weimerskirch, H. (1984). Incubation routine, body mass regulation and egg neglect in the blue petrel *Halobaena caerulea*. *Ibis*, **136**, 285–90.

Cheng, M.-F. and Silver, R. (1975). Estrogen-progesterone regulation of nest building and incubation behavior in ovariectomized ring doves (*Streptopelia risoria*). *Journal of Comparative Physiology and Psychology*, **88**, 256–63.

Cherel, Y., Fréby, F., Gilles, J. and J.-P. Robin. (1993). Comparative fuel metabolism in gentoo and king penguins: adaptation to brief versus prolonged fasting. *Polar Biology*, **13**, 263–9.

Cherel, Y., Gilles, J., Handrich, Y. and Le Maho, Y. (1994). Nutrient reserve dynamics and energetics during long-term fasting in the king penguin (*Aptenodytes patagonicus*). *Journal of Zoology, London*, **234**, 1–12.

Christian, K. A. and Weavers, B. W. (1996). Thermoregulation of monitor lizards in Australia: an evaluation of methods in thermal biology. *Ecological Monographs*, **66**, 139–57.

Cichon, M. (2000). Costs of incubation and immunocompetence in the collared flycatcher. *Oecologia*, **125**, 453–7.

Cichon, M. and Kozlowski, M. (2000). Ageing and typical survivorship curves result from optimal resource allocation. *Evolutionary Ecology Research*, **2**, 857–70.

Clark, A. B. and Wilson, D. S. (1981). Avian breeding adaptations: Hatching asynchrony, brood reduction, and nest failure. *Quarterly Review of Biology*, **56**, 253–77.

Clark, A. B. and Wilson, D. S. (1985). The onset of incubation in birds. *American Naturalist*, **125**, 603–11.

Clark, G. A. (1960). Notes on the embryology and evolution of the Megapodes (Aves: Galliformes). *Yale Peabody Museum Postilla*, **45**, 1–7.

Clark, G. A. (1964). Life histories and the evolution of megapodes. *Living Bird*, **3**, 149–67.

Clark, J. M., Norell, M. A. and Chiappe, L. M. (1999). An oviraptorid skeleton from the Late Cretaceous of Ukhaa Tolgod, Mongolia, preserved in an avianlike brooding position over and Oviraptorid nest. *American Museum Novitates*, **3265**, 1–36.

Clark, L. (1990). Starlings as herbalists: countering parasites and pathogens. *Parasitology Today*, **6**, 358–60.

Clark, L. and Mason, J. R. (1985). Use of nest material as insecticidal and antipathogenic agents by the European starling. *Oecologia*, **67**, 169–76.

Clark, L. and Mason, J. R. (1986). Olfactory discrimination of plant volatiles by the European starling. *Chemical Senses*, **11**, 590–9.

Clark, L. and Mason, J. R. (1987). Olfactory discrimination of plant volatiles by the European starling. *Animal Behaviour*, **35**, 227–35.

Clark, L. and Mason, J. R. (1988). Effect of biologically active plants used as nest material and the derived benefit to starling nestlings. *Oecologia*, **77**, 174–80.

Clark, L. and Smeraski, C. A. (1990). Seasonal shifts in odor acuity by starlings. *Journal of Experimental Zoology*, **255**, 22–9.

Clemens, D. T. (1990). Interspecific variation and effects of altitude on blood properties of rosy finches (*Leucosticte arctoa*) and house finches (*Carpodacus mexicanus*). *Physiological Zoology*, **63**, 288–307.

Clotfelter, E. D. and Yasukawa, K. (1999). The function of onset of nocturnal incubation in red-winged blackbirds. *Auk*, **116**, 417–26.

Cloues, R., Ramos, C. and Silver, R. (1990). Vasoactive intestinal polypeptide-like immunoreactivity during reproduction in doves: influence of experience and number of offspring. *Hormones and Behavior*, **24**, 215–31.

Clutton-Brock, T. H. (1991). *The Evolution of Parental Care*. Princeton University Press, Princeton.

Cobley, N. D. and Shears, J. R. (1999). Breeding performance of gentoo penguins (*Pygoscelis papua*) at a colony exposed to high levels of human disturbance. *Polar Biology*, **21**, 355–60.

Coleman, R. M. and Whittall, R. D. (1988). Clutch size and the cost of incubation in the Bengalese finch (*Lonchura striata*). *Behavioural Ecology and Sociobiology*, **23**, 367–72.

Collias, N. (1964). The evolution of nests and nest building in birds. *American Zoologist*, **4**, 175–90.

Collias, N. E. (1997). On the origin and evolution of nest building by passerine birds. *Condor*, **99**, 253–70.

Collias, N. E. and Collias, E. C. (1964). The evolution of nest building in weaverbirds (Ploceidae). *University of California Publications in Zoology*, 73, 1–162.

Collias, N. E. and Collias, E. C. (1984). *Nest Building and Bird Behaviour*. Princeton University Press, Princeton.

Conder, P. J. (1948). The breeding biology and behaviour of the continental goldfinch *Carduelis carduelis carduelis*. *Ibis*, **90**, 493–525.

Conway, C. J. and Martin, T. E. (2000a). Effects of ambient temperature on avian incubation behavior. *Behavioral Ecology*, **11**, 178–88.

Conway, C. J. and Martin, T. E. (2000b). Evolution of passerine incubation behaviour: influence of food, temperature and nest predation. *Evolution*, **54**, 670–85.

Cooper, J. A. (1979). Trumpeter swan nesting behaviour. *Wildfowl*, **30**, 55–71.

Corley Smith, G. T. (1969). A high altitude hummingbird on the Volcano Cotopaxi. *Ibis*, 111, 17–22.

Cott, H. B. (1940). *Adaptive Coloration in Animals*. Methuen and Co. Ltd., London.

Cott, H. B. (1948). Edibility of the eggs of birds. *Nature* ,**161**, 8–11.

Cott, H. B. (1951). The palatability of the eggs of birds: Illustrated by experiments on the food preferences of the hedgehog (*Erinaceus eropaeus*). *Proceedings of the Zoological Society of London*, **121**, 1–41.

Cott, H. B. (1952). The palatability of the eggs of birds: Illustrated by three seasons' experiments (1947, 1948 and 1950) on the food preferences of the rat (*Rattus norvegicus*); and with special reference to the protective adaptations of eggs considered in relation to vulnerability. *Proceedings of the Zoological Society of London*, **122**, 1–54.

Cott, H. B. (1953). The palatability of the eggs of birds: Illustrated by experiments on the food preferences of the ferret (*Putorius furo*) and cat (*Felis catus*); with notes on other egg-eating Carnivora. *Proceedings of the Zoological Society of London*, **123**, 123–41.

Cott, H. B. (1954). The palatability of the eggs of birds: Mainly based upon observations of an egg panel. *Proceedings of the Zoological Society of London*, **124**, 335–463.

Cracraft, J. (1988). The major clades of birds. In: *The Phylogeny and classification of the Tetrapods*. The Systematic Association Special Volume 35(A) (M. J. Benton, Ed.) pp. 339–61, Claredon Press, London.

Craig, J. L. and Jamieson, I. G. (1990). Pukeko: different approaches and some different answers. In: *Cooperative Breeding in Birds: Long-term studies of Ecology and Behaviour* (eds P. B. Stacey and W. D. Koenig), pp. 387–412. Cambridge University Press, Cambridge.

Cramp, S. (1977). *Handbook of the Birds of Europe, the Middle East and North Africa, The Birds of the Western Palearctic, Vol. II Ostrich to Ducks*. Oxford University Press, Oxford.

Cramp, S. (1980). *Handbook of the Birds of Europe, the Middle East and North Africa, The Birds of the Western Palearctic, Vol. II Hawks to Bustards.* Oxford University Press, Oxford.

Cramp, S. (1983). *Handbook of the Birds of Europe, the Middle East and North Africa, The Birds of the Western Palearctic, Vol. III Waders to Gulls.* Oxford University Press, Oxford.

Cramp, S. (1992). *Handbook of the Birds of Europe, the Middle East and North Africa. The Birds of the Western Palearctic. Volume VI. Warblers.* Oxford University Press, Oxford.

Cramp, S. and Perrins, C. M. (1993). *Handbook of the Birds of Europe, the Middle East and North Africa. The Birds of the Western Palearctic. Volume VII. Flycatchers to Shrikes.* Oxford University Press, Oxford.

Cramp, S. and Perrins, C. M. (1994). *Handbook of the Birds of Europe, the Middle East and North Africa, The Birds of the Western Palearctic, Vol. VIII Crows to Finches.* Oxford University Press, Oxford.

Crisóstomo, S., D. Guémené, D., Garreau-Mills, M. and Zadworny, D. (1997). Prevention of the expression of incubation behaviour using passive immunisation against prolactin in turkey hens (*Meleagris gallopavo*). *Reproduction, Nutrition, and Development*, **37**, 253–66.

Crome, F. H. J. and Brown, H. E. (1979). Notes on social organization and breeding of the orange-footed scrubfowl *Megapodius reinwardt. Emu*, **79**, 111–9.

Crook, J. H. (1960). Nest form and construction in certain African weaver birds. *Ibis*, **102**, 1–25.

Crook, J. H. (1963). A comparative analysis of nest structure in the weaver birds (Ploceinae). *Ibis*, **105**, 238–62.

Croxall, J. P. (1982). Energy costs of incubation and moult of petrels and penguins. *Journal of Animal Ecology*, **51**, 177–94.

Cruz, A. and Wiley, J. W. (1989). The decline of an adaptation in the absence of a presumed selection pressure. *Evolution*, **43**, 55–62.

Curson, D. R., Goguen, C. B. and Mathews, N. E. (2000). Long-distance commuting by brow-headed cowbirds in New Mexico. *Auk*, **117**, 795–9.

Daan, S., Deerenberg, C. and Dijkstra, C. (1996). Increased daily work precipitates natural death in the Kestrel. *Journal of Animal Ecology*, **65**, 539–44.

Daan, S., Dijkstra, C. and Tinbergen, J. M. (1990). Family planning in the Kestrel (*Falco tinnunculus*): the ultimate control of variation in laying date and clutch size. *Behaviour*, **114**, 83–116.

Daniel, J. C. (1957). An embryological comparison of the domestic fowl and the red-winged blackbird. *Auk*, **74**, 340–58.

Dareste, C. (1891). *Recherche Sur La Production Artificielle Des Monstruostities Ou Essais De Teratogenie Experimentale*, 2nd Edition. Paris.

Davies, A. K. and Baggott, G. K. (1989a). Clutch size and nesting sites of the Mandarin Duck *Aix galericulata. Bird Study*, **36**, 32–6.

Davies, A. K. and Baggott, G. K. (1989b). Egg-laying, incubation and intraspecific nest parasitism by the Mandarin Duck *Aix galericulata. Bird Study*, **36**, 115–22.

Davies, J. C. and Cooke, F. (1983). Intraclutch hatch synchronization in the lesser snow goose. *Canadian Journal of Zoology*, **61**, 1398–401.

Davies, N. B. (2000). *Cuckoos, Cowbirds and Other Cheats.* T and A D Poyser, London.

Davies, N. B. and Brooke, M. de L. (1988). Cuckoos versus reed warblers: Adaptations and counter adaptations. *Animal Behaviour*, **36**, 262–84.

Davies, N. B. and Brooke, M. de L. (1989a). An experimental study of co-evolution between the cuckoo, *Cuculus canorus*, and its hosts. I. Host egg discrimination. *Journal of Animal Ecology*, **58**, 207–24.

Davies, N. B. and Brooke, M. de L. (1989b). An experimental study of co-evolution between the cuckoo, *Cuculus canorus*, and its hosts. II. Host egg markings, chick discrimination and general discussion. *Journal of Animal Ecology*, **58**, 225–36.

Davies, N. B. and Brooke, M. de L. (1998). Cuckoos versus hosts. Experimental evidence for coevolution. In: *Parasitic Birds and Their Hosts: Studies in Coevolution* (eds S. I. Rothstein and S. K. Robinson), pp. 59–79. Academic Press, New York.

Davies, S. J. J. F. (1962). Nest building of the magpie goose *Anseranus semipalmata*. *Ibis*, **104**, 147–57.

Davis, L. S. and McCaffrey, F. T. (1986). Survival analysis of eggs and chicks in Adelie Penguins (*Pygoscelis adeliae*). *Auk*, **103**, 379–88.

Davis, L. S. and McCaffrey, F. T. (1989). Recognition and parental investment in Adélie penguins. *Emu*, **89**, 155–8.

Davis, S. D., Williams, J. B., Adams, W. J. and Brown, S. L. (1984a). The effect of egg temperature on attentiveness in the Belding's savannah sparrow. *Auk*, **101**, 556–66.

Davis, T. D., Platter-Reiger, M. F. and Ackerman, R. A. (1984b). Incubation water loss by pied-billed grebe eggs: adaption to a hot, wet nest. *Physiological Zoology*, **57**, 384–91.

Davis, W. A. and Russell, S. M. (1984). *Birds in Southeastern Arizona.* Tucson Audubon Society, Tucson, Arizona.

Dawson, A. and Goldsmith, A. R. (1982). Prolactin and gonadotrophin secretion in wild starlings (*Sturnus vulgaris*) during the annual cycle and in relation to nesting, incubation, and rearing young. *General and Comparative Endocrinology*, **48**, 213–21.

Dawson, A. and Goldsmith, A. R. (1983). Modulation of gonadotropin and prolactin secretion by daylength and breeding behavior in free-living starlings, *Sturnus vulgaris*. *Journal of Zoology, London*, **206**, 241–52.

Dawson, W. R. and Bartholomew, G. A. (1968). Temperature regulation and water economy of desert birds. In: *Desert Biology* (ed. G. W. Brown, Jr.), pp. 357–94. Academic Press, New York.

Dawson, W. R. and Hudson, J. W. (1970). Birds, In: *Comparative Physiology of Thermoregulation, Volume 1* (ed. G. C. Whittow), pp. 224–310. Academic Press, New York.

Dawson, W. R. and O'Connor, T. P. (1996). Energetic features of avian thermoregulatory responses. In: *Avian Energetics and Nutritional Ecology* (ed. C. Carey), pp. 85–124. Chapman and Hall, New York.

Deeming, D. C. (1987). The effect of cuticle removal on the water vapour conductance of eggshells of several species of domestic bird. *British Poultry Science*, **28**, 231–7.

Deeming, D. C. (1989a). Characteristics of unturned eggs: A critical period, retarded embryonic growth and poor albumen utilisation. *British Poultry Science*, **30**, 239–49.

Deeming, D. C. (1989b). Failure to turn eggs during incubation: Development of the area vasculosa and embryo growth. *Journal of Morphology*, **201**, 179–85.

Deeming, D. C. (1989*c*). Importance of sub-embryonic fluid and albumen in the embryo's response to turning of the egg during incubation. *British Poultry Science*, **30**, 591–606.

Deeming, D. C. (1991). Reasons for the dichotomy in the need for egg turning during incubation in birds and reptiles. In: *Egg Incubation: Its Effects on Embryonic Development in Birds and Reptiles* (eds D. C. Deeming and M. W. J. Ferguson), pp. 307–23. Cambridge University Press, Cambridge.

Deeming, D. C. (1995). The hatching sequence of ostrich (*Struthio camelus*) embryos with notes on development as observed by candling. *British Poultry Science*, **36**, 67–78.

Deeming, D. C. and Ar, A. (1999). Factors affecting the success of commercial incubation. In: *The Ostrich: Biology, Production and Health* (ed. D. C. Deeming), pp. 159–90. CAB International, Wallingford.

Deeming, D. C. and Ferguson, M. W. J. (1989). In the heat of the nest. *New Scientist*, **No. 1657**, 33–8.

Deeming, D. C. and Ferguson, M. W. J. (1991a). *Egg Incubation: Its Effects On Embryonic Development In Birds And Reptiles*. Cambridge University Press, Cambridge.

Deeming, D. C. and Ferguson, M. W. J. (1991b). Physiological effects of incubation temperature on embryonic development in reptiles and birds. In: *Egg Incubation: Its Effects on Embryonic Development in Birds and Reptiles* (eds D. C. Deeming and M. W. J. Ferguson), pp. 147–71. Cambridge University Press, Cambridge.

Deeming, D. C. and Ferguson, M. W. J. (1991c) Egg turning has no effect on growth and development of embryos of *Alligator mississippiensis*. *Acta Zoologica*, **72**, 125–8.

Deeming, D. C., Rowlett K., and Simkiss, K. (1987). Physical influences on embryo development. *Journal of Experimental Zoology*, **Supplement 1:** 341–5.

Deeming, D. C. and Thompson, M. B. (1991). Gas exchange across reptilian eggshells. In: *Egg Incubation: Its Effects on Embryonic Development in Birds and Reptiles* (eds D. C. Deeming and M. W. J. Ferguson), pp. 277–84. Cambridge University Press, Cambridge.

Dekker, R. W. R. J. (1988). Notes on ground temperatures at nesting sites of the Maleo *Macrocephalon maleo* (Megapodiidae). *Emu*, **88**, 124–7.

Dekker, R. W. R. J. (1989). Predation and western limits of megapode distribution. *Journal of Biogeography*, **6**, 317–21.

Dekker, R. W. R. J. and Brom, T. G. (1990). Maleo eggs and the amount of yolk in relation to different incubation strategies in megapodes. *Australian Journal of Zoology*, **38**, 19–24.

Derksen, D. V. (1977). A quantitative analysis of the incubation behaviour of the Adelie penguin. *Auk*, **94**, 552–66.

Derrickson, K. C. and Breitwisch, R. (1992). Northern Mockingbird. In: T*he Birds of North America, No.* 7 (eds A. Poole, P. Stettenheim and F. Gill). The Academy of Natural Sciences, Philadelphia: The American Ornithologists' Union, Washington, DC.

Deviche, P. (1997). Seasonal reproductive pattern of White-winged Crossbills in interior Alaska. *Journal of Field Ornithology*, **68**, 613–21.

Diaz, J. A. (1994). Field thermoregulatory behavior in the western Canarian lizard *Gallotia galloti*. *Journal of Herpetology*, **28**, 325–33.

Dijkstra, C., Bijlsma, A., Daan, S., Meijer, T. and Zijltra, M. (1990). Brood size manipulations in the kestrel (*Falco tinnunculus*): effects on offspring and parent survival. *Journal of Animal Ecology*, **59**, 269–85.

Dingus, L. and Rowe, T. (1998). The Mistaken Extinction. Dinosaur Evolution and the Origin of Birds. W. H. Freeman and Co., New York.

Dittami, J. (1981). Seasonal changes in the behavior and plasma titers of various hormones in barheaded geese, *Anser indicus*. *Zeitshrift für Tierpsychologie*, **55**, 289–324.

Dolan, P. M. and Wright, P. L. (1984). Damaged western flycatcher eggs in nests containing brown-headed cowbird chicks. *Condor*, **86**, 483–5.

Dong, Z.-M. and Currie, P. J. (1996). On the discovery of an oviraptorid skeleton on a nest of eggs at Bayan Mandahu, Inner Mongolia, People's Republic of China. *Canadian Journal of Earth Sciences*, **33**, 631–6.

Dorward, D. F. (1963). The fairy tern *Gygis alba* on Ascension Island. *Ibis*, **103**, 365–78.

Downs, C. T. and Ward, D. (1997). Does shading behavior of incubating shorebirds in hot environments cool the eggs or the adults? *Auk*, **114**, 717–24.

Dragon, S., Carey, C., Martin, K. and Baumann, R. (1999). Effect of high altitude and *in vivo* adenosine/ß-adrenergic receptor blockade on ATP and 2,3BPG concentrations in red blood cells of avian embryos. *Journal of Experimental Biology*, **202**, 2787–95.

Dreisig, H. (1984). Control of body temperature in shuttling ectotherms. *Journal of Thermal Biology*, **9**, 229–33.

Dreisig, H. (1985). A time budget model of thermoregulation in shuttling ectotherms. *Journal of Arid Environments*, **8**, 191–205.

Drent, R. H. (1970). Functional aspects of incubation in the Herring gull (*Larus argentatus* Pont.). *Behaviour*, **Supplement 17**, 1–132.

Drent, R. (1972). Adaptive aspects of the physiology of incubation. *Proceedings of the XVth International Ornithological Congress*, pp. 255–80.

Drent, R. H. (1973). The natural history of incubation. In: *Breeding Biology of Birds* (ed. D. S. Farner), pp. 262–322. National Academy of Sciences, Washington D.C.

Drent, R. H. (1975). Incubation. In: *Avian Biology, Volume 5* (eds D. S. Farner and J. R. King), pp. 333–420. Academic Press, New York.

Drent, R. H. and Daan, S. (1980). The prudent parent: energetic adjustments in avian breeding. *Ardea*, **68**, 225–52.

Drent, R. H. and Piersma, T. (1990). An exploration of the energetics of leap-frog migration in arctic-breeding waders. In: *Bird migration: Physiology and Ecophysiology* (ed. E. Gwinner), pp. 399–412. Springer Verlag Berlin.

Drent, R. H., Postuma, K. and Joustra, T. (1970). The effect of egg temperature on incubation behaviour in the herring gull. *Behaviour*, **Supplement 17**, 237–61.

Drent, R. H., Tinbergen, J. M. and Biebach, H. (1985). Incubation in the starling, *Sturnus vulgaris*: resolution of the conflict between egg care and foraging. *Netherlands Journal of Zoology*, **35**, 103–23.

Driver, P. M. (1965). 'Clicking' in the egg-young of nidifugous birds. *Nature*, **206**, 315.
Driver, P. M. (1967). Notes on the clicking of avian egg-young, with comments on its mechanisms and function. *Ibis*, **109**, 434–7.
Duckworth, J. W. (1992). Effects of mate removal on the behaviour and reproductive success of Reed Warblers *Acrocephalus scirpaceus*. *Ibis*, **134**, 164–70.
Dufty, A. M., Jr., Goldsmith, A. R. and Wingfield, J. C. (1987). Prolactin secretion in a brood parasite: The brown-headed cowbird, *Molothrus ater*. *Journal of Zoology, London*, **212**, 669–75.
Duncker, H.-R. (1978). Development of the avian respiratory and circulatory systems. In: *Respiratory Function in Birds, Adult and Embryonic* (ed. J. Piiper), pp. 260–73. Springer-Verlag, Heidelberg.
Dunning, J. B. (1993). *CRC Handbook of Avian Body Masses*. CRC Press, Boca Raton.
Durant, J. M., Massemun, S., Thouzeau, C. and Handrich, Y. (2000). Body reserves and nutritional needs during laying preparation in barn owls. *Journal of Comparative Physiology B*, **170**, 253–60.
Eastzer, D., Chu, P. R. and King, A. P. (1980). The young cowbird: Average or optimal nestling? *Condor*, **82**, 417–25.
Eaton, S. W. (1992). Wild Turkey. In: *The Birds of North America, No. 22* (eds A. Poole, P. Stettenheim and F. Gill). The Academy of Natural Sciences, Philadelphia and, The American Ornithological Union, Washington DC.
Eberhard, J. (1998). The evolution of nest building behaviour in *Agapornis* parrots. *Auk*, **115**, 455–64.
Ebling, F., Goldsmith, A. R. and Follett, B. (1982). Plasma prolactin and luteinizing hormone during photoperiodically induced testicular growth and regression in starlings (*Sturnus vulgaris*). *General and Comparative Endocrinology*, **48**, 485–90.
Edmunds, M. (1990). The evolution of cryptic coloration. In: *Insect Defenses. Adaptive Mechanisms and Strategies of Prey and Predators* (ed. D. L. Evans and J. O. Schmidt), pp. 3–21. State University of New York Press, Albany.
Ehrlich, P. R., Breedlove, D. E., Brussard, P. F. and Sharp, M. A. (1972). Weather and the 'regulation' of subalpine populations. *Ecology*, **53**, 243–7.
El Halawani, M. E., Burke, W. H. and Dennison, P. T. (1980). Effects of nest-deprivation on serum prolactin level in nesting female turkeys. *Biology of Reproduction*, **23**, 118–23.
El Halawani, M. E., Burke, W. H., Millam, J. R., Fehrer, S. C. and Hargis, B.M. (1984). Regulation of prolactin and its role in gallinaceous bird reproduction. *Journal of Experimental Zoology*, **232**, 521–9.
El Halawani, M. E., Silsby, J. L., Rozenboim, I. and Pitts, G. R. (1986). Hormonal induction of incubation in ovariectomized female turkeys (*Meleagris gallopavo*). *Biology of Reproduction*, **35**, 59–67.
Elias, St. (1964). The subembryonic fluid in the hen's egg. *Review of Roumaine Embryology and Cytology*, **1**, 165–92.
Ellison, W. G. (1992). Blue-gray Gnatcatcher. In: *Birds of North America, No. 23* (eds A. Poole and P. Stettenheim). American Ornithologists' Union, Washington, D.C. and Academy of Natural Sciences, Philadelphia.
El Wailly, A. (1966). Energy requirements for egg-laying and incubation in the zebra finch, *Taeniopygia castanotis*. *Condor*, **68**, 582–94.

Elzanowski, A. (1981). Embryonic bird skeletons from the Late Cretaceous of Mongolia. *Acta Palaeontologica Polonica*, **42**, 147–79.
Emlen, J. T. and Miller, D. E. (1969). Pace-setting mechanisms of the nesting cycle in the ring-billed gull. *Behaviour*, **33**, 237–61.
Emlen, S. T. and Wrege, P. H. (1986). Forced copulations and intra-specific parasitism: Two costs of social living in the white-fronted bee-eater. *Ethology*, **71**, 2–29.
Enderson, J. H., Temple, S. A. and Swartz, L. G. (1972). Time-lapse photographic records of nesting peregrine falcons. *Living Bird*, **11**, 113–28.
Enemar, A. and Arheimer, O. (1989). Developmental asynchrony and onset of incubation among passerine birds in a mountain birch forest of Swedish lapland. *Ornis Fennica*, **66**, 32–40.
Enkerlin-Hoeflich, E. C. and Hogan, K. M. (1997). Red-crowned parrot (*Amazona viridigenalis*). In: *The Birds of North America, No.* 292 (eds A. Poole and F. Gill). The Academy of Natural Sciences, Philadelphia: The American Ornithologists' Union, Washington, DC.
Epple, W. and Bühler, P. (1981). Eiwenden, eirollen und positionswechsel der brütenden schleier-eule *Tyto alba. Okologie der Vogel*, **3**, 203–11.
Erikstad, K. E. and Tveraa, T. (1995). Does the cost of incubation set limits to clutch size in Common Eiders *Somateria mollissima*? *Oecologia*, **103**, 270–4.
Etches, R. J., Garbutt, A. and Middleton, A. L. (1979). Plasma concentrations of prolactin during egg laying and incubation in the ruffed grouse (*Bonasa umbellus*). *Canadian Journal of Zoology*, **57**, 1624–7.
Evans, M. E. (1975). Breeding behaviour of captive Bewick's swans. *Wildfowl*, **26**, 117–30.
Evans, R. M. (1988a). Embryonic vocalizations and the removal of foot webs from pipped eggs in the American white pelican. *Condor*, **90**, 721–3.
Evans, R. M. (1988b). Embryonic vocalizations as care soliciting signals, with particular reference to the American White Pelican. In: *Acta XIX Congressus Internationalis Ornithologici* (ed. H. Ouellet), pp. 1467–75. University of Ottawa Press, Ottawa, Canada.
Evans, R. M. (1989). Egg temperatures and parental behavior during the transition from incubation to brooding in the American White Pelican. *Auk*, **106**, 26–33.
Evans, R. M. (1990a). Terminal egg neglect in the American White Pelican. *Wilson Bulletin*, **102**, 684–92.
Evans, R. M. (1990b). Effects of low incubation temperatures during the pipped egg stage on hatchability and hatching times in domestic chickens and Ring-billed Gulls. *Canadian Journal Zoology*, **68**, 836–40.
Evans, R. M. (1990c). Terminal-egg chilling and hatching intervals in the American White Pelican. *Auk*, **107**, 431–4.
Evans, R. M. (1990d). Vocal regulation of temperature by avian embryos: a lab study with pipped eggs of the American white pelican. *Animal Behaviour*, **40**, 969–79.
Evans, R. M. (1990e). Embryonic fine tuning of pipped egg temperatures in the American White Pelican. *Animal Behaviour*, **40**, 963–8.
Evans, R. M. (1992). Embryonic and neonatal vocal elicitation of parental brooding and feeding responses in American White Pelicans. *Animal Behaviour*, **44**, 667–75.

Evans, R. M. (1994). Cold-induced calling and shivering in young American white pelicans: Honest signalling of offspring need for warmth in a functionally integrated thermoregulatory system. *Behaviour*, **129**, 13–34.

Evans, R. M. (1995). Incubation temperature in the Australasian gannnet *Morus serrator*. *Ibis*, **137**, 340–4.

Evans, R. M. and Knopf, F. L. (1993). American White Pelican (*Pelecanus erythrorhynchos*). In: *The Birds of North America, No. 57* (eds A. Poole and F. Gill). The Academy of Natural Sciences, Philadelphia and, The American Ornithological Union, Washington DC.

Evans, R. M. and Lee, S. C. (1991). Terminal-egg neglect: Brood reduction strategy or cost of asynchronous hatching? *Proceedings of the International Ornithological Congress*, **20**, 1734–40.

Evans, R. M., Whitaker, A. and Wiebe, M. O. (1994). Development of vocal regulation of temperature by embryos in pipped eggs of ring-billed gulls. *Auk*, **111**, 596–604.

Evans, R. M., Wiebe, M. O., Lee, S. C. and Bugden, S. C. (1995). Embryonic and parental preferences for incubation temperature in herring gulls: implications for parent-offspring conflict. *Behavioral Ecology and Sociobiology*, **36**, 17–23.

Eycleshymer, A. C. (1906). Some observations and experiments on the natural and artificial incubation of the egg of the common fowl. *Biological Bulletin*, **12**, 360–74.

Facemire, C. F., Facemire, M. E. and Facemire, M. C. (1990). Wind as a factor in the orientation of entrance of cactus wren nests. *Condor,* **92**, 1073–5.

Faraci, F. M. (1991). Adaptations to hypoxia in birds: how to fly high. *Annual Review of Physiology,* **53**, 59–70.

Fautin, R. W. (1941). Incubation studies of the yellow-headed blackbird. *Wilson Bulletin*, **53**, 107–22.

Feduccia, A. (1999). *The Origin and Evolution of Birds, 2nd Edition*. Yale University Press, New Haven.

Ferguson, M. W. J. (1985). Reproductive biology and embryology of the crocodilians. In: *Biology of the Reptilia. Volume 14, Development A,* (eds C. Gans, F. Billet and P. F. A. Maderson), pp. 329–491. John Wiley and Sons, New York.

Fetterolf, P. M. and Blokpoel, H. (1983). Reproductive performance of Caspian terns at a new colony on Lake Ontario, 1979–1981. *Journal of Field Ornithology*. **54**, 170–86.

Finch, D. M. (1983). Brood parasitism of the Abert's towhee: Timing, frequency, and effects. *Condor*, **85**, 355–9.

Finkler, M. S., Van Orman, J. B. and Sotherland, P. R. (1998). Experimental manipulation of egg quality in chickens: influence of albumen and yolk on the size and body composition of near-term embryos in a precocial bird. *Journal of Comparative Physiology B*, **168**, 17–24.

Fisher, J. and Peterson, R. T. (1964). *Birds, An Introduction to General Ornithology.* Aldus Books, London.

Fitzpatrick, S. (1996). Male and female incubation in Pied Wagtails (*Motacilla alba*): shared costs or increased parental care? *Ornis Fennica*, **73**, 88–96.

Fivizzani, A. and Oring, L. (1986). Plasma steroid hormones in relation to behavioral sex role reversal in the spotted sandpiper, *Actitis macularia*. *Biology of Reproduction*, **35**, 1195–201.

Fleischer, R. C. (1985). A new technique to identify and assess the dispersion of eggs of individual brood parasites. *Behavioral Ecology and Sociobiology*, **17**, 91–9.

Fleischer, R. C., Smyth, A. P. and Rothstein, S. I. (1987). Temporal and age-related laying rate of the brown-headed cowbird in the eastern Sierra Nevada, California. *Canadian Journal of Zoology*, **65**, 2724–30.

Fleischer, R. C. and Smith, N. G. (1992). Giant cowbird eggs in the nests of two icterid hosts: The use of morphology and electrophoretic variants to indentify individuals and species. *Condor*, **94**, 572–8.

Flint, P. L. and Grand, J. B. (1999). Incubation behavior of spectacled eiders on the Yukon-Kuskokwim Delta, Alaska. *Condor*, **101**, 413–6.

Flint, P. L., Lindberg, M. S., MacCluskie, M. S. and Sedinger, J. S. (1994). The adaptive significance of hatching synchrony of waterfowl eggs. *Wildfowl*, **45**, 248–54.

Forbes, M. R. L. and Ankney, C. D. (1987). Hatching asynchrony and food allocation within broods of Pied-billed Grebes, *Podilymbus podiceps*. *Canadian Journal of Zoology*, **65**, 2872–7.

Forsythe, D. M. (1971). Clicking in the egg-young of the Long-billed Curlew. *Wilson Bulletin*, **83**, 441–2.

Fowler, G. S., Wingfield, J. C., Boersma, P. D. and Sosa, R. A. (1994). Reproductive endocrinology and weight change in relation to reproductive success in the Magellanic penguin (*Spheniscus magellanicus*). *General and Comparative Endocrinology*, **94**, 305–15.

Fraga, R. M. (1983). The eggs of the parasitic screaming cowbird (*Molothrus rufoaxillaris*) and its host, the Baywinged Cowbird (*M. badius*): Is there evidence for mimicry? *Journal für Ornithologie*, **124**, 187–93.

Fraga, R. M. (1985). Host-parasite interactions between chalk-browed mockingbirds and shiny cowbirds. *Ornithological Monographs*, **36**, 829–44.

Fraga, R. M. (1986). *The Bay-winged Cowbird* (Molothrus badius) *and Its Brood Parasites: Interactions, Coevolution and Comparative Efficiency*. Ph.D. Thesis, University of California, Santa Barbara.

Fraga, R. M. (1998). Interactions of the parasitic screaming and shiny cowbirds (*Molothrus rufoaxillaris* and *M. bonariensis*) with a shared host, the bay-winged cowbird (*M. badius*). In: *Parasitic Birds and Their Hosts: Studies in Coevolution* (eds S. I. Rothstein and S. K. Robinson), pp. 173–93. Oxford University Press, Oxford.

Franks, E. C. (1967). The responses of incubating ringed turtle doves (*Streptopelia risoria*) to manipulated egg temperatures. *Condor*, **69**, 268–76.

Frederickson, L. H. and Weller, M. W. (1972). Responses of Adelie penguins to colored eggs. *Wilson Bulletin*, **84**, 309–14.

Freed, L. A. (1981). Loss of mass in breeding wrens: stress or adaptation? *Ecology*, **62**, 1179–86.

Freeman, B. M. (1964). The emergence of the homeothermic-metabolic response in the fowl (*Gallus domesticus*). *Comparative Biochemistry and Physiology*, **13** , 413–22.

Freeman, B. M. and Vince, M. A. (1974). *Development of the Avian Embryo*. Chapman and Hall, London.

Freeman, S. (1988). Egg variability and conspecific nest parasitism in the *Ploceus* weaverbirds. *Ostrich*, **59**, 49–53.

French, N. R. and Hodges, R. W. (1959). Torpidity in cave-roosting hummingbirds. *Condor*, **61**, 223.

Fretwell, S. (1973). Why do robins lay blue eggs? *The Bird Watch*, **1**, 1, 4.

Freund, L. (1926). Besondere Bildungen im mikroskopischen Aufbau der Vogelhaut. *Verh. dt zool. Ges.*, **31**, 153–8.

Friedmann, H. (1929). *The Cowbirds: A Study in the Biology of Social Parasitism.* C. C. Thomas, Springfield.

Friedmann, H. (1955). The honey-guides. *United States National Museum Bulletin*, **208**, 1–273.

Friedmann, H. (1963). Host relations of the parasitic cowbirds. *United States National Museum Bulletin*, **233**, 1–292.

Friedmann, H. and Kiff, L. F. (1985). The parasitic cowbirds and their hosts. *Proceedings of the Western Foundation of Vertebrate Zoology*, **2**, 225–302.

Frith, H. J. (1956a). Temperature regulation in the nesting mounds of the mallee-fowl, *Leiopa ocellata* Gould. *CSIRO Wildlife Research,* **1**, 79–95.

Frith, H. J. (1956b). Breeding habits of the family Megapodiidae, *Ibis*, **98**, 620–40.

Frith, H. J. (1957). Experiments on the control of temperature in the mound of the mallee-fowl, *Leipoa ocellata* Gould (Megapodiidae). *CSIRO Wildlife Research*, **2**, 101–10.

Frith, H. J. (1959a). Breeding in the malleefowl, *Leiopa ocellata* Gould (Megapodiidae). *CSIRO Wildlife Research*, **4**, 31–60.

Frith, H. J. (1959b). Incubator birds. *Scientific American* **201**, 52–8.

Frith, H. J. (1962). *The Mallee Fowl.* Angus and Robertson, Sydney.

Frith, H. J. (1982). *Complete Book of Australian Birds.* Reader's Digest Pty. Ltd., Surrey, N.S.W.

Frost, P. G. H. and Siegfried, W. R. (1974). The cooling rate of eggs of moorhen *Gallinula chloropus* in single and multi-egg clutches. *Ibis*, **119**, 77–80.

Fujimaki, Y., Hanawa, S., Ozaki, K., Yuonoki, O., Nisijima, F., Khrabryi, V. M., Sarikov, Y. B. and Shibnev, Y. B. (1989). Breeding status of the hooded crane, *Grus moncha*, along the Bikin River in the Far East of the USSR. *Strix*, **8**, 199–217. [Data cited by Winter, S. W., Andryushchenko, Y. A. and Gorlov, P. I. (1996).]

Gabrielsen, G. W. and Steen, J. B. (1979). Tachycardia during egg-hypotnermia in incubating ptarmigan (*Lagopus lagopus*). *Acta Physiologica Scandinavia*, **107**, 273–7.

Gales, R. and Green, B. (1990). The annual energetics cycle of Little Penguins (*Eudyptula minor*). *Ecology*, **71**, 2297–312.

Garcia, V., Jouventin, P. and Mauget, R. (1996). Parental care and the prolactin secretion pattern in the king penguin: An endogenously timed mechanism? *Hormones and Behavior*, **30**, 259–65.

Gärtner, K. (1981). Das Wegnehmen von Wirtsvogeleiern durch den Kuckuck (*Cuculus canorus*). *Ornithologische Mittlengunen*, **33**, 115–31.

Gaston, A. J. (1976). Brood parasitism by the pied crested cuckoo *Clamator jacobinus*. *Journal of Animal Ecology*, **45**, 331–48.

Gaston, A. J. and Nettleship, D. N. (1981). *The Thick-billed Murres of Prince Leopold Island.* Canadian Wildlife Service Monograph Series Number 6.

Gaston, A. J., De Forest, L. N. and Noble, D. G. (1993). Egg recognition and egg stealing in murres (*Uria* spp.). *Animal Behaviour*, **45**, 301–6.

Gee, G. F., Hatfield, J. S. and Howey, P. W. (1995). Remote monitoring of parental incubation conditions in the greater sandhill crane. *Zoo Biology*, **14**, 159–72.

Geers, R., Michels, H., Nackaerts, G. and Konings, F. (1983). Metabolism and growth of chickens before and after hatch in relation to incubation temperatures. *Poultry Science*, **62**, 1869–75.

Genelly, R. E. (1955). Annual cycle in a population of California quail. *Condor*, **57**, 263–85.

Gibbons, D. W. (1986). Brood parasitism and cooperative nesting in the moorhen, *Gallinula chloropus*. *Behavioral Ecology and Sociobiology*, **19**, 221–32.

Gibbs, H. L., Sorenson, M. D., Marchetti, K., Brooke, M. de L., Davies, N. B. and Nakamura, H. (2000). Genetic evidence for female host-specific races of the common cuckoo. *Nature*, **407**, 183–6.

Gibson, F. (1971). The breeding biology of the American avocet (*Recurvirostra americana*) in central Oregon. *Condor*, **73**, 444–54.

Giese, M. and Riddle, M. (1999). Disturbance of emperor penguin *Aptenodytes forsteri* chicks by helicopters. *Polar Biology*, **22**, 366–71.

Gifford, M. M. (1993). *Inter- and Intraclutch Egg Variation in Eastern Kingbirds (*Tyrannus tyrannus*): A Test of Kingbirds' Ability to Recognize Foreign Eggs*. M. Sc. Thesis, University of Manitoba, Winnipeg.

Gilbert, A. B. (1971). The egg: its physical and chemical aspects. In: *Physiology and Biochemistry of the Domestic Fowl, Volume 3* (eds D. J. Bell and B. M. Freeman), pp. 1379–99. Academic Press, London.

Gill, B. J. (1998). Behavior and ecology of the shining cuckoo *Chrysococcyx lucidus*. In: *Parasitic Birds and Their Hosts: Studies in Coevolution* (eds S. I. Rothstein and S. K. Robinson), pp. 143–51. Academic Press, New York.

Gill, F. B. (1990). *Ornithology*. Freeman, New York.

Girard, H. and Visschedijk, A. H. J. (1987). Altitude hypocapnia at 2,800 m does not affect development of the chicken embryo. *Journal of Experimental Zoology*, **Supplement 1**, 365–70.

Goad, M. S. and Mannan, R. W. (1987). Nest site selection by elf owls in Saguaro National Monument, Arizona. *Condor*, **89**, 659–62.

Gochfeld, M. (1977). Intraclutch egg variation: The uniqueness of the common tern's third egg. *Bird-Banding*, **48**, 325–32.

Goldsmith, A. R. (1991). Prolactin and avian reproductive strategies. *Acta Congressus Internationalis Ornithologici XX*, **4**, 2063–71.

Goldsmith, A. R., Burke, S. and Prosser, J. M. (1984). Inverse changes in plasma prolactin and LH concentrations in female canaries after disruption and reinitiation of incubation. *Journal of Endocrinology*, **103**, 251–6.

Goldsmith, A. R. and Williams, D. M. (1980). Incubation in mallards (*Anas platyrhynchos*): changes in plasma levels of prolactin and luteinizing hormone. *Journal of Endocrinology*, **86**, 371–80.

Golet, G. H., Irons, D. B. and Estes, J. A. (1998). Survival costs of chick rearing in black-legged kittiwakes. *Journal of Animal Ecology*, **67**, 827–41.

Goodwin, D. (1953). Observations on the voice and behaviour of the red-legged partridge *Alectoris rufa. Ibis*, **95**, 581–614.

Goodwin, D. (1983). *Pigeons and Doves of the World.* British Museum (Natural History), London and Cornell University Press, Comstock Publishing Associates.

Gordon, M. H. and Warren, R. (1895). On the response of the chick, before and after hatching, to changes of external temperature. *Journal of Physiology*, **17**, 331–48.

Gosler, A. (1993). *The Great Tit.* Hamlyn, London.

Göth, A. (1995). Summary of Masters thesis 'Zur ontogenese des Polynesischen Grossfusshuhnes (*Megapodius pritchardii*, Megapodiidae)'. *Megapode Newsletter*, **9**, 4–7.

Göth, A. (2000). *Survival, habitat selectivity and behavioural development of Australian brush turkey Alectura lathami chicks.* PhD Thesis, Giffith University.

Göth, A. and Vogel, U. (1997). Egg laying and incubation of the Polynesian megapode. *Annual Review of the World Pheasant Association*, **1996/97**, 43–54.

Göth, A, and Vogel, U. (1999). Notes on breeding and conservation of birds on Niuafo'ou Island, Kingdom of Tonga. *Pacific Conservation Biology*, **5**, 103–14.

Götmark, F. (1992). Blue eggs do not reduce nest predation in the song thrush, *Turdus philomelos. Behavioral Ecology and Sociobiology*, **30**, 245–52.

Götmark, F. (1993). Conspicuous nests may select for non-cryptic eggs: A comparative study of avian families. *Ornis Fennica*, **70**, 102–5.

Gottlieb, G. (1968). Prenatal behavior in birds. *Quarterly Review of Biology*, **43**, 148–74.

Gottlieb, G. (1988). Development of species identification in ducklings: XV. Individual auditory recognition. *Developmental Psychobiology*, **21**, 509–22.

Gottlieb, G. and Vandenbergh, J. G. (1968). Ontogeny of vocalization in duck and chick embryos. *Journal of Experimental Zoology*, **168**, 307–26.

Grant, G. S. (1982). Avian incubation. Egg temperature, nest humidity and behavioral thermoregulation in a hot environment. *Ornithological Monographs*, **30**, 1–75.

Grant, G. S., Pettit, T. N., Rahn, H., Whittow, G. C. and Paganelli, C. V. (1982). Water loss from Laysan and black-footed albatross eggs. *Physiological Zoology*, **55**, 405–14.

Grant, N. D. (1998). *Host choice of a generalist brood parasite, the brown-headed cowbird (*Molothrus ater*).* M.Sc. Thesis, Laurentian University, Sudbury.

Grant, P. R. (1983). The relative size of Darwin's finch eggs. *Auk*, **100**, 228–9.

Gratto-Trevor, C. L., Oring, L. W., Fivizzani, A. J., El Halawani, M. E. and Cooke, F. (1990). The role of prolactin in parental care in a monogamous and a polyandrous shorebird. *Auk*, **107**, 718–29.

Grau, C. R. (1976). The ring structure of avian egg yolk. *Poultry Science*, **55**, 1418–22.

Grau, C. R. (1984). Egg formation. In: *Seabird Energetics* (eds G. C. Whittow and H. Rahn), pp. 35–57. Plenum, New York.

Graul, W. D. (1975). Breeding biology of the mountain plover. *Wilson Bulletin*, **87**, 6–31.

Grenier, J. L. and Beissinger, S. R. (1999). Variation in the onset of incubation in a neotropical parrot. *Condor*, **101**, 752–61.

Grieef, P. M. and Sealy, S. G. (2000). Simulated host activity does not attract parasitism by brown-headed cowbirds (*Molothrus ater*). *Bird Behavior*, **13**, 69–78.

Grigorescu, D., Weishampel, D., Norman, D., Seclamen, M., Rusu, M., Baltres, A. and Teodorescu, V. (1994). Late Maastrichtian dinosaur eggs from the Hateh Basin (Romania). In: *Dinosaur Eggs and Babies* (eds K. Carpenter, K. F. Hirsch and J. R. Horner), pp. 75–87. Cambridge University Press, Cambridge.

Groscolas, R. (1988). The use of body mass loss to estimate metabolic rate in fasting sea birds: a critical examination based on emperor penguins (*Aptenodytes forsteri*). *Comparative Biochemistry and Physiology*, **90A**, 361–6.

Groscolas, R., Decrock, F., Thil, M. A., Fayolle, C., Boissery, C. and Robin, J. P. (2000). Refeeding signal in fasting-incubating king penguins: changes in behavior and egg temperature. *American Journal of Physiology-Regulatory Integrative and Comparative Physiology*, **279**, R2104–12.

Grover, R. M., Anderson, D. L. and Damon, R. A. Jr. (1980). The correlation between egg shell color and specific gravity as a measure of shell strength. *Poultry Science*, **59**, 1335–6.

Grzybowski, J. A. (1995). Black-capped vireo (*Vireo atricapillus*). In: *The Birds of North America, No.* 181 (eds A. Poole and F. Gill). The Academy of Natural Sciences, Philadelphia: The American Ornithologists' Union, Washington, DC.

Gwinner, H., Oltrogge, M., Trost, L. and Nienaber, U. (2000). Green plants in starling nests: effects on nestlings. *Animal Behaviour*, **59**, 301–09.

Haftorn, S. (1966). Egg laying and incubation in tits based on temperature recordings and direct observations. *Sterna*, 7, 49–102.

Haftorn, S. (1978). Egg-laying and regulation of egg temperature during incubation in the goldcrest *Regulus regulus*. *Ornis Scandinavia*, **9**, 2–21.

Haftorn, S. (1981). Incubation during the egg-laying period in relation to clutch-size and other aspects of reproduction in the great tit *Parus major*. *Ornis Scandinavica*, **12**, 169–85.

Haftorn, S. (1983). Egg temperature during incubation in the Great Tit *Parus major*, in relation to ambient temperature, time of day and other factors. *Fauna Norvegica Series C, Cinclus*, **6**, 22–38.

Haftorn, S. (1984). The behaviour of an incubating coal tit *Parus ater* in relation to environmental regulation of nest temperature. *Fauna Norvegica Series* C, Cindus, 7, 12–20.

Haftorn, S. (1988). Incubating female passerines do not let the egg temperature fall below the 'physiological zero temperature' during their absences from the nest. *Ornis Scandinavica*, **19**, 97–110.

Haftorn, S. (1994). The act of tremble-thrusting in tit nests, performance and possible function. *Fauna Norvegica, Series C, Cinclus*, **17**, 55–74.

Haftorn, S., and Reinertsen, R. E. (1982). Regulation of body temperature and heat transfer to eggs during incubation. *Ornis Scandinavia*, **13**, 1–10.

Haftorn, S. and Reinertsen, R. E. (1985). The effect of temperature and clutch size on the energetic cost of incubation in a free-living Blue Tit (*Parus caeruleus*). *Auk*, **102**, 470–8.

Hahn, D. C. and Fleischer, R. C. (1995). DNA fingerprint similarity between female and juvenile brown-headed cowbirds trapped together. *Animal Behaviour*, **49**, 1577–80.

Hahn, D. C., Sedgwick, J. A., Painter, I. S. and Casna, N. J. (1999). A spatial and genetic analysis of cowbird host selection. In: *Research and Management of the*

Brown-headed Cowbird in Western Landscapes (eds M. L. Morrison, L. S. Hall, S. K. Robinson, S. I. Rothstein, D. C. Hahn and T. D. Rich), pp. 204–17. Studies in Avian Biology, Number 18.

Hainsworth, F. R. (1995). Optimal body temperatures with shuttling: desert antelope ground squirrels. *Animal Behavior*, **49**, 107–16.

Hainsworth, F. R., Collins, B. G. and Wolf, L. L. (1977). The function of torpor in hummingbirds. *Physiological Zoology*, **50**, 215–22.

Hainsworth, F. R., Moonan, T., Voss, M. A., Sullivan, K. A. and Weathers, W. W. (1998). Time and heat allocations to balance conflicting demands during intermittent incubation by Yellow-eyed Juncos. *Journal of Avian Biology*, **29**, 113–20.

Håland, A. (1986). Intraspecific brood parasitism in fieldfares *Turdus pilaris* and other passerine birds. *Fauna Norvegica, Series C, Cinclus*, **9**, 62–7.

Hales, J. R. S., Foldes, A., Fawcett, A. A. and King, R. B. (1982). The role of adrenergic mechanisms in thermoregulatory control of blood flow through capillaries and arteriovenous anastomoses in sheep hind limb. *Pflugers Archives*, **395**, 93–8.

Hall, M. R. (1986). Plasma concentrations of prolactin during the breeding cycle in the cape gannet (*Sula capensis*): a foot incubator. *General and Comparative Endocrinology*, **64**, 112–21.

Hall, M. R. (1987). External stimuli affecting incubation behavior and prolactin secretion in the duck (*Anas platyrhynchos*). *Hormones and Behavior*, **21**, 269–87.

Hall, M. R. and Goldsmith, A. R. (1983). Factors affecting prolactin secretion during breeding and incubation in the domestic duck (*Anas platyrhynchos*). *General and Comparative Endocrinology*, **49**, 270–6.

Hall, T., Harvey, S. and Chadwick, A. (1986). Control of prolactin secretion in birds: a review. *General and Comparative Endocrinology*, **62**, 171–84.

Halupka, K. (1994). Incubation feeding in meadow pipit *Anthus pratensis* affects female time budget. *Journal of Avian Biology*, **25**, 251–3.

Hamas, M. J. (1994). Belted Kingfisher (*Ceryle alcyon*). In: *The Birds of North America, No.* 84 (eds A. Poole and F. Gill). The Academy of Natural Sciences, Philadelphia: The American Ornithologists' Union, Washington, DC.

Hamburger, V. and Hamilton, H. L. (1951). A series of normal stages in the development of the chick embryo. *Journal of Morphology*, **88**, 49–92.

Hamilton, W. J., III and Orians, G. H. (1965). Evolution of brood parasitism in altricial birds. *Condor*, **67**, 361–82.

Handrich, Y. (1989a). Incubation water loss in King penguin egg. I. Change in egg and brood pouch parameters. *Physiological Zoology*, **62**, 96–118.

Handrich, Y. (1989b). Incubation water loss in king penguin egg. II. Does the brood patch interfere with eggshell conductance? *Physiological Zoology*, **62**, 119–32.

Hanka, L. R., Packard, G. C., Sotherland, P. R., Taigen, T. L., Boardman, T. J. and Packard, M. J. (1979). Ontogenetic changes in water-vapor conductance of eggs of yellow-headed blackbirds (*Xanthocephalus xanthocephalus*). *Journal of Experimental Zoology*, **210**, 183–8.

Hann, H. C. (1947). An oven-bird incubates a record number of eggs. *Wilson Bulletin*, **59**,173–4.

Hansell, M. H. (2000). *Bird Nests and Construction Behaviour*. Cambridge University Press, Cambridge.

Hanson, H. L. (1962). The incubation patch of wild geese; its recognition and significance. *Arctic*, **12**, 139–50.

Hardy, J. W. (1957). The least tern in the Mississippi valley. *Michigan State University Museum Biological Series*, **1**, 1–60.

Harper, R. G., Juliano, S. A. and Thompson, C. F. (1994). Intrapopulation variation in hatching synchrony in house wrens: test of the individual optimization hypothesis. *Auk*, **111**, 516–24.

Harrison, H. H. (1975). *A Field Guide to Birds' Nests*. In the United States east of the Mississippi River. Houghton Mifflin Company, Boston.

Hartshorne, J. M. (1962). Behaviour of the eastern bluebird at the nest. *Living Bird*, **1**, 131–49.

Harvey, R. L. (1993). *Practical Incubation*. Hancock House Publishers, Surrey, British Colombia.

Haskell, D. G. (1995). Forest fragmentation and nest predation: Are experiments with Japanese quail eggs misleading? *Auk*, **112**, 767–70.

Hatchwell, B. J., Fowlie, M. K., Ross, D. J. and Russell, A. F. (1999). Incubation behaviour of Long-tailed Tits: why do males provision incubating females? *Condor*, **101**, 681–6.

Hauber, M. E. (2000). Nest predation and cowbird parasitism in song sparrows. *Journal of Field Ornithology*, **71**, 389–98.

Hawkins, L. L. (1986). Nesting behaviour of male and female whistling swans and implications of male incubation. *Wildfowl*, **37**, 5–27.

Haycock, K. A. and Threlfall, W. (1975). The breeding biology of the herring gull in Newfoundland. *Auk*, **92**, 678–97.

Haywood, S. (1993a). Sensory and hormonal control of clutch size in birds. *Quarterly Review of Biology*, **68**, 33–60.

Haywood, S. (1993b). Sensory control of clutch size in the zebra finch (*Taeniopygia guttata*). *Auk*, **110**, 778–86.

Heaney, V. and Monaghan, P. (1995). A within-clutch trade-off between egg production and rearing in birds. *Proceedings of the Royal Society of London B*, **261**, 361–5.

Heaney, V. and Monaghan, P. (1996). Optimal allocation of effort between reproductive phases: the trade-off between incubation costs and subsequent brood rearing capacity. *Proceedings of the Royal Society of London B*, **263**, 1719–24.

Heath, M. and Hansell, M.H. (2001). Weaving techniques in two species of Icterini, the yellow oriole and the crested oropendola. In *Studies in Trinidad and Tobago: Ornithology Honouring Richard ffrench* (ed. F. Hayes). Occasional Papers of The Department of Life Sciences. University of the West Indies, St Augustine, Trinidad, in press.

Hébert, P. N. (1993). An experimental study of brood reduction and hatching asynchrony in yellow warblers. *Condor*, **95**, 362–71.

Hébert, P. N. (1999). Seasonal variation in hatching spreads in mountain bluebirds (*Sialia currucoides*): a test of the nest failure hypothesis. *Bird Behavior*, **13**, 15–21.

Hébert, P. N. and McNeil, R. (1999a). Hatching asynchrony and food stress in ring-billed gulls: an experimental study. *Canadian Journal of Zoology*, **77**, 515–23.

Hébert, P. N. and McNeil, R. (1999b). Nocturnal activity of ring-billed gulls at and away from the colony. *Waterbirds*, **22**, 445–51.

Hébert, P. N. and Sealy, S. G. (1992). Onset of incubation in yellow warblers: A test of the hormonal hypothesis. *Auk*, **109**, 249–55.

Hébert, P. N. and Sealy (1993a). Hatching asynchrony in the yellow warbler: A test of the sexual conflict hypothesis. *American Naturalist*, **142**, 881–92.

Hébert, P. N. and Sealy, S. G. (1993b). Hatching asynchrony in yellow warblers: A test of the nest-failure hypothesis. *Ornis Scandinavica*, **24**, 10–4.

Hector, J. A. L. and Goldsmith, A. R. (1985). The role of prolactin during incubation: Comparative studies of three *Diomedea* albatrosses. *General and Comparative Endocrinology*, **60**, 236–43.

Hegner, R. and Wingfield, J. (1987). Effects of experimental manipulation of testosterone levels on parental investment and breeding success in male house sparrows. *Auk*, **104**, 462–9.

Hegyi, Z. and Sasvari, L. (1998). Parental condition and breeding effort in waders. *Journal of Animal Ecology*, **67**, 41–53.

Heij, C. J., Rompas, C. F. E., and Moeliker, C. W. (1997). The biology of the Moluccan megapode *Eulipoa wallacei* (Aves, Megapodiidae) on Haruku and other Moluccan Islands; part 2: final report. *Deinsea*, **3**, 1–123.

Heinroth, O. (1922). Die beziehungen zwischen vogelgewicht, eigewicht, gelegegewicht und brutdauer. *Journal für Ornithologie*, **70**, 172–285.

Hendricks, P. and Norment, C. J. (1992). Effects of a severe snowstorm on subalpine and alpine populations of nesting American pipits. *Journal of Field Ornithology*, **63**, 331–8.

Hepp, G. R., Kennamer, R. A. and Harvey, W. F. (1990). Incubation as a reproductive cost in female Wood Ducks. *Auk*, **107**, 756–64.

Herbert, R. A. and Herbert, K. G. S. (1965). Behavior of peregrine falcons in the New York City region. *Auk*, **82**, 62–94.

Hertz, P. E., Huey, R. B. and Stevenson, R. D. (1993). Evaluating temperature regulation by field active ectotherms: the fallacy of the inappropriate question. *American Naturalist*, **142**, 796–818.

Hewitson, W. C. (1846). *Coloured Illustrations of the Eggs of British Birds, Accompanied with Descriptions of the Eggs, Nests, etc., Volume 1*. J. Van Voorst, London.

Hiatt, E., Goldsmith, A. R. and Farner, D. (1987). Plasma levels of prolactin and gonadotropins during the reproductive cycle of white-crowned sparrows (*Zonotrichia leucophrys*). *Auk*, **104**, 208–17.

Hiebert, S. M. (1990). Energy costs and temporal organization of torpor in the rufous hummingbird (*Selasphorus rufous*). *Physiological Zoology*, **63**, 1082–97.

Hiebert, S. M. (1991). Seasonal differences in the response of hummingbirds to food restrictions: Body mass and the use of torpor. *Condor*, **93**, 526–37.

Hiebert, S. M. (1992). Time-dependent thresholds for torpor initiation in the rufous hummingbird. *Journal of Comparative Physiology B*, **162**, 249–55.

Hiebl, I., Braunitzer, G. and Schneeganss, D. (1987). The primary structure of the major and minor hemoglobin-components of adult Andean goose *(Chloephaga melanoptera,* Anatidae): The mutation Leu $\Rightarrow$ Ser in position 55 of the β-chains. *Biological Chemistry Hoppe-Seyler,* **368**, 1559–69.

Higuchi, H. (1989). Responses of the bush warbler *Cettia diphone* to artificial eggs of *Cuculus canorus* in Japan. *Ibis*, **131**, 94–8.

Hill, D. P. and Sealy, S. G. (1994). Desertion of nests parasitized by cowbirds: Have clay-coloured sparrows evolved an anti-parasite defence? *Animal Behaviour*, **48**, 1063–70.

Hillström, L. (1995). Body mass reduction during reproduction in the Pied Flycatcher *Ficedula hypoleuca*: physiological stress or adaptation for lowered costs of locomotion? *Functional Ecology*, **9**, 807–17.

Hinde, R. A. (1962). Temporal relations of brood patch development in domesticated canaries. *Ibis*, **104**, 90–7.

Hinde, R. A. (1965). Interaction of internal and external factors in integration of canary reproduction. In: *Sex and Behaviour* (ed. F. A. Beach), pp. 381–415. John Wiley and Sons, New York.

Hinde, R. A., Bell, R. Q. and Steel, E. (1963). Changes in sensitivity of the canary brook patch during the natural breeding season. *Animal Behaviour*, **11**, 553–60.

Hinde, R. A. and Steel, E. (1964). Effect of exogenous hormones on the tactile sensitivity of the canary brood patch. *Journal of Endocrinology*, **30**, 355–60.

Hirota, S. (1894). On the sero-amniotic connection and the foetal membranes in the chick. *Journal of the College of Science, University of Tokyo*, **6**, 337–97.

Hochachka, W. and Smith, J. N. M. (1991). Determinants and consequences of nestling condition in Song Sparrows. *Journal of Animal Ecology*, **60**, 995–1008.

Hockey, P. A. R. (1982). Adaptiveness of nest site selection and egg coloration in the African black oystercatcher *Haematopus moquini*. *Behavioral Ecology and Sociobiology*, **11**, 117–23.

Hofslund, P. B. (1957). Cowbird parasitism of the northern yellow-throat. *Auk*, **74**, 42–8.

Höhn, E. O. (1962). A possible endocrine basis of brood parasitism. *Ibis*, **104**, 418–21.

Höhn, O. (1959). Prolactin in the cowbird's pituitary in relation to avian brood parasitism. *Nature*, **184**, 2030.

Höhn, O. (1972). Prolactin lack in a brood parasite: A summary report and appeal for material. *Ibis*, **114**, 108.

Hoi, H., Schleicher, B. and Valera, F. (1994). Female mate choice and nest desertion in penduline tits, *Remiz pendulinus*: the importance of nest quality. *Animal Behaviour*, **48**, 743–6.

Holcomb, L. C. (1967). Goldfinch accept young after long and short incubation. *Wilson Bulletin*, **79**, 348.

Holcomb, L. C. (1969). Egg turning behaviour of birds in response to color-marked eggs. *Bird-Banding*, **40**, 105–13.

Holford, K. C. and Roby, D. D. (1993). Factors limiting fecundity of captive brown-headed cowbirds. *Condor*, **95**, 536–45.

Holmes, R. T. and Pitelka, F. A. (1998). Pectoral sandpiper (*Calidris melanotos*). In: *The Birds of North America, No.* 348 (eds A. Poole and F. Gill). The Academy of Natural Sciences, Philadelphia: The American Ornithologists' Union, Washington, DC.

Holyoak, D. (1969). The function of the pale egg colour of the jackdaw. *Bulletin of the British Ornithologists' Club*, **89**, 159.

Holyoak, D. (1970). The relation of egg colour to laying sequence in the carrion crow. *Bulletin of the British Ornithologists' Club*, **90**, 40–2.

Horváth, L. (1955). Red-footed falcons in Ohat-Woods, near Hortobágy. *Acta Zoologica Academiae Scientarum*, **1**, 245–87.

Horváth, O. (1964). Seasonal differences in rufous hummingbird nest height and their relation to nest climate. *Ecology*, **45**, 235–41.

Hötker, H. (2000). Conspecific nest parasitism in the pied avocet *Recurvirostra avosetta*. *Ibis*, **142**, 280–8.

Hötker, H., Kölsch, G. and Visser, G. H. (1996). Die energieumsatz brütender Säbelschnäbler *Recurvirostra avocetta*. *Journal für Ornithologie*, **137**, 203–12.

Houston, A. I. and Carbone, C. (1992). The optimal allocation of time during the diving cycle. *Behavioral Ecology*, **3**, 255–65.

Houston, A. I. and McNamara, J. M. (1985). A general theory of central place foraging for single-prey loaders. *Theoretical Population Biology*, **28**, 233–62.

Howard, E. (1957). Ontogenetic changes in the freezing point and sodium and potassium content of the subgerminal fluid and blood plasma of the chick embryo. *Journal of Cellular and Comparative Physiology*, **50**, 451–70.

Howe, H. F. (1976). Egg size, hatching asynchrony, sex and brood reduction in the common grackle. *Ecology*, **57**, 1195–207.

Howe, S. and Kilgore, D. L., Jr. (1987). Convective and diffusive gas exchange in nest cavities of the northern flicker (*Colaptes auratus*). *Physiological Zoology*, **60**, 707–12.

Howe, S., Kilgore, D. L., Jr. and Colby, C. (1987). Respiratory gas concentrations and temperature within the nest cavities of the northern flicker (*Colaptes auratus*). *Canadian Journal of Zoology*, **65**, 1541–7.

Howell, J. C. (1942). Notes on the nesting habits of the American robin (*Turdus migratorius* L.). *American Midland Naturalist*, **28**, 529–603.

Howell, T. R. (1979). Breeding biology of the Egyptian plover, *Pluvianus aegyptius*. *University of California Publications in Zoology*, **113**, 1–76.

Howell, T. R., Araya, B. and Millie, W.R. (1974). Breeding biology of the gray gull, *Larus modestus*. *University of California Publications in Zoology*, **104**, 1–57.

Howell, T. R. and Dawson, W. R. (1954). Nest temperatures and attentiveness in the Anna hummingbird. *Condor*, **56**, 93–7.

Howey, P., Board, R. G., Davis, D. H. and Kear, J. (1984). The microclimate of the nests of waterfowl. *Ibis*, **126**, 16–32.

Hoy, G. and Ottow, J. (1964). Biological and oological studies of the Molothrine cowbirds (Icteridae) of Argentina. *Auk*, **81**, 186–203.

Hoyt, D. F. (1979). Practical methods of estimating volume and fresh weight of bird eggs. *Auk*, **96**, 73–7.

Hoyt, D. F. and Rahn, H. (1980). Respiration of avian embryos- a comparative analysis. *Respiration Physiology*, **39**, 255–64.

Hubalek, Z. (1978). Coincidence of fungal species associated with birds. *Ecology*, **59**, 438–42.

Hubalek, Z. and Balat, F. (1974). The survival of microfungi in the nests of tree-sparrow [*Passer montanus* L.] in the nest-boxes over the winter season. *Mycopathologia et Mycologia applicata*, **54**, 517–30.

Hubalek, Z., Juricova, Z. and Halouzka, J. (1995). A survey of free-living birds as hosts and 'lessors' of microbial pathogens. *Folia Zoologica*, **44**, 1–11.

Hubalek, Z., Balat, F., Touskova, I. and Vlk, J. (1973). Mycoflora of birds' nests in nest-boxes. *Mycopatholgia et Mycologia applicata*, **49**, 1–12.

Huggins, R. A. (1941). Egg temperatures of wild birds under natural conditions. *Ecology*, **22**, 148–57.

Hughes, J. M. (1997). Taxonomic significance of host-egg mimicry by facultative brood parasites of the avian genus *Coccyzus* (Cuculidae). *Canadian Journal of Zoology*, **75**, 1380–6.

Hughes, J. M. (1999). Yellow-billed cuckoo (*Coccyzus americanus*). In: *Birds of North America*, number 418 (eds A. Poole and F. G. Gill). The Birds of North America, Inc., Philadelphia.

Humphrey, T. J. (1994). Contamination of eggs with potential human pathogens. In: *Microbiology of the Avian Egg* (eds R. G. Board and R. Fuller), pp. 93–116. Chapman and Hall, London.

Hunt, K. E., Hahn, T. P. and Wingfield, J. C. (1999). Endocrine influences on parental care during a short breeding season: testosterone and male parental care in Lapland Longspurs (*Calcarius lapponicus*). *Behavioural Ecology and Sociobiology*, **45**, 360–9.

Hussell, D. J. T. (1972). Factors affecting clutch size in arctic passerines. *Ecological Monographs*, **42**, 317–64.

Hussell, D. J. T. (1985). On the adaptive basis for hatching asynchrony: Brood reduction, nest failure and asynchronous hatching in snow buntings. *Ornis Scandinavica*, **16**, 205–12.

Hutchison, R. E. (1975). Effects of ovarian steroids and prolactin on the sequential development of nesting behavior in female budgerigars. *Journal of Endocrinology*, **67**, 29–39.

Ignarro, L. J. (1981). Endothelium-derived nitric oxide: action and properties. *The FASEB Journal*, **3**, 31–6.

Immelman, K. (1971). Ecological aspects of periodic reproduction. In: *Avian Biology, Volume 1* (eds D. S. Farner, J. R. King and K. C. Parkes), pp. 342–89. Academic Press, New York.

Impekoven, M. (1970). Prenatal experience of parental calls and pecking in the laughing gull (*Larus atricilla* L.). *Animal Behaviour*, **19**, 475–80.

Impekoven, M. (1973). The response of incubating laughing gulls (*Larus atricilla* L.) to calls of hatching chicks. *Behaviour*, **46**, 94–113.

Impekoven, M. (1976). Prenatal parent-young interactions in birds and their long-term effects. *Advances in the Study of Behavior*, **7**, 201–53.

Impekoven, M. and Gold, P. S. (1973). Prenatal origins of parent-young interactions in birds: A naturalistic approach. In: *Behavioral Embryology* (ed. G. Gottlieb), pp. 325–56. Academic Press, New York.

Ingold, J. L. and R. Galati (1997). Golden Crowned Kinglet. In: *Birds of North America, No. 301* (eds A. Poole and P. Stettenheim). American Ornithologists' Union, Washington, D.C. and Academy of Natural Sciences, Philadelphia.

Inouye, R. S., Huntly, N. J. and Inouye, D. W. (1981). Non-random orientation of Gila woodpecker nest entrances in saguaro cacti. *Condor*, **83**, 88–9.

Irving, L. (1972). *Arctic Life of Birds and Mammals Including Man*. Springer-Verlag, Heidelberg.

Irving, L. and Krog, J. (1956). Temperature during the development of birds in arctic nests. *Physiological Zoology*, **29**, 195–205.

Jackson, J. A. (1976). A comparison of some aspects of the breeding ecology of red-headed and red-bellied woodpeckers in Kansas. *Condor*, **78**, 67–76.

Jackson, N. H. and Roby, D. D. (1992). Fecundity and egg-laying patterns of captive yearling brown-headed cowbirds. *Condor*, **94**, 585–89.

Jackson, W. M. (1992a). Estimating conspecific nest parasitism in the northern masked weaver based on within female variability in egg appearance. *Auk*, **109**, 435–43.

Jackson, W. M. (1992b). Relative importance of parasitism by *Chrysococcyx* cuckoos versus conspecific nest parasitism in the northern masked weaver *Ploceus taeniopterus*. *Ornis Scandinavica*, **23**, 203–6.

Jackson, W. M. (1998). Egg discrimination and egg-color variability in the northern masked weaver. In: *Parasitic Birds and Their Hosts* (eds S. I. Rothstein and S. K. Robinson), pp. 407–16. Oxford University Press, Oxford.

Jacob, J. (1978). Uropygial Gland secretions and feather waxes. In: *Chemical Zoology X, Aves* (eds M. Florkin, B. T. Scheer and A. H. Brush), pp. 165–211. Academic Press, New York.

Jacob, J., Balthazart, J. and Schoffeniels, E. (1979). Sex differences in the chemical composition of uropygial gland waxes in domestic ducks. *Biochemical Systematics and Ecology*, **7**, 149–53.

Jacob, J., Eigener, U. and Hoppe, U. (1997). The structure of preen gland waxes from pelecaniform birds containing 3,7-dimethyloctan-1-ol – An active ingredient against dermatophytes. *Zeitschrift Fur Naturforschung C*, **52**, 114–23.

Jacob, J. and Ziswiler, V. (1982). The Uropygial gland. In: *Avian Biology*, Volume 6 (eds D. S. Farner, J. R. King and K. C. Parkes), pp. 199–324. Academic Press, New York.

Janik, D. S. and Buntin, J. D. (1985). Behavioural and physiological effects of prolactin in incubating ring doves. *Journal of Endocrinology*, **105**, 201–9.

Janzen, D. (1978). Predation intensity on eggs on the ground in two Costa Rican forests. *American Midland Naturalist*, **100**, 467–70.

Jaramillo, A. P. (1993). *Parasite-host coevolution in the cowbirds* Molothrus rufoaxillaris *and* Molothrus badius: *Egg mimicry in shape and size*. M.Sc. Thesis, University of Toronto.

Jensen, R. A. C. (1966). Genetics of cuckoo egg polymorphism. *Nature*, **209**, 827.

Jessen, T.-H., Weber, R. E., Fermi, G., Tame, J. and Braunitzer, G. (1991). Adaptation of bird hemoglobins to high altitudes: demonstration of molecular mechanism by protein engineering. *Proceedings of the National Academy of Science USA*, **88**, 6519–22.

Jobin, B. and Picman, J. (1997). The effect of egg coloration on predation of artificial ground nests. *Canadian Field-Naturalist*, **111**, 591–4.

Johns, J. E. and Pfeiffer, E. W. (1963). Testosterone-induced incubation patches of phalarope birds. *Science, New York*, **140**, 1225–6.

Johnsgard, P. A. (1993). *Cormorants, Darters and Pelicans of the World*. Smithsonian Institution Press, Washington DC.

Johnsgard, P. A. (1997a). *The Hummingbirds of North America, 2nd edition*. Smithsonian Institution Press, Washington, DC.

Johnsgard, P. A. (1997b). *The Avian Brood Parasites. Deception at the Nest.* Oxford University Press, Oxford.

Johnsgard, P. A. (1983). *Cranes of the World.* Croom Helm, London.

Johnson, R. G. and Temple, S. A. (1990). Nest predation and brood parasitism of tallgrass prairie birds. *Journal of Wildlife Management*, **54**, 106–11.

Johnson, S. R. and Campbell, R. W. (1995). Crested Mynah (*Acridotheres cristatellus*). In: *The Birds of North America, No.* 157 (eds A. Poole and F. Gill). The Academy of Natural Sciences, Philadelphia: The American Ornithologists' Union, Washington, DC.

Johnston, P. M. and Comar, C. L. (1955). Distribution and contribution of calcium from the albumen, yolk and shell to the developing chick embryo. *American Journal of Physiology*, **183**, 365–70.

Johnston, R. F. (1992). Rock Dove. In: *The Birds of North America, No.* 13 (eds A. Poole, P. Stettenheim and F. Gill). The Academy of Natural Sciences, Philadelphia: The American Ornithologists' Union, Washington, DC.

Jones, D. N. (1988a). Hatching success of the Australian Brush-turkey *Alectura lathami* in South-east Queensland. *Emu*, **88**, 260–2.

Jones, D. N. (1988b). Construction and maintenance of the incubation mounds of the Australian brush-turkey *Alectura lathami. Emu*, **88**, 210–8.

Jones, D. N. (1989). Modern megapode research a post-Frith review. *Corella*, **13**, 145–54.

Jones, D. N. (1999). What we don't know about megapodes. In: *Proceedings of the Third International Megapode Symposium, Nhill, Australia* (eds R. W. R. J. Dekker, D. N. Jones and J. Benshemesh), *Zoologische Verhandelingen* **327**, 159–68.

Jones, D. N., Dekker, R. W. R. J. and Roselaar, C. S. (1995). *The Megapodes Megapodiidae.* Oxford University Press, Oxford.

Jones, D. N., and Everding, S. E. (1991). Australia Brush turkeys in a suburban environment: implications for conflict and conservation. *Australian Wildlife Research*,**18**, 260–3.

Jones, G. (1987). Time and energy constraints during incubation in free-living Swallows (*Hirundo rustica*): an experimental study using precision electronic balances. *Journal of Animal Ecology*, **56**, 229–45.

Jones, I. L. (1992). Factors affecting survival of adult Least Auklets (*Aethia pusilla*) at St. Paul Island, Alaska. *Auk*, **109**, 576–84.

Jones, R. E. (1968). *The role of prolactin and gonadal steroids in the reproduction of California quail,* Lophortyx californicus. Doctoral dissertation, University of California, Berkeley.

Jones, R. E. (1969a). Hormonal control of incubation patch development in the California quail, *Lophortyx californicus. General and Comparative Endocrinology*, **13**, 1–13.

Jones, R. E. (1969b). Epidermal hyperplasia in the incubation patch of the California quail, *Lophortyx californicus*, in relation to pituitary prolactin content. *General and Comparative Endocrinology*, **12**, 498–502.

Jones, R. E. (1971). The incubation patch of birds. *Biological Reviews*, **46**, 315–39.

Jones, R. E., Kreider, J. W. and Criley, B. B. (1970). Incubation patch of the chicken: response to hormones *in situ* and transplanted to a dorsal site. *General and Comparative Endocrinology*, **15**, 398–403.

Jouventin, P. and Mauget, R. (1996). The endocrine basis of the reproductive cycle in the king penguin (*Aptenodytes patagonicus*). *Journal of Zoology, London*, **238**, 665–78.

Joyce, F. J. (1993). Nesting success of rufous-naped wrens (*Campylorhynchos rufinucha*) is greater near wasp nests. *Behaviour Ecology and Sociobiology*, **32**, 71–7.

Kacelnik, A. and Houston, A. I. (1984). Some effects of energy costs on foraging strategies. *Animal Behaviour*, **32**, 609–14.

Kattan, G. H. (1993). *Reproductive strategy of a generalist brood parasite, the shiny cowbird, in the Cauca Valley, Colombia*. Ph.D. Thesis, University of Florida, Gainesville.

Kattan, G. H. (1995). Mechanisms of short incubation period in brood-parasitic cowbirds. *Auk*, **112**, 335–42.

Kemal, R. E. and Rothstein, S. I. (1988). Mechanisms of avian egg recognition: Adaptive responses to eggs with broken shells. *Animal Behaviour*, **36**, 175–83.

Kemp, M. (1995). *The Hornbills Bucerotiformes*. Oxford University Press, Oxford.

Kempenaers, B., Pinxten, R. and Eens, M. (1995). Intraspecific brood parasitism in two tit *Parus* species: Occurrence and responses to experimental parasitism. *Journal of Avian Biology*, **26**, 114–20.

Kendeigh, S. C. (1952). *Parental Care and its Evolution in Birds*. Illinois Biological Monographs, **XXII**, The University of Illinois Press, Urbana.

Kendeigh, S. C. (1963). Thermodynamics of incubation in the house wren, *Troglodytes aedon*. *Proceedings of the XIIIth International Ornithological Congress, Ithaca*, 884–904.

Kendeigh, S. C. (1973). The natural history of incubation-discussion. In: *Breeding Biology of Birds* (ed. D. S. Farner), pp. 311–20. National Academy of Sciences, Washington, DC.

Kendeigh, S. C., Dol'nik, V. R. and Gavrilov, V. M. (1977). Avian energetics. In: *Granivorous Birds In Ecosystems* (eds J. Pinowski and S. C. Kendeigh), pp. 127–204. Cambridge University Press, Cambridge UK.

Kendeigh, S. C., Kramer, T. C. and Hamerstrom, F. (1956). Variations in egg characteristics of the house wren. *Auk*, **73**, 42–65.

Kendra, P. E., Roth, R. R. and Tallamay, D. W. (1988). Conspecific brood parasitism in the house sparrow. *Wilson Bulletin*, **100**, 80–90.

Kennamer, R. A., Harvey, IV, W. F. and Hepp, G. R. (1990). Embryonic development and the nest attentiveness of wood ducks during egg laying. *Condor*, **92**, 587–92.

Kennedy, E. D. (1991). Determinate and indeterminate egg-laying patterns: A review. *Condor*, **93**, 106–24.

Kennedy, G. Y. and Vevers, H. G. (1976). A survey of avian eggshell pigments. *Comparative Biochemistry and Physiology*, **55B**, 117–23.

Kepler, A. K. (1977). Comparative study of todies (Todidae): With emphasis on the Puerto Rican Tody, *Todus mexicanus*. *Publications of the Nuttall Ornithological Club*, **16**. Cambridge, Massachusetts.

Kern, M. (1984). Racial differences in nests of white crowned sparrows. *Condor*, **86**, 455–66.

Kern, M. D. (1986). Changes in water-vapor conductance of common canary eggs during the incubation period. *Condor*, **88**, 390–3.

Kern, M. D. and Bushra, A. (1980). Is the incubation patch required for the construction of a normal nest? *Condor*, **82**, 328–34.

Kern, M. D. and Coruzzi, L. (1979). The structure of the canary's incubation patch. *Journal of Morphology,* **162**, 425–52.

Kern, M. D. and Cowie, R. J. (2000). Female Pied Flycatchers fail to respond to variations in nest humidity. *Comparative Biochemistry and Physiology,* **127A**, 113–9.

Kern, M. D., Cowie, R. J. and Yeager, M. (1992). Water loss, conductance, and structure of eggs of pied flycatchers during egg laying and incubation. *Physiological Zoology*, **65**, 1162–87.

Kern, M. D. and Ferguson, M. W. J. (1997). Gas permeability of American alligator eggs and its anatomical basis. *Physiological Zoology*, **70**, 530–46.

Kern, M. and van Riper, C. (1984). Altitudinal variations in nests of the Hawaiian honeycreeper, *Hemignathus virens virens*. *Condor*, **86**, 443–54.

Kerpez, T. A. and Smith, N. S. (1990). Nest-site selection and nest-cavity characteristics of Gila woodpeckers and northern flickers. *Condor,* **92**, 193–8.

Kersten, M. (1996). Time and energy budgets of oystercatchers *Haematopus ostralegus* occupying territories of different quality. *Ardea*, **84A**, 57–72.

Kersten, M. and Piersma, T. (1987). High levels of energy expenditure in shorebirds: metabolic adaptations to an energetically expensive way of life. *Ardea*, 75, pp. 175–187.

Ketterson, E. D., Nolan, V. J. and Wolf, L. (1990). Effect of sex, stage of reproduction, season, and mate removal on prolactin in dark-eyed juncos. *Condor*, **92**, 922–30.

Ketterson, E. D., Nolan, V. J., Wolf, L. and Ziegenfus, C. (1992). Testosterone and avian life histories: effects of experimentally elevated testosterone on behavior and correlates of fitness in the dark-eyed junco (*Junco hyemalis*). *American Naturalist*, **140**, 980–99.

Kilpi, M. and Byholm, P. (1995). The odd colour of the last laid egg in herring gull *Larus argentatus* clutches: Does it reflect egg quality? *Ornis Fennica*, **72**, 85–8.

Kim, C.-H., Yamagishi, S. and Won, P.-O. (1995). Egg-color dimorphism and breeding success in the crow tit (*Paradoxornis webbiana*). *Auk*, **112**, 831–9.

King, J. R. (1973). Energetics of reproduction in birds. In: *Breeding Biology Of Birds* (ed. D. S. Farner), pp. 78–107. National Academy of Sciences , Washington DC.

Klandorf, H., Probert, I. L. and Iqbal, M. (1999). In the defense against hyperglycemia: an avian strategy. *Worlds Poultry Science Journal*, **55**, 251–68.

Kleindorfer, S., Fessl, B. and Hoi, H. (1995). More is not always better: male incubation in two acrocephalus warblers. *Behaviour*, **132**, 607–25.

Kleven, O., Moksnes, A., Røskaft, E. and Honza, M. (1999). Host species affects the growth rate of cuckoo (*Cuculus canorus*) chicks. *Behavioral Ecology and Sociobiology*, **47**, 41–6.

Kloska, C. and Nicolai, J. (1988). Fortpflanzungsverhalten des Kamm-Talegalla (*Aepypodius arakianius* Salvad.). *Journal für Ornithologie*, 129, 185–204.

Kluivjer, H. N. (1950). Daily routines of the great tit *Parus m. major* L. *Ardea*, **38**, 99–135.

Koenig, W. D., Stacey, P. B., Stanback, M. T. and Mumme, R. L. (1995). Acorn woodpecker (*Melanerpes formicivorus*). In: *The Birds of North America, No.* 194 (eds A. Poole and F. Gill). The Academy of Natural Sciences, Philadelphia: The American Ornithologists' Union, Washington, DC.

Koepcke, M. (1972). Über die Resistenzformen der Vogelnester in einem begrenzten Gebeit des tropischen Regenwaldes in Peru. *Journal für Ornithologie*, **113**, 138–60.

Koford, R. R., Bowen, B. S. and Vehrencamp, S. L. (1990). Groove-billed Anis: joint-nesting in a tropical cuckoo. In: *Cooperative Breeding in Birds: Long-term studies of Ecology and Behaviour* (eds P. B. Stacey and W. D. Koenig), pp. 335–55. Cambridge University Press, Cambridge.

Kossack, C. W. (1947). Incubation temperatures of Canada geese. *Journal of Wildlife Management*, **11**, 119–26.

Koutnik, J. (1927). Die Hautveranderungen der Brütfleckbildung beim Haushuhn. *Prage. Arch. Tiermed.* **7** (Teil A), 129–41.

Kozlovic, D. R., Knapton, R. W. and Barlow, J. C. (1996). Unsuitability of the house finch as a host of the brown-headed cowbird. *Condor*, **98**, 253–58.

Kramer, D. L. (1988). The behavioral ecology of air breathing by aquatic animals. *Canadian Journal of Zoology*, **66**, 89–94.

Krebbs, K. (1992). Four years of hummingbird breeding at the Arizona-Sonora Desert Museum. *AAZK Animal Keepers Forum*. November.

Krebbs, K. (1999). The Arizona-Sonora Desert Museum's hummingbird propagation program. *Proceedings 22nd Annual Wildlife Conference, International Wildlife Rehabilitation Conference*, Tucson, AZ.

Krementz, D. G. and Ankney, C. D. (1986). Bioenergetics of egg production by female house sparrows. *Auk*, **103**, 299–305.

Kruger, K., Prinzinger, R. and Schuchmann, K.-L. (1982). Torpor and metabolism in hummingbirds. *Comparative Biochemistry and Physiology*, **73A**, 679–89.

Kruijt, J. P. (1958). Speckling of the herring gull egg in relation to brooding behaviour. *Archives Néerlandaises de Zoologie*, **12**, 565–7.

Kuiper, J. W. and Ubbels, P. (1951). A biological study of natural incubation and its application to artificial incubation. *9th World's Poultry Congress, Edinburgh*: 105–12.

Kyle, P. D. and Kyle, G. Z. (1993). An evaluation of the role of microbial flora in the saliva transfer technique for hand-rearing chimney swifts (*Chaetura pelagica*). *Wildlife rehabilitation*, **8**, 65–71.

Labisky, R. F. and Jackson, G. L. (1966). Characteristics of egg-laying and eggs of yearling pheasants. *Wilson Bulletin*, **78**, 379–99.

Lack, D. (1947). The significance of clutch size. *Ibis*, **89**, 302–52.

Lack, D. (1954). *The Natural Regulation of Animal Numbers*. Oxford Press, London.

Lack, D. (1956). A review of the genera and nesting habits of swifts. *Auk*, 73, 1–32.

Lack, D. (1958). The significance of the colour of turdine eggs. *Ibis*, **100**, 145–66.

Lack, D. (1968). *Ecological Adaptations for Breeding in Birds*. Methuen, London.

Lancaster, D. A. (1964). Life history of the boucard tinamou in British Honduras. Part II: Breeding biology. *Condor*, **66**, 253–76.

Lange, B. (1928). Die Brutflecke der Vögel und die für sie wichtigen Hauteigentümlichkeiten. *Gegenbaurs Jahrb.* **59**, 601–712.

Lanier, G. A. Jr. (1982). A test for conspecific egg discrimination in three species of colonial passerine birds. *Auk*, **99**, 519–25.

Lanyon, W. E. (1988a). A phylogeny of the thirty-two genera in the *Elaenia* assemblage of tyrant flycatchers. *American Museum Novitates*, **2914**, 1–57.

Lanyon, W. E. (1988b). A phylogeny of the flatbill and tody-tyrant assemblage of tyrant flycatchers. *American Museum Novitates*, **2923**, 1–41.

Lasiewski, R. C. (1963). Oxygen consumption of torpid, resting, active, and flying hummingbirds. *Physiological Zoology*, **36**, 122–40.

Lasiewski, R. C. and Dawson, W. R. (1967). A re-examination of the relation between standard metabolic rate and body weight in birds. *Condor*, **69**, 13–23.

Laskey, A. R. (1944). A study of the Cardinal in Tennessee. *Wilson Bulletin*, **56**, 27–44.

Latter, G. V. and Baggott, G. K. (1996). Effect of egg turning and fertility upon the sodium concentration of albumen of the Japanese quail. *British Poultry Science*, **37**, 301–8.

Latter, G. V. and Baggott, G. K. (2000a). Ion and fluid transport by the blastoderm of the Japanese quail. *British Poultry Science*, **41**, S43–4.

Latter, G. V. and Baggott, G. K. (2000b). The effect of egg turning and fertility upon the potassium concentration of albumen and yolk of the Japanese quail. *British Poultry Science*, **41**, S44–5.

Lawes, M. J. and Kirkman, S. (1996). Egg recognition and interspecific brood parasitism rates in red bishops (Aves: Ploceidae). *Animal Behaviour*, **52**, 553–63.

Laybourne, R. C. (1974). Collision between a vulture and an aircraft at an altitude of 37,000 feet. *Wilson Bulletin*, **86**, 461–2.

Lea, R. W., Dods, A. S. M., Sharp, P. J. and Chadwick, A. (1981). The possible role of prolactin in the regulation of nesting behaviour and the secretion of luteinising hormone in broody bantams. *Journal of Endocrinology*, **91**, 89–97.

Lea, R. W., Richard-Yris, M. A. and Sharp, P. J. (1996). The effect of ovariectomy on concentrations of plasma prolactin and LH and parental behavior in the domestic fowl. *General and Comparative Endocrinology*, **101**, 115–21.

Lea, R. W. and Sharp, P. J. (1989). Concentrations of plasma prolactin and luteinizing hormone following nest deprivation and renesting in ring doves (*Streptopelia risoria*). *Hormones and Behavior*, **23**, 279–89.

Lea, R. W., Talbot, R. T. and Sharp, P. J. (1991). Passive immunization against chicken vasoactive intestinal polypeptide suppresses plasma prolactin and crop sac development in incubating ring doves. *Hormones and Behavior*, **25**, 283–94.

Lea, R. W., Vowles, D. M. and Dick, H. R. (1986). Factors affecting prolactin secretion during the breeding cycle of the ring dove (*Streptopelia risoria*) and its possible role in incubation. *Journal of Endocrinology*, **110**, 447–58.

Leboucher, G., Richard-Yris, M.-A., Chadwick, A. and Guémené, D. (1996). Influence of nest deprivation on behaviour, hormonal concentrations and on the ability to resume incubation in domestic hens (*Gallus domesticus*). *Ethology*, **102**, 660–71.

Leboucher, G., Richard-Yris, M. A., William, J. and Chadwick, A. (1990). Incubation and maternal behavior in domestic hens: influence of the presence of chicks on circulating luteinizing hormone, prolactin, and estradiol and on behavior. *British Poultry Science*, **31**, 851–62.

Lee, S. C., Evans, R. M. and Bugden, S. C. (1993). Benign neglect of terminal eggs in herring gulls. *Condor*, **95**, 507–14.

Lehrman, D. (1964). The reproductive behavior of ring doves. *Scientific American*, **211**, 48–54.

Lehrman, D. S. (1955). The physiological basis of parental feeding behaviour in the ring dove (*Streptopelia risoria*). *Behaviour* 7, 241–86.

Le Maho, Y. (1977). The emperor penguin: a strategy to live and breed in the cold. *American Scientist,* **65**, 680–93.

Leon-Velarde, F., Whittembury, J., Carey, C. and Monge, C. (1984). Shell characteristics of eggs laid at 2800 m by hens transported from sea level 24 hours after hatching. *Journal of Experimental Zoology,* **230**, 137–9.

Leopold, A. S., Erwin, M., Oh, J. and Browning. B. (1976). Phytoestrogens: adverse effects on reproduction of the California Quail. *Science,* **191**, 98–100.

Leroi, A. (2001). Molecular signals versus the Loi de Balancement. *Trends in Ecology and Evolution*, **16**, 24–9.

Lessells, C. M. (1991). The Evolution of Life Histories. In: *Behavioural Ecology* (eds, J. R. Krebs and N. B. Davies), pp. 32–68. Blackwell, Oxford.

Lichtenstein, G. (1998). Parasitism by shiny cowbirds of rufous-bellied thrushes. *Condor*, **100**, 680–7.

Lichtenstein, G. and Sealy, S. G. (1998). Nestling competition, rather than supernormal stimulus, explains the success of parasitic brown-headed cowbird chicks in yellow warbler nests. *Proceedings of the Royal Society of London B*, **265**, 249–54.

Lifjeld, J. T. and Slagsvold, T. (1986). The function of courtship feeding during incubation in the pied flycatcher *Ficedula hypoleuca. Animal Behaviour,* **34**, 1441–53.

Lifjeld, J. T., Slagsvold, T. and Stenmark, G. (1987). Allocation of incubation feeding in a polygynous mating system: a study on pied flycatchers *Ficedula hypoleuca. Animal Behaviour*, **35**, 1663–9.

Lill, A. (1979). Nest inattentiveness and its influence on the development of the young in the superb lyrebird. *Condor*, **81**, 225–31.

Lill, A. (1986). Time-energy budgets during incubation during reproduction and the evolution of single parenting in the Superb Lyrebird. *Australian Journal of Zoology*, **34**, 351–71.

Limpert, R. J. and Earnst, S. L. (1994). Tundra Swan (*Cygnus columbianus*). In: *The Birds of North America, No. 89* (eds A. Poole and F. Gill). The Academy of Natural Sciences, Philadelphia and, The American Ornithological Union, Washington DC.

Lin, X., Lee, C.-G., Casale, E. S. and Shih, J. C. H. (1992). Purification and characterization of a keratinase from a feather-degrading *Bacillus licheniformis* strain. *Applied and Environmental Microbiology*, **58**, 3271–5.

Lincoln, G. A. (1974). Predation of incubator birds (*Megapodius fregcinet*) by Komodo dragons (*Varanus komodoensis*). *Journal of Zoology London*, **174**, 419–28.

Lind, H. (1961). Studies on the behaviour of the Black-tailed Godwit (*Limosa limosa* (L)). *Meddelelse Naurfredningsradets Reservatudvalg*, **66**, 11–157.

Lindell, C. (1996). Patterns of nest usurpation: When should species converge on nest niches? *Condor*, **98**, 464–73.

Lindén, M. and Møller, A. P. (1989). Cost of reproduction and covariation of life history traits in birds. *Trends in Ecology and Evolution*, **4**, 367–71.

Lindholm, A. K. (1997). *Evolution of Host Defences Against Brood Parasitism*. Ph. D. Thesis, University of Cambridge.

Lindström, A. and Kvist, A. (1995). Maximal energy intake is proportional to basal metabolic rate. *Proceedings Of The Royal Society Of London B*, **261**, 337–43.

Lindstrom, J. (1999). Early development and fitness in birds and mammals. *Trends in Ecology and Evolution*, **14**, 343–8.

Liversidge, R. (1961). Pre-incubation development of *Clamator jacobinus*. *Ibis*, **103a**, 624.

Livesey, T. R. (1936). Cuckoo problems. *Journal of the Bombay Natural History Society*, **38**, 734–58.

Livesey, T. R. (1938). Egg stealing by Kashia Hills cuckoo (*Cuculus canorus bakeri* Hartert). *Journal of the Bombay Natural History Society*, **40**, 561–4.

Lloyd, J. A. (1965a). Seasonal development of the incubation patch in the starling (*Sturnus vulgaris*). *Condor*, **67**, 67–72.

Lloyd, J. A. (1965b). Effects of environmental stimuli on the development of the incubation patch in the European starling (*Sturnus vulgaris*). *Physiological Zoology*, **38**, 121–8.

Lloyd, P., Plagány, É., Lepage, D., Little, R. M. and Crowe, T. M. (2000). Nest-site selection, egg pigmentation and clutch predation in the ground-nesting Namaqua sandgrouse *Pterocles namaqua*. *Ibis*, **142**, 123–31.

Lombardo, M. P., Power, H. W., Stouffer, P. C., Romagnano, L. C. and Hoffenberg, A. S. (1989). Egg removal and intraspecific brood parasitism in the European starling (*Sturnus vulgaris*). *Behavioral Ecology and Sociobiology*, **24**, 217–23.

Lombardo, M. P., Thorpe, P. A., Cichewicz, R., Henshaw, M., Millard, C., Steen, C. and Zeller, T. K. (1996). Communities of cloacal bacteria in tree swallow families. *Condor*, **89**, 167–72.

Lomholt, J. P. (1974). Relationship of weight loss to ambient humidity of birds' eggs during incubation. *Journal of Comparative Physiology*, **105**, 189–96.

Lomholt, J. P. (1984). A preliminary study of local oxygen tensions inside bird eggs and gas exchange during early stages of embryonic development. In: *Respiration and Metabolism of Embryonic Vertebrates* (ed. R. S. Seymour), pp. 289–309. Dr W. Junk, Dordrecht.

Long, C. A. (1969). The origin and evolution of mammary glands. *Bioscience*, **19**, 519–23.

Longo, L. D., Hill, E. P. and Power, G. G. (1972). Theoretical analysis of factors affecting placental O_2 transfer. *American Journal of Physiology*, **222**, 730–9.

Lorenzana, J. C. and Sealy, S. G. (1998). Adult brood parasites feeding nestlings and fledglings of their own species: A review. *Journal of Field Ornithology*, **69**, 364–75.

Lorenzana, J. C. and Sealy, S. G. (1999). A meta-analysis of the impact of parasitism by the Brown-headed Cowbird on its hosts. In: *Research and Management of the*

Brown-headed Cowbird in Western Landscapes (eds M. L. Morrison, L. S. Hall, S. K. Robinson, S. I. Rothstein, D. C. Hahn and T. D. Rich), pp. 241–53. *Studies in Avian Biology*, Number 18.

Lormée, H., Jouventin, P., Chastel, O. and Mauget, R. (1999). Endocrine correlates of parental care in an Antarctic winter breeding seabird, the emperor penguins, *Aptenodytes forsteri*. *Hormones and Behavior*, **35**, 9–17.

Lormée, H., Jouventin, P., Lacroix, A., Lallemand, J. and Chastel, O. (2000). Reproductive endocrinology of tropical seabirds: Sex-specific patterns in LH, steroids, and prolactin secretion in relation to parental care. *General and Comparative Endocrinology*, **117**, 413–26.

Lotem, A. and Nakamura, H. (1998). Evolutionary equilibria in avian brood parasitism: An alternative to the 'arms race-evolutionary lag' concept. In: *Parasitic Birds and Their Hosts: Studies in Coevolution* (eds S. I. Rothstein and S.K. Robinson), pp. 223–35. Academic Press, New York.

Lowther, P. E. (1988). Spotting pattern of the last laid egg of the house sparrow. *Journal of Field Ornithology*, **59**, 51–4.

Lowther, P. E. and Cink, C. L. (1992). House Sparrow (*Passer domesticus*). In: *The Birds of North America, No. 12* (eds A. Poole, P. Stettenheim and F. Gill). The Academy of Natural Sciences, Philadelphia: The American Ornithologists' Union, Washington, DC.

Lowther, P. E. and Nocedal, J. (1997). Olive Warbler (*Peucedramus taeniatus*). In: *The Birds of North America, No. 310* (eds A. Poole and F. Gill). The Academy of Natural Sciences, Philadelphia and, The American Ornithological Union, Washington DC.

Lundy, H. (1969). A review of the effects of temperature, humidity, turning and gaseous environment in the incubator on the hatchability of the hen's egg. In: *The Fertility and Hatchability of the Hen's Egg* (eds T. C. Carter and B. M. Freeman), pp. 143–76. Oliver and Boyd, Edinburgh, United Kingdom.

Lyon, B. E. (1993). Tactics of parasitic American coots: Host choice and the pattern of egg dispersion among host nests. *Behavioral Ecology and Sociobiology*, **33**, 87–100.

Lyon, B. E. and Hamilton, L. D. (1992). The frequency of conspecific brood parasitism and the pattern of laying determinancy in yellow-headed blackbirds. *Condor*, **94**, 590–7.

Lyon, B. E. and Montgomerie, R. D. (1985). Incubation feeding in snow buntings: female manipulation or indirect male parental care? *Behavioural Ecology and Sociobiology*, **17**, 279–84.

Lyon, B. and Montgomerie, R. (1995). Snow bunting (*Plectrophenax nivalis*). In: *The Birds of North America, No.* 198 (eds A. Poole and F. Gill). The Academy of Natural Sciences, Philadelphia: The American Ornithologists' Union, Washington, DC.

MacKinnon, J. (1978). Sulawasi Megapodes. *World Pheasant Association Journal*, **3**, 96–103.

Maclean, G. L. (1974). Egg-covering in the Charadrii. *Ostrich*, **45**, 167–74.

Madge, S. G. (1970). Nest of the long-billed spiderhunter *Arachnothera robusta*. *Malay Naturalists Journal*, **23**, 125.

Magrath, M. J. L. (1999). Breeding ecology of the fairy martin. *Australian Journal of Zoology*, **47**, 463–77.

Magrath, M. J. L. and Elgar, M. A. (1997). Paternal care declines with increased opportunity for extra-pair matings in fairy martins. *Proceedings of the Royal Society of London B*, **264**, 1731–6.

Magrath, R. D. (1989). Hatching asynchrony in altricial birds: nest failure and adult survival. *American Naturalist*, **131**, 893–900.

Magrath, R. D. (1990). Hatching asynchrony in altricial birds. *Biological Review*, **65**, 587–622.

Magrath, R. D. (1991). Nestling weight and juvenile survival in the Blackbird (*Turdus merula*). *Journal of Animal Ecology*, **60**, 335–51.

Magrath, R. D. (1992). Roles of egg mass and incubation pattern in establishment of hatching hierarchies in the blackbird (*Turdus merula*). *Auk*, **109**, 474–87.

Maher, W. J. (1962). Breeding biology of the snow petrel near Cape Hallett, Antarctica. *Condor*, **64**, 488–99.

Mahoney, S. P. and Threlfall, W. (1981). Notes on the eggs, embryos and chick growth of common guillemots *Uria aalge* in Newfoundland. *Ibis*, **123**, 211–8.

Major, R. E., Gowing, G. and Kendal, C. E. (1996). Nest predation in Australian urban environments and the role of the pied currawong, *Strepera graculina. Australian Journal of Ecology*, **21**, 399–409.

Major, R. E. and Kendal, C. E. (1996). The contribution of artificial nest experiments to understanding avian reproductive success: A review of methods and conclusions. *Ibis*, **138**, 298–307.

Mallory, M. L. and Weatherhead, P. J. (1990). Effects of nest parasitism and nest location on eggshell strength in waterfowl. *Condor*, **92**, 1031–9.

Mallory, M. L. and Weatherhead, P. J. (1993). Incubation rhythms and mass loss of Common Goldeneyes. *Condor*, **95**, 849–59.

Manuwal, D. A. (1974). The incubation patches of Cassin's Auklet. Condor, **76**, 481–4.

March, J. B., Sharp, P. J., Wilson, P. W. and Sang, H. M. (1994). Effect of active immunization against recombinant-derived chicken prolactin fusion protein on the onset of broodiness and photoinduced egg laying in bantam hens. *Journal of Reproduction and Fertility*, **101**, 227–33.

Marchetti, K. (2000). Egg rejection in a passerine bird: Size does matter. *Animal Behaviour*, **59**, 877–83.

Marder, J. and Gavrieli-Levin, I. (1986). Body and egg temperature regulation in incubating pigeons exposed to heat stress: the role of skin evaporation. *Physiological Zoology*, **59**, 532–8.

Marks, J. S., Evans, D. L. and Holt, D. W. (1994). Long-eared owl (*Asio otus*). In: *The Birds of North America, No. 133* (eds A. Poole and F. Gill). The Academy of Natural Sciences, Philadelphia: The American Ornithological Union, Washington DC.

Marsh, R. L. and Dawson, W. R. (1989). Avian adjustments to cold. In: *Advances in Comparative and Environmental Physiology. Volume 4. Animal Adaptation to Cold* (ed. L. C. H. Wang), pp. 206–53. Springer-Verlag, Berlin.

Marti, C. M. (1991). Barn owl. In: *The Birds of North America, No. 1* (eds A. Poole, P. Stettenheim and F. Gill). The Academy of Natural Sciences, Philadelphia: The American Ornithologists' Union, Washington, DC.

Martin, K., Holt, R. F. and Thomas, D. F. (1993). Getting by on high: ecological energetics of arctic and alpine grouse. In: *Life in the Cold. Ecological, Physiological, and Molecular Mechanisms* (ed. C. Carey), pp. 33–41. Westview Press, Boulder, Colorado.

Martin, S. G. (1974). Adaptations for polygynous breeding in the bobolink, *Dolichonyx oryzivorus*. *American Zoologist*, **14**, 109–19.

Martin, T. E. (1987). Artificial nest experiments: Effects of nest appearance and type of predator. *Condor*, **89**, 925–8.

Martin, T. E. (1995). Avian life history evolution in relation to nest sites, nest predation, and food. *Ecological Monographs*, **65**, 101–27.

Martin, T. E. and Ghalambor, C. K. (1999). Males feeding females during incubation I: required by microclimate or constrained by nest predation? *American Naturalist*, **153**, 131–9.

Martinez, J. G., Soler, J. J., Soler, M. and Burke, T. (1998). Spatial patterns of egg laying and multiple parasitism in a brood parasite: A non-territorial system in the great spotted cuckoo (*Clamator glandarius*). *Oecologia*, **117**, 286–94.

Masman, D., Daan, S. and Beldhuis, H. J. A. (1988). Ecological energetics of the Kestrel: daily energy expenditure throughout the year based on time-energy budget, food intake and doubly labelled water methods. *Ardea*, **76**, 64–81.

Mason, P. (1980). *Ecological and Evolutionary Aspects of Host Selection in Cowbirds*. Ph. D. Thesis, University of Texas, Austin.

Mason, P. (1986). Brood parasitism in a host generalist, the shiny cowbird: I. The quality of different species as hosts. *Auk*, **103**, 52–60.

Mason, P. and Rothstein, S. I. (1986). Coevolution and avian brood parasitism: Cowbird eggs show evolutionary response to host discrimination. *Evolution*, **40**, 1207–14.

Mason, P. and Rothstein, S. I. (1987). Crypsis versus mimicry and the color of shiny cowbird eggs. *American Naturalist*, **130**, 161–7.

Massoni, V. and Reboreda, J. C. (1999). Egg puncture allows shiny cowbirds to assess host egg development and suitability for parasitism. *Proceedings of the Royal Society of London B*, **266**, 1871–4.

Mauck, R. A. and Grubb, T. C. (1995). Petrel parents shunt all experimentally increased reproductive costs to their offspring. *Animal Behaviour*, **49**, 999–1008.

Maunder, J. E. and Threlfall, W. (1972). The breeding biology of the black-legged kittiwake in Newfoundland. *Auk*, **89**, 789–816.

Mayfield, H. F. (1961). Vestiges of a proprietary interest in nests by the brown-headed cowbird parasitizing the Kirtland's warbler. *Auk*, **78**, 162–6.

Mazen, M. B., Moubasher, A. H. and Bagy, M. M. K. (1994). Seasonal distribution of fungi in bird nests in Egypt. *Microbiological research*, **149**, 429–34.

McAldowie, A. M. (1886). Observations on the development and the decay of the pigment layer on birds' eggs. *Journal of Anatomy and Physiology*, **20**, 225–37.

McCluskie, M. C. and Sedinger, J. S. (2000). Nutrient reserves and clutch-size regulation of northern shovelers in Alaska. *Auk*, **117**, 971–9.

McCracken, K. G., Afton, A. D. and Alisauskas, R. T. (1997). Nest morphology and body size of Ross' geese and lesser snow geese. *Auk*, **114**, 610–8.

McFarland, K. P. and Rimmer, C. C. (1996). Horsehair fungus, *Marasmius androsaceus*, used as nest lining by birds of the subalpine spruce-fir community in the Northeastern United States. *Canadian Field-Naturalist*, **110**, 541–3.

McKinney, D. F. (1952). Incubation and hatching behaviour in the mallard. *Wildfowl Trust Annual Report*, **5**, 68–70.

McLaren, C. M. (2000). *Patterns of host nest use by Brown-headed Cowbirds parasitizing Song Sparrows and Yellow Warblers*. M.Sc. Thesis, University of Manitoba, Winnipeg.

McLaren, C. M. and Sealy, S. G. (2000). Are nest predation and brood parasitism correlated in yellow warblers? A test of the cowbird predation hypothesis. *Auk*, **117**, 1056–60.

McLennan, J. A. (1988). Breeding of North Island brown kiwi, *Apteryx australis mantelli*, in Hawke's Bay, New Zealand. *New Zealand Journal of Ecology*, **11**, 89–97.

McLennan, J. A. and McCann, A. J. (1988). Incubation temperatures of great spotted kiwi, *Apteryx haastii*. *New Zealand Journal of Ecology*, **15**, 163–6.

McMaster, D. G. (1997). *An experimental investigation of strategies used by Brown-headed Cowbirds to optimize parental care*. Ph.D. Thesis, University of Manitoba, Winnipeg.

McMaster, D. G. and Sealy, S. G. (1997). Host-egg removal by brown-headed cowbirds: A test of the host incubation limit hypothesis. *Auk*, **114**, 212–20.

McMaster, D. G. and Sealy, S. G. (1998a). Short incubation periods of brown-headed cowbirds: How do cowbird eggs hatch before Yellow Warbler eggs? *Condor*, **100**, 102–11.

McMaster, D. G. and Sealy, S. G. (1998b). Red-winged blackbirds (*Agelaius phoeniceus*) accept prematurely hatching brown-headed cowbirds (*Molothrus ater*). *Bird Behavior*, **12**, 67–70.

McMaster, D. G. and Sealy, S. G. (1999). Do brown-headed cowbird hatchlings alter adult yellow warbler behavior during the hatching period? *Journal of Field Ornithology*, **70**, 365–73.

McNamara, J. M. and Houston, A. I. (1996). State-dependent life histories. *Nature*, **380**, 215–21.

McRae, S. B. (1995). Temporal variation in responses to intraspecific brood parasitism in the moorhen. *Animal Behaviour*, **49**, 1073–88.

McRae, S. B. (1997). Identifying eggs of conspecific brood parasites in the field: A cautionary note. *Ibis*, **139**, 701–4.

Mead, P. S. and Morton, M. L. (1985). Hatching asynchrony in the mountain white-crowned sparrow (*Zonotrichia leucophyrs oriantha*): A selected or incidental trait? *Auk*, **102**, 781–92.

Meanley, B. (1992). King Rail. In: *The Birds of North America, No. 3* (eds A. Poole, P. Stettenheim and F. Gill). The Academy of Natural Sciences, Philadelphia: The American Ornithologists' Union, Washington, DC.

Medway, L. (1961). Domestic pigeon. The stimulus provided by the egg in the nest. *Journal of Endocrinology*, **23**, 9–12.

Mehmke, U., Gerlach, H., Kosters, J. and Hausmann, S. (1992). Studies of the Aerobic Bacterial-Flora On the Nesting-Material of Singing Birds. *Deutsche Tierarztliche Wochenschrift*, **99**, 478–82.

Meijer, T. (1990). Incubation development and clutch size in the starling. *Ornis Scandinavica*, **21**, 163–8.

Meijer, T. (1995). Importance of tactile and visual stimuli of eggs and nest for termination of egg laying of red junglefowl. *Auk*, **112**, 483–8.

Meijer, T., Daan, S. and Hall, M. (1990). Family planning in the kestrel (*Falco tinnunculus*): the proximate control of covariation of laying date and clutch size. *Behaviour*, **114**, 117–36.

Meijer, T. and Siemers, I. (1993). Incubation development and asynchronous hatching in junglefowl. *Behaviour*, **127**, 309–22.

Meir, M. and Ar, A. (1986). Further increasing hatchability of turkey eggs and poult quality by correcting water loss and incubator oxygen pressure. *Proceedings of the 7th European Poultry Conference,* **2**, 749–53.

Meir, M. and Ar, A. (1990). Gas pressures in the air cell of the ostrich egg prior to pipping as related to oxygen consumption, eggshell gas conductance and egg temperature. *Condor*, **92**, 556–63.

Meir, M., Nir, A. and Ar, A. (1984). Increasing hatchability of turkey eggs by matching incubator humidity to shell conductance of individual eggs. *Poultry Science,* **63**, 1489–96.

Melin, E., Wiken, T. and Oblom, K. (1947). Antibiotic agents in the substrates from cultures of the genus *Marasmius*. *Nature*, **159**, 840–1.

Merilä, J. and Wiggins, D. A. (1997). Mass loss in breeding Blue Tits: the role of energetic stress. *Journal of Animal Ecology*, **66**, 452–60.

Merkle, M. S. and Barclay, R. M. R. (1996). Body mass variation in breeding mountain bluebirds *Sialia currucoides*: evidence of stress or adaptation for flight? *Journal of Animal Ecology*, **65**, 401–13.

Mermoz, M. E. and Reboreda, J. C. (1994). Brood parasitism of the shiny cowbird, *Molothrus bonariensis*, on the brown-and-yellow marshbird, *Pseudoleistes virescens*. *Condor*, **96**, 716–21.

Mermoz, M. E. and Reboreda, J. C. (1999). Egg-laying behaviour by shiny cowbirds parasitizing brown-and-yellow marshbirds. *Animal Behaviour*, **58**, 873–82.

Merola-Zwartjes, M. and Ligon, J. D. (2000). Ecological energetics of the Puerto Rican tody: Heterothermy, torpor, and intra-island variation. *Ecology*, **81**, 990–1003.

Mersten-Katz, C. (1997). Gas composition in the nest cavity of the Syrian woodpecker, *Dendrocopus syriacus*. *Israel Journal of Zoology,* **43**, 111.

Mertens, J. A. L. (1977). The energy requirements for incubation in great tits, *Parus major* L. *Ardea*, **65**, 184–96.

Mertens, J. A. L. (1980), The energy requirements for incubation in Great Tits and other bird species. *Ardea*, **68**, 185–92.

Mertens, J. A. L. (1987). The influence of temperature on the energy reserves of female Great Tits during the breeding season. *Ardea*, **75**, 73–80.

Metcalfe, J., Stock, M. K. and Ingermann, R. L. (1987). Development of the avian embryo. *Journal of Experimental Zoology*, **Supplement 1**.

Metcalfe, N. B. and Monaghan, P. (2001). Compensation for a bad start: grow now – pay later. *Trends in Ecology and Evolution*, **16**, 254–60.

Mewaldt, L. R. (1956). Nesting behaviour of the Clark nutcracker. *Condor*, **58**, 3–23.

Mico, M. M. (1998). *Yellow Warbler Nests: Structure, Building Materials and Cowbird Parasitism.* M.Sc. Thesis, University of Manitoba.

Middleton, A. L. A. (1991). Failure of brown-headed cowbird parasitism in nests of the American goldfinch. *Journal of Field Ornithology*, **62**, 200–3.

Midtgård, U. (1988). Innervation of arteriovenous anastomoses in the brood patch of the domestic fowl. *Cell Tissue Research*, **252**, 207–10.

Midtgård, U., Sejrsen, P. and Johansen, K. (1985). Blood flow in the brood patch of bantam hens. Evidence of cold vasodilatation. *Journal of Comparative Physiology B*, **155**, 703–10.

Mikhailov, K. E., Sabath, K. and Kurzanov, S. M. (1994). Eggs and nests from the Cretaceous of Mongolia. In: *Dinosaur Eggs and Babies* (eds K. Carpenter, K. F. Hirsch and J. R. Horner), pp. 88–115. Cambridge University Press, Cambridge.

Miller, D. E. (1972). Parental acceptance of young as a function of incubation time in the ring-billed gull. *Condor*, **74**, 482–4.

Mills, A. M. (1987). Size of host egg and egg size in brown-headed cowbirds. *Wilson Bulletin*, **99**, 490–1.

Mills, T. K., Lombardo, M. P. and Thorpe, P. A. (1999). Microbial colonization of the cloacae of nestling tree swallows. *Auk*, **116**, 947–56.

Minguez, E. (1998). The costs of incubation in the British Storm Petrel: an experimental study in a single-egg layer. *Journal of Avian Biology*, **29**, 183–9.

Mirosh, L. W. and Becker, W. A. (1974). Storage and incubation temperature effects on hatching time of Coturnix Quail eggs. *Poultry Science*, **53**, 432–4.

Mitchell, D. E. (2000). Allen's Hummingbird (*Selasphorus sasin*). In: *Birds of North America, No. 501* (eds A. Poole and P. Stettenheim). The Academy of Natural Sciences, Philadelphia: The American Ornithologists' Union, Washington, DC.

Mitchell, C. D. (1994). Trumpeter swan (*Cygnus buccinator*). In: *The Birds of North America, No. 105* (eds A. Poole and F. Gill). The Academy of Natural Sciences, Philadelphia: The American Ornithologists' Union, Washington, DC.

Moksnes, A. (1992). Egg recognition in chaffinches and bramblings. *Animal Behaviour*, **44**, 993–5.

Moksnes, A. and Røskaft, E. (1987). Cuckoo host interactions in Norwegian mountain areas. *Ornis Scandinavica*, **18**, 168–72.

Moksnes, A. and Røskaft, E. (1989). Adaptations of meadow pipits to parasitism by the common cuckoo. *Behavioral Ecology and Sociobiology*, **24**, 25–30.

Moksnes, A. and Røskaft, E. (1992). Responses of some rare cuckoo hosts to mimetic model cuckoo eggs and to foreign conspecific eggs. *Ornis Scandinavica*, **23**, 17–23.

Moksnes, A. and Røskaft, E. (1995). Egg-morphs and host preference in the common cuckoo (*Cuculus canorus*): An analysis of cuckoo and host eggs from European museum collections. *Journal of Zoology, London*, **236**, 625–48.

Moksnes, A., Røskaft, E. and Braa, A. T. (1991). Rejection behavior by common cuckoo hosts towards artificial brood parasite eggs. *Auk*, **108**, 348–54.

Moksnes, A., Røskaft, E., Braa, A. T., Korsnes, L., Lampe, H. M. and Pedersen, H. C. (1990). Behavioural responses of potential hosts towards artificial cuckoo eggs and dummies. *Behaviour*, **116**, 64–89.

Moksnes, A., Røskaft, E., Hagen, L. G., Honza, M., Mørk, C. and Olsen, P. H. (2000). Common cuckoo *Cuculus canorus* and host behaviour at reed warbler *Acrocephalus scirpaceus* nests. *Ibis*, **142**, 247–58.

Moksnes, A., Røskaft, E. and Korsnes, L. (1993). Rejection of cuckoo (*Cuculus canorus*) eggs by meadow pipits (*Anthus pratensis*). *Behavioral Ecology*, **4**, 120–7.
Moksnes, A., Røskaft, E. and Tysse, T. (1995). On the evolution of the blue cuckoo eggs in Europe. *Journal of Avian Biology*, **26**, 13–9.
Møller, A. P. (1984). On the use of feathers in birds' nests: predictions and tests. *Ornis Scandinavica*, **15**, 38–42.
Møller, A. P. (1987a). Intraspecific nest parasitism and anti-parasite behaviour in swallows, *Hirundo rustica*. *Animal Behaviour*, **35**, 247–54.
Møller, A. P. (1987b). Nest lining in relation to the nesting cycle in the Swallow *Hirundo rustica*. *Ornis Scandinavica*, **18**, 148–9.
Møller, A. P. (1990). Nest predation selects for small size in the nest of the blackbird. *Oikos*, **57**, 237–40.
Møller, A. P. and Petrie, M. (1991). Evolution of intraspecific variability in birds' eggs: Is intraspecific nest parasitism the selective agent? *Proceedings of the International Ornithological Congress*, **20**, 1041–8.
Monaghan, P. and Nager, R. G. (1997). Why don't birds lay more eggs? *Trends* in *Ecology and Evolution*, **12**, 270–3.
Montevecchi, W. A. (1976). Field experiments on the adaptive significance of avian eggshell pigmentation. *Behaviour*, **58**, 26–39.
Moreau, R. E. (1941). A contribution to the breeding biology of the palm swift *Cypselus parvus*. *Journal of the East Africa and Uganda Natural History Society*, **15**, 154–70.
Morel, M.-Y. (1973). Contribution à l'étude dynamique de la population de *Lagonosticta senegala* L. (Estrildides) à Richard-Toll (Sénégal). Interrelations avec le parasite *Hypochera chalybeata* (Müller) (Viduines). *Mémoires du Muséum National d'Histoire Naturelle, série A, Zoologie*, **78**, 1–156.
Moreno, E., Moreno, J. and A. de León. (1999a). The effect of nest size on stone-gathering behavior in the chinstrap penguin. *Polar Biology*, **22**, 90–2.
Moreno, J. (1989a). Energetic constraints on uniparental incubation in the wheatear *Oenanthe oenanthe L.*. *Ardea*, **77**, 107–15.
Moreno, J. (1989b). Strategies of mass change in breeding birds. *Biological Journal of the Linnean Society*, **37**, 297–310.
Moreno, J., Barbosa, A., Potti, J. and Merino, S. (1997). The effects of hatching date and parental quality on chick growth and creching age in the Chinstrap Penguin (*Pygoscelis antarctica*): A field experiment. *Auk*, **114**, 47–54.
Moreno, J. and Carlson, A. (1989). Clutch size and the costs of incubation in the Pied Flycatcher *Ficedula hypoleuca*. *Ornis Scandinavica*, **20**, 123–8.
Moreno, J., Gustafsson, L., Carlson, A. and Pärt, T. (1991). The cost of incubation in relation to clutch size in the collared flycatcher (*Ficedula albicollis*). *Ibis*, **133**, 186–93.
Moreno, J. and Hillstrom, L. (1992). Variation in time and energy budgets of breeding Wheatears. *Behaviour*, **120**, 11–39.
Moreno, J. and Sanz, J. J. (1994). The relationship between the energy expenditure during incubation and clutch size in the Pied Flycatcher *Ficedula hypoleuca*. *Journal of Avian Biology*, **25**, 125–30.

Moreno, J., Veiga, J. P., Cordero, P. J. and Minguez, E. (1999b). Effects of parental care on reproductive success in the polygynous spotless starling (*Sturnus unicolor*). *Behavioural Ecology and Sociobiology*, **47**, 47–53.

Morgan, K. R., Paganelli, C. V. and Rahn, H. (1978). Egg weight loss and nest humidity during incubation in two Alaskan gulls. *Condor,* **80**, 272–5.

Morton, M. L. (1976). Adaptive strategies of *Zonotrichia* breeding at high latitude or high altitude. *Proceedings of the 16th International Ornithological Congress,* **1974**, 322–36.

Morton, M. L. (1978). Snow conditions and the onset of breeding in the mountain white-crowned sparrow. *Condor,* **80**, 285–9.

Morton, M. L. and Carey, C. (1971). Growth and the development of endothermy in the mountain white-crowned sparrow (*Zonotrichia leucophrys oriantha*). *Physiological Zoology,* **44**, 177–89.

Morton, M. L., Horstmann, J. and Osborn, J. (1972). Reproductive schedule and nesting success of the mountain white-crowned sparrow (*Zonotrichia leucophrys oriantha*) in the central Sierra Nevada. *Condor,* **74**, 152–63.

Morton, M. L. and Pereyra, M. E. (1985). The regulation of egg temperatures and attentiveness patterns in the dusky flycatcher (*Empidonax oberholseri*). *Auk*, **102**, 25–37.

Morton, M. L., Sockman, K. W. and Peterson, L. E. (1993). Nest predation in the mountain white-crowned sparrow. *Condor,* **95**, 72–82.

Moskát, C. and Fuisz, T. I. (1999). Reactions of red-backed shrikes *Lanius collurio* to artificial cuckoo *Cuculus canorus* eggs. *Journal of Avian Biology*, **30**, 175–81.

Mueller, W. J. (1958). Shell porosity of chicken eggs 1. CO_2 loss and CO_2 content of infertile eggs during storage. *Poultry Science*, **37**, 437–44.

Mugaas, J. N. and King, J. R. (1981). Annual variation of daily energy expenditure by the Black-billed Magpie. A study of thermal and behavioral energetics. *Studies in Avian Biology*, **5**, 1–78.

Murphy, E. C. and Haukioja, E. (1986). Clutch size in nidicolous birds. In: *Current Ornithology, Volume 4* (ed. R. F. Johnston), pp. 141–80. Plenum press, New York.

Murton, R. K. and Westwood, N. J. (1977). *Avian Breeding Cycles.* Clarendon Press, Oxford.

Nager, R. G. and von Noordwijk, A. J. (1992). Energetic limitation in the egg-laying period of Great Tits. *Proceedings of the Royal Society of London B*, **249**, 259–63.

Nair, G. and Dawes, C. M. (1980). The effects of cooling the egg on the respiratory movements of the hatching quail (*Coturnix c. japonica*). *Comparative Biochemistry and Physiology,* **67**, 587–92.

Nakumara, S., Hashimoto, H. and Sootome, O. 1984. Breeding ecology of *Motacilla alba* and *M. grandis* and their interspecific relationship. *Journal of the Yamashima Institute of Ornithology*, **16**, 114–35.

Nakamura, T. and Cruz, A. (2000). The ecology of egg-puncture behavior by the shiny cowbird in southwestern Puerto Rico. In: *Ecology and Management of Cowbirds and Their Hosts. Studies in the Conservation of North American Passerine Birds* (eds J. M. N. Smith, T. L. Cook, S. I. Rothstein, S. K. Robinson and S. G. Sealy), pp. 178–86. University of Texas Press, Austin.

Naylor, B. J., Szuba, K. J. and Bendell, J. F. (1988). Nest cooling and recess length of incubating spruce grouse. *Condor*, **90**, 489–92.

Nelson, J. B. (1965). The behaviour of the gannet. *British Birds*, **58**, 313–36.

Nelson, J. B. (1978). *The Sulidae, Gannets and Boobies*. Oxford University Press, Oxford.

Neudorf, D. L. and Sealy, S. G. (1994). Sunrise nest attentiveness in cowbird hosts. *Condor*, **96**, 162–9.

Neufeld, L. J. (1998). *Egg Colour Variation, Nest Depredation, and Egg Size Variation in Brewer's Blackbirds (*Euphagus cyanocephalus*)*. BSc. Thesis, University of Manitoba, Winnipeg.

New, D. A. T. (1957). A critical period for the turning of hens' eggs. *Journal of Embryology and Experimental Morphology*, **5**, 293–9.

Nice, M. M. (1937). Studies in the life history of the song sparrow, Volume 1. A population study of the song sparrow. *Transactions of the Linnaean Society of New York*, **4**, 1–247.

Nice, M. M. (1954). Incubation periods throughout the ages. *Centaurus*, **3**, 311–59.

Nice, M. M. (1962). Development of behavior in precocial birds. *Transactions of the Linnean Society of New York*, **8**, 1–211.

Nicolaus, L. K. (1987). Conditioned aversions in a guild of egg predators: Implications for aposematism and prey defense mimicry. *American Midland Naturalist*, **117**, 405–19.

Nicolaus, L. K., Cassel, J. F., Carlson, R. B. and Gustavson, C. R. (1983). Taste-aversion conditioning of crows to control predation on eggs. *Science*, **220**, 212–4.

Niethammer, K. R. and Patrick-Caistilaw, L. B. (1998). White Tern (*Gygis alba*). In: *The Birds of North America, No. 371* (eds A. Poole and F. Gill). The Academy of Natural Sciences, Philadelphia and, The American Ornithological Union, Washington DC.

Nilsson, J.-A. (1993a). Energetic constraints on hatching asynchrony. *American Naturalist*, **141**, 158–66.

Nilsson, J.-A. (1993b). Bisexual incubation facilitates hatching asynchrony. *American Naturalist*, **142**, 712–7.

Nilsson, J. Å. (1994). Energetic bottle-necks during breeding and the reproductive cost of being too early. *Journal of Animal Ecology*, **63**, 200–8.

Nilsson, J. Å. and Smith, H. G. (1988). Incubation feeding as a male tactic for early hatching. *Animal Behaviour*, **36**, 641–7.

Nilsson, S. G. (1984). The evolution of nest-site selection among hole nesting birds: the importance of nest predation and competition. *Ornis Scandinavica*, **15**, 167–75.

Noble, D. G. (1995). *Coevolution and Ecology of Seven Sympatric Cuckoo Species and Their Hosts in Namibia*. Ph. D. Thesis, University of Cambridge.

Noble, G. K. and Lehrman, D. S. (1940). Egg recognition by the laughing gull. *Auk*, **57**, 22–43.

Nol, E. and Blanken, M. S. (1999). Semipalmated Plover (*Charadrius semipalmatus*). In: T*he Birds of North America, No. 444* (eds A. Poole and F. Gill). The Academy of Natural Sciences, Philadelphia: The American Ornithologists' Union, Washington, DC.

Nolan, V. (1978). The ecology and behavior of the prairie warbler *Dendroica discolor*. *Ornithological Monographs*, **26**. American Ornithologist's Union.

Norell, M. A., Clark, J. M., Chiappe, L. M. and Dashzeveg, D. (1995). A nesting dinosaur. *Nature*, **378**, 774–6.

Norell, M. A., Clark, J. M., Dashzeveg, D., Barsbold, R., Chiappe, L. M., Davidson, A. R., Mckenna, M. C., Perle, A. and Novacek M. J. (1994). A theropod dinosaur embryo and the affinities of the Flaming Cliffs dinosaur eggs. *Science, New York*, **266**, 779–82.

Norton-Griffiths, M. (1969). The organisation, control, and development of parental feeding in the Oystercatcher (*Haematopus ostralegus*). *Behaviour*, **34**, 55–114.

Nur, N. (1984). The consequences of brood size for breeding Blue Tits II. Nestling weight, offspring survival and optimal brood size. *Journal of Animal Ecology*, **53**, 497–517.

Nur, N. (1988). The consequences of brood size for breeding Blue Tits. III measuring the cost of reproduction: survival, future fecundity and differential dispersal. *Evolution*, **42**, 351–62.

Nuttall, P. A. (1997). Viruses, bacteria and fungi of birds. In: *Host-parasite evolution: General principles and avian models* (eds D. H. Clayton and J. Moore), pp. 271–302. Oxford University Press, New York.

O'Connor, M. P. (1999). Physiological and ecological implications of a simple model of heating and cooling in reptiles. *Journal of Thermal Biology*, **24**, 113–36.

Øien, I. J., Honza, M., Moksnes, A. and Røskaft, E. (1996). The risk of parasitism in relation to the distance from reed warbler nests to cuckoo perches. *Journal of Animal Ecology*, **65**, 147–53.

Øien, I. J., Moksnes, A. and Røskaft, E. (1995). Evolution of variation in egg color and marking pattern in European passerines: Adaptations in a coevolutionary arms race with the cuckoo, *Cuculus canorus*. *Behavioral Ecology*, **6**, 166–74.

Oniki, Y. (1979). Nest-egg combinations: Possible antipredatory adaptations in Amazonian birds. *Revista Brasileira de Biologia*, **39**, 747–67.

Oniki, Y. (1985). Why robin eggs are blue and birds build nests: Statistical tests for Amazonian birds. *Ornithological Monographs*, **36**, 536–45.

Oosterwoud, A. (1987). Effect of egg handling on egg quality. In: *Egg Quality – Current Problems and Recent Advances* (eds R. G. Wells and C. G. Belyavin), pp. 283–91. Butterworths, London.

Opel, H. and Proudman, J. A. (1988). Effects of poults on plasma concentrations of prolactin in turkey hens incubating without eggs or a nest. *British Poultry Science*, **29**, 791–800.

Opel, H. and Proudman, J. A. (1989). Plasma prolactin levels in incubating turkey hens during pipping of the eggs and after introduction of poults into the nest. *Biology of Reproduction*, **40**, 981–7.

Oppenheim, R. W. (1972). Prehatching and hatching behaviour in birds: a comparative study of altricial and precocial species. *Animal Behaviour*, **20**, 644–55.

Oppenheim, R. W. and Levin, H. L. (1974). Short-term changes in incubation temperature: Behavioral and physiological effects in the chick embryo from 6 to 20 days. *Developmental Psychobiology*, **8**, 103–15.

Orians, G. H. and Janzen, D. H. (1974). Why are embryos so tasty? *American Naturalist*, **108**, 581–92.

Orians, G. H. and Pearson, N. E. (1979). On the theory of central place foraging. In: *Analysis of Ecological Systems* (eds D. J. Horn, D. R. Mitchell and G. R. Stairs), pp. 155–77. Ohio State University Press, Columbus, Ohio.

Oring, L., Fivizzani, A., Colwell, M. and El Halawani, M. E. (1988). Hormonal changes associated with natural and manipulated incubation in the sex-role reversed Wilson's phalarope. *General and Comparative Endocrinology*, **72**, 247–56.

Oring, L., Fivizzani, A. and El Halawani, M. E. (1986b). Changes in plasma prolactin associated with laying and hatch in the Spotted Sandpiper. *Auk*, **103**, 820–2.

Oring, L., Fivizzani, A., El Halawani, M. E. and Goldsmith, A. (1986a). Seasonal changes in prolactin and luteinizing hormone in the polyandrous spotted sandpiper, *Actitis macularia*. *General and Comparative Endocrinology*, **62**, 394–403.

Oring, L., Fivizzani, A. J. and El Halawani, M. E. (1989). Testosterone-induced inhibition of incubation in the spotted sandpiper (*Actitus macularia*). *Hormones and Behavior*, **23**, 412–23.

Oring, L. W. and Fivizzani, A. J. (1991). Reproductive endocrinology of sex-role reversal. In: *Acta Congressus Internationalis Ornithologici XX* Vol. 4 (eds B. D. Bell, R. O. Cossee, J. E. C. Flux, B. D. Heather, R. A. Hitchmough, C. J. R. Robertson and M. J. Williams), pp. 2072–80. New Zealand Ornithological Congress Trust Board, Wellington.

Ortega, C. P. (1998). *Cowbirds and Other Brood Parasites*. University of Arizona Press, Tucson.

Ortega, C. P. and Cruz, A. (1991). A comparative study of cowbird parasitism in yellow-headed blackbirds and red-winged blackbirds. *Auk*, **109**, 16–24.

Ortega, C. P., Ortega, J. C., Rapp, C. A. and Backensto, S. A. (1998). Validating the use of artificial nests in predation experiments. *Journal of Wildlife Management*, **62**, 925–32.

Osuga, D. T. and Feeney, R. E. (1968). Biochemistry of the egg-white proteins of the ratite group. *Archives of Biochemistry and Biophysics*, **124**, 560–74.

Osborn, H. F. (1924). Three new theropods, *Protoceratops* zone, central Mongolia. *American Museum Novitates*, **144**, 1–12.

Packard, G. C., Sotherland, P. R. and Packard, M. J. (1977). Adaptive reduction in permeability of avian eggshells to water vapour at high altitudes. *Nature*, **266**, 255–6.

Packard, M. J. (1996). Patterns of mobilization and deposition of calcium in embryos of oviparous, amniotic vertebrates. *Israel Journal of Zoology*, **40**, 481–92.

Paganelli, C. V. (1991). The avian eggshell as a mediating barrier: respiratory gas fluxes and pressures during development. In: *Egg Incubation: Its Effects on Embryonic Development in Birds and Reptiles* (eds D. C. Deeming and M. W. J. Ferguson), pp. 261–75. Cambridge University Press, Cambridge.

Paganelli, C. V., Ackerman, R. A. and Rahn, H. (1978). The avian egg: *in vivo* conductances to oxygen, carbon dioxide, and water vapor in late development. In: *Respiratory Function in Birds, Adult and Embryonic* (ed. J. Piiper), pp. 212–8. Springer-Verlag, Berlin.

Paganelli, C. V., Olszowka, A. and Ar, A. (1974). The avian egg: surface area, volume and density. *Condor*, **76**, 319–25.

Paganelli, C. V., Ar, A., Rahn, H. and Wangensteen, O. D. (1975). Diffusion in the gas phase: the effects of ambient pressure and gas composition. *Respiration Physiology*, **25**, 247–58.

Paganelli, C. V., Sotherland, P. R., Olszowka, A. and Rahn, H. (1988). Regional differences in diffusive conductance/perfusion ratio in the shell of the hen's egg. *Respiration Physiology*, **71**, 45–56.

Pages, T., Fuster, J. F. and Palacios, L. (1991). Thermal responses of the fresh water turtle *Mauremys caspica* to step-function changes in the ambient temperature. *Journal of Thermal Biology*, **16**, 337–43.

Palmer, C., Christian, K. A. and Fisher, A. (2000). Mound characteristics and behaviour of the orange-footed scrubfowl in the seasonal tropics of Australia. *Emu*, **100**, 54–63.

Palmer, R. S. (1962). *Handbook of North American Birds. Volume 1*. Yale University Press, New Haven.

Palmgren, M. and Palmgren, P. (1939). Uber die Warmeisolierungskapazitatverschiedener kleinvogelnester. *Ornis Fennica*, **16**, 1–6.

Parker, J. W. (1999). Mississippi Kite (*Ictinia mississippiensis*). In: *The Birds of North America, No.* 402 (eds A. Poole and F. Gill). The Academy of Natural Sciences, Philadelphia: The American Ornithologists' Union, Washington, DC.

Parker, T. A., Parker, S. A. and Plenge, M. A. (1982). *An Annotated Checklist of Peruvian Birds*. Buteo Books, Vermillion, South Dakota.

Parsons, J. (1976). Factors determining the number and size of eggs laid by the herring gull. *Condor*, **78**, 481–92.

Partridge, L. (1992). Measuring reproductive costs. *Trends in Ecology and Evolution*, **7**, 99–100.

Patel, M. D. (1926). The physiology of the pigeon's milk. *Physiological Zoology*, **9**, 129–31.

Payne, R. B. (1966). The absence of brood patch in Cassin auklets. *Condor*, **68**, 209–10.

Payne, R. B. (1973a). Behavior, mimetic songs and song dialects, and relationships of the parasitic indigobirds (*Vidua*) of Africa. *Ornithological Monographs*, **11**, 1–333.

Payne, R. B. (1973b). Individual laying histories and the clutch size and numbers of eggs of parasitic cuckoos. *Condor*, **75**, 414–38.

Payne, R. B. (1974). The evolution of clutch size and reproductive rates in parasitic cuckoos. *Evolution*, **28**, 169–81.

Payne, R. B. (1976). The clutch size and numbers of eggs of brown-headed cowbirds: Effects of latitude and breeding season. *Condor*, **78**, 337–42.

Payne, R. B. (1977a). The ecology of brood parasitism in birds. *Annual Review of Ecology and Systematics*, **8**, 1–28.

Payne, R. B. (1977b). Clutch size, egg size, and the consequences of single vs. multiple parasitism in parasitic finches. *Ecology*, **58**, 500–13.

Payne, R. B. (1980). Seasonal incidence of breeding, moult and local dispersal of Red-billed Firefinches *Lagonosticta senegala* in Zambia. *Ibis*, **122**, 43–56.

Payne, R. B. (1988). Clutch size, laying periodicity and behaviour in the honeyguides *Indicator indicator* and *I. minor*. *Proceedings of the Pan-African Ornithological Congress*, **5**, 537–47.

Payne, R. B. (1989). Egg size of African honeyguides (Indicatoridae): Specialization for brood parasitism? *Tauraco*, **1**, 201–10.

Payne, R. B. (1997). Avian brood parasitism. In: *Host-Parasite Evolution. General Principles and Avian Models* (ed. D. H. Clayton and J. Moore), pp. 338–69. Oxford University Press, Oxford.

Payne, R. B. and Payne, L. L. (1998). Nestling eviction and vocal begging behaviors in the Australian glossy cuckoos *Chysococcyx basalis* and *C. lucidus*. In: *Parasitic Birds and Their Hosts: Studies in Coevolution* (eds S. I. Rothstein and S.K. Robinson), pp. 152–69. Academic Press, New York.

Peakall, D. B. (1960). Nest records of the yellowhammer. *Bird Study*, 7, 94–103.

Pearson, J. T., Haque, M. A., Hou, P.-C. L. and Tazawa, H. (1996). Developmental patterns of O_2 consumption, heart rate and O_2 pulse in unturned eggs. *Respiration Physiology*, **103**, 83–7.

Pearson, T. G. and Burroughs, J. (1936). *Birds of America*. Garden City Publishing Co. Ltd., New York.

Pedersen, H. C. (1989). Effects of exogenous prolactin on parental behaviour in free-living female willow ptarmigan *Lagopus l. lagopus*. *Animal Behaviour*, **38**, 926–34.

Peer, B. D. (1998). *An Experimental Investigation of Egg Rejection Behavior in the Grackles (Quiscalus)*. Ph.D. Thesis, University of Manitoba, Winnipeg.

Peer, B. D. and Bollinger, E. K. (1997). Explanations for the infrequent cowbird parasitism on common grackles. *Condor*, **99**, 151–61.

Peer, B. D. and Bollinger, E. K. (1998). Rejection of cowbird eggs by mourning doves: A manifestation of nest usurpation? *Auk*, **115**, 1057–62.

Peer, B. D. and E. K. Bollinger. (2000). Why do female brown-headed cowbirds remove host eggs? A test of the incubation efficiency hypothesis. In: *Ecology and Management of Cowbirds and Their Hosts. Studies in the Conservation of North American Passerine Birds* (eds J. M. N. Smith, T. L. Cook, S. I. Rothstein, S. K. Robinson and S. G. Sealy), pp. 187–92. University of Texas Press, Austin.

Peer, B. D., Robinson, S. K. and Herkert, J. K. (2000). Egg rejection by cowbird hosts in grasslands. *Auk*, **117**, 892–901.

Peer, B. D. and Sealy, S. G. (1999a). Laying time of the bronzed cowbird. *Wilson Bulletin*, **111**, 137–9.

Peer, B. D. and Sealy, S. G. (1999b). Parasitism and egg puncture behavior by bronzed cowbirds and brown-headed cowbirds in sympatry. In: *Research and Management of the Brown-headed Cowbird in Western Landscapes* (eds M. L. Morrison, L. S. Hall, S. K. Robinson, S. I. Rothstein, D. C. Hahn and T. D. Rich), *Studies in Avian Biology*, **18**, 235–40.

Peer, B. D. and Sealy, S. G. (2000a). Conspecific brood parasitism and egg rejection in great-tailed grackles. *Journal of Avian Biology*, **31**, 271–7.

Peer, B. D. and Sealy, S. G. (2000b). Responses of scissor-tailed flycatchers (*Tyrannus forficatus*) to experimental cowbird parasitism. *Bird Behavior*, **13**, 63–7.

Perrins, C. M. (1967). The short apparent incubation period of the cuckoo. *British Birds*, **60**, 51–2.

Perrins, C. M. (1977). The role of predation in the evolution of clutch size. In: *Evolutionary Biology* (eds B. Stonehouse and C. M. Perrins), pp. 181–92. Macmillan Press, London.

Perrins, C. M. and McCleery, R. H. (1985). The effect of age and pair bond on the breeding success of great tits, *Parus major*. *Ibis*, **127**, 306–315.

Perrins, C. R. and Middleton, A. L. A. (1985). *The Encyclopaedia of Birds*. George Allen and Unwin, London.

Persson, I. and Goransson, G. (1999). Nest attendance during egg laying in pheasants. *Animal Behaviour*, **58**, 159–64.

Peters, R. H. (1983). *The Ecological Implications of Body Size*. Cambridge University Press, Cambridge.

Peterson, A. J. (1955). The breeding cycle in the bank swallow. *Wilson Bulletin*, **67**, 235–86.

Peterson, K. L. and Best, L. B. (1985a). Nest-site selection by sage sparrows. *Condor*, **87**, 217–21.

Peterson, K. L. and Best, L. B. (1985b). Brewer's sparrow nest-site characteristic in a sagebrush community. *Journal of Field Ornithology*, **56**, 23–7.

Peterson, M. R., Schmutz, J. A. and Rockwell, R. F. (1994). Emperor Goose (*Chen canagica*). In: *The Birds of North America, No.* 97 (eds A. Poole and F. Gill). The Academy of Natural Sciences, Philadelphia: The American Ornithologists' Union, Washington, DC.

Petit, L. J. (1991). Adaptive tolerance of cowbird parasitism by prothonotory warblers: A consequence of nest-site limitation? *Animal Behaviour*, **41**, 425–32.

Pettifor, R. A., Perrins, C. M. and McCleery, R. H. (1988). Individual optimisation of clutch size in Great Tits. *Nature*, **336**, 160–2.

Pettingill, O. S., Jr. (1985). *Ornithology in Laboratory and Field*, 5th edition. Academic Press, Orlando.

Picman, J. (1989). Mechanism of increased puncture resistance of eggs of brown-headed cowbirds. *Auk*, **106**, 577–83.

Picman, J. (1997). Are cowbird eggs unusually strong from the inside? *Auk*, **114**, 66–73.

Picman, J. and Pribil, S. (1997). Is greater eggshell density an alternative mechanism by which parasitic cuckoos increase the strength of their eggs? *Journal für Ornithologie*, **138**, 531–41.

Picman, J., Pribil, S. and Picman, A. K. (1996). The effect of intraspecific egg destruction on the strength of marsh wren eggs. *Auk*, **113**, 599–607.

Piersma, T., Lindström, A., Drent, R. H., Tulp, I., Jukema, J., Morrisson, R. I. G., Reneerkens, J., Schekkerman, H. and Visser, G. H. (2001). Daily energy expenditure of high arctic shorebirds: a circumpolar data set confirms high metabolic costs during incubation? *Journal of Animal Ecology*, in press.

Pijanowski, B. C. (1992). A revision of Lack's brood reduction hypothesis. *American Naturalist*, **139**, 1270–92.

Pinxten, R., Eens, M. and Verheyen, R. F. (1993). Male and female nest attendance during incubation in the facultatively polygynous European Starling. *Ardea*, **81**, 125–33.

Pitelka, F. A. (1940). Breeding behavior of the black-throated green warbler. *Wilson Bulletin*, **52**, 3–18.

Polis, G. A. ed. (1991). *The Ecology of Desert Communities*. University of Arizona Press, Tucson.

Ponomareva, T. (1971). Role of environmental factors and nest properties in incubating birds behavior. *Zoologicheskii Zhurnal*, **50**, 1709–19.

Porter, W. P. and Gates, D. M. (1969). Thermodynamic equilibria of animals with environment. *Ecological Monographs*, **39**, 227–44.

Portman, A. (1961). Sensory organs: skin, taste and olfaction. In: *Biology and Comparative Physiology of Birds, Volume 2* (ed. A. J. Marshall), pp. 37–48. New York, Academic Press.

Potter, E. F. (1980). Notes on nesting Yellow-billed cuckoos. *Journal of Field Ornithology*, **51**, 17–29.

Potter, E. F. (1989). Egg-turning by Northern Cardinal prior to onset of incubation. *The Chat*, **53**, 4–7.

Potti, J. (1998). Variation in the onset of incubation in the pied flycatcher (*Ficedula hypoleuca*): fitness consequences and constraints. *Journal of Zoology, London*, **245**, 335–44.

Potts, P. L. and Washburn, K. W. (1974). Shell evaluation of white and brown egg strains by deformation, breaking strength, shell thickness and specific gravity. 1. Relationship to egg characteristics. *Poultry Science*, **53**, 1123–8.

Poulsen, H. (1953). A study of incubation responses and some other behaviour patterns in birds. *Vidensk. Medd. Fra Dansk naturh. Foren.*, **115**, 1–131.

Poulton, E. B. (1890). *The Colours of Animals*. Kegan Paul, Trench, Trübner, and Co. Ltd., London.

Poussart, C., Larochelle, J. and Gauthier, G. (2000). The thermal regime of eggs during laying and incubation in greater snow geese. *Condor*, **102**, 292–300.

Power, H. W., Litovich, E. and Lombardo, M. P. (1981). Male Starlings delay incubation to avoid being cuckolded. *Auk*, **98**, 386–9.

Powers, D. R. (1996). Magnificent Hummingbird. In: *Birds of North America, No. 221* (eds A. Poole and P. Stettenheim). American Ornithologists' Union and Academy of Natural Sciences, Philadelphia.

Powers, D. R. and Wethington, S. M. (1999). Broad-billed Hummingbird. In: *Birds of North America, No. 430* (eds A. Poole and P. Stettenheim). American Ornithologists' Union and Academy of Natural Sciences, Philadelphia.

Pratt, H. M. (1970). Breeding biology of Great blue herons and common egrets in central California. *Condor*, **72**, 407–16.

Preston, F. W. (1957). Pigmentation of eggs: Variation in the clutch sequence. *Auk*, **74**, 28–41.

Priddel, D. and Wheeler, R. (1990). Survival of Malleefowl chicks in the absence of ground-dwelling predators. *Emu*, **90**, 81–7.

Prinzinger, R. (1974). Untersuchungen über das verhalten des Schwarzhalstauchers *Podiceps n. nigricollis*, Brehm (1831). *Anzeiger Ornithologischen Gesellschaft in Bayern*, **13**, 1–34.

Prinzinger, R. (1979). Der Schwarzhalstaucher *Podiceps nigricollis*. Die Neue Brehm-Bucherei 521, Ziemsen Verlag, Wittenberg-Lutherstadt.

Prinzinger, R. (1992). Energy costs of incubation in the Blackbird *Turdus merula*. *Der Ornithologische Beobachter*, **89**, 111–25.

Prinzinger, R., Lübben, I. and Schuchmann, K.-L. (1989). Energy metabolism and body temperature in 13 sunbird species (Nectariniidae). *Comparative Biochemistry and Physiology*, **92A**, 393–402.

Proctor, H. and Owens, I. (2000). Mites and birds: diversity, parasitism and coevolution. *Trends in Ecology and Evolution*, **15**, 358–64.

Pugh, G. F. J. (1972). The contamination of birds' feathers by fungi. *Ibis*, **114**, 172–7.

Pugh, G. J. F. and Evans, M. D. (1970a). Keratinophilic fungi associated with birds. I. Fungi isolated from feathers, nests and soils. *Transactions of the British Mycological Society*, **54**, 233–40.

Pugh, G. J. F. and Evans, M. D. (1970b). Keratinophilic fungi associated with birds II. Physiological studies. *Transactions of the British Mycological Society*, **54**, 241–50.

Pulliainen, E. (1971). Behaviour of a nesting capercaille (*Tetrao urogallus*) in north-eastern Lapland. *Ann. Zool. Fennici*, **8**, 456–62.

Pulliainen, E. (1978). Behaviour of a willow grouse (*Lagopus l. lagopus*) at the nest. *Ornis Fennica*, **55**, 141–8.

Punnett, R. C. (1933). Inheritance of egg-colour in the 'parasitic' cuckoos. *Nature*, **132**, 892–3.

Putnam, L. S. (1949). The life history of the Cedar Waxwing. *Wilson Bulletin*, **61**, 141–82.

Rahn, H. (1977). Adaptation of the avian embryo to altitude: the role of gas diffusion through the eggshell. In: *Respiratory Adaptations, Capillary Exchange, and Reflex Mechanisms* (eds A. S. Paintal and P. Gill-Kumar), pp. 94–105. Vallabhbhai Patel Chest Institute, University of Delhi, Delhi.

Rahn, H. (1984). Factors controlling the rate of incubation water loss in bird eggs. In: *Respiration and Metabolism of Embryonic Vertebrates* (ed. R. S. Seymour), pp. 271–88. Dr. W. Junk, Dordrecht.

Rahn, H. (1991). Why birds lay eggs. In: *Egg Incubation. Its Effects on Embryonic Development in Birds and Reptiles* (eds D. C. Deeming and M. W. J. Ferguson) pp. 345–60. Cambridge University Press, Cambridge.

Rahn, H., Ackerman, R. A. and Paganelli, C. V. (1977a). Humidity in the avian nest and egg water loss during incubation. *Physiological Zoology*, **50**, 269–83.

Rahn, H. and Ar, A. (1974). The avian egg: incubation time and water loss. *Condor*, **76**, 147–52.

Rahn, H. and Ar, A. (1980). Gas exchange of the avian egg: time, structure, and function. *American Zoologist*, **20**, 477–84.

Rahn, H., Ar, A. and Paganelli, C. V. (1979). How bird eggs breathe. *Scientific American*, **240**, 46–55.

Rahn, H., Carey, C., Balmas, K., Bhatia, B. and Paganelli, C. V. (1977b). Reduction of pore area of the avian eggshell as an adaptation to altitude. *Proceedings of the National Academy of Science USA*, **74**, 3095–8.

Rahn, H., Curran-Everett, L. and Booth, D. T. (1988). Eggshell differences between parasitic and nonparasitic Icteridae. *Condor*, **90**, 962–4.

Rahn, H. and Dawson, W. R. (1979). Incubation water loss in eggs of Heermann's and western gulls. *Physiological Zoology*, **52**, 451–60.

Rahn, H. and Hammel, H. T. (1982). Incubation water loss, shell conductance, and pore dimensions in Adélie penguin eggs. *Polar Biology*, **1**, 91–7.

Rahn, H., Krog, J. and Mehlum, F. (1983). Microclimate of the nest and egg water loss of the eider *Somateria mollissima* and other waterfowl in Spitzsbergen. *Polar Research*, **1**, 171–83.

Rahn, H., Ledoux, T., Paganelli, C. V. and Smith, A. H. (1982). Changes in eggshell conductance after transfer of hens from an altitude of 3,800 m to 1,200 m. *Journal of Applied Physiology*, **53**, 1429–31.

Rahn, H. and Paganelli, C. V. (1988a). Frequency distribution of egg mass of passerine and non-passerine birds based on Schönwetter's tables. *Journal für Ornithologie*, **129**, 236–9.

Rahn, H. and Paganelli, C. V. (1988b). Length, breadth and elongation of avian eggs from the tables of Schönwetter. *Journal für Ornithologie*, **129**, 366–9.

Rahn, H. and Paganelli, C. V. (1989). Shell mass, thickness and density of avian eggs derived from the tables of Schönwetter's. *Journal für Ornithologie*, **130**, 59–68.

Rahn, H. and Paganelli, C. V. (1990). Gas fluxes in avian eggs: driving forces and the pathway for exchange. *Comparative Biochemistry and Physiology*, **95A**, 1–15.

Rahn, H., Paganelli, C. V. and Ar, A. (1974). The avian egg: air-cell gas tension, metabolism and incubation time. *Respiration Physiology*, **22**, 297–309.

Rahn, H., Paganelli, C. V. and Ar, A. (1975). Relation of avian egg weight to body weight. *Auk*, **92**, 750–65.

Rahn, H., Paganelli, C. V., Nisbet, I. C. T. and Whittow, G. C. (1976). Regulation of incubation water loss in eggs of seven species of terns. *Physiological Zoology*, **49**, 245–59.

Ramesh, R., Proudman, J. A. and Kuenzel, W. J. (1995). Changes in pituitary somatotrophs and lactotrophs associated with ovarian regression in the turkey hen (*Meleagris gallopavo*). *Comparative Biochemistry and Physiology*, **112C**, 327–34.

Ramsay, S. L. (1997). *Nutritional constraints on egg production in the blue tit*. PhD thesis, University of Glasgow, Scotland.

Ramsey, S. M., Goldsmith, A. R. and Silver, R. (1985). Stimulus requirements for prolactin and LH secretion in incubating ring doves. *General and Comparative Endocrinology*, **59**, 246–56.

Rand, A. L. and Rand, R. M. (1943). Breeding notes on the Phainopepla. *Auk*, **60**, 333–40.

Rauter, C. and Reyer, H.-U. (1997). Incubation pattern and foraging effort in female water pipit *Anthus spinoletta*. *Ibis*, **139**, 441–6.

Rawles, M. E. (1960). The integumentary system. In: *Biology and Comparative Physiology of Birds, Volume 1* (ed. A. J. Marshall), pp. 189–240. New York, Academic Press.

de Reamur, R. A. F. (1751). *Konst om Tamme-vogelen van Allerhandesoort in alle Jaartyden Uittebroeijen en Opiebrengen zp door 't Middle van Mest als van't Gewone Vuur, Vol. 1*, 1–384, Pieter de Hondt, s'Gravenhage.

Regel, J. and Pütz, K. (1997). Effect of human disturbance on body temperature and energy expenditure in penguins. *Polar Biology*, **18**, 246–53.

Reid, J. M., Monaghan, P. and Ruxton, G. D. (1999). The effect of clutch cooling rate on Starling incubation strategy. *Animal Behaviour*, **58**, 1161–7.

Reid, J. M., Monaghan, P. and Ruxton, G. D. (2000a). The consequences of clutch size for incubation conditions and hatching success in starlings. *Functional Ecology*, **14**, 560–5.

Reid, J. M., Monaghan, P. and Ruxton, G. D. (2000b). Resource allocation between reproductive phases: the importance of thermal conditions in determining the cost of incubation. *Proceedings of the Royal Society of London B*, **267**, 37–41.

Reid, J. M., Ruxton, G. D. Monaghan, P. and Hilton, G. M. (2002). Energetic consequences of clutch temperature and clutch size for a uniparental intermittent incubator. *Auk*, in press.

Remmert, H. (1980). *Arctic Animal Ecology*. Springer-Verlag, Berlin.

Rensch, B. (1924). Zur Entstehung der Mimikry der Kuckuckseier. *Journal für Ornithologie*, **72**, 461–72.

Rensch, B. (1925). Verhalten von Singvögeln bei Aenderung des Geleges. *Ornithologische Monatsberichte*, **33**, 169–73.

Reznick, D. (1985). Costs of reproduction: an evaluation of the empirical evidence. *Oikos*, **44**, 257–67.

Reznick, D. (1992). Measuring the costs of reproduction. *Trends in Ecology and Evolution*, **7**, 42–5.

Reznick, D., Perry, E. and Travis, J. (1986). Measuring the cost of reproduction – a comment. *Evolution*, **40**, 1338–44.

Richard-Yris, M. A., Chadwick, A., Guémené, D., Grillou-Schuelke, H. and Leboucher, G. (1995). Influence of the presence of chicks on the ability to resume incubation behavior in domestic hens (*Gallus domesticus*). *Hormones and Behavior*, **29**, 425–41.

Richard-Yris, M. A., Guemene, D., Lea, R. W., Sharp, P. J., Bedecarrats, G., Foraste, M. and Wauters, A. M. (1998b). Behaviour and hormone concentrations in nest deprived and renesting hens. *British Poultry Science*, **39**, 309–17.

Richard-Yris, M. A., Leboucher, G., Chadwick, A. and Garnier, D. H. (1987). Induction of maternal behavior in incubating and non-incubating hens: Influence of hormones. *Physiology and Behavior*, **40**, 193–200.

Richard-Yris, M. A., Sharp, P. J., Wauters, A. M., Guémené, D., Richard, J. P. and Forasté, M. (1998a). Influence of stimuli from chicks on behavior and concentrations of plasma prolactin and luteinizing hormone in incubating hens. *Hormones and Behavior*, **33**, 139–48.

Richards, P. D. G. and Deeming, D. C. (2001). Correlation between shell colour and ultrastructure in pheasant eggs. *British Poultry Science*, **42**, 338–42.

Richardson, M. K., Deeming, D. C. and Cope, C. (1998). Morphology of the distal tip of the upper beak of the ostrich (*Struthio camelus*) embryo during hatching. *British Poultry Science*, **39**, 575–8.

Richardson, M. K., Hanken, J., Gooneratne, M. L., Pieau, C., Raynaud, A., Selwood, L. and Wright, G. M. (1997). There is no highly conserved embryonic stage in the vertebrates: implications for current theories of evolution and development. *Anatomy and Embryology*, **196**, 91–106.

Ricklefs, R. E. (1969). An analysis of nesting mortality in birds. *Smithsonian Contributions to Zoology*, **9**, 1–48.

Ricklefs, R. E. (1974). Energetics of reproduction in birds. In: *Avian Energetics, Number 15* (ed. R. A. Paynter, Jr), pp. 152–292. Publications of the Nuttall Ornithology Club, Cambridge.

Ricklefs, R. E. (1992). Embryonic development period and the prevalence of avian blood parasites. *Proceedings of the National Academy of Sciences USA*, **89**, 4722–5.

Ricklefs, R. E. (1993). Sibling competition, hatching asynchrony, incubation period, and lifespan in altricial birds. *Current Ornithology*, **11**, 199–276.

Ricklefs, R. E. and Hainsworth, F. R. (1969a). Temperature regulation in nestling cactus wrens: the nest environment. *Condor,* **71**, 32–7.

Ricklefs, R. E. and Hainsworth, F. R. (1969b). Temperature dependent behavior of the cactus wren. *Ecology,* **49**, 227–33.

Ricklefs, R. E. and Hussell, D. J. T. (1984). Changes in adult mass associated with the nesting cycle in the European Starling. *Ornis Scandinavica*, **15**, 155–61.

Ricklefs, R. E. and Smeraski, C. A. (1983). Variation in incubation period within a population of the European starling. *Auk*, **100**, 926–31.

Ricklefs, R. E. and Starck, J. M. (1998). Embryonic growth and development. In: *Avian Growth and Development. Evolution within the Altricial-Precocial Spectrum.* (eds J. M. Starck and R. E. Ricklefs), pp. 31–58. Oxford University Press, Oxford.

Ricklefs, R. E. and Williams, J. B. (1984). Daily energy expenditure and water-turnover rate of adult European Starlings (*Sturnus vulgaris*) during the nesting cycle *Auk*, **101**, 707–16.

Riggs, D. S. (1963). *The Mathematical Approach to Physiological Problems*. Williams and Wilkens, Baltimore.

Robert, M. and Sorci, G. (1999). Rapid increase of host defence against brood parasites in a recently parsitized area: The case of village weavers in Hispaniola. *Proceedings of the Royal Society of London B*, **266**, 941–6.

Robertson, I. S. (1961). Studies on the effect of humidity on the hatchability of hens eggs I. The determination of optimum humidity for incubation. *Journal of Agricultural Science,* **57**, 185–95.

Robertson, S. L and Smith, E. N. (1981). Thermal conductance and its relation to thermal time constants. *Journal of Thermal Biology*, **6**, 129–43.

Robin, J. P., Frain, M., Sardet, C., Groscolas, R. and Le Maho, Y. (1988). Protein and lipid utilization during long term fasting in emperor penguins. *American Journal of Physiology* **254** *(Regulatory Integrative Comparative Physiology* **23**), R61–8.

Robinson, T. R., Sargent, R. R. and Sargent, M. B. (1996). Ruby-throated Hummingbird. In: *Birds of North America, No. 204* (eds A. Poole and P. Stettenheim). American Ornithologists' Union and Academy of Natural Sciences, Philadelphia.

Roff, D. A. (1992). The Evolution Of Life Histories; Theory And Analysis. Chapman and Hall, London.

Rofstad, G. and Sandvik, F. (1985). Variation in egg size of the hooded crow *Corvus corone cornix*. *Ornis Scandinavica*, **16**, 38–44.

Rohwer, F. C. and Freeman, S. (1989). The distribution of conspecific nest parasitism in birds. *Canadian Journal of Zoology*, **67**, 239–53.

Romanoff, A. L. (1960). *The Avian Embryo*. Macmillan, New York.

Romanoff, A. L. (1967). *Biochemistry of the Avian Embryo*. John Wiley and Sons, New York.

Romanoff, A. L. and Romanoff, A. J. (1949). *The Avian Egg*. John Wiley and Sons, New York.

Roper, D. S. (1983). Egg incubation and laying behaviour of the incubator bird *Megapodius freycinet* on Savo. *Ibis*, **125**, 384–9.

Rothstein, S. I. (1970). *An Experimental Investigation of the Defenses of the Hosts of the Parasitic Brown-headed Cowbird (*Molothrus ater*)*. Ph.D. Thesis, Yale University, New Haven.

Rothstein, S. I. (1975a). Mechanisms of avian egg-recognition: Do birds know their own eggs? *Animal Behaviour*, **23**, 268–78.

Rothstein, S. I. (1975b). An experimental and teleonomic investigation of avian brood parasitism. *Condor*, **77**, 250–71.

Rothstein, S. I. (1977). Cowbird parasitism and egg recognition of the northern oriole. *Wilson Bulletin*, **89**, 21–32.

Rothstein, S. I. (1982a). Successes and failures in avian egg and nestling recognition with comments on the utility of optimality reasoning. *American Zoologist*, **22**, 547–60.

Rothstein, S. I. (1982b). Mechanisms of avian egg recognition: Which egg parameters elicit responses by rejecter species? *Behaviorial Ecology and Sociobiology*, **11**, 229–39.

Rothstein, S. I. (1990). A model system for coevolution: Avian brood parasitism. *Annual Review of Ecology and Systematics*, **21**, 481–508.

Rothstein, S. I. (1992). Brood parasitism, the importance of experiments and host defences of avifaunas on different continents. *Proceedings of the Pan-African Ornithological Congress*, **7**, 521–35.

Rothstein, S. I. and Robinson, S. K. (1998). *Parasitic Birds and Their Hosts: Studies in Coevolution*. Oxford University Press, Oxford.

Rowan, M. K. (1983). *The Doves, Parrots, Louries and Cuckoos of Southern Africa*. David Philip, Cape Town.

Rowe, B. (1978). Incubation temperatures of the North Island brown kiwi (*Apteryx australis mantelli*). *Notornis*, **25**, 213–7.

Rowley, I. (1970). The use of mud in nest-building – A review of the incidence and taxonomic importance. *Ostrich*, **Supplement 8**, 139–48.

Royle, N. J., Surai, P. F., McCartney, R. J. and Speake, B. K. (1999). Parental investment and egg yolk lipid composition in gulls. *Functional Ecology*, **13**, 298–306.

Russell, S. M. (1969). Regulation of egg temperatures by incubating white-winged doves. In: *Physiological Systems in Semiarid Environments* (eds C. C. Hoff and M. L. Riedesel), pp. 107–12. University of New Mexico Press.

Russell, S. M. (1996). Anna's Hummingbird. In: *Birds of North America, No. 226* (eds A. Poole and F. Gill). The Academy of Natural Sciences, Philadelphia: The American Ornithologists' Union, Washington, D.C.

Sabbath, K. (1991). Upper Cretaceous amniotic eggs from the Gobi desert. *Acta Palaeontologica Polonica*, **36**, 151–92.

Sæther, B. E., Andersen, R. and Pedersen, H. C. (1993). Regulation of parental effort in a long-lived seabird: an experimental manipulation of the cost of reproduction in the antarctic petrel, *Thalassoica antarctica*. *Behavioral Ecology and Sociobiology*, **33**, 147–50.

Sahni, A., Tandon, S. K., Jolly, A., Bajpai, S., Sood, A. and Srinivasan, S. (1994). Upper Cretaceous dinosaur eggs and nesting sites from the Deccan volcano-sedimentary province of penisular India. In: *Dinosaur Eggs and Babies* (eds K. Carpenter, K. F. Hirsch and J. R. Horner), pp. 204–26. Cambridge University Press, Cambridge.

Saino, N. and Fasola, M. (1996). The function of embryonic vocalizations in the little tern (*Sterna albifrons*). *Ethology*, **102**, 265–71.

Salonen, V. and Penttinen, A. (1988). Factors affecting nest predation in the great crested grebe: Field observations, experiments and their statistical analysis. *Ornis Fennica*, **65**, 13–20.

Sandercock, B. K. (1997). Incubation capacity and clutch size determination in two calidrine sandpipers: a test of the four-egg threshold. *Oecologia*, **110**, 50–9.

Sankaran, R, and Sivakumar, K. (1999). Preliminary results of an ongoing study of the Nicobar megapode *Megapodius nicobariensis* Blyth. In: *Proceedings of the Third International Megapode Symposium, Nhill, Australia* (eds R.W.R.J. Dekker, D.N. Jones and J. Benshemesh), *Zoologische Verhandelingen* **327**, 75–90.

Sanz, J. J. (1996). Effect of food availability on incubation period in the Pied Flycatcher (*Ficedula hypoleuca*). *Auk*, **113**, 249–53.

Sanz, J. J. and Moreno, J. (1995). Experimentally induced clutch size enlargements affect reproductive success in the Pied Flycatcher. *Oecologia*, **103**, 358–64.

Sanz, J. J. and Tinbergen, J. M. (1999). Energy expenditure, nestling age and brood size: an experimental study of parental behavior in the Great Tit *Parus major*. *Behavioral Ecology*, **10**, 598–606.

Sauer, E. G. F and Sauer, E. M. (1966). The behaviour and ecology of the South African ostrich. *Living Bird*, **5**, 45–75.

Savage, D. L. (1897). Observations on the cowbird. *Iown Ornithologist*, **3**, 4–7.

Saylor, R. D. (1992). Ecology and evolution of brood parasitism in waterfowl. In: *Ecology and Management of Breeding Waterfowl* (eds B. D. J. Batt, A. D. Afton, M. G. Anderson, C. D. Ankney, D. H. Johnson, J. A. Kadlec and G. L. Krapu), pp. 290–322. University of Minnesota Press, Minneapolis.

Schaefer, V. (1976). Geographic variation in the placement and structure of oriole nests. *Condor*, **78**, 443–8.

Schaefer, V. H. (1980). Geographic variation in the insulative qualities of nests of the northern oriole. *Wilson Bulletin*, **92**, 466–74.

Schaffner, F. C. (1990). Egg recognition by elegant terns (*Sterna elegans*). *Colonial Waterbirds*, **13**, 25–30.

Schantz, W. E. (1944). All day record of an incubating robin *Turdus migratorius*. *Wilson Bulletin*, **56**, 118.

Schardien, B. J. and Jackson, J. A. (1979). Belly-soaking as a thermoregulatory mechanism in nesting killdeers. *Auk*, **96**, 604–6.

Schmidt-Nielson, K. (1964). *Desert Animals. Physiological Problems of Heat and Water*. Oxford University Press, Oxford.

Schmidt-Nielsen, K. (1984). *Scaling. Why Is Animal Size So Important?* Cambridge University Press, Cambridge.

Schoech, S. J., Mumme, R. L. and Wingfield, J. C. (1996). Prolactin and helping behaviour in the cooperatively breeding Florida scrub-jay, *Aphelocoma c. coerulescens*. *Animal Behaviour*, **52**, 445–56.

Schönwetter M. (1960–1985). *Handbuch der oologie* (ed. W. Meise). Akademie-Verlag, Berlin.

Schuchmann, K. L. (1999). Family Trochilidae (Hummingbirds). In: *Handbook of Birds of the World, Volume 5* (eds J. del Hoyo, A. Elliott and J. Sargatal), pp. 468–535. Lynx Edicions, Barcelona, Spain.

Schwagmeyer, P. L., Mock, D. W., Lamey, T. C., Lamey, C. S. and Beecher, M. D. (1991). Effects of sibling contact on hatch timing in an asynchronously hatching bird. *Animal Behaviour*, **41**, 887–94.

Schwagmeyer, P. L., St. Clair, R. C., Moodie, J. D., Lamey, T. C., Schnell, G. D. and Moodie, M. N. (1999). Species differences in male parental care in birds: a reexamination of correlates with paternity. *Auk*, **116**, 487–503.

Scott, D. M. (1991). The time of day of egg laying by the brown-headed cowbird and other icterines. *Canadian Journal of Zoology*, **69**, 2093–9.

Scott, D. M. and Ankney, C. D. (1980). Fecundity of the brown-headed cowbird in southern Ontario. *Auk*, **97**, 677–83.

Scott, D. M. and Ankney, C. D. (1983). Do Darwin's finches lay small eggs? *Auk*, **100**, 226–7.

Scott, D. M. and Lemon, R. E. (1996). Differential reproductive success of brown-headed cowbirds with Northern Cardinals and three other hosts. *Condor*, **98**, 259–71.

Sealy, S. G. (1984). Interruptions extend incubation by ancient murrelets, crested auklets, and least auklets. *Murrelet*, **65**, 53–6.

Sealy, S. G. (1992). Removal of yellow warbler eggs in association with cowbird parasitism. *Condor*, **94**, 40–54.

Sealy, S. G. (1995). Burial of cowbird eggs by parasitized yellow warblers: An empirical and experimental study. *Animal Behaviour*, **49**, 877–89.

Sealy, S. G. (1996). Evolution of host defenses against brood parasitism: Implications of puncture-ejection by a small passerine. *Auk*, **113**, 346–55.

Sealy, S. G. and Bazin, R. C. (1995). Low frequency of observed cowbird parasitism on eastern kingbirds: Host rejection, effective nest defense, or parasite avoidance? *Behavioral Ecology*, **6**, 140–5.

Sealy, S. G., Hobson, K. A. and Briskie, J. V. (1989). Responses of yellow warblers to experimental intraspecific brood parasitism. *Journal of Field Ornithology*, **60**, 224–9.

Sealy, S. G., Neudorf, D. L. and Hill, D. P. (1995). Rapid laying by brown-headed cowbirds *Molothrus ater* and other parasitic birds. *Ibis*, **137**, 76–84.

Sealy, S. G., Neudorf, D. L., Hobson, K. A. and Gill, S. A. (1998). Nest defense by potential hosts of the brown-headed cowbird: Methodological approaches, benefits of defense, and coevolution. In: *Parasitic Birds and Their Hosts: Studies in Coevolution* (eds S. I. Rothstein and S. K. Robinson), pp. 194–211. Oxford University Press, Oxford.

Seebohm, H. (1896). *Coloured Figures of the Eggs of British Birds*. Pawson and Brailsford, Sheffield.

Seel, D. C. (1968). Clutch-size, incubation and hatching success in the house sparrow and tree sparrow *Passer* spp. at Oxford. *Ibis*, **110**, 270–82.

Seel, D. C. and Davis, P. R. K. (1981). Cuckoos reared by unusual hosts in Britain. *Bird Study*, **28**, 242–3.

Selander, R. K. (1960). Failure of estrogen and prolactin treatment to induce brood patch formation in Brown-headed Cowbirds. *Condor*, **62**, 65.

Selander, R. K. and Kuich, L. L. (1963). Hormonal control and development of the incubation patch in icterids, with notes on behavior of cowbirds. *Condor*, **65**, 73–90.

Selander, R. K. and Yang, S. Y. (1966). The incubation patch of the house sparrow, *Passer domesticus* Linnaeus, *General and Comparative Endocrinology*, **6**, 325–33.

Senar, J. C. (1989). Do Cardueline finches help their young to hatch? *Le Gerfaut*, **79**, 185–7.

Serreze, M. C., Walsh, J. E., Chapini, F. S., Osterkamp, T., Dyurgerov, M., Romanovsky, V., Oechel, W. C., Morison, J., Zhang, T. and Barry, R. G. (2000). Observational evidence of recent change in the northern high-latitude environment. *Climatic Change*, **46**, 159–207.

Serventy, D. L. (1971). Biology of desert birds. In: *Avian Biology, Volume 1* (eds D. S. Farner, J. R. King and K. C. Parkes), pp. 287–339. Academic Press, New York.

Seviour, E. M., Sykes, F. R. and Board, R. G. (1972). A microbiological survey of the incubated eggs of chickens and waterfowl. *British Poultry Science*, **13**, 549–56.

Seymour, R. S. (1979). Dinosaur eggs: gas conductance through the shell, water loss during incubation and clutch size. *Palaeobiology*, **5**, 1–11.

Seymour, R. S. (1984a). *Respiration and Metabolism of Embryonic Vertebrates*. Dr W. Junk Publishers, Dordrecht.

Seymour, R. S. (1984b). Patterns of lung aeration in the perinatal period of domestic fowl and brush turkey. In: *Respiration and Metabolism of Embryonic Vertebrates* (ed. R. S. Seymour), pp. 319–32. Junk, Dordrecht.

Seymour, R. S. (1985). Physiology of megapode eggs and incubation mounds. *Acta XVIII Congressus Internationalis Ornithologici*, **2**, 254–63.

Seymour, R. S. (1991). The Brush turkey. *Scientific American*, **265(6)**, 68–74.

Seymour, R. S. and Ackerman, R. A. (1980). Adaptations to underground nesting in birds and reptiles. *American Zoologist*, **20**, 437–47.

Seymour, R. S. and Bradford, D. F. (1992). Temperature regulation in the incubation mounds of the Australian brush-turkey. *Condor*, **94**, 134–50.

Seymour, R. S., Vleck, D. and Vleck, C. M. (1986). Gas exchange in the incubation mounds of megapode birds. *Journal of Comparative Physiology B*, **156**, 773–82.

Seymour, R. S., Vleck, D., Vleck, C. M. and Booth, D. T. (1987). Water relations of buried eggs of mound building birds. *Journal of Comparative Physiology B*, **157**, 413–22.

Sharp, P. J. (1997). Immunological control of broodiness. *World's Poultry Science Journal*, **53**, 23–31.

Sharp, P. J., Dawson, A. and Lea, R. W. (1998). Control of luteinizing hormone and prolactin secretion in birds. *Comparative Biochemistry and Physiology*, **119C**, 275–82.

Sharp, P. J., Macnamee, M. C., Sterling, R. J., Lea, R. W. and Pedersen, H. C. (1988). Relationships between prolactin, LH and broody behaviour in bantam hens. *Journal of Endocrinology*, **118**, 279–86.

Sharp, P. J., Scanes, C. G., Williams, J. B., Harvey, S. and Chadwick, A. (1979). Variations in the concentrations of prolactin, luteinizing hormone, growth hormone and progesterone in the plasma of brooding bantams (*Gallus domesticus*). *Journal of Endocrinology*, **80**, 51–7.

Sharp, P. J., Sterling, R. J., Talbot, R. T. and Huskisson, N. S. (1989). The role of hypothalamic vasoactive intestinal polypeptide in the maintenance of prolactin secretion in incubating bantam hens: observations using passive immunization, radioimmunoassay and immunohistochemistry. *Journal of Endocrinology*, **122**, 5–13.

Sherry, T. W. and Holmes, R. T. (1997). American Redstart (*Setophaga ruticilla*). In: *The Birds of North America, No. 277* (eds A. Poole and F. Gill). The Academy of Natural Sciences, Philadelphia, and The American Ornithological Union, Washington DC.

Shine, R. (1988). Parental care in Reptiles. In: *Biology of the Reptilia. Volume 16, Ecology B Defense and Life History* (eds C. Gans and R. B. Huey), pp. 276–329. Alan R. Liss, Inc., New York.

Shoffner, R. N., Shuman, R., Otis, J. S. and Bitgood, J. J. (1982). The effect of a protoporphyrin mutant on some economic traits of the chicken. *Poultry Science*, **61**, 817–20.

Shugart, G. W. (1987). Individual clutch recognition by Caspian terns, *Sterna caspia*. *Animal Behaviour*, **35**, 1563–5.

Sibley, C. G. and Ahlquist, J. E. (1972). *A Comparative Study of the Egg-White Proteins of Passerine Birds*. Peabody Museum of Natural History (Yale University), New Haven.

Sibley, S. G. and Monroe, B. L. (1990). *Distribution and Taxonomy of Birds of the World*. Yale University Press, New Haven.

Sidis, Y., Zilberman, R. and Ar, A. (1994). Thermal aspects of nest placement in the orange-tufted sunbird (*Nectarinia osea*). *Auk*, **111**, 1001–5.

Siegfried, W. R. and Frost, P. G. H. (1974). Egg temperature and incubation behaviour of the ostrich. *Madoqua*, **Series 1(3)**, 63–6.

Siegfried, W. R. and Frost, P. G. H. (1975). Continuous breeding and associated behaviour in the moorhen *Gallinula chloropus*. *Ibis*, **117**, 102–9.

Siikamäki, P. (1995). Are large clutches costly to incubate – the case of the Pied Flycatcher. *Journal of Avian Biology*, **26**, 76–80.

Silver, R. and Gibson, M. J. (1984). Termination of incubation in doves: influence of egg fertility and absence of mate. *Hormones and Behavior*, **14**, 93–106.

Silverin, B. (1980). Effects of long-acting testosterone treatment on free-living pied flycatchers, *Ficedula hypoleuca*, during the breeding period. *Animal Behaviour*, **28**, 906–12.

Silverin, B. (1991). Annual changes in plasma levels of LH, and prolactin in free-living female great tits (*Parus major*). *General and Comparative Endocrinology*, **83**, 425–31.

Silverin, B. (1995). Reproductive adaptations to breeding in the north. *American Zoologist*, **35**, 191–202.

Silverin, B. and Goldsmith, A. R. (1983). Reproductive endocrinology of free-living pied flycatchers *(Ficedula hypoleuca)*: Prolactin and FSH secretion in relation to incubation and clutch size. *Journal of Zoology (London)*, **200**, 119–30.

Silverin, B. and Goldsmith, A. (1984). The effects of modifying incubation on prolactin secretion in free-living pied flycatchers. *General and Comparative Endocrinology*, **55**, 239–44.

Silverin, B. and Goldsmith, A. (1997). Natural and photoperiodically induced changes in plasma prolactin levels in male great tits. *General and Comparative Endocrinology*, **105**, 145–54.

Silverin, B., Viebke, P. A. and Westin, J. (1989). An artificial simulation of the vernal increase in day length and its effects on the reproductive system in three species of tits (*Parus* spp.), and modifying effects of environmental factors – a field experiment. *Condor*, **91**, 598–608.

Simmons, K. E. L. (1955). Studies on Great Crested Grebes. *Aviculture Magazine*, **61**, 294–316.

Sinclair, J. R. (2000). The Behaviour, Ecology and Conservation of Three Species of Megapode in Papua New Guinea. M Sc thesis, University of Otago, New Zealand.

Sinclair, J. R., O'Brien, T. and Kinnaird, M.F. (1999). Observations on the breeding biology of the Philippine Megapode (*Megapodius cumingii*) in North Sulawesi, Indonesia. *Tropical Biodiversity* **6**, 87–97.

Singleton, D. R. and Harper, R. G. (1998). Bacteria in old House Wren nests. *Journal of Field Ornithology*, **69**, 71–4.

Skagen, S. K. (1987). Hatching asynchrony in American goldfinches: an experimental study. *Ecology*, **68**, 1747–59.

Skowron, C. and Kern, M. D. (1980). The insulation in nests of selected North American songbirds. *Auk*, **97**, 816–24.

Skutch, A. F. (1957). The incubation patterns of birds. *Ibis*, **99**, 69–93.

Skutch, A. F. (1960). *Life Histories of Central American Birds, Volume 2 Families Fringllidae, Vireonidae, Sylviidae, Turdidae, Trogloditidae, Paridae, Corvidae, Hirundinidae and Tyrannidae, No. 34*. Cooper Ornithological Society. Berkeley, California.

Skutch, A. F. (1961). The nest as a dormitory. *Ibis*, **103**, 50–70.

Skutch, A. F. (1962). The constancy of incubation. *Wilson Bulletin*, **74**, 115–52.

Skutch, A. F. (1969). *Life Histories of Central American Birds, Volume 3: Families Cotingidae, Pipridae, Formicariidae, Furnariidae, Dendrocolaptidae and Picidae, No. 35*. Cooper Ornithological Society, Berkeley, California.

Skutch, A. F. (1976). *Parent Birds and Their Young*. University of Texas Press, Austin and London.

Skutch, A. F. (1997). *Life of the Flycatcher*. University of Oklahoma Press, Norman and London.

Slagsvold, T. (1980). Egg predation in woodland in relation to the presence and density of breeding fieldfares *Turdus pilaris*. *Ornis Scandinavica*, **11**, 92–8.

Slagsvold, T. (1982a). Clutch size variation in passerine birds: The nest predation hypothesis. *Oecologia*, **54**, 159–69.

Slagsvold, T. (1982b). Clutch size, nest size and hatchling asynchrony in birds: experiments with the fieldfare *(Turdus pilaris)*. *Ecology*, **63**, 1389–99.

Slagsvold, T. (1986). Asynchronous versus synchronous hatching in birds: experiments with the pied flycatcher. *Journal of Animal Ecology*, **55**, 1115–34.

Slagsvold, T. (1989a). On the evolution of clutch size and nest size in passerine birds. *Oecologia*, **79**, 300–5.

Slagsvold, T. (1989b). Experiments on clutch size and nest size in passerine birds. *Oecologia*, **80**, 297–302.

Slagsvold, T. (1997). Brood division in birds in relation to offspring size: Sibling rivalry and parental control. *Animal Behaviour*, **54**, 1357–68.

Slagsvold, T. (1998). On the origin and rarity of interspecific nest parasitism in birds. *American Naturalist*, **152**, 264–72.

Slagsvold, T. and Johansen, M. A. (1998). Mass loss in female Pied Flycatchers *Ficedula hypoleuca* during late incubation: supplementation fails to support the reproductive stress hypothesis. *Ardea*, **86**, 203–10.

Slagsvold, T. and Lifjeld, J. T. (1989). Hatching asynchrony in birds: the hypothesis of sexual conflict over parental investment. *American Naturalist*, **134**, 239–53.

Slater, P. J. B. (1967). External stimuli and readiness to incubate in the Bengalese finch. *Animal Behaviour*, **15**, 520–6.

Smart, I. H. M. (1991). Egg-shape in birds. In: *Egg Incubation: Its Effects on Embryonic Development in Birds and Reptiles* (eds D. C. Deeming and M. W. J. Ferguson), pp. 101–16. Cambridge University Press, Cambridge.

Smith, A. H., Burton, R. R. and Besch, E. L. (1969). Development of the chick embryo at high altitude. *Federation Proceedings*, **28**, 1092–8.

Smith, H. G. (1989). Larger clutches take longer to incubate. *Ornis Scandinavica*, **20**, 156–8.

Smith, H. G. (1995). Experimental demonstration of a trade-off between mate attraction and paternal care. *Proceedings of the Royal Society of London B*, **260**, 45–51.

Smith, H. G., Kallander, H., Hultman, J. and Sanzén, B. (1989). Female nutritional state affects the rate of male incubation feeding in the pied flycatcher *Ficedula hypoleuca*. *Behavioral Ecology and Sociobiology*, **24**, 417–20.

Smith, H. G. and Montgomerie, R. (1992). Male incubation in barn swallows: the influence of nest temperature and sexual selection. *Condor*, **94**, 750–9.

Smith, H. G., Sandell, M. I. and Bruun, M. (1995). Paternal care in the European starling: incubation. *Animal Behaviour*, **50**, 323–31.

Smith, N. G. (1968). The advantage of being parasitized. *Nature*, **219**, 690–4.

Smith, K. G. (1988). Clutch-size dependent asynchronous hatching and brood reduction in *Junco hyemalis*. *Auk*, **105**, 200–3.

Smith, W. K., Roberts, S. W. and Miller, P.C. (1974). Calculating the nocturnal energy expenditure of an incubating Anna's hummingbird. *Condor* **76**, 176–83.

Snyder, N. R. R., Enkerlin-Hoeflich, E. C. and Cruz-Nieto, M. A. (1999). Thick-billed Parrot (*Rhynchopsitta pachyrhyncha*). In: *The Birds of North America, No.* 406 (eds A. Poole and F. Gill). The Academy of Natural Sciences, Philadelphia: The American Ornithologists' Union, Washington, DC.

Sockman, K. W. and Schwabl, H. (1998). Hypothermic tolerance in an embryonic American Kestrel *Falco sporverius*. *Canadian Journal of Zoology,* **76**, 1399–402.

Sockman, K. W. and Schwabl, H. (1999). Daily estradiol and progesterone levels relative to laying and onset of incubation in canaries. *General and Comparative Endocrinology*, **114**, 257–68.

Sockman, K. W., Schwabl, H. and Sharp, P. J. (2000). The role of prolactin in the regulation of clutch size and onset of incubation behavior in the American kestrel. *Hormones and Behavior*, **38**, 168–76.

Soler, J. J., Martínez, J. G., Soler, M. and Møller, A. P. (1999). Genetic and geographic variation in rejection behavior of cuckoo eggs by European magpie populations: An experimental test of rejecter-gene flow. *Evolution*, **53**, 947–56.

Soler, J. J. and Møller, A. P. (1996). A comparative analysis of the evolution of variation in appearance of eggs of European passerines in relation to brood parasitism. *Behavioral Ecology*, **7**, 89–94.

Soler, J. J., Soler, M., Møller, A. P. and Martinez, J. G. (1995). Does the great spotted cuckoo choose magpie hosts according to their parenting ability? *Behavioral Ecology and Sociobiology*, **36**, 201–6.

Soler, M. (1990). Relationships between great spotted cuckoo *Clamator glandarius* and its corvid hosts in a recently colonized area. *Ornis Scandinavica*, **21**, 212–23.

Soler, M., Soler, J. J. and Martinez, J. C. (1997). Great spotted cuckoos improve their reproductive success by damaging magpie host eggs. *Animal Behaviour*, **54**, 1227–33.

Soler, M., Soler, J. J., Martínez, J. G. and Møller, A. P. (1994). Micro-evolutionary change in host response to a brood parasite. *Behavioral Ecology and Sociobiology*, **35**, 295–301.

Solís, J. C. and Lope, F. de. (1995). Nest and egg crypsis in the ground-nesting stone curlew *Burhinus oedicnemus*. *Journal of Avian Biology*, **26**, 135–8.

Solomon, S. E. (1987). Egg shell pigmentation. In: *Egg Quality – Current Problems and Recent Advances* (ed. R. G. Wells and C. G. Belyarin), pp. 147–57. Butterworths and Co. Ltd., London.

Solomon, S. E., Bain, M. M., Cranstoun, S. and Mascimento, V. (1994). Hen's egg shell structure and function. In: *Microbiology of the Avian Egg* (eds R. G. Board and R. Fuller), pp. 1–24. Chapman and Hall, London.

Sorenson, M. D. (1991). The functional significance of parasitic egg laying and typical nesting in redhead ducks: An analysis of individual behaviour. *Animal Behaviour*, **42**, 771–96.

Sossinka, R. (1980). Ovarian development in an opportunistic breeder, the Zebra Finch *Poephila* [*Taenopygia*] *guttata castanotis*. *Journal of Experimental Zoology*, **211**, 225–30.

Sotherland, P. R., Ashen, M. D., Shuman, R. D., and Tracy, C. R. (1984). The water balance of bird eggs incubated in water. *Physiological Zoology*, **57**, 338–48.

Sotherland, P. R., Packard, G. C., Taigen, T. L. and Boardman, T. J. (1980). An altitudinal cline in conductance of cliff swallow (*Petrichelidon pyrrhonota*) eggs to water vapor. *Auk*, **97**, 177–85.

Sotherland, P. R. and Rahn, H. (1987). On the composition of bird eggs. *Condor*, **89**, 48–65.

Sparks, N. H. C. (1994). Shell accessory materials: structure and function. In: *Microbiology of the Avian Egg* (eds R. G. Board and R. Fuller), pp. 25–42. Chapman and Hall, London.

Sparks, N. H. C. and Board, R. G. (1984). Cuticle, shell porosity and water uptake through hens' eggshells. *British Poultry Science*, **25**, 267–76.

Sparks, N. H. C. and Board, R. G. (1985). Bacterial penetration of the recently oviposited shell of hens' eggs. *Australian Veterinary Journal*, **62**, 169–70.

Spaw, C. D. and Rohwer, S. (1987). A comparative study of eggshell thickness in cowbirds and other passerines. *Condor*, **89**, 307–18.

Speake, B. K., Decrock, F., Surai, P. F. and Groscolas, R. (1999a). Fatty acid composition of the adipose tissue and yolk lipids of a bird with a marine-based diet, the Emperor penguin (*Aptenodytes forsteri*). *Lipids*, **34**, 283–90.

Speake, B. K., Murray, A. M. B. and Noble, R. C. (1998). Transport and transformations of yolk lipids during development of the avian embryo. *Progress in Lipid Research*, **37**, 1–32.

Speake, B. K., Surai, P. F., Noble, R. C., Beer, J. V. and Wood, N. A. R. (1999b). Differences in egg lipid and anti-oxidant composition between captive and wild pheasants and geese. *Comparative Biochemistry and Physiology*, **B124**, 101–7.

Speake, B. K. and Thompson, M. B. (1999). Comparative aspects of yolk lipid utilisation in birds and reptiles. *Poultry and Avian Biology Reviews*, **10**, 181–211.

Speakman, J. R. (1997). *Doubly Labelled Water: Theory and Practice*. Chapman Hall, New York.

Spellerberg, I. F. (1969). Incubation temperatures and thermoregulation in the McCormick skua. *Condor,* **71,** 59–67.

Spiers, D. E. and Baummer, S. C. (1990). Embryonic development of Japanese quail (*Coturnix coturnix japonica*) as influenced by periodic cold exposure. *Physiological Zoology*, **63**, 516–35.

St. Clair, C. C. (1992). Incubation behaviour, brood patch formation and obligate brood reduction in Fiordland crested penguins. *Behavorial Ecology and Sociobiology*, **31**, 409–16.

Stacey, P. B. and Koenig, W. D. (1991). Cooperative breeding in the acorn woodpecker. In: *Behavior and Evolution of Birds* (ed. D. W. Mock), pp. 95–106. W. H. Freeman and Company, New York.

Starck, J. M. (1996). Comparative morphology and cytokinetics of skeletal growth in hatchlings of altricial and precocial birds. *Zoologischer Anzeiger*, **235**, 53–75.

Starck, J. M. and Ricklefs, R. E. (1998). *Avian Growth and Development. Evolution within the Altricial-Precocial Spectrum*. Oxford University Press, Oxford.

Steadman, D. W. (1999). The biogeography and extinction of megapodes in Oceania. In: *Proceedings of the Third International Megapode Symposium, Nhill, Australia* (eds R.W.R.J. Dekker, D.N. Jones and J. Benshemesh), *Zoologische Verhandelingen* **327**, 7–21.

Stearns, S. C. (1992). *The Evolution of Life Histories*. Oxford University Press, Oxford.

Steel, E. A. and Hinde, R. A. (1963). Hormonal control of brood patch and oviduct development in domesticated canaries. *Journal of Endocrinology*, **26**, 11–24.

Steel, E. A. and Hinde, R. A. (1964). Effect of exogenous oestrogen on brood patch development of intact and ovariectomized canaries. *Nature*, **202**, 718–9.

Steen, J. B. and Parker, J. (1981). The egg 'numerostat'. A new concept in the regulation of clutch-size. *Ornis Scandinavia*, **12**, 109–10.

Stenning, M. J. (1996). Hatching asynchrony, brood reduction and other rapidly reproducing hypotheses. *Trends in Ecology and Evolution*, **11**, 243–6.

Stetten, G., Koontz, F., Sheppard, C. and Koontz, C. (1980). Telemetric egg for monitoring nest microclimate of endangered birds. *Instrument Society of America Proceedings*, 321–7.

Stettenheim, P. (1972). The integument of birds. In: *Avian Biology, Volume 2* (eds D. S. Farner, J. R. King and K. C. Parkes), pp. 1–63. Academic Press, New York.

Stewart, P. A. (1971). Egg turning by an incubating wood duck. *Wilson Bulletin*, **83**, 97–9.

Stewart, P. A. (1974). A nesting of black vultures. *Auk*, **91**, 595–600.

Stewart, R. and Rambo, T. B. (2000). Cloacal microbes in House Sparrows. *Condor*, **102**, 679–84.

Steyn, P. and Howells, W. W. (1975). Supplementary notes on the breeding biology of the striped cuckoo. *Ostrich*, **46**, 258–60.

Stokke, B. G., Moksnes, A., Røskaft, E., Rudolfsen, G. and Honza, M. (1999). Rejection of artificial cuckoo (*Cuculus canorus*) eggs in relation to variation in egg appearance among reed warblers (*Acrocephalus scirapeus*). *Proceedings of the Royal Society of London B*, **266**, 1483–8.

Stoleson, S. H. and Beissinger, S. R. (1995). Hatching asynchrony and the onset of incubation in birds, revisited: When is the critical period? *Current Ornithology*, **12**, 191–270.

Stoleson, S. H. and Beissinger, S. R. (1999). Egg viability as a constraint on hatching asynchrony at high ambient temperatures. *Journal of Animal Ecology*, **68**, 951–62.

Stouffer, P. C., Kennedy, E. D. and Power, H. R. (1987). Recognition and removal of intraspecific eggs by starlings. *Animal Behaviour*, **35**, 1583–4.

Strickland, D., and Oulett, H. (1993). Gray Jay. In: *The Birds of North America*, No. 40 (eds A. Poole and F. Gill), Academy of Natural Sciences, Philadelphia: The American Ornithologists Union, Washington, DC.

Stiles, F. G. (1973). Food supply and the annual cycle of the Anna hummingbird. *University of California Publications in Zoology*, **97**.

Sturkie, P. D. (1976). *Avian Physiology*. Springer-Verlag, New York.

Sturm, L. (1945). A study of the nesting activity of the American Redstart. *Auk*, **62**, 189–206.

Suarez, R. K. (1992). Hummingbird flight: Sustaining the highest mass-specific metabolic rates among vertebrates. *Experientia*, **48**, 565–70.

Swart, D. and Rahn, H. (1988). Microclimate of ostrich nests: measurements of egg temperature and nest humidity using egg hygrometers. *Journal of Comparative Physiology B*, **157**, 845–53.

Swennen, C. (1968). Nest protection of eiderducks and shovelers by means of faeces. *Ardea*, **56**, 248–58.

Swynnerton, C. F. M. (1916). On the coloration of the mouths and eggs of birds. II. On the coloration of eggs. *Ibis*, **Series 10, 4**, 529–606.

Swynnerton, C. F. M. (1918). Rejections by birds of eggs unlike their own: With remarks on some of the cuckoo problems. *Ibis*, **Series 10, 6**, 127–54.

Sydeman, W. J. and Emslie, S. D. (1992). Effects of parental age on hatching asynchrony, egg size and third-chick disadvantage in western gulls. *Auk*, **109**, 242–48.

Sykes, P. W. (1993). Male Kirkland's warbler with incubation patch. *Wilson Bulletin*, **105**, 356–9.

Székely, T., Karsai, I. and Williams, T. D. (1994). Determination of clutch-size in the Kentish Plover *Charadrius alexandrinus*. *Ibis*, **136**, 341–8.

Tacha, T. C., Nesbitt, S. A. and Vohs, P. A. (1992). Sandhill Crane. In: *The Birds of North America, No. 31* (eds A. Poole, P. Stettenheim and F. Gill). The Academy of

Natural Sciences, Philadelphia: The American Ornithologists' Union, Washington, DC.

Taigen, T. L., Packard, G. C., Sotherland, P. R., Boardman, T. J. and Packard M. J. (1980). Water-vapour conductance of black-billed magpie (*Pica pica*) eggs collected along an altitudinal gradient. *Physiological Zoology,* **53**, 163–9.

Tarburton, M. K. and Minot, E. O. (1987). A novel strategy of reproduction in birds. *Animal Behaviour*, **35**, 1898–9.

Tatner, P. and Bryant, D. M. (1993). Interspecific variation in daily energy expenditure during avian incubation. *Journal of Zoology, London*, **231**, 215–32.

Taylor, L. W. and Kreutziger, G. O. (1965). The gaseous environment of the chick embryo in relation to its development and hatchability. 2. Effect of carbon dioxide and oxygen levels during the period of the fifth through the eighth days of incubation. *Poultry Science,* **44**, 98–106.

Taylor, L. W. and Kreutziger, G. O. (1966). The gaseous environment of the chick embryo in relation to its development and hatchability. 3. Effect of carbon dioxide and oxygen levels during the period of the ninth through the twelfth days of incubation. *Poultry Science,* **45**, 867–84.

Taylor, L. W. and Kreutziger, G. O. (1969). The gaseous environment of the chick embryo in relation to its development and hatchability. 4. Effect of carbon dioxide and oxygen levels during the period of the thirteenth through the sixteenth days of incubation. *Poultry Science,* **48**, 871–7.

Taylor, L. W., Kreutziger, G. O. and Abercrombie, G. L. (1971). The gaseous environment of the chick embryo in relation to its development and hatchability. 5. Effect of carbon dioxide and oxygen levels during the terminal days of incubation. *Poultry Science,* **50**, 66–78.

Taylor, L. W., Sjodin, R. A. and Gunns, C. A. (1956). The gaseous environment of the chick embryo in relation to its development and hatchability. 1. Effect of carbon dioxide and oxygen levels during the first four days of incubation upon hatchability. *Poultry Science,* **35**, 1206–15.

Tazawa, H. (1980a). Oxygen and CO_2 exchange and acid-base regulation in the avian embryo. *American Zoologist,* **20**, 395–404.

Tazawa, H. (1980b). Adverse effect of failure to turn the avian egg on the embryo oxygen exchange. *Respiration Physiology,* **41**, 137–42.

Tazawa, H. and Mochizuki, M. (1977). Oxygen analyses of chicken embryo blood. *Respiration Physiology,* **31**, 203–15.

Tazawa, H., Turner, J. S. and Paganelli, C. V. (1988a). Cooling rates of living and killed chicken and quail eggs in air and helium-oxygen gas mixture. *Comparative Biochemistry and Physiology*, **90A**, 99–102.

Tazawa, H., Visschedijk, A. H. J., Wittman, J. and Piiper, J. (1983). Gas exchange, blood gases and acid-base status in the chick before, during and after hatching. *Respiration Physiology*, **53**, 173–85.

Tazawa, H., Wakayama, H., Turner, J. S. and Paganelli, C. V. (1988b). Metabolic compensation for gradual cooling in developing chick embryos. *Comparative Biochemistry and Physiology*, **89A**, 125–9.

Telfair, R., C. II. and Morrison, M. C. (1995). Neotropic Cormorant (*Phalacrocorax brasilianus*). In: *The Birds of North America, No. 137* (eds A. Poole and F. Gill). The

Academy of Natural Sciences, Philadelphia: The American Ornithologists' Union, Washington, DC.

Templeton, R. K. (1983). Why do Herring Gull chicks vocalise in the shell? *Bird Study*, **30**, 73–4.

Teuschl, Y., Taborsky, B. and Taborsky, M. (1998). How do cuckoos find their hosts? The role of habitat imprinting. *Animal Behaviour*, **56**, 1425–33.

Thomas, C. J., Thompson, D. B. A. and Galbraith, H. (1989). Physiognomic variation in dotterel *Charadrius morinellus* clutches. *Ornis Scandinavica*, **20**, 145–50.

Thomas, D. H. and Robin, A. P. (1977). Comparative studies of thermoregulatory and osmoregulatory behavior and physiology of five species of sandgrouse (Aves: Pterocliidae) in Morocco. *Journal of Zoology, London*, **183**, 229–49.

Thomas, L. C. (1980). *Fundamentals of Heat Transfer*. Prentice-Hall, Englewood Cliffs, New Jersey.

Thompson, M. B. (1989). Patterns of metabolism in embryonic reptiles. *Respiration Physiology*, **76**, 243–56.

Thompson, M. B. and Goldie, K. N. (1990). Conductance and structure of eggs of Adelie penguins, *Pygoscelis adeliae*, and its implications for incubation. *Condor*, **92**, 304–12.

Thomson, D. L., Furness, R. W. and Monaghan, P. (1998a). Field metabolic rates of Kittiwakes *Rissa tridactyla* during incubation and chick rearing. *Ardea*, **86**, 169–75.

Thomson, D. L., Monaghan, P. and Furness, R. W. (1998b). The demands of incubation and avian clutch size. *Biological Reviews*, **73**, 293–304.

Tideman, S. C. and Marples, T. G. (1988). Selection of nest sites by three species of fairy-wrens (*Malurus*). *Emu*, **88**, 9–15.

Tinbergen, J. M. and Boerlijst, M. C. (1990). Nestling weight and survival in individual Great Tits (*Parus major*). *Journal of Animal Ecology*, **59**, 1113–28.

Tinbergen, J. M. and Dietz, M. W. (1994). Parental energy expenditure during brood rearing in the Great Tit (*Parus major*) in relation to body mass, temperature, food availability and clutch size. *Functional Ecology*, **8**, 563–72.

Tinbergen, N. (1953). *The Herring Gull's World*. Collins, London.

Tinbergen, N., Broekhuysen, G. J., Feekes, F., Houghton, J. C. W., Kruuk, H. and Szulc, E. (1962). Egg shell removal by the black-headed gull, *Larus ridibundus* L.; A behaviour component of camouflage. *Behaviour*, **19**, 74–117.

Todd, D. (1983). Pritchard's megapode on Niuafo'os Island, Kingdom of Tonga. *Journal of the World Pheasant Association*, **8**, 69–88.

Tøien, Ø. (1989). Effect of clutch size on efficiency of heat transfer to cold eggs in incubating bantam hens. In: *Physiology of Cold Adaptation in Birds* (eds C. Bech and R. E. Reinertsen), pp. 305–13. NATO ASI Series A: Life Sciences, Volume 173, Plenum Press, New York.

Tøien, O. (1993a). Control of shivering and heart rate in incubating bantam hens upon sudden exposure to cold eggs. *Acta Physiologica Scandinavica*, **149**, 205–14.

Tøien, O. (1993b). Dynamics of heat transfer to cold eggs in incubating bantam hens and a black grouse. *Journal of Comparative Physiology B*, **163**, 182–8.

Tøien, O., Aulie, A. and Steen, J. (1986). Thermoregulatory responses to egg cooling in incubating bantam hens. *Journal of Comparative Physiology B*, **156**, 303–7.

Tombre, I. M. and Erikstad, K. E. (1996). An experimental study of incubation effort in high-Arctic Barnacle Geese. *Journal of Animal Ecology*, **65**, 325–31.

Tomialojc, L. (1992). Colonisation of dry habitats by the song thrush *Turdus philomelos*: is the type of nest material an important nest constraint? *Bulletin of the British Ornithologist's Club*, **112**, 27–34.

Tranter, H. S. and Board, R. G. (1982). The antimicrobial defense of avian eggs: biological perspective and chemical basis. *Journal of Applied Biochemistry*, **4**, 295–338.

Trimmer, J. D. (1950). *Response of Physical Systems*. J. Wiley and Sons, New York.

Trine, C. L. (2000). Effects of multiple parasitism on cowbird and wood thrush nesting success. In: *Ecology and Management of Cowbirds and Their Hosts: Studies in the Conservation of North American Passerine Birds* (eds J. N. M. Smith, T. L. Cook, S. I. Rothstein, S. K. Robinson and S. G. Sealy), pp. 135–44. University of Texas Press, Austin.

Trivers, R. L. (1974). Parent-offspring conflict. *American Zoologist*, **14**, 249–64.

Trost, C. H. and Webb, C. L. (1986). Egg moving by two species of corvid. *Animal Behaviour*, **34**, 294–5.

Tschanz, B. (1959). Zur Brutbiologie der Trottellumme (*Uria aalge aalge* Pont.). *Behaviour*, **14**, 1–100.

Tschanz, B. (1968). Trottellummen: Die Entstehung der persönlichen Beziehungen zwischen Jungvogel und Eltern. *Zeitschrift für Tierpsychologie*, **4**, 4–100.

Tschanz, B., Ingold, P. and Lengacher, H. (1969). Eiform und bruterflog bei trottellumen *Uria aalge aalge* Pnt. *Der Ornithologische Beobachter*, **66**, 25–42.

Tullett, S. G. (1984). The porosity of avian eggshells. *Comparative Biochemistry and Physiology*, **78A**, 5–13.

Tullett, S. G. (1987). Egg shell formation and quality. In: *Egg Quality – Current Problems and Recent Advances* (eds R. G. Wells and C. G. Belyavin), pp. 123–46. Butterworths, London.

Tullett, S. G. and Board, R. G. (1976). Oxygen flux across the integument of the avian egg during incubation. *British Poultry Science,* **17**, 441–50.

Tullett, S. G. and Burton, F. G. (1982). Factors affecting the weight and water status of the chick at hatch. *British Poultry Science*, **23**, 361–9.

Tullett, S. G. and Deeming, D. C. (1982). The relationship between eggshell porosity and oxygen consumption in the domestic fowl. *Comparative Biochemistry and Physiology*, **72A**, 529–33.

Turner, D. A. and Gerhart J. (1971). 'Foot-wetting' by incubating African skimmers *Rynchops flavirostris*. *Ibis*, **113**, 244.

Turner, J. S. (1985). Cooling rate and size of birds' eggs – a natural isomorphic body. *Journal of Thermal Biology*, **10**, 101–4.

Turner, J. S. (1987a). Blood circulation and the flows of heat in an incubated egg. *Journal of Experimental Zoology*, **Supplement 1**, 99–104.

Turner, J. S. (1987b). On the transient temperatures of ectotherms. *Journal of Thermal Biology*, **12**, 207–14.

Turner, J. S. (1990). The thermal energetics of an incubated chicken egg. *Journal of Thermal Biology,* **15**, 211–6.

Turner, J. S. (1991). The thermal energetics of incubated bird eggs. In: *Egg Incubation: Its Effects on Embryonic Development in Birds and Reptiles* (eds D. C. Deeming and M. W. J. Ferguson), pp. 117–45. Cambridge University Press, Cambridge.

Turner, J. S. (1994a). Time and energy in the intermittent incubation of birds' eggs. *Israel Journal of Zoology*, **40**, 519–40.

Turner, J. S. (1994b). Transient-state thermal properties of contact-incubated chicken eggs. *Physiological Zoology*, **67**, 1426–47.

Turner, J. S. (1994c). Thermal impedance of a contact-incubated bird's egg. *Journal of Thermal Biology*, **19**, 237–43.

Turner, J. S. (1997). On the thermal capacity of a bird's egg warmed by a brood patch. *Physiological Zoology*, **70**, 470–80.

Tyler, C. (1965). A study of the egg shells of the Sphenisciformes. *Journal of Zoology, London*, **147**, 1–19.

Utter, J. M. (1971). *Daily energy expenditure of free-living Purple Martins (*Progne subis*) and Mockingbirds (*Mimus polyglottos*) with a comparison of two nothern populations of Mockingbirds.* Doctoral thesis, Rutgers University, New Brunswick, New Jersey.

Valanne, K. (1966). Incubation behaviour and temperature of Capercaille (*Tetrao urogallus*) and willow Grouse (*Lagopus lagopus*). *Suomen Riista*, **19**, 30–41.

Van Ee, C. A. (1966). Notes on the breeding behaviour of the blue crane *Tetrapteryx paradisea. Ostrich*, **37**, 23–9.

van Kampen, H. S. (1996). A framework for the study of filial imprinting and the development of attachment. *Psychonomic Bulletin and Review*, **3**, 3–20.

Verbeek, N. A. M. (1973). The exploitation system of the yellow-billed magpie. *University of California at Berkley Publications in Zoology*, **99**, 1–58.

Verbeek, N. A. M. (1986). Aspects of the breeding biology of an expanded population of glaucous-winged gulls in British Columbia. *Journal of Field Ornithology*, **57**, 22–33.

Verbeek, N. A. M. (1988). Differential predation of eggs in clutches of glaucous-winged gulls *Larus glaucescens. Ibis*, **130**, 512–8.

Verbeek, N. A. M. (1990). Differential predation on eggs in clutches of northwestern crows: The importance of egg color. *Condor*, **92**, 695–701.

Verboven, N. and Visser, M. E. (1998). Seasonal variation in local recruitment in Great Tits: the importance of being early. *Oikos*, **81**, 511–24.

Verhulst, S. and Tinbergen, J. M. (1997). Clutch size and parental effort in the Great Tit *Parus major. Ardea*, **85**, 111–26.

Vernon, C. J. (1970). Pre-incubation embryonic development and egg 'dumping' by the jacobin cuckoo. *Ostrich*, **41**, 259–60.

Viega, J. P. (1992). Hatching asynchrony in the house sparrow: a test of the egg-viability hypothesis. *American Naturalist*, **139**, 669–75.

Viega, J. P. and Viñuela, J. (1993). Hatching asynchrony and hatching success in the house sparrow: evidence for the egg viability hypothesis. *Ornis Scandinavica*, **24**, 237–42.

Victoria, J. K. (1972). Clutch characteristics and egg discriminative ability of the African village weaverbird *Ploceus cucullatus. Ibis*, **114**, 367–76.

Village, A. (1990). *The Kestrel.* T and A D Poyser, London.

Vince, M. A. (1964). Social facilitation of hatching in the bobwhite quail. *Animal Behaviour*, **12**, 531–4.
Vince, M. A. (1966a). Potential stimulation produced by avian embryos. *Animal Behaviour*, **14**, 34–40.
Vince, M. A. (1966b). Artificial acceleration of hatching in quail embryos. *Animal Behaviour*, **14**, 389–94.
Vince, M. A. (1968). Retardation as a factor in the synchronization of hatching. *Animal Behaviour*, **16**, 332–5.
Vince, M. A. (1969). Embryonic communication, respiration and the synchronization of hatching. In: *Bird Vocalizations* (ed. R. A. Hinde), pp. 233–60. Cambridge University Press, Cambridge.
Vince, M. A. (1972). Communication between quail embryos and the synchronisation of hatching. *Proceedings of the International Ornithological Congress*, **15**, 357–62.
Vince, M. A., Ockleford, E. and Reader, M. (1984). The synchronisation of hatching in quail embryos: Aspects of development affected by a retarding stimulus. *Journal of Experimental Zoology*, **229**, 273–82.
Viñuela, J. (2000). Opposing selective pressures on hatching asynchrony: egg viability, brood reduction, and nestling growth. *Behavioral Ecology and Sociobiology*, **48**, 333–43.
Vispo, C. R. and Bakken, G. S. (1993). The influence of thermal conditions on the surface activity of thirteen-lined ground squirrels. *Ecology*, **74**, 377–89.
Visschedijk, A. H. J. (1968a). The air space and embryonic respiration 3. The balance between oxygen and carbon dioxide in the air space of the incubating chicken egg and its role in stimulating pipping. *British Poultry Science*, **9**, 197–210.
Visschedijk, A. H. J. (1968b). The air space and embryonic respiration 2. The times of pipping and hatching as influenced by an artificially changed permeability of the shell over the air space. *British Poultry Science*, **9**, 185–96.
Visschedijk, A. H. J. (1980). Effects of barometric pressure and abnormal gas mixtures on gaseous exchange by the avian embryo. *American Zoologist*, **20**, 469–76.
Visschedijk, A. H. J., Ar, A., Rahn, H. and Piiper, J. (1980). The independent effects of atmospheric pressure and oxygen partial pressure on gas exchange of the chicken embryo. *Respiration Physiology*, **39**, 33–44.
Visschedijk, A. H. J., Girard, H. and Ar, A. (1988). Gas diffusion in the shell membranes of the hen's egg: Lateral diffusion in situ. *Journal of Comparative Physiology B*, **158**, 567–74.
Visser, M. E. and Verboven, N. (1999). Long-term fitness effects of fledging date in Great Tits. *Oikos*, **85**, 445–50.
Vleck, C. M. (1981a). Energetic cost of incubation in the zebra finch. *Condor*, **83**, 229–37.
Vleck, C. M. (1981b). Hummingbird incubation: female attentiveness and egg temperature. *Oecologia*, **51**, 199–205.
Vleck, C. M. (1993). Hormones, reproduction, and behavior in birds of the Sonoran desert. In: *Avian Endocrinology* (ed. P. J. Sharp), pp. 73–86. Journal of Endocrinology Ltd., Bristol.
Vleck, C. M., Bucher, T. L., Reed, W. L. and Kristmundsdottir, A. Y. (1999). Changes in reproductive hormones and body mass through the reproductive cycle in the Adélie

Penguin (*Pygoscelis adeliae*), with associated data on courting-only individuals. In: *Proceedings of the 22nd International Ornithological Congress* (eds N. Adams and R. Slotow), pp. 1210–23. University of Natal, Durban.

Vleck, C. M. and Hoyt, D. F. (1980). Patterns of metabolism and growth in avian embryos. *American Zoologist*, **20**, 405–16.

Vleck, C. M., Mays, N. A., Dawson, J. W. and Goldsmith, A. R. (1991). Hormonal correlates of parental and helping behavior in cooperatively breeding Harris' Hawks (*Parabuteo unicinctus*). *Auk*, **108**, 638–48.

Vleck, C. M. and Priedkalns, J. (1985). Reproduction in zebra finches: hormone levels and effect of dehydration. *Condor,* **87**, 37–46.

Vleck, C. M., Ross, L. L., Vleck, D. and Bucher, T. L. (2000). Prolactin and parental behavior in Adélie penguins: effects of absence from nest, incubation length, and nest failure. *Hormones and Behavior*, **38**, 149–58.

Vleck, C. M., Vertalino, N., Vleck, D. and Bucher, T. L. (2000). Stress, corticosterone, and heterophil to lymphocyte ratios in free-living Adélie penguins. *Condor,* **102,** 392–400.

Vleck, C. M. and Vleck, D. (1987). Metabolism and energetics of avian embryos. *Journal of Experimental Zoology*, **Supplement 1**, 111–25.

Vleck, C. M. and Vleck, D. (2001). Physiological condition and reproductive consequences in Adelie penguins. *American Zoologist*, (in press).

Vleck, C. M., Vleck, D., Rahn, H. and Paganelli, C. V. (1983). Nest microclimate, water-vapor conductance, and water loss in heron and tern eggs. *Auk,* **100**, 76–83.

Vleck, D. Vleck, C. M. and Hoyt, D. F. (1980). Metabolism of avian embryos: ontogeny of oxygen consumption in the Rhea and Emu. *Physiological Zoology*, **53**, 125–35.

Vleck, D., Vleck C. M. and Seymour, R. S. (1984). Energetics of embryonic development in the megapode birds, Mallee Fowl *Leipoa ocellata* and Brush Turkey *Alectura lathami. Physiological Zoology*, **57**, 444–56.

von Haartman, L. (1956). Der einfluss der temperatur auf den brutrhythmus experimentell nachgewiesen. *Ornis Fennica*, **33**, 100–7.

von Haartman, L. (1957). Adaptation in hole-nesting birds. *Evolution*, **11**, 339–47.

Voss, M. A. and Hainsworth, F. R. (2001). Relatively simple, precise methods to analyze temperature transients in ectotherms. *Journal of Thermal Biology*, in press.

Vowles, D. M. and Lea, R. W. (1986). External factors affecting the duration of broody behavior in the ring dove (*Streptopelia risoria*). *Hormones and Behavior*, **20**, 249–62.

Wagner, H. O. (1955). Einfluss der Poikolothermie bei kolibris auf ihre brutbiologie. *Journal für Ornithologie,* **96**, 361–8.

Wakayama, H. and Tazawa, H. (1988). The analysis of PO_2 difference between air space and arterialized blood in chicken eggs with respect to widely altered shell conductance. In: *Oxygen Transport to Tissue, X* (eds M. Mochizuki, C. R. Honig, T. Koyama, T. K. Goldstick and D. F. Bruley), pp. 699–708. Plenum Press, New York.

Walkinshaw, L. H. (1951). Nesting of white-naped crane in Detroit Zoological Park, Michigan. *Auk*, **68**, 194–202.

Wallace, A. R. (1889). *Darwinism.* Macmillan and Co., London.

Walsberg, G. E. (1980). The gaseous microclimate of the avian nest during incubation. *American Zoologist*, **20**, 363–72.

Walsberg, G. E. (1981). Nest-site selection and the radiative environment of the Warbling Vireo. *Condor*, **83**, 86–8.

Walsberg, G. E. (1983a). A test for regulation of nest humidity in two bird species. *Physiological Zoology*, **56**, 231–5.

Walsberg, G. E. (1983b). Avian ecological energetics. In: *Avian Biology, Volume 7* (eds D. S. Farner, J. R. King and K. C. Parkes), pp. 161–220. Academic Press, New York.

Walsberg, G. E. (1985). Physiological consequences of microhabitat selection. In: *Habitat Selection in Birds* (ed. M. L. Cody), pp. 389–413. Academic Press, Orlando, Florida.

Walsberg, G. E. and King, J. R. (1978a). The energetic consequences of incubation for two passerine species. *Auk*, **95**, 644–55.

Walsberg, G. E. and King, J. R. (1978b). The relationship of the external surface area of birds to skin surface area and body mass. *Journal of Experimental Biology*, **76**, 185–9.

Walsberg, G. E. and King, J. R. (1978c). The heat budget of incubating mountain White-crowned Sparrows (*Zonotrichia leucophrys oriantha*) in Oregon. *Physiological Zoology*, **51**, 92–103.

Walsberg, G. E. and Voss-Roberts, K. A. (1983). Incubation in desert-nesting doves: mechanisms for egg cooling? *Physiological Zoology*, **56**, 88–93.

Walton, P. R., Ruxton, G. D. and Monaghan, R. (1998). Avian diving, respiratory physiology and the marginal value theorem. *Animal Behaviour*, **56**, 165–74.

Wang, Q. and Buntin, J. D. (1999). The roles of stimuli from young, previous breeding experience, and prolactin in regulating parental behavior in ring doves (*Streptopelia risoria*). *Hormones and Behavior*, **35**, 241–53.

Wangensteen, O. D. and Rahn, H. (1970/71). Respiratory gas exchange by the avian embryo. *Respiration Physiology*, **11**, 31–45.

Ward, D. (1988). Belly-soaking in the blacksmith plover *Vanellus armatus*. *Ostrich*, **59**, 142.

Ward, D. (1990). Incubation temperatures and behavior of crowned, black-winged, and lesser black-winged plovers. *Auk*, **107**, 10–7.

Ward, S. (1996). Energy expenditure of female Barn Swallows *Hirundo rustica* during egg formation. *Physiological Zoology*, **69**, 930–51.

Warham, J. (1990). *The Petrels, Their Ecology and Breeding Systems*. Academic Press, London.

Waser, N. M. and Real, L. A. (1979). Effective mutualism between sequentially flowering plant species. *Nature*, **281**, 670–2.

Weatherhead, P. J. (1989). Sex ratios, host-specific reproductive success, and impact of brown-headed cowbirds. *Auk*, **106**, 358–66.

Weathers, W. W. (1985). Energy cost of incubation in the canary. *Comparative Biochemistry and Physiology*, **81A**, 411–3.

Weathers, W. W. and K. A. Sullivan. (1989a). Nest attentiveness and egg temperature in the Yellow-eyed Junco. *Condor*, **91**, 628–33.

Weathers, W. W. and Sullivan, K. A. (1989b). Juvenile foraging proficiency, parental effort, and avian reproductive success. *Ecological Monographs*, **59**, 223–46.

Weathers, W. W. and Sullivan, K. A. (1993). Seasonal patterns of time and energy allocation by birds. *Physiological Zoology*, **66**, 511–36.

Weaver, R. L. and West, F. H. (1943). Notes on the breeding of the pine siskin (*Spinus pinus pinus*). *Auk*, **60**, 492–504.

Webb, D. R. (1987). Thermal tolerance of avian embryos: A review. *Condor*, **89**, 874–98.

Webb, D. R. and King, J. (1983a). An analysis of the heat budgets of the eggs and nest of the white-crowned sparrow, *Zonotrichia leucophrys*, in relation to parental attentiveness. *Physiological Zoology*, **56**, 495–505.

Webb, D. and King, J. (1983b). Heat transfer relations of avian nestlings. *Journal of Thermal Biology*, **8**, 301–10.

Webb, G. J. W. and Cooper-Preston, H. (1989). Effects of incubation temperature on crocodiles and the evolution of reptilian oviparity. *American Zoologist*, **29**, 953–71.

Webster, M. D. (1999). Verdin. In: *Birds of North America, No. 470* (eds A. Poole and P. Stettenheim). Academy of Natural Sciences, Philadelphia: American Ornithologists' Union, Washington, DC.

Webster, M. D. and Weathers, W. W. (2000). Seasonal changes in energy and water use by verdins, *Auriparus flaviceps*. *Journal of Experimental Biology*, **203**, 3333–44.

Wedral, E. M., Vadehra, D. V. and Baker, R. C. (1974). Chemical composition of the cuticle, and the inner and outer shell membranes from eggs of *Gallus gallus*. *Comparative Biochemistry and Physiology*, **47B**, 631–40.

Weeden, J. S. (1966). Diurnal rhythm of attentiveness of incubating female tree sparrows (*Spizella arborea*) at a northern latitude. *Auk*, **83**, 368–88.

Weeks, H. P. Jr. (1994). Eastern Phoebe (*Sayornis phoebe*). In: *The Birds of North America, No. 94* (eds A. Poole and F. Gill). The Academy of Natural Sciences, Philadelphia: The American Ornithologists' Union, Washington, DC.

Weimerskirch, H., Chastel, O. and Ackermann, L. (1995). Adjustment of parental effort to manipulated foraging ability in a pelagic seabird, the thin-billed prion *Pachyptila belcheri*. *Behavioural Ecology and Sociobiology*, **36**, 11–6.

Weir, D. G. (1973). Status and habits of *Megapodius prichardii*. *Wilson Bulletin*, **85**, 79–82.

Weller, M. W. (1958). Observations on the incubation behavior of a Common Nighthawk. *Auk*, **75**, 48–59.

Weller, M. W. (1959). Growth, weight and plumages of the redhead (*Aythya americana*). *Wilson Bulletin*, **69**, 5–38.

Weller, M. W. (1961). Breeding biology of the least bittern. *Wilson Bulletin*, **73**, 11–35.

Weller, M. W. (1968). The breeding biology of the parasitic black-headed duck. *Living Bird*, **7**, 169–207.

Welty, J. C. (1975). *The Life of Birds, 2nd Edition*. Saunders, Philadelphia.

Welty, J. C. (1982). *The Life of Birds, 3rd edition*. Saunders College Publishing, Philadelphia.

Welty, J. C. and Baptista, L. (1988). *The Life of Birds, 4th Ed*. Harcourt Brace Jovanovich College Publishers, Fort Worth.

Wendeln, H. and Becker, P. H. 1999. Effects of parental quality and effort on the reproduction of common terns. *Journal of Animal Ecology*, **68**, 205–14.

West, G. C. (1968). Bioenergetics of captive Willow Ptarmigan under natural conditions. *Ecology*, **49**, 1035–45.

Westerterp, K. and Bryant, D. M. (1984). Energetics of free existence in swallows and martins (Hirundinidae) during breeding: a comparative study using doubly labelled water. *Oecologia*, **62**, 376–81.

Westerterp, K. and Drent, R. H. (1985). Flight energetics of the Starling *Sturnus vulgaris* during the parental period. In: *Proceedings of the 18th International Ornithological Congress*, pp. 392–8.

Westmoreland, D. and Best, L. B. (1986). Incubation continuity and the advantage of cryptic egg coloration to mourning doves. *Wilson Bulletin*, **98**, 297–300.

Westmoreland, D. and Kiltie, R. A. (1996). Egg crypsis and clutch survival in three species of blackbirds (Icteridae). *Biological Journal of the Linnean Society*, **58**, 159–72.

White, D. H., Mitchell, C. A. and Cromartie, E. (1982). Nesting ecology of roseate spoonbills at Neuces Bay, Texas. *Auk*, **99**, 275–84.

White, F. N., Bartholomew, G. A. and Kinney, J. L. (1978). Physiological and ecological correlates of tunnel nesting in the European bee-eater, *Merops apiaster*. *Physiological Zoology*, **51**, 140–54.

White, F. N. and Kinney, J. L. (1974). Interactions among behavior, environment, nest and eggs result in regulation of egg temperature. *Science*, **189**, 107–15.

White, H. B. (1991). Maternal diet, maternal proteins and egg quality. In: *Egg Incubation: Its Effects on Embryonic Development in Birds and Reptiles* (eds D. C. Deeming and M. W. J. Ferguson), pp. 1–15. Cambridge University Press, Cambridge.

White, N. R. (1984). Effects of embryonic auditory stimulation on hatch time in the domestic chick. *Bird Behaviour*, **5**, 122–6.

Whitfield, D. P. and Brade, J. J. (1991). The breeding behaviour of the Knot (*Calidris canutus*). *Ibis*, **133**, 246–55.

Whittow, G. C. (1980). Physiological and ecological correlates of prolonged incubation in sea birds. *American Zoologist*, **20**, 427–36.

Whittow, G. C. and Berger, A. J. (1977). Heat loss from the nest of the Hawaiian honeycreeper, 'Amakihi'. *The Wilson Bulletin*, **89**, 480–3.

Whittow, G. C. Pettit, T. N. Ackerman, R. A. and Paganelli, C. V. (1990). The regulation of water loss from the eggs of the red-footed Booby (*Sula sula*). *Comparative Biochemistry and Physiology*, **93A**, 807–10.

Whittow, G. C. and Tazawa, H. (1991). The early development of thermoregulation in birds. *Physiological Zoology*, **64**, 1371–90.

Wickler, S. J. and Marsh, R. L. (1981). Effects of nestling age and burrow depth on CO_2 and O_2 concentrations in the burrows of bank swallows (*Riparia riparia*). *Physiological Zoology*, **54**, 132–6.

Wiebe, K. L. and Bortolotti, G. R. (1993). Brood patches of American Kestrels: an ecological and evolutionary perspective. *Ornis Scandinavica*, **24**, 197–204.

Wiebe, K. L. and Bortolotti, G. R. (1994). Food supply and hatching spans of birds: energy constraints or facultative manipulation? *Ecology*, **75**, 813–23.

Wiebe, K. L. and Martin, K. (1995). Ecological and physiological effects on egg laying intervals in ptarmigan. *Condor*, **97**, 708–17.

Wiebe, K. L. and Martin, K. (2000). The use of incubation behaviour to adjust avian reproductive costs after egg laying. *Behavioural Ecology and Sociobiology*, **48**, 463–70.

Wiebe, K. L., Wiehn J. and Korpimäki, E. (1998). The onset of incubation in birds: Can females control hatching patterns? *Animal Behaviour*, **55**, 1043–52.
Wijnandts, H. (1984). Ecological energetics of the Long-eared Owl (*Asio otus*). *Ardea*, **72**, 1–92.
Wiklund, C. G. (1985). Fieldfare, *Turdus pilaris*, breeding strategy: The importance of asynchronous hatching and resources needed for egg formation. *Ornis Scandinavica*, **16**, 213–21.
Wiles, G. J and Conry, P. J. (2001). Characteristics of nest mounds of Micronesian megapodes in Palau. *Journal of Field Ornithology*, in press.
Wiley, J. W. (1988). Host selection by the shiny cowbird. *Condor*, **90**, 289–303.
Williams, A. J. (1980). Variation in weight of eggs and its effect on the breeding biology of the great skua. *Emu*, **80**, 198–202.
Williams, C. M., Richter, C. S., Mackenzie, J. M. and Shih, J. C. H. (1990). Isolation, identification, and characterization of a feather degrading bacterium. *Applied and Environmental Microbiology*, **56**, 1509–15.
Williams, D. L. G., Seymour, R. S. and Kerourio, P. (1984). Structure of fossil dinosaur eggshell from the Aix Basin, France. *Palaeogeography, Palaeoclimatology, Palaeoecology*, **45**, 23–37.
Williams, G. C. (1966). Natural selection, the cost of reproduction, and a refinement of Lack's principle. *American Naturalist*, **100**, 687–90.
Williams, J. B. (1987). Field metabolism and food consumption of Savannah Sparrows during the breeding season. *Auk*, **104**, 277–89.
Williams, J. B. (1988). Field metabolism of Tree Swallows during the breeding season. *Auk*, **105**, 706–14.
Williams, J. B. (1991). On the importance of energy considerations to small birds with gyneparental intermittant incubation. In: *Acta XX Congressus Internationalis Ornithologici* (ed. B. D. Bell), pp. 1964–1975. New Zealand Ornithological Congress Trust Board, Wellington, New Zealand.
Williams, J. B. (1993a). Nest orientation of Orangebreasted Sunbirds in South-Africa. *Ostrich*, **64**, 40–2.
Williams, J. B. (1993b). On the importance of energy considerations to small birds with gynelateral intermittent incubation. *Proceedings of the 20th International Ornithological Congress*, pp. 1964–75.
Williams, J. B. (1993c). Energetics of incubation in free-living orange-breasted sunbirds in South Africa. *Condor*, **95**, 115–26.
Williams, J. B. (1996). Energetics of avian incubation. In: *Avian Energetics and Nutritional Ecology* (ed. C. Carey), pp. 375–416. Chapman and Hall, New York.
Williams, T. D. (1994). Intraspecific variation in egg size and egg composition in birds: effects on offspring fitness. *Biological Reviews*, **68**, 35–59.
Williams, J. B. (2001). The energy expenditure and water flux of free-living Dune Larks in the Namib: A test of the re-allocation hypothesis on a desert bird. *Functional Ecology*, in press.
Williams, J. B. and Dwinnel, B. (1990). Field metabolism of free-living Savannah Sparrows during incubation: a study using doubly labelled water. *Physiological Zoology*, **63**, 353–72.

Williams, J. B. and Nagy, K. A. (1985). Water flux and energetics of nesting Savannah Sparrows in the field. *Physiological Zoology*, **58**, 515–25.

Williams, J. B. and Tieleman, B. I. (1999). Lizard burrows provide thermal refugia for larks in the Arabian desert. *Condor,* **101**, 714–7.

Williams, J. B. and Tieleman, B. I. (2001). Physiological ecology and behavior of desert birds. In: *Current Ornithology, Volume 16* (eds V. Nolan, E. Ketterson and J. Thompson), Kluwer Academic/Plenum Publishers, New York, in press.

Williams, T. D. (1995). *The Penguins Spheniscidae*. Oxford University Press, Oxford.

Williamson, F. S. L. and Emison, W. B. (1971). Variation in the timing of breeding and molt of the Lapland longspur (*Calcarius lapponicus*) in Alaska, with relation to differences in latitude. *BioScience,* **21**, 701–7.

Williamson, S. (2000). Blue-throated Hummingbird. In: *Birds of North America, No. 531* (eds A. Poole and P. Stettenheim). Academy of Natural Sciences, Philadelphia: The American Ornithologists' Union, Washington, DC.

Willis, E. (1961). A study of nesting ant-tanagers in British Honduras. *Condor*, **63**, 479–503.

Wilson, H. R. (1991). Physiological requirements of the developing embryo: temperature and turning. In: *Avian Incubation* (ed. S. G. Tullett), pp. 145–56. Butterworths-Heinemann, London.

Wilson, R. P., Culik, B. M., Dannfeld, R. and Adelung, D. (1991). People in Antarctica–how much to Adelie penguins *Pygoscelis adeliae* care? *Polar Biology,* **11**, 363–70.

Wilson, R. T., Wilson, M. P. and Durkin, J. W. (1986). Breeding biology of the barn owl *Tyto alba* in central Mali. *Ibis*, **128**, 81–90.

Wimberger, P. H. (1984). The use of green plant-material in bird nests to avoid ectoparasites. *Auk*, **101**, 615–18.

Wingfield, J., Hegner, R., Dufty, A. J. and Ball, G. (1990). The 'challenge hypothesis': theoretical implications for patterns of testosterone secretion, mating systems and breeding strategies. *American Naturalist*, **136**, 829–46.

Wingfield, J. C. and Goldsmith, A. R. (1990). Plasma levels of prolactin and gonadal steroids in relation to multiple-brooding and renesting in free-living populations of the song sparrows, *Melospiza melodia. Hormones and Behavior*, **24**, 89–103.

Wingfield, J. C., Hahn, T. P., Wada, M. and Schoech, S. J. (1997). Effects of day length and temperature on gonadal development, body mass and fat deposits in White-Crowned Sparrows, *Zonotrichia leucophrys pugetensis. General and Comparative Endocrinology,* **107**, 44–62.

Wingfield, J. C. O'Reilly, K. M. and Astheimer, L. B. (1995). Modulation of adrenocortical responses to acute stress in arctic birds: a possible ecological basis. *American Zoologist,* **35**, 285–94.

Wingfield, J. C., Vleck, C. M. and Moore, M. C. (1992). Seasonal changes of the adrenocortical response to stress in birds of the Sonoran desert. *Journal of Experimental Zoology*, **264**, 419–28.

Wink, M., Ristow, D. and Wink, C. (1985). Biology of Eleonora's falcon (*Falco eleonorae*): 7. Variability of clutch size, egg dimensions and egg coloring. *Raptor Research*, **19**, 8–14.

Winkler, D. W. and Sheldon, F. H. (1993). Evolution of nest construction in swallows (Hirundinidae): a molecular phylogenetic perspective. *Proceedings of the National Academy of Sciences USA*, **90**, 5705–7.

Winter, S. W., Andryushchenko, Y. A. and Gorlov, P. I. (1996). The behavior of breeding Hooded Crane. *Proceedings of the 3rd European Crane Workshop* (eds H. Prange, G. Nowald and W. Mewes), pp. 293–320. European Crane Working Group, Halle, Germany.

Winter, S. W., Gorlov, P. I. and Andryushchenko, Y. A. (1999). Neues aus der forschung an paläarktischen kranichen. *Vogelwelt*, **120**, 367–76.

Wishart, G. J. and Staines, H. J. (1999). Measuring sperm : egg interaction to assess breeding efficiency in chickens and turkeys. *Poultry Science*, **78**: 428–36.

Witmer, M. C., Mountjoy, D. J. and Elliot, L. (1997). Cedar Waxwing (*Bombycilla cedrorum*). In: *The Birds of North America, No. 309* (eds A. Poole and F. Gill). The Academy of Natural Sciences, Philadelphia: The American Ornithologists' Union, Washington DC.

Witschi, E. (1949). Utilisation of the egg albumen by the avian fetus. *Contributions to Ornithological Biology, Wissenesch*, pp. 111–22. Carl Winter, Heidelberg.

Wittman, J. and Kaltner, H. (1988). Formation and changes of the subembryonic liquid from turned, unturned and cultured Japanese Quail eggs. *Biotechnology and Applied Biochemistry*, **10**, 338–45.

Wood, D. R. and Bollinger, E. K. (1997). Egg removal by brown-headed cowbirds: A field test of the host incubation efficiency hypothesis. *Condor*, **99**, 851–7.

Woolf, N. K., Bixby, J. L. and R. R. Capranica. (1976). Prenatal experience and avian development: Brief auditory stimulation accelerates the hatching of Japanese quail. *Science*, **194**, 959–60.

Woolfenden, B. (2000). *Demography and breeding behaviour of brown-headed cowbirds: An examination of host use, individual mating patterns and reproductive success using microsatellite DNA markers*. Ph.D. Thesis, McMaster University, Hamilton.

Wyllie, I. (1975). Study of cuckoos and reed warblers. *British Birds*, **68**, 369–78.

Wyllie, I. (1981). *The Cuckoo*. Batsford, London.

Yahner, R. H. and DeLong, C. A. (1992). Avian predation and parasitism on artificial nests and eggs in two fragmented landscapes. *Wilson Bulletin*, **104**, 162–8.

Yahner, R. H. and Mahan, C. G. (1996). Effects of egg type on depredation of artificial ground nests. *Wilson Bulletin*, **108**, 129–36.

Yasukawa, K. and Searcy, W. A. (1995). Red-winged blackbird (*Agelaius phoeniceus*). In: *The Birds of North America, No.184* (eds A. Poole and F. Gill). The Academy of Natural Sciences, Philadelphia: The American Ornithologists' Union, Washington, DC.

Yerkes, T. (1998). The influence of female age, body mass, and ambient conditions on Redhead incubation constancy. *Condor*, **100**, 62–8.

Yom-Tov, Y. (1980a). Intraspecific nest parasitism in birds. *Biological Reviews*, **55**, 93–108.

Yom-Tov, Y. (1980b). Intraspecific nest parasitism among Dead Sea sparrows *Passer moabiticus*. *Ibis*, **122**, 234–7.

Yom-Tov, Y. and Ar, A. (1993). Incubation and fledging durations of woodpeckers. *Condor*, **95**, 282–7.

Yom-Tov, Y., Ar, A. and Mendelssohn, H. (1978). Incubation behavior of the Dead Sea sparrow. *Condor*, **80**, 341–3.

Yom-Tov, Y. and Hilborn, R. (1981). Energetic constraints on clutch size and time of breeding in temperate zone birds. *Oecologia*, **48**, 234–43.

Yom-Tov, Y., Wilson, R. and Ar, A. (1986). Water loss from Jackass Penguin *Spheniscus demersus* eggs during natural incubation. *Ibis,* **128**, 1–8.

Youngren, O. M., El Halawani, M. E., Silsby, J. L. and Phillips, R. E. (1991). Intracranial prolactin perfusion induces incubation behavior in turkey hens. *Biology of Reproduction*, **44**, 425–31.

Zaan, R. A. (1996). *The Zebra Finch. A Synthesis of Field and Laboratory Studies.* Oxford University Press, Oxford.

Zaan, R. A., Morton, S. R., Jones, K. R. and Burley, N. T. (1994). The timing of breeding by Zebra finches in relation to rainfall in central Australia. *Emu,* **95**, 208–22.

Zack, R. (1982). Hatching asynchrony, egg size, growth, and fledging in tree swallows. *Auk*, **99**, 695–700.

Zadworny, D. and Etches, R. J. (1987). Effects of ovariectomy or force feeding on the plasma concentrations of prolactin and luteinizing hormone in incubating turkey hens. *Biology of Reproduction*, **36**, 81–8.

Zadworny, D., Walton, J. and Etches, R. (1985). Effect of feed and water deprivation or force-feeding on plasma prolactin concentration in turkey hens. *Biology of Reproduction*, **32**, 241–7.

Zann, R. and Rossetto, M. (1991). Zebra finch incubation: brood patch, egg temperature and thermal properties of the nest. *Emu*, **91**, 107–20.

Zerba, E. and Morton, M. L. (1983a). The rhythm of incubation from egg laying to hatching in mountain white-crowned sparrows. *Ornis Scandinavica* **14**, 188–97.

Zerba E. and Morton, M. L. (1983b). Dynamics of incubation in mountain white-crowned sparrows. *Condor,* **85**, 1–11.

Zimmerman, J. L. (1983). Cowbird parasitisn of dickcissels in different habitats and at different nest densities. *Wilson Bulletin*, **95**, 7–22.

Zyskowski, K. and Prum, R. O. (1999). Phylogenetic analysis of the nest architecture of neotropical ovenbirds (Furnariidae). *Auk*, **116**, 891–911.

Index

air space (cell) 48–9, 163, 239, 251
albumen 28, 30, 32, 35, 159, 179
  changes during incubation 50–2, 159–60
  changes during storage 50–1
  proteins 30–1, 51
  roles 31, 50
  sac 47
  turning 163, 165–7, 171, 176
  water content 32, 51, 159–60
allantoic fluid 46, 47, 159, 202–3
allantois 47
altitude 213–4, 238, 242–7
  gas pressure 244, 246
altricial 8, 31, 43, 47, 51–2, 55, 73, 74, 75, 89, 91, 95, 98, 145, 165–7, 171–4, 176–7, 274
amnion 45–6
amniotic fluid 45, 46, 51, 159, 171
arthropods 12, 19, 20, 23, 183, 184
attentiveness 13, 54, 69, 75–6, 78–80, 82–3, 86–7, 91, 94, 97, 131, 148, 153, 155, 157, 215, 223, 240, 310, 319
  climate 77, 87, 157, 240
  clutch mass 83–5
  day-length 76
  distribution 73–6
  during laying 276
  egg mass 79–80, 83
  female-only 79–80, 83, 215, 223
  foraging 76–7
  hatching 272
  male-feeding 76, 300, 301–3, 319
  male-only 79–80, 83
  nests 77
  observations 76
  shared incubation 73–4, 79–80, 83, 223
  stage of incubation 77
  weight loss 76

bacteria 179–80
  control 184–6
  in nests 183–4
  on eggshells 186–9, 190
  pathogenic 183, 184
  water vapour conductance 188–9
barometric pressure 243–4, 245
behaviour
  clutch size 83–4, 85
  egg size 78–83
  evolution 78
  hummingbird 209
  incubation 113–4, 176–7, 215–6, 270–1, 320–1
  mating 320
  microbial control 184–6
  nesting 110
  non-incubating 72–3
  off nest 71, 72–3, 147
  on nest 67–71, 147
  time-budget 67–8, 71, 310
bird–nest incubation unit 1–2, 8, 28, 87, 139, 155
blastoderm 29, 44
breeding, timing 240, 250
broodiness 54, 55, 58, 110
brooding 55
brood parasite 60, 62, 66, 109, 254–69
  acceptance of egg 256–7
  behaviour 259–60
  egg characteristics 256, 257, 261–2
  egg numbers 257–8, 259
  egg recognition 285–8
  host behaviour 256, 259–60
  hosts 255–7
  multiple parasitism 262–3
  nests 256, 260, 262
  occurrence 255
brood patch 58, 59, 60, 81, 87, 119, 131, 140, 148, 149–50, 217, 264, 270, 315
  absence 107–8
  actions, direct 115–8
  actions, indirect 113–5
  anaesthesia 59, 112, 113, 116
  artificial 128, 132, 160
  blood flow 116–7, 119, 150
  characteristics 100, 109
  de-feathering 102, 104–5, 108, 109, 110
  denervation 59, 60, 103, 112, 113, 114
  dermis 103–4
  environmental factors 111–2
  epidermis 102–3
  evolution 100
  females 106–7
  heat production 116–7
  heat transfer 104, 116, 119, 120, 127, 131–5
  males 78, 105–7, 116
  morphology 101–5
  number 105
  oedema 100, 103–4, 111
  pattern 101
  score 110–1
  size 106, 117, 150, 158
  skin 102–4
  species differences 104–5
  stimulation 112
  tactile sensitivity 11
  temperature 146, 147, 150, 227–8, 230
  vascularisation 103, 104, 108, 110, 116

brood patch (*cont.*):
  vasoconstriction 117
  vasodilation 117, 150
  water loss 117
brood pouch 9, 108, 117, 127

carbon dioxide *see* respiratory gases
chorio-allantoic membrane (CAM) 46, 47, 48, 52, 91, 147, 150, 160, 171, 172, 202
chorion 46
climate 11, 69, 82, 83, 105, 119, 151, 243, 303, 323
  behaviour 69, 77
  cold 239–40, 243, 323
  humidity 143, 155, 238, 242, 247, 250–1
  rainfall 77, 168, 169, 218, 250
  snow 243
  solar radiation 151, 243, 251
  temperature 11, 54, 69, 77–8, 95–7, 112, 120, 153, 214–5, 238, 242, 247, 248, 276–8, 303, 306, 309–10
  wind 138, 143, 157, 239, 249
clutch size 83–5, 86, 299, 303, 304, 319
  brood parasite 261–2
  cooling 152, 318
  energetics 318
  hatching 271–2, 274
  hummingbirds 210–2, 221
  incubation 318
  turning 168

dinosaur 2–5
  eggs 3
  eggshells 3, 4, 6
  embryo 2–5
  nest 3, 4, 5
developmental modes 43, 44, 47, *see also* altricial, precocial

ecological
  constraints 270
  factors 274–9, 321
  trade-offs 270, 278, 303, 314, 317
ectoderm 44, 45
egg
  broken 268
  brood patch contact 58, 59
  burying 6, 69
  chilling 240–1
  colour 71, 280–97
  composition 30–1, 32, 35, 163–5, 177, 200
  cooling 127, 129, 136–7, 151–2, 224, 225, 227–8, 232–3, 238, 249, 276
  covering 69, 71
  dimensions 33, 34, 36, 208, 211, 256, 261
  ejection 260
  electronic 155, 161, 163, 167, 168
  energy balance 120
  formation 28–9, 294–5
  heat flow 119, 128, 132, 137, 138
  heating 225, 227–8, 230
  laying 55, 57, 59, 114–5, 258, 259–60
  maculation 280–97
  mass 33, 34, 35, 79–80, 86, 129, 131, 163–5, 172–5, 208, 210–2, 258, 261, 266, 278
  mimicry 254, 256, 258, 286, 288–92, 289, 290
  microbial penetration 179–80
  morphology 29–30
  movement 267
  neglect 277
  overheating 248, 293
  palatability 293–4
  pecking 260–1
  predation 260, 280–81, 295, 296–7, 317
  recognition 71–2, 284–8
  removal 115, 260, 262, 318
  scaling 210–3
  shading 148, 248
  shape 33–5, 257
  surface area 36, 210–2
  temperature 69, 98, 107–8, 116, 120–1, 125, 128, 129–130, 135–6, 138–9, 141–2, 144, 145–52, 153, 154, 225, 228–9, 231, 233, 234–6, 248, 262
  thermal characteristics 81–3, 123–34, 137
  turning *see* turning
  volume 36, 261
  water content 32, 33
  weight loss 38, 154–5, 212–3, 244, 251–2
  wetting 69, 148–9, 249
eggshell 28–9, 35–6, 48–9, 52–3
  accessory material 37
  characteristics 36, 38, 50, 212, 239, 257, 264
  cuticle 41, 188–9, 190, 242
  intraclutch variability 294–6
  megapode 200–1
  membranes 28–9, 171, 176
  pigment 294–5
  pores 36–7, 40, 179, 200–1, 242
  porosity 49, 239, 242 *see also* water vapour conductance
  position 163–4
  role 35–6
  solar filter 292–3
  strength 257, 264, 293
  structure 35–7, 38, 52–3, 179
  surface microbes 186–9
  temperature 82
embryo
  brood parasite 266–7
  circulation 118, 125, 127, 128, 147
  clicking noises 89–90, 267
  communication 50
  development 43–4, 266–8, 271, 316
  differentiation 43–4, 47
  Embryo–embryo interaction 88–91, 267, 271
  growth 31, 35, 44–5, 48, 171
  heat production 120, 150
  location in egg 145

metabolism 39, 128
morphometrics 45
mortality 179
orientation 133
parent–embryo interaction 91–9, 158
pre-incubation development 266–7
stages 43–4, 47–8
temperature 90, 141, 201
thermal tolerance 248–9
viability 277
vocalisations 90–1, 92–93, 95–6, 252
endoderm 44, 46
energetics 84, 129, 212, 216–7, 220, 241, 299–313
constraints 303–4
costs 304–13
limitations 316–20
evolution 2, 6, 23–5, 175–7, 300
nesting behaviour in megapodes 192–3
extra-embryonic membranes 43, 45–7

feathers 16, 21, 27, 100, 180–1, 186
bacteria 186–8
loss 102, 104–5, 108
pigmentation 292–3
re-growth 108
wax 185–6
feeding 65
fertilisation 28
fitness 314–23, 324–5
food availability 278–9
foot webs 94, 107
fossils 2
fungi 16, 180–3, 186
control 184–6
in nests 180–3
on eggshells 187–9
pathogenic 181–2
temperature 180–1

hatchability 204–5, 304
brood parasites 261–2, 263
hatching 48–50, 98–9, 168, 252, 267–8
acceleration 50, 90, 91, 267
asynchronous 97–8, 265, 271–2, 274
breeding season 274
fluid loss 202–3
megapodes 202–3
parental age 275
parental assistance 98–9
parental neglect 95–6
retardation 91
spread 265, 274–5
synchrony 58, 89, 90, 95, 271, 274
turning 168
vocalisations 93–4, 252, 267
hatchling 204
mass 45
water content 33
hatchling maturity *see* altricial, precocial

hormones 54, 106, 107
androgens 109
brood patch 108–12
corticosterone 242
oestrogens 57–58, 102, 108, 109, 111, 112, 254, 270
progesterone 58, 108
prolactin 55–62, 106, 108, 109, 111, 112, 113, 114, 254, 270
sex steroids 58, 110
testosterone 58–9, 106

incubation
artificial 1, 171, 178
behaviour 54, 57–60, 63–7, 67–70, 87, 113–4, 209, 215–6
by nestlings 315
constancy 73–76 *see also* attentiveness
costs 316–23
cycles 225, 227, 228, 230–3, 234
during laying 276
environment 263–4
female-only 63–5, 67, 74, 75, 209, 223, 300, 311
initiation 57
intermittent 116, 130, 131, 148, 223, 224, 299, 300, 304–5, 322
male-only 58–9, 63–5, 67, 74, 78
model 225–35
onset 270–9
period 36, 39, 44, 60–1, 165–7, 172, 174, 175, 176–7, 208, 209, 210, 265–6, 270, 300
recesses 65, 67, 77–8, 79, 81, 84, 86, 151, 227
sessions 63–5, 67, 77–8, 79, 81, 84, 86, 227
shared 63–5, 66, 73–4, 75, 278, 300, 308, 311, 323
study 161, 170
temperature 95, 176–7, 218, 224, 227, 262
weight loss during 241–2, 317–9
incubators 171

latitude 243
life-history theory 299–300, 314

megapodes 6, 40, 48, 66, 160, 169, 172, 174
burrow nesters 194–7, 198, 199
egg 200–1
embryonic development 200–5
hatching 202–5
heat sources 195, 199
incubation behaviour 193–4
incubation mounds 193–4, 198
incubation temperature 195, 199–200
laying 196
mound builders 193–7, 198
nest conditions 197–200
nest location 195–7, 198
mesoderm 44, 45, 46, 103

metabolism 207, 212, 241, 299, 301, 302, 304–7
  body mass 301–2, 306–7, 309
  gender 308
  seabirds 311–3
  terminology 300–1
microbial
  growth 41, 179
  respiration 193
molecular genetics 24, 25, 255, 257–8, 263, 329
mud 11, 16, 22

nest 2, 14, 18, 19, 140, 143, 145, 214, 219, 240–1, 255
  artificial 281
  attachment 17–9
  bacteria 183–4
  box 181–2
  burrow 6, 13, 14
  cavity 296
  conductance 154
  construction 15–23, 69, 93–4, 102, 151, 182–3, 213, 270, 315
  defence 68, 70
  deprivation 114
  function 25–7
  fungi 180–3
  heating 319
  humidity 9, 38, 144, 154–6, 198, 263
  hygiene 185
  insulation 21, 27, 82, 151, 152–4, 213, 219, 241
  location 10–13, 17, 70, 147, 213–4, 216, 248–9, 275
  loss 61
  material 10, 15–16, 20
  microclimate 144, 145, 146, 159, 193, 250
  orientation 11, 249–50
  parasitised 259–60
  phylogeny 24
  protection 5, 12
  recognition 285
  relief 70–1
  respiratory gas composition 156–7, 199
  role 8, 9, 20, 151
  shape 13–5
  size 20–1, 27
  structure 16–21, 26, 264
  temperature 9, 13, 146, 154, 217–8
  thermal properties 219, 315
  tunnel 144, 156
  usurpation 288
  ventilation 156, 157, 170
  water content 180, 181
  weaving 23
  wet 251–2
  zones 17, 25
nestling phase 307–9, 310, 321–3
nitric oxide 91–2, 140, 150, 158, 159
  embryonic hyperthermia 158
  embryonic hypoxia 158, 159
  synthase 150

Oviraptor 2–5
oxygen *see* respiratory gases
oxygen consumption
  embryo 158, 201
  parental 153

phylogeny 24, 67, 175
physiological zero (PZT) 141, 229, 232, 236, 276, 277
pipping 48–9, 251
polygyny 209
precocial 8, 31, 43, 47, 51, 55, 73, 74, 75, 89, 91, 98, 145, 165–7, 171–4, 176–7
predation 9, 10, 27, 143, 176, 205, 275–6, 317

reproduction costs 303, 314–23
reptiles 2, 5, 6, 143, 155, 159, 160, 176, 316
respiratory gases 36, 39, 48–9, 117, 156–8, 197–8, 201, 239, 245–7, 249, 263–4
  altitude 245
  conductance 158, 201, 239, 246–7

sero-amniotic connection 45, 51–2
sub-embryonic fluid 46, 47, 51, 159, 171

temperature field 121–128, 131–2
  eggs 122–8
  heat flow 121–3, 218–9
thermal
  capacity 130, 132
  conductance 120, 124, 128, 154, 218–9
  conductivity 121, 217
  impedance 81, 84, 132–5, 138
  mutualism 139–42
  resistance 81, 124, 125, 128, 133
thermoregulation 69, 95–6, 140, 148–9, 203–4, 247, 248
  water balance 247–8
torpor 207, 216–7
turning 51, 67, 68, 72, 93, 94, 97, 159–60, 218, 252
  absence 165, 167, 169–70, 175
  angle 163–4
  albumen 165–7, 172, 174
  axis 163
  before incubation 161–2
  behaviour 162–3
  by beak 162
  by feet 162–3, 169, 216
  by hand 172
  captivity 164
  clutch mass 168, 170
  critical period 159
  diffusion gradients 171
  diurnal cycle 164, 167–8
  egg mass 165–7, 172, 173
  egg position 163–4, 168

evolution 175–7
frequency 164–9, 172–5, 176
gender 164
hatching 168
incubation period 165–7, 168, 170, 172, 173, 215–6
marked eggs 170
reasons 171
study 170
timing 163
tremble-thrusting 162

vegetation 16, 21, 26, 251–2
antimicrobial actions 184–5

water
absorption 252
loss 154, 239, 250
movements 51

water vapour conductance 3, 4, 5, 6, 36, 37–41, 154–5, 213, 242, 244, 250
brood patch 117
changes 39–41, 188–9, 190, 201, 242, 247
effective 244, 246
Fick's law 38
metabolism 39
water loss 154–6, 212–3

yolk 28, 29, 35, 46, 51, 52, 263
composition 263
deposition 28–9
energy content 32
fatty acid profile 30, 31, 52
morphology 29–30, 46
sac membrane 46–7, 52
sac retraction 46, 52
water content 32

zygote 44